An Introduction to TRANSPORT PHENOMENA in MATERIALS ENGINEERING

DAVID R. GASKELL

School of Materials Engineering
Purdue University

Macmillan Publishing Company
NEW YORK

Maxwell Macmillan Canada
TORONTO

Maxwell Macmillan International
NEW YORK   OXFORD   SINGAPORE   SYDNEY

Editor: David Johnstone
Senior Production Supervisor: J. Edward Nève
Production Manager: Sandra Moore
Text Designer: Jane Edelstein
Cover Designer: Jane Edelstein
Illustrations: Academy Art Works, Inc.

This book was set in 10/12 Times Roman by Waldman Graphics, Inc., printed by Book Press, Inc., and bound by Book Press, Inc. The cover was printed by Lehigh Press Lithographers, Inc.

Macmillan Publishing Company
866 Third Avenue, New York, New York 10022

Macmillan Publishing Company is
part of the Maxwell Communication
Group of Companies.

Maxwell Macmillan Canada, Inc.
1200 Eglinton Avenue, E.
Suite 200
Don Mills, Ontario M3C 3N1

Library of Congress Cataloging in Publication Data

Gaskell, David R.,
      An introduction to transport phenomena in materials engineering /
   David R. Gaskell.
         p.   cm.
      Includes bibliographical references and index.
      ISBN 0-02-340720-4
      1. Materials—Fluid dynamics.   2. Mass transfer.   3. Heat—
   Transmission.   I. Title.
   TA418.5.G37   1992
   620.1'1296—dc20   91-9470
                                                                CIP

Printing: 1 2 3 4 5 6 7 8   Year: 2 3 4 5 6 7 8 9 0 1

In memory of
Thomas Skehan
and
Jeffrey Weiss

Preface

In their classic text, *Transport Phenomena,* Bird, Stewart, and Lightfoot state their opinion that the subject of transport phenomena should rank along with thermodynamics, mechanics, and electromagnetism as one of the "key engineering sciences." This thought was not shared by many traditional metallurgists, and diffusion in the solid state was the only aspect of transport phenomena included in many traditional university metallurgy curricula. However, as metallurgists transformed themselves into materials scientists and engineers, and the artificial barriers between the various engineering disciplines were lowered, the materials engineers began to see the truth in the opinion of Bird, Stewart, and Lightfoot. To begin with, undergraduate students of materials engineering were sent off to study classical transport phenomena in single courses offered by departments of chemical engineering, which was reasonably successful until the chemical engineers realized that the scope of the subject required at least a two-course sequence. With the proliferation of courses on polymeric materials, ceramics, and so on, in the evolving materials engineering curricula, there was simply not enough room to accommodate a two-course sequence in transport phenomena, and hence the materials engineering students disappeared from the departments of chemical engineering and transport phenomena disappeared from many materials engineering curricula. In 1984 a decision was made by the faculty of the School of Materials Engineering at Purdue to initiate a required course on transport phenomena in materials engineering, and the present text evolved from lecture notes written for the new course.

In presenting an introduction to transport phenomena to students who have not had any previous exposure to fluid dynamics or heat transfer or mass transport, a careful balance must be made between an explanation of the fundamentals that govern the dynamics of fluid flow and the transport of heat and mass, on the one hand,

and on the other, illustration of the application of the fundamentals to specific systems of interest in materials engineering. As the present text is introductory, emphasis is placed on discussion of the properties of fluids that determine the nature of their flow, and in the belief that an understanding of phenomena is best achieved by first considering commonplace examples, the flow of familiar fluids such as air, water, and oils is used to illustrate the relative influences of the properties and geometry on fluid flow.

The text is organized in discrete sections on fluid flow, heat transfer, and mass transport, as this allows the development of the discussion of heat transfer to employ analogies of heat flow with fluid flow and allows the discussion of mass transport to employ analogies of mass transport with heat transfer. SI units are used throughout the book except in a few equations that were originally derived using cgs units and contain recognizable numerical constants. Conversion of any unit in the British System to SI can be made by memorizing three conversions: 1 inch equals 2.54 centimeters, 1 kilogram equals 2.2 pounds mass (lb_m), and 1 kelvin equals $\frac{9}{5}$ of a Fahrenheit degree. A table of factors for the conversion of British units to SI is provided as Appendix C.

I wish to express my gratitude to Cheryl Waller for her considerable assistance in the preparation of the text.

David R. Gaskell
West Lafayette
Indiana

Contents

3 Equations of Continuity and Conservation of Momentum and Fluid Flow Past Submerged Objects 102

4 Turbulent Flow 135

5 Mechanical Energy Balance and Its Application to Fluid Flow 185

6   Transport of Heat by Conduction   235

7   Transport of Heat by Convection   295

11   Mass Transport in Fluids   522

Symbols

A	area	m^2
B	mobility	s/kg
Bi	Biot number	Eq. (8.7)
$\bar{c}$	average speed of an atom in a gas phase	m/s
C	concentration	g mol/m^3
C_D	discharge coefficient	dimensionless
C_p	constant pressure heat capacity	J/(kg·K)
C_p	pitot tube coefficient	dimensionless
C_v	constant volume heat capacity	J/(kg·K)
d	atomic diameter	m
D_{A-B}	interdiffusion coefficient in the system $A-B$	m^2/s
D_e	equivalent diameter	m
D_i	chemical diffusion coefficient of species i	m^2/s
D_i^*	self-diffusion coefficient of species i	m^2/s
e_f	friction loss factor	dimensionless
E	total emissive power of radiation	W/m^2
E_a	activation energy	J
$E_b(T)$	emissive power of a blackbody at the temperature T	W/m^2
E_f	friction loss	m^2/s^2 (J/kg)
E_k	kinetic energy	J
E_p	potential energy	J
f	friction factor	dimensionless
F	force	N
F_{ij}	view factor	dimensionless
F_K	drag force	N
Fo	Fourier number	Eq. (8.11)
G	total irradiation	W/m^2
Ga	Galileo number	Eq. (4.71)
Gr	Grashof number	Eq. (7.116)
Gr_m	mass transfer Grashof number	Eq. (11.89)
h	height	m
h	head	m
h	heat transfer coefficient	W/(m^2·K)
h_L	loss of head	m
h_m	mass transfer coefficient	m/s
H	enthalpy	J
ΔH_{evap}	latent heat of evaporation	J
ΔH_s	heat of solidification	J
I	electric current	A
$I_{\lambda,e}$	spectral intensity of radiation	W/(m^2·sr·μm)
j	mass diffusion flux relative to mixture mass average velocity	kg/(m^2·s)
J	molar diffusion flux	g·mol/(m^2·s)
J	total radiosity	W/m^2
J^*	molar diffusion flux relative to mixture molar average velocity	g mol/(m^2·s)
k	thermal conductivity	W/(m·K)
k_m	mean thermal conductivity	W/(m·K)

K	flow coefficient	dimensionless
K	kinetic energy per unit volume	J/m^3
K	equilibrium constant	dimensionless
L	length	m
L_c	characteristic length	m
L_e	equivalent length	m
Le	Lewis number	Eq. (11.67)
L_E	entry length	m
$L_{E,th}$	thermal entry length	m
m	mass	kg
$\dot{M}$	mass flowrate	kg/s
M	molecular weight	kg/g mol
$\dot{n}$	mass flux relative to fixed coordinates	$kg/(m^2 \cdot s)$
n_i	number of gram moles of species i	g mol
$\dot{N}$	molar flux relative to fixed coordinates	$g\ mol/(m^2 \cdot s)$
Nu	Nusselt number	Eq. (7.7)
p_i	partial pressure of species i	Pa
p_i^o	saturated vapor pressure of species i	Pa
P	pressure	Pa
P	specific permeability	D'Arcy $(10^{-8}\ cm^2)$
Pr	Prandtl number	Eq. (7.4)
P_w	wetted perimeter	m
q	rate of heat flow	W
$\dot{q}$	rate of generation of heat	W/m^3
q'	heat flux	W/m^2
Q	thermal energy, heat	J
r	radius	m
$\dot{r}$	reaction rate	$g\ mol/(m^2 \cdot s)$ or $g\ mol/(m^3 \cdot s)$
R	radius	m
R	electric resistance	Ω
R_h	resistance to heat flow by convection	$K^2 \cdot m/W$
R_h	hydraulic radius	m
R_k	resistance to heat flow by conduction	$K^2 \cdot m/W$
Ra	Rayleigh number	Eq. (7.122)
Re	Reynolds number	Eq. (2.2)
Sc	Schmidt number	Eq. (11.66)
Sh	Sherwood number	Eq. (11.64)
St	Stanton number	Eq. (7.27)
St_m	mass transfer Stanton number	Eq. (11.82)
t	time	s
T	temperature	K
T_s	surface temperature	K
T_∞	ambient temperature	K
U	internal energy	J
v	velocity	m/s
v_x	x-directed component of velocity	m/s
V	volume	m^3
V	electric voltage	V
$\dot{V}$	volume flowrate	m^3/s

w	weight	N
w	work	J
W	width	m
$\dot{W}$	molar flowrate	g mol/s
X_i	mole fraction of species i	dimensionless

Constants

c_0	speed of light in a vacuum	2.998×10^8 m/s
C_1	first radiation constant	3.742×10^8 W·μm⁴/m²
C_2	second-radiation constant	1.439×10^4 μm·K
C_3	third radiation constant	$2.897.6$ μm·K
g	acceleration due to gravity	9.8067 m/s²
h_0	Planck's constant	6.6256×10^{-34} J·s
$\hbar$	Boltzmann's constant	1.38054×10^{-23} J/K
N_0	Avogadro's number	6.0232×10^{23}/g mol
R	Universal gas constant	8.3144 J/(g mol·K)
σ	Stefan–Boltzmann constant	5.670×10^{-8} W/(m²·K⁴)

Greek Symbols

α	thermal diffusivity	m²/s
α	adsorptivity	dimensionless
α	isothermal compressibility	Pa⁻¹
β	isothermal expansivity	K⁻¹
β'	coefficient of change of density w/composition	dimensionless
γ	ratio of C_p to C_v	dimensionless
γ	activity coefficient	dimensionless
δ	thickness of momentum boundary layer	m
δ_C	thickness of concentration boundary layer	m
δ_T	thickness of thermal boundary layer	m
ϵ	emissivity	dimensionless
η	viscosity	Pa·s [= kg/(m·s)]
η	efficiency	dimensionless
Θ	mean residence time	s
λ	shape factor	dimensionless
λ	mean free path of an atom in a gas phase	m
λ	wavelength	μm
ν	kinematic viscosity	m²/s
ρ	density	kg/m³
ρ	specific electric resistance	Ω·m
ρ	reflectivity	dimensionless
ρ_i	mass concentration of species i	kg/m³
σ	collision diameter	Å
τ	shear stress	Pa
τ	shear stress at a wall	Pa

$\tau_{y,x}$	rate of transport of momentum in the y direction due to fluid flow in the x-direction	Pa
ϕ	dissipation function	s^{-2}
Ω	collision integral	dimensionless
ω	void fraction	dimensionless
ω	solid angle	sr
ω_i	mass fraction of species i	dimensionless

1

Engineering Units and Pressure in Static Fluids

1.1 Origins of Engineering Units

Standards of weights and measures were originally developed for the purposes of trade. Linear measurements were related to dimensions of the human body and weights were determined by what a human being or an animal could carry; and the earliest measures of weight and length were devised in Babylonia and Egypt in 3000 B.C. The Babylonian unit of mass was the *mina* (which varied in the range 700 to 900 g), and the more familiar *shekel*, as the biblical Hebrew standard coin and mass, was originally a Babylonian unit. The Egyptian unit of length was the *cubit*, which is the length of an arm from the elbow to the extended fingertips. The cubit was standardized as the royal master cubit (1 cubit = 0.523 m) and contained 28 digits, the digit being the breadth of a finger. The cubit had a marvellous set of subunits as follows:

$$
\begin{array}{rcl}
4 \text{ digits} & = & 1 \text{ palm} \\
5 \text{ digits} & = & 1 \text{ hand} \\
12 \text{ digits} & = & 1 \text{ small span} \\
14 \text{ digits} & = & 1 \text{ large span} \\
16 \text{ digits} & = & 1 \text{ t'ser} \\
24 \text{ digits} & = & 1 \text{ small cubit} \\
28 \text{ digits} & = & 1 \text{ royal cubit}
\end{array}
$$

A cubit stick was divided into 28 digits, the fourteenth of which was subdivided into 16 equal lengths, the fifteenth of which was subdivided into 15 equal lengths, and so on until the twenty-eighth, which was subdivided into 2 equal lengths. The Babylonian unit of length was the *kus*, approximately 0.530 m, also known as the

Babylonian cubit, and the Babylonian unit of liquid measure was the *ka*, the volume of a cube of one hand breadth. The cube, however, had to contain a mass of 1 great mina of water.

With the passage of time, definition of the units of weights and measures passed from the Babylonians and Egyptians to the Hittites, Assyrians, Phoenicians, and the Hebrews, from them to the Greeks and from the Greeks to the Romans. The Greek unit of length was the finger (0.0193 m), with 16 fingers equaling 1 foot and 24 fingers equaling 1 Olympic cubit. The Roman unit of length was the inch, with 12 inches per foot, 5 feet per pace, and 1000 paces per Roman mile. Although the Roman pace has vanished, the mile has survived. The Roman unit of mass was the *libra*, which has retained its identity as the pound (lb) in the British system.

The British system of units developed in a haphazard way from Roman units and improvisations during the Middle Ages. Although the units of pounds, feet, and gallons were widely used, their values varied with time and location. Attempts at standardization by the monarchy led to the development of the British Imperial System of Units, which was used officially in the English-speaking world until 1965. In the tenth century, King Edgar established the Winchester Standard, named after the ancient capital of England, and kept the royal bushel measure. In the fourteenth century the yard was defined in terms of the girth of the king, and this led to the definitions of 3 feet to the yard and 12 inches to the foot. One inch was the length of three barley corns. In the sixteenth century Queen Elizabeth defined the rod as 5.5 yards, being the length of the left feet of 16 men lined up heel to toe as they emerged from church. In the seventeenth century the acre, which has been based on the area that one man and one horse could plow in a day, was defined as 4840 square yards, the furlong (furrow long) was established as 660 feet, and the several trade pounds in use were decreased to two: the troy (or apothecary) pound for weighing precious metals, jewels, and drugs, with the pound avoirdupois used for everything else sold by weight. In 1824 the imperial gallon was defined as the volume of 10 lb avoirdupois of distilled water at 62°F and a barometric pressure of 30 in., and this replaced the differing gallons used for wine, ale, and corn.

Ironically, while the British were reforming their weights and measures during the nineteenth century, the Americans were in the process of adopting units that had been discarded in Britain by a parliamentary act in 1824. Consequently, the U.S. gallon, which is based on the Queen Anne wine gallon of volume 231 cubic inches, is 17% smaller than the British Imperial gallon, and the U.S. bushel, of volume 2150.42 cubic inches, which is based on the abandoned Winchester bushel, is approximately 3% smaller than the British Imperial bushel. In the United States a hundredweight is 100 lb, which makes a short ton contain 2000 lb, and in the British Imperial system a hundredweight is 112 lb, which makes a long ton contain 2240 lb. To confuse matters further, a metric tonne, which is 1000 kg, contains 2200 lb.

The metric system, which was adopted in the major English-speaking nations in the world in 1959, has its origins in the French Revolution. In 1670, Gabriel Monton, who was the vicar of St. Paul's Church in Lyon, suggested that the unit of length be defined as the arc of 1 minute of longitude, and, in 1790, after the French Revolution, when the political climate was favorable, the French Academy of Sciences defined the meter as 1 ten millionth of the length of the meridian from the north pole to Paris. The gram was defined as the mass of 1 cubic centimeter of water at 4°C, the liter was defined as the volume of a cube of length 10 centimeters, and a platinum cylinder was fabricated and declared the standard for 1000 grams, or 1 kilogram. The widespread use of the metric system in Europe was a direct result of Napoleon's military power and influence.

Force, which is any action that tends to change the position or shape of a body, was examined experimentally by Galileo and theoretically by Sir Isaac Newton in the seventeenth century. Galileo, who introduced the experimental method to physical science, dropped objects from the leaning tower of Pisa and measured the acceleration due to gravity as being approximately 980 centimeters per second per second. Newton, who is supposed to have begun thinking about gravitation when an apple fell on his head, explained the action of force in his famous three laws of action and established the classical theory of gravitational force in his *Principia*, which was published in 1687. Gravitation is a universal force of attraction that acts between any two bodies, and, on the Earth, the weight of a body is its downward force of gravity, which is equal to the product of its mass and the acceleration due to gravity. A source of confusion as to the distinction between mass and weight was introduced in the British Imperial System of Units by the selection of the pound (lb_m) as the unit of mass and the pound force (lb_f) as the unit of force. When a force of 1 lb_f is applied to a mass of 1 lb_m it causes an acceleration of 32.17 feet per second per second and thus a mass of 1 lb_m has a weight of 32.17 lb_f in a standard gravitational field. The confusion was not diminished by the definition of the slug as a mass of 32.17 lb_m, or the definition of the poundal as the force which when applied to a mass of 1 lb_m causes an acceleration of 1 foot per second per second. Also, a hundredweight is actually a mass. Pressure, which is force per unit area, probably has more defined units than any other physical phenomenon. These include the standard atmosphere (atm), pascal (Pa), pounds per square inch absolute (psia), pounds per square inch gauge (psig), millimeters of mercury (mm Hg), inches of mercury, feet of water, torr, bar, and what is possibly the most ridiculous of all defined units, the electron-volt per cubic angstrom.

The concept of energy developed very slowly in the physical sciences and, indeed, the discipline of classical mechanics was developed without use of the notion of energy. Energy, as the ability to do work, was recognized in the seventeenth century by Gallileo, who noted that when a mass is lifted by a pulley system, the product of the force applied to the pulley and the distance through which the pulley moves is equal to the product of the weight of the mass raised and the distance through which it is raised. In the British Imperial System of Units the unit of energy is the foot-pound force; the amount of work done when a force of 1 lb_f moves through a distance of 1 foot, and in the currently used International System of Units (SI) the unit of energy is the joule, which is the work done when 1 newton of force moves through a distance of 1 meter. The notion of heat as a form of energy was not recognized until the eighteenth century and was not proven until the nineteenth century. The Greek philosophers considered heat to be an invisible fluid that could reside within matter. The temperature of a body was determined by the pressure of the fluid, caloric, which it contained, and heat transfer between bodies at different temperatures was caused by the flow of caloric from the hotter body (containing caloric at the higher pressure) to the colder body (containing caloric at the lower pressure) until equalization of the pressures was attained. In 1798, while supervising the boring of cannon at the military arsenal in Munich, Benjamin Thompson, Count von Rumford, noticed that the increase in the temperature of the cannon was proportional to the amount of work performed during the boring and provided the first numerical equivalence of heat and work. The equivalence was finally determined accurately in 1840 by James Joule, who performed work by a variety of means in a quantity of adiabatically contained water and noted that the increase in the temperature of the water was directly proportional to the amount of work performed. He determined the equivalence to be 778 foot-pounds force of mechanical energy per British thermal unit of thermal energy, where the British thermal unit (Btu) is the amount of heat required to increase the temperature of 1 lb_m of water from 60°F to 61°F.

A definition of a unit of power, the rate at which work is performed, became necessary in the eighteenth century when steam engines replaced horses for pumping water out of deep coal and ore mines in Great Britain. James Watt, who had invented the first efficient steam engine in 1763, conducted experiments with strong dray horses and defined the unit of power as the horsepower, with 1 horsepower (hp) being work performed at the rate of 33,000 foot-pounds force per hour. Watt's definition is actually 50% more than the rate at which an average horse can perform during a working day. The SI unit of power is the watt, which is work done at the rate of 1 joule per second, and in everyday use, the units of horsepower and watts are used comfortably in different applications by the average person who might not know the definition of either unit. For example, the reader of an automobile magazine would not be surprized to read that a particular four-cylinder, 24-valve double-overhead-cam automobile engine is capable of delivering 190 hp, but might be confused by the information that the engine can deliver 140 kilowatts (kW). Similarly, a person, comfortable buying a 100-watt lamp bulb might be confused if the label on the bulb rated it at 0.0134 hp.

Temperature scales were defined in the eighteenth century by the astronomer Anders Celsius and the physicist Daniel Fahrenheit. In the *Celsius (centigrade) scale* the reference points are the normal freezing and normal boiling temperatures of water, and the scale between these points is divided into 100 Celsius or centigrade degrees, with 0°C being the normal freezing temperature of water and 100°C being the normal boiling temperature. In the *Fahrenheit scale* the reference points are the eutectic in the $H_2O-NaCl$ system and the body temperature of a healthy horse, and the scale between these two points is divided into 100 Fahrenheit degrees. The eutectic temperature in the salt−water system is 0°F and a healthy horse has a body temperature of 100°F. A healthy human being, who is slightly cooler than a healthy horse, has a body temperature of 98.6°F. As with the units of power, the units of temperature are used differently in different applications. For example, would a person who is comfortable with the thermostat in a room set at 75°F be comfortable if the temperature in the room were 45°C, and is the eutectic temperature in the Ag−Cu system, read off the phase diagram as 780°C, greater or less than 1400°F?

In 1802, Joseph Gay-Lussac observed that the coefficient of thermal expansion of what he called "permanent gases" had a constant value of 1/267. The *coefficient of thermal expansion* is defined as the fractional increase, with increasing temperature at constant pressure, of the volume of the gas at 0°C, and thus, if, on cooling, a "permanent gas" did not condense or sublime, its volume would decrease to zero at a temperature of $-267°C$. This observation indicated that an absolute zero of temperature exists and lead to the definition of an absolute temperature scale called the *ideal gas temperature scale*. More refined measurement of the coefficient of thermal expansion of permanent gases, made by Regnault in 1847, gave a value of 1/273, and hence the absolute zero of temperature is $-273°C$. In 1848, William Thomson, Lord Kelvin, showed that the ideal gas temperature scale is identical with the thermodynamic temperature scale derived from considering the efficiencies with which reversible thermodynamic engines convert heat to work. In SI the unit of temperature is the kelvin (K), which has the same magnitude as the Celsius degree, and the triple point of water is defined as occurring at the temperature 273.1600 K. The normal freezing temperature of water is thus 273.15 K.

In an attempt to eliminate the the proliferation of units (such as electron-volts, barns, Rydbergs, Bohr magnetons, angstoms, etc.) improvised in the increasing number of subdisciplines of science and engineering that emerged during the twentieth century, and to increase the precision of defined units, the 11th General Conference on Weights and Measure, meeting in Paris in 1960 devised a new International System of Units (SI) which defined six base units as follows.

1. *Length: meter*. The meter is defined as 1,650,763.73 wavelengths in a vacuum of the orange-red line of the spectrum of krypton 86.
2. *Mass: kilogram*. The standard for the unit of mass, kilogram, is a platinum–iridium alloy cylinder kept by the International Bureau of Weights and Measures in Sèvres, which is near Paris.
3. *Time: second*. The second is defined as the duration of 9,192,631,770 cycles of radiation associated with a specified transition, or change in energy level, in the cesium atom.
4. *Electric current: ampere*. The ampere is defined as the magnitude of the current which, when flowing through each of two long parallel wires separated by 1 meter in free space, causes a force between the two wires (due to their magnetic fields) of 2×10^{-7} newtons per meter of length.
5. *Thermodynamic temperature: kelvin*. The thermodynamic temperature scale has as reference points the hypothetical absolute zero of temperature, 0 K, and the triple point of water, defined as 273.1600 K.
6. *Light intensity: candela*. The candela is defined as the luminous intensity of 1/600,000 of a square meter of a cavity radiating at the melting temperature of platinum, 2042 K.

The elementary and derived SI units and their symbols are listed in Appendix A, and the prefixes for multiples and submultiples of the units, and their symbols, are listed in Appendix B. The original British Imperial System of units and their equivalents in SI are listed in Appendix C.

1.2 Concept of Pressure

The force exerted by a gas on the walls of the vessel containing the gas is caused by collisions of the individual atoms or molecules of the gas with the wall. An individual atom of mass m approaches the wall in some direction with the velocity **v**, and after an elastic collision with the wall, rebounds in another direction at the velocity **v**. As momentum, $m\mathbf{v}$, is a vector quantity that depends on direction as well as magnitude, the change in the direction of motion of the atom causes a change in its momentum. For example, if the atom of mass m is traveling in the $+x$ direction normal to the wall, its momentum before the collision is $m\mathbf{v}$, and after the collision is $-m\mathbf{v}$. The change in the momentum of the atom as a result of the collision is thus $(m\mathbf{v}) - (-m\mathbf{v}) = 2m\mathbf{v}$. The population of atoms in the gas is continuously colliding with the walls of the containing vessel, and hence there is a continuous rate of change of momentum, which has the units

$$\frac{\text{momentum}}{\text{time}} = \frac{\text{mass} \times \text{velocity}}{\text{time}} = \text{kg}\,\frac{\text{m}}{\text{s}^2}$$

These are also the units of force in newtons, and thus the force exerted by the gas on the walls of the container arises from, and is equal to, the rate of change of momentum of the gas atoms or molecules caused by their collisions with the walls. The force exerted by a liquid on the walls of its container arises from the same phenomenon, and the fluid pressure is this force per unit area. Any imaginary surface within the body of the fluid experiences the same force in the direction normal to the plane of the imaginary surface.

The pressure at any point in a static fluid is the same in all directions. This can be seen by considering the element of fluid with sides of lengths Δa, Δb, and Δc and unit depth, located within a body of fluid, as shown in Fig. 1.1. Mechanical equilibrium requires that the algebraic sum of the forces acting in the x-direction be zero,

$$P_a \sin \phi \, \Delta a = P_b \sin \theta \, \Delta b \tag{1.1}$$

and the algebraic sum of the forces acting in the y-direction be zero,

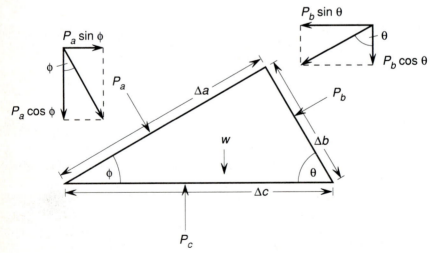

FIGURE 1.1. Pressures acting on a volume element of a fluid.

$$P_a \cos \phi \, \Delta a + P_b \cos \theta \, \Delta b + w = P_c \, \Delta c \qquad (1.2)$$

In Eq. (1.2), w is the body force, or the weight of the element of fluid, which arises from the influence of the gravitational field on the mass of fluid in the element. For a mass of fluid m of density ρ in an element of volume V,

$$w = mg = \rho g V = \tfrac{1}{2}\rho g \, \Delta a \, \Delta b \qquad (1.3)$$

From the geometry of the triangular cross section of the element,

$$\sin \theta = \frac{\Delta y}{\Delta b}$$

$$\cos \theta = \frac{\Delta c_2}{\Delta b}$$

$$\sin \phi = \frac{\Delta y}{\Delta a}$$

$$\cos \phi = \frac{\Delta c_1}{\Delta a}$$

Thus in Eq. (1.1),

$$P_a = P_b \qquad (1.4)$$

and in Eq. (1.2),

$$P_a \, \Delta c_1 + P_b \, \Delta c_2 + \tfrac{1}{2}\rho g \, \Delta a \, \Delta b = P_c \, \Delta c$$

or, as

$$P_a = P_b \qquad \text{and} \qquad \Delta c_1 + \Delta c_2 = \Delta c$$

$$P_a + \frac{1}{2} \rho g \, \frac{\Delta a \, \Delta b}{\Delta c} = P_c \qquad (1.5)$$

As the element is made infinitesimally small, the body force approaches zero, and hence in the limit of $\Delta a = \Delta b = \Delta c \to 0$,

$$P_a = P_c$$

and thus

$$P_a = P_c = P_c \qquad (1.6)$$

that is, the pressure at any point in the static fluid is the same in all directions.

The only forces acting in a static fluid are normal forces (such as those giving rise to the pressures P_a, P_b, and P_c in Fig. 1.1) and the body forces that arise from the influence of gravity on the mass of the fluid. This influence of gravity is such that the pressure on any plane in a static fluid normal to the gravitational field arises from the weight of fluid above the plane, and thus the pressure at a point in a static fluid in a gravitational field varies with position in the direction of the field.

Consider the column of fluid of cross-sectional area A in a gravitational field above the level z shown in Fig. 1.2(a). The body force exerted by the column of fluid above the level z on the fluid below the level z equals $P|_z A$, where $P|_z$ is the normal pressure in the fluid at the level z. In Fig. 1.2(b) the body force exerted by the column of fluid above the level $z + \Delta z$ on the fluid below the level $z + \Delta z$ equals $P|_{z+\Delta z} A$, where $P|_{z+\Delta z}$ is the normal pressure in the fluid at the level $z + \Delta z$. Thus in Fig. 1.2(c) the difference between the normal force at level z and the normal force at level $z + \Delta z$ equals the body force arising from the weight of the fluid between the two levels:

$$P|_{z+\Delta z} A + \rho g A \, \Delta z = P|_z A$$

or

$$\frac{P|_{z+\Delta z} - P|_z}{\Delta z} = -\rho g$$

which in the limit of $\Delta z \to 0$ becomes

$$\frac{dP}{dz} = -\rho g \tag{1.7}$$

Consider atmospheric air above the surface of the Earth. The standard atmospheric pressure at sea level is 1.01325×10^5 Pa (1 standard atmosphere) at a temperature of 15°C (288 K) and the standard acceleration due to gravity is $g = 32.174$ ft/s² (which is defined in terms of the pull of the Earth's gravitational field on a mass of 1 lb at sea level and 45°N latitude). In SI units the standard acceleration due to gravity is thus

$$g = 32.174 \, \frac{\text{ft}}{\text{s}^2} \times 12 \, \frac{\text{in.}}{\text{ft}} \times 2.54 \, \frac{\text{cm}}{\text{in}} \times 0.01 \, \frac{\text{m}}{\text{cm}} = 9.8067 \text{ m/s}^2$$

Atmospheric air behaves as an ideal gas, in which case it obeys the ideal gas law,

$$PV = \frac{RT}{M} \tag{1.8}$$

where P = pressure of the gas (Pa)
  V = volume of the gas per unit mass ($= 1/\rho$ in m³/kg)
  R = universal gas constant (8.3144 J·g mol^{-1} · K^{-1})
  T = absolute temperature of the gas (K)
  M = molecular weight of the gas (kg/g mol)

Rearrangement of Eq. (1.8) as

$$\rho = \frac{1}{V} = \frac{PM}{RT}$$

and substitution into Eq. (1.7) gives

$$\frac{dP}{dz} = -\frac{PMg}{RT}$$

or

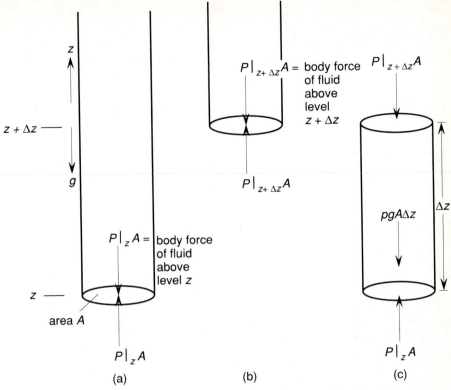

FIGURE 1.2. Derivation of the barometric formula from consideration of the body force of a fluid in a gravitational field.

$$\frac{dP}{P} = -\frac{Mg}{RT}\,dz \tag{1.9}$$

Assuming that T and g are independent of z, integration of Eq. (1.9) between the limits

$$P = P_0 \text{ (standard atmospheric pressure)} \quad \text{at } z = 0 \text{ (sea level)}$$

and

$$P = P \quad \text{at } z = z$$

gives

$$\ln\frac{P}{P_0} = -\frac{Mgz}{RT}$$

or

$$P = P_0 e^{-Mgz/RT} \tag{1.10}$$

and Eq. (1.10) is known as the *barometric formula*.

Air is a mixture of 21 volume percent oxygen (of molecular weight 0.032 kg/g mol) and 79 volume percent nitrogen (of molecular weight 0.028 kg/g mol), and thus the molecular weight of air is

$$
\begin{aligned}
M &= (0.21 \times 0.032) + (0.79 \times 0.028) \text{ kg/g mol} \\
&= 0.02884 \text{ kg/g mol}
\end{aligned}
$$

Thus with

$$M = 0.02884 \text{ kg/g mol}$$

$$g = 9.807 \text{ m/s}^2$$

$$R = 8.3144 \text{ J/g mol·K} \ (\text{kg·m}^2\text{·s}^{-1}\text{·g mol}^{-1}\text{·K}^{-1})$$

$$T = 288 \text{ K}$$

$$z = \text{height in meters}$$

$$\frac{Mg}{RT} = \frac{0.02884 \times 9.807}{8.3144 \times 288} = 1.181 \times 10^{-4} \text{ m}^{-1}$$

and hence Eq. (1.10) becomes

$$P = P_0 e^{-1.181 \times 10^{-4}z} \tag{1.11}$$

which is shown as line A in Fig. 1.3.

By convention, the international standard atmosphere is defined on the basis of a uniform temperature gradient of $-6.5°C$ per 1000 m of altitude from sea level to a height of 11,000 m and a constant temperature of $-56.5°C$ for heights greater than 11,000 m. Thus in the range $z = 0$ to $z = 11,000$ m,

$$T = 288 - 0.0065z \tag{1.12}$$

which when substituted into Eq. (1.9) gives

$$\frac{dP}{P} = \frac{-Mg}{R(188 - 0.0065z)} \, dz \tag{1.13}$$

integration of which gives

$$
\begin{aligned}
\ln \frac{P}{P_0} &= \frac{Mg}{0.0065R} \ln \frac{288 - 0.0065z}{288} \\
&= 5.245 \ln \frac{288 - 0.0065z}{288}
\end{aligned}
$$

or

$$P = P_0 \left(\frac{288 - 0.0065z}{288} \right)^{5.245} \tag{1.14}$$

Equation (1.14) is shown as line B in Fig. 1.3.

From Eq. (1.14), it is calculated that the air pressure at the summit of Pike's Peak in Colorado (elevation 14,110 ft, or 4301 m) is 0.59 atm, which illustrates why

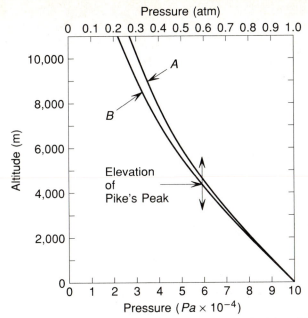

FIGURE 1.3. Variation of atmospheric pressure with altitude. Line *A*, which is given by Eq. (1.11), is drawn for a constant air temperature of 288 K. Line *B*, which is given by Eq. (1.14), is drawn for an air temperature that decreases at the rate of 6.5°C per 1000 m to an altitude of 11,000 m.

tourists have difficulty breathing at the summit. At 17,000 ft (5182 m), which is the maximum altitude at which passengers without oxygen masks can fly in an unpressurized airplane, the air pressure is 0.52 atm.

For completeness, it must be pointed out that the gravitational constant itself varies with distance above sea level as

$$g = 9.8067 \times \left(\frac{6388}{6388 + z}\right)^2$$

where z is in kilometers. The gravitation acceleration at the summit of Pike's Peak is thus 9.79 m/s^2.

1.3 Measurement of Pressure

In 1643, Torricelli filled a 4-foot-long glass tube with mercury and immersed the open end of the tube in a bath of mercury as shown in Fig. 1.4. He observed that the mercury did not run out of the tube and concluded that a vacuum had been created above the mercury in the tube and that the column of mercury was supported by the atmospheric pressure acting on the free surface of the mercury in the bath.

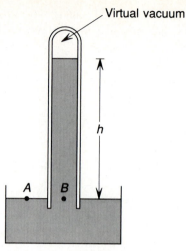

FIGURE 1.4.   Torricelli vacuum created in an inverted flask containing a liquid.

At equilibrium the body force, per unit area, on the surface of the mercury at point
A exerted by the column of air above the surface is equal to the body force exerted,
per unit area, at point B exerted by the column of mercury above B. The body force
F exerted by the column of mercury of mass m, height h, and cross-sectional area
A is

$$F = mg$$
$$= \rho Ahg$$

which gives rise to a static pressure at point B of

$$P = \frac{F}{A} = \rho gh$$

and hence measurement of the height h allows determination of the atmospheric
pressure at point B. Mercury is considered to be an incompressible liquid, in which
case its density is independent of position within the column. Thus the static pressure
in the mercury column decreases linearly with z from the value ρgh at point B to
virtually zero at the free surface at the height h. Standard atmospheric pressure at
0°C corresponds to $h = 760$ mm and is referred to as "760 mm Hg" or "29.92 in
Hg." If the long cylindrical tube contained water, standard atmospheric pressure
would correspond to $h = 33.93$ ft and would be referred to as "33.93 ft of water."
The arrangement shown in Fig. 1.4 gave rise to the development of the barometer,
which provides a means of measuring the atmospheric pressure.

Most devices that measure pressure actually measure the difference between the
pressures of two fluids. For example, the pressure gauge that is used to measure the
air pressure in an inflated rubber tire actually measures the difference between the
pressure of the air in the tire and the pressure of the atmospheric air outside the tire.
This gives rise to the distinction between absolute pressure and gauge pressure.
Inflating a tire to a measured pressure of 35 psi means that the difference between

the pressures inside and outside the tire is 35 psi. If the atmospheric pressure outside the tire is 14.7 psi, the actual pressure inside the tire is $35 + 14.7 = 49.7$ psi. The actual pressure is called the *absolute pressure* and is the value that is determined by the rate of change of momentum of the atoms or molecules of the gas caused by their collisions with the walls of the container, and the difference between the absolute pressure and atmospheric pressure is called the *gauge pressure*. Thus

absolute pressure = gauge pressure + atmospheric pressure

With reference to the inflated tire

49.7 psia (lb_f per square inch absolute pressure)

= 35 psig (lb_f per square inch gauge pressure)

+ 14.7 psia (absolute atmospheric pressure)

Figure 1.5 shows a simple U-tube manometer, which allows measurement of the difference between the pressure P' of a fluid and the atmospheric pressure P_0. The U-tube is partially filled with an incompressible fluid of density ρ (which is immiscible with the fluid, the pressure of which is being measured), and at equilibrium the absolute pressure at level A is the same in both legs of the manometer. At level A in the left leg the pressure is that of the fluid, P', and at level A in the right leg the pressure is that due to the body force of the column of manometric fluid of height h plus the atmospheric pressure, P_0, exerted on the surface of the manometric fluid. Thus

$$P' = \rho h g + P_0$$

The barometer shown in Fig. 1.4 is also a manometer, in that it measures the difference between the atmospheric pressure at A and the pressure of the virtual vacuum above the column of liquid in the closed end of the tube. The pressure of

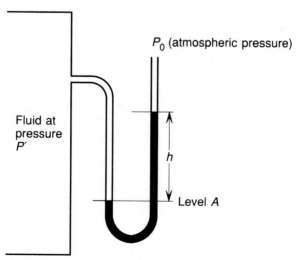

FIGURE 1.5. Simple U-tube manometer for measuring differences in pressure.

the virtual vacuum is actually the saturated vapor pressure of the liquid in the barometer, which with mercury at 298 K as the liquid is 0.76 Pa. This is negligible in comparison with an atmospheric pressure of approximately 10^5 Pa.

Figure 1.6 shows a U-tube manometer containing two immiscible liquids, 1 and 2, of densities ρ_1 and ρ_2 (where $\rho_1 > \rho_2$). The cross-sectional area of the tubing connecting the two reservoirs is a and the cross-sectional areas of the reservoirs is A. With atmospheric pressure P_0 exerted on the free surfaces of both liquids, the equilibrium state is as shown in Fig. 1.6(a). The pressure at level 1 is the same in both legs of the manometer,

$$P_0 + \rho_1 h_1 g = P_0 + \rho_2 h_2 g$$

or

$$\rho_1 h_1 = \rho_2 h_2 \qquad (1.15)$$

In Fig. 1.6(b) the pressure exerted on the surface of liquid 2 has been increased to P', which causes the level of the surface of liquid 2 to fall through the distance Z, the position of contact of the two liquids in the right leg of the U-tube to fall through the distance z, and the level of the surface of liquid 1 to rise the distance Z. At the new state of equilibrium the pressures in the two legs are equal at level 2 and hence

$$P_0 + \rho_1 g x_1 = P' + \rho_2 g x_2$$

or

$$P' - P_0 = g(\rho_1 x_1 - \rho_2 x_2) \qquad (1.16)$$

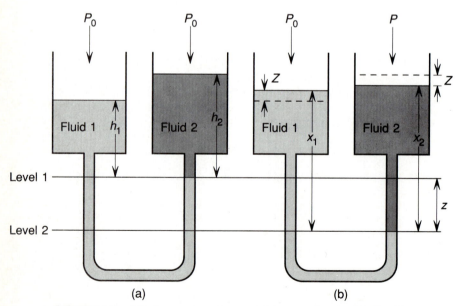

FIGURE 1.6.   U-tube manometer containing two immiscible liquids.

From geometric considerations

$$x_1 = h_1 + Z + z$$

and

$$x_2 = h_2 - Z + z$$

substitution of which into Eq. (1.16) gives

$$P' - P_0 = g[\rho_1(h_1 + Z + z) - \rho_2(h_2 - Z + z)]$$
$$= g[\rho_1 h_1 - \rho_2 h_2 + z(\rho_1 - \rho_2) + Z(\rho_2 + \rho_1)]$$

which in view of the equality in Eq. (1.15) becomes

$$P' - P_0 = g[z(\rho_1 - \rho_2) + z(\rho_2 + \rho_1)] \tag{1.17}$$

Conservation of mass and geometry also give

$$ZA = za$$

substitution of which into Eq. (1.17) gives

$$P' - P_0 = g \left[z(\rho_1 - \rho_2) + \frac{za}{A} (\rho_2 + \rho_1) \right]$$
$$= gz \left[(\rho_1 - \rho_2) + \frac{a}{A} (\rho_2 + \rho_1) \right] \tag{1.18}$$

As z is the measured quantity, the sensitivity of the manometer is increased by decreasing the difference between ρ_1 and ρ_2 and decreasing the ratio a/A.

1.4 Pressure in Incompressible Fluids

In the case of incompressible fluids it is often necessary to know the pressure at some depth below the free surface of the liquid, in which case it is convenient to consider that h increases in the $-z$ direction (which is in the direction of the gravitational field). Thus, taking P_0 to be pressure at the free surface, the pressure at a depth h below the surface is obtained from integration of Eq. (1.7) as

$$P = P_0 + \rho gh \tag{1.19}$$

that is, in a fluid of constant density, the pressure increases linearly with increasing depth below the free surface.

Figure 1.7(a) shows a plate of length L and width W immersed vertically in a liquid of density ρ. From Eq. (1.19) the pressure exerted by the liquid on each face of the plate varies linearly with depth below the free surface from a gauge pressure of zero at the free surface to a gauge pressure of ρgL at the bottom of the plate. The total force, F_{total}, exerted on each side of the plate is obtained from the definition

$$F_{total} = \int_A P \, dA$$

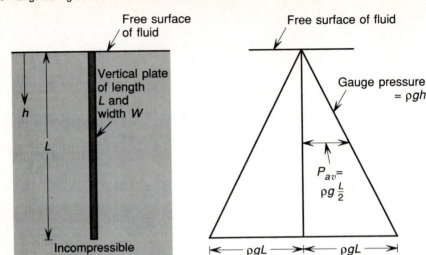

(a)

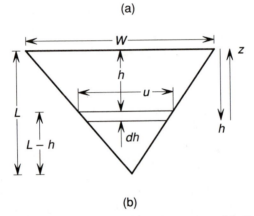

(b)

FIGURE 1.7. (a) Vertical plate immersed in an incompressible fluid and the variation, with depth, of the pressures exerted by the fluid on both sides of the plate; (b) diagram for consideration of the pressure exerted by an incompressible fluid on a triangle immersed vertically in the fluid.

in which A is area. For a plate of constant width W, $dA = W\,dh$ and hence

$$F_{total} = \int_0^L \rho g h W\,dh = \frac{\rho g L^2 W}{2}$$

The average pressure exerted on each side of the plate is thus

$$P_{av} = \frac{F_{total}}{WL} = \frac{\rho g L}{2}$$

which is the local gauge pressure exerted halfway down the plate.

Consider the case of an immersed triangular plate as shown in Fig. 1.7(b). The pressure exerted on the plate again increases linearly with depth, and for the triangle,

$$dA = u \, dh$$

From similar triangles,

$$\frac{u}{W} = \frac{L - h}{L}$$

and hence

$$u = (L - h)\frac{W}{L}$$

and

$$dA = \frac{(L - h)W \, dh}{L}$$

Therefore,

$$F_{total} = \int_A P \, dA = \int_0^L \frac{\rho g z (L - h) W \, dh}{L}$$

$$= \rho g W \left(\int_0^L h \, dh - \int_0^L \frac{h^2}{L} \, dh \right)$$

$$= \rho g W \left(\frac{L^2}{2} - \frac{L^2}{3} \right)$$

$$= \frac{\rho g W L^2}{6}$$

and the average pressure, P_{av}, is obtained as

$$P_{av} = \frac{F_{total}}{A} = \frac{\rho g W L^2 / 6}{W L / 2} = \frac{\rho g L}{3}$$

which is the local gauge pressure exerted one-third of the way down the plate.

These are two examples of the general case in which the average pressure exerted on a vertical plate is equal to the local value exerted at the centroid (or center of gravity) of the plate.

EXAMPLE 1.1 ───────────────────────────────────────

The rectangular metal water tank shown in Fig. 1.8 has a glass window of dimensions 2 m × 2 m located at a distance of 1 m below the waterline in the tank. What force does the water, of density 1000 kg/m³, exert on the window?

From Eq. (1.19) the gauge pressure is

$$P - P_0 = \rho g h$$
$$= 1000 \times 9.81 h$$
$$= 9810 h \text{ Pa}$$

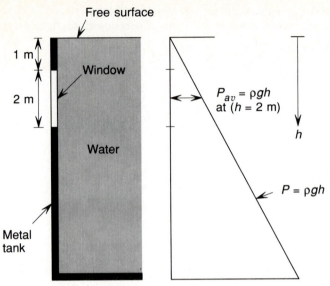

FIGURE 1.8. Variation, with depth, of the pressure exerted by an incompressible fluid on the walls of its container.

Thus the local pressure exerted at the centroid of the window (at $h = 2$ m) is 19,620 Pa and the total force exerted on the window by the water is

$$F_{total} = 19,620 \times 4 = 78,480 \text{ N}$$

EXAMPLE 1.2

A cylindrical tank of radius 2 m is placed horizontally and is half-filled with oil of density 888 kg/m³. Calculate the force exerted by the oil on one end of the tank.

$$F = \int_{area} P \, dA$$

where in Fig. 1.9

$$P = \rho g y$$

and

$$dA = 2x \, dy$$

From the equation of a circle,

$$x^2 + y^2 = R^2$$

$$x = (R^2 - y^2)^{1/2}$$

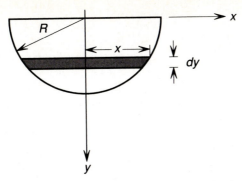

FIGURE 1.9. Consideration of the pressure exerted by a fluid on the ends of a cylindrical tank.

Thus

$$dA = 2(R^2 - y^2)^{1/2} \, dy$$

and

$$\begin{aligned}
F &= \int \rho g y 2(R^2 - y^2)^{1/2} \, dy \\
&= 2\rho g \left[-\tfrac{1}{3}(R^2 - y^2)^{3/2} \right]_0^R \\
&= 2\rho g \left(0 + \frac{R^3}{3} \right)
\end{aligned}$$

or

$$\begin{aligned}
F &= \tfrac{2}{3}\rho g R^3 \\
&= \tfrac{2}{3} \times 888 \times 9.807 \times 2^3 \\
&= 46{,}450 \text{ N}
\end{aligned}$$

In this example,

$$P_{av} = \frac{F}{A} = \frac{\tfrac{2}{3}\rho g R^3}{\tfrac{1}{2}\pi R^2} = \frac{4}{3} \frac{\rho g R}{\pi}$$

which is the local pressure at the centroid of the semicircle,

$$y = \frac{4}{3}\frac{R}{\pi}, \qquad x = 0$$

   Figure 1.10(a) shows a vertical plate acting as a dam separating two bodies of water, each of which have the same depth, and hence the free surfaces of which are at the same level. As the variations of pressure with depth on both sides are equal and opposite, there is no net force on the dam. In Fig. 1.10(b) the depths of the

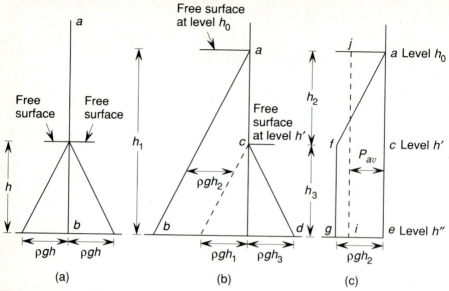

FIGURE 1.10. (a) Variation, with depth of immersion, of the pressures exerted on both sides of a dam when the free surfaces of the fluid are at the same level on each side of the dam; (b) variation, with depth of immersion, of the pressures exerted on both sides of a dam when the free surfaces of the fluid are at different levels on each side of the dam; (c) variation, with depth of immersion, of the net pressure exerted on the dam shown in (b).

liquids, h_1 and h_3, differ from one another and hence the free surfaces occur at different levels, at h_0 on the left side and at h' on the right. The variation of gauge pressure with depth below the free surface at the left side is given by the line ab and the corresponding variation on the right is given by the line cd. The average gauge pressure on the left side is the local value at the depth $h_1/2$,

$$P_{av}(\text{left}) = \frac{\rho g h_1}{2}$$

and thus for a plate width of W, the force exerted on the left face is

$$F = \frac{\rho g h_1}{2} \times h_1 W = \frac{\rho g h_1^2 W}{2}$$

Similarly, the average gauge pressure on the right side is the local value at the depth $h_3/2$,

$$P_{av}(\text{right}) = \frac{\rho g h_3}{2}$$

and hence the force exerted on the right face is

$$F = \frac{\rho g h_3^2 W}{2}$$

The net force acting on the dam is thus

$$F_{net} = \frac{\rho g W}{2} (h_1^2 - h_3^2) \tag{1.20}$$

Alternatively, in Fig. 1.10(b), the gauge pressure exerted by the liquid on the left increases linearly from zero at the free surface at level h_0 to the value $\rho g h_2$ at the level h'. Below the level h' the pressure exerted on the right face, cd, is nullified by that part of the pressure exerted on the left given by ce, and hence the net gauge pressure on the dam between the levels h' and h'' is constant at the value $\rho g h_2$. The variation of net gauge pressure on the dam with depth below the level h_0 is thus as shown in Fig. 1.10(c). The average net pressure is then obtained as the value of P that makes the area of the rectangle $ieaji$ equal to the area $afgea$,

$$
\begin{aligned}
P_{av}(h_2 + h_3) &= \int_{h_0}^{h''} P \, dh \\
&= \int_{h_0}^{h'} P \, dh + \int_{h'}^{h''} P \, dh \\
&= \int_{h_0}^{h'} \rho g h \, dh + \int_{h'}^{h''} \rho g h_2 \, dh \\
&= \frac{\rho g h^2}{2} \Bigg|_{h_0}^{h'} + \rho g h_2 \int_{h'}^{h''} dh \\
&= \frac{\rho g h_2^2}{2} + \rho g h_2 h_3
\end{aligned}
$$

or

$$P_{av} = \frac{\rho g}{2} \frac{h_2^2 + 2h_2 h_3}{h_2 + h_3}$$

in which case the net force on the dam is

$$
\begin{aligned}
F_{net} &= P_{av} \times \text{area} \\
&= \frac{\rho g}{2} (h_2^2 + 2h_2 h_3) W
\end{aligned} \tag{1.21}
$$

which, as $h_1 = h_2 + h_3$, is identical with Eq. (1.20).

1.5 Buoyancy

The phenomenon of buoyancy was first considered the third century B.C. by Archimedes, who enunciated what is known as *Archimedes' principle:* A body immersed in or floating in a fluid is acted upon by an upward buoyant force equal to

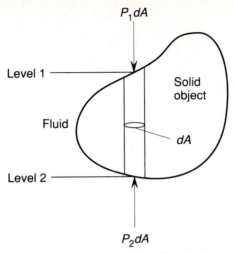

$P_1 dA$

Level 1

Solid object

Fluid

dA

Level 2

$P_2 dA$

FIGURE 1.11.   Illustration of Archimedes' principle.

the weight of fluid displaced, which acts through the center of gravity of the displaced volume. Figure 1.11 shows a solid object of density ρ_s immersed in a fluid of density ρ_f. Consider the cylindrical volume element of cross-sectional area dA located in the object. The differential upward force $dF_{(up)}$ exerted on the element by the fluid is given by

$$dF_{(up)} = P_2\, dA - P_1\, dA$$

But this is also the body force of an identical volume element of fluid given by

$$\rho_f g(z_2 - z_1)\, dA$$

Thus, integrating over the entire surface of solid object gives the total upward force exerted on the object by the fluid as

$$F_{(up)} = \rho_f g V$$

where V is the volume of the object.

EXAMPLE 1.3

In air an object weighs 3 N, and when immersed in water, weighs 1.5 N. If the density of water is 1000 kg/m³, what is the density of the object? The force balance is illustrated in Fig. 1.12.

The decrease in the weight of the object when immersed in water equals the buoyancy force, which is the weight of displaced water,

decrease in weight = buoyancy = weight of displaced water
$$= \rho_f g \text{ (volume of object)}$$

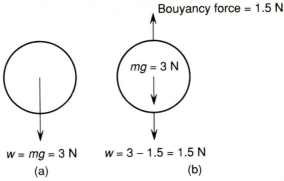

FIGURE 1.12. Difference between the weights obtained when an object is (a) weighed in air and (b) weighed in water.

Therefore,

$$\text{volume of object} = \frac{\text{buoyancy}}{\rho_f g}$$

But

$$\text{weight of the object in air} = \rho_s g \text{ (volume of object)}$$
$$= \frac{\rho_s \times \text{ buoyancy}}{\rho_f}$$

Therefore,

$$\rho_s = \frac{\rho_f \times \text{ weight of the object in air}}{\text{buoyancy}}$$
$$= \frac{1000 \times 3}{3 - 1.5} = 2000 \text{ kg/m}^3$$

In this example the influence of the buoyancy force of air on the measurement of the weight of the object in air is ignored.

EXAMPLE 1.4

A rod consisting of a 0.1-m length of hollow glass tubing of radius 0.005 m and composite density 600 kg/m^3, to one end of which is attached a 0.001-m length of iron rod of radius 0.005 m and density 7870 kg/m^3, is immersed in water. What fraction of the glass tube lies below the surface of the water?

With reference to Fig. 1.13,

$$\text{weight of the rod} = (\rho g V)_{\text{glass}} + (\rho g V)_{\text{iron}}$$
$$= \pi g(600 \times 0.1 \times 0.005^2)$$
$$+ \pi g(7870 \times 0.001 \times 0.005^2)$$
$$= 0.001697 \pi g$$

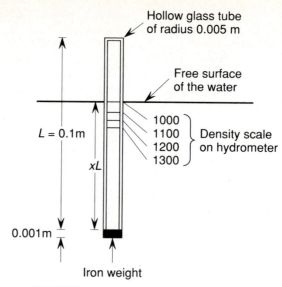

FIGURE 1.13. Principle of a hydrometer.

and

weight of the water displaced $= \pi g (0.001 \times 0.005^2 + 0.1 \times 0.005^2 x)1000$

At equilibrium the weight of the rod equals the weight of displaced water, and hence

$$x = \frac{0.001697 - 0.005^2}{0.1 \times 1000 \times 0.005^2} = 0.67$$

Therefore, approximately two-thirds of the length of the glass tube is below the surface of the water. As the density of the fluid in which the composite rod floats is increased, the volume of displaced fluid decreases and hence x decreases. The variation of x with fluid density is listed in Table 1.1.

TABLE 1.1

ρ (kg/m^3)	1000	1100	1200	1300
x	0.667	0.605	0.554	0.511

The composite rod is the basis for construction of a hydrometer, on which a scale placed on the stem permits the measurement of the density of a fluid by observing the length of stem protruding through the surface of the fluid in which the fluid is immersed.

EXAMPLE 1.5

A balloon filled with 60 kg of helium gas at a pressure of 2.25×10^5 Pa at a temperature of 15°C is tethered to the Earth at sea level. Calculate the force exerted on the tether and the altitude to which the balloon ascends when released.

The atomic weight of helium is 4×10^{-3} kg/g mol. Thus the volume V of the balloon is

$$V = \frac{mRT}{MP} = \frac{60 \times 8.3144 \times 288}{4 \times 10^{-3} \times 2.25 \times 10^5} = 159.6 \text{ m}^3$$

The density of air at standard atmospheric pressure at 288 K is 1.233 kg/m³, and hence the upward buoyant force on the ballon is

$$\text{buoyancy} = \rho g V = 1.233 \times 9.81 \times 159.6 = 1931 \text{ N}$$

The upward force on the tether is thus the buoyancy force minus the weight of the balloon,

$$1931 - 60 \times 9.81 = 1342 \text{ N}$$

When released, the balloon rises to an altitude at which the upward buoyancy force equals the weight of the balloon,

$$\text{buoyancy} = \rho_{air} g V = 60g$$

or the density of the atmospheric air is

$$\rho_{air} = \frac{60}{159.6} = 0.376 \text{ kg/m}^3$$

(which is also the density of the helium in the balloon).

Designating ρ_1, P_1, and T_1 as the density, pressure, and temperature of air at standard atmosphere pressure and 288 K at sea level and ρ_2, P_2, and T_2 as the corresponding values at the altitude to which the balloon rises, the ideal behavior of atmospheric air gives

$$\frac{P_1}{T_1 \rho_1} = \frac{P_2}{T_2 \rho_2} \tag{i}$$

However, from Eq. (1.12),

$$T_2 = 288 - 0.0065z$$

and from Eq. (1.14),

$$P_2 = P_1 \left(\frac{288 - 0.0065z}{288} \right)^{5.245}$$

substitution of which into Eq. (i) gives

$$\frac{1}{T_1 \rho_1} = \left(\frac{288 - 0.0065z}{288} \right)^{5.245} \times \frac{1}{\rho_2 (288 - 0.0065z)}$$

or

$$\frac{p_2}{T_1 \rho_1} \times 288^{5.245} = (288 - 0.0065z)^{4.245}$$

or

$$\frac{0.376 \times (288)^{4.245}}{1.233} = 8.40 \times 10^9 = (288 - 0.0065z)^{4.245}$$

which has the solution

$$z = 10,810 \text{ m}$$

As a check, at $z = 10,810$ m,

$$P_2 = P_1 \left(\frac{288 - 0.0065z}{288} \right)^{5.245}$$

$$= 101,325 \times \left(\frac{288 - 0.0065 \times 10,810}{288} \right)^{5.245}$$

$$= 23,369 \text{ Pa}$$

and

$$T_2 = 288 - 0.0065 \times 10,810 = 218 \text{ K}$$

Therefore, from Eq. (i),

$$\rho_2 = \frac{P_2 T_1}{P_1 T_2} \rho_1 = \frac{23,369 \times 288}{101,325 \times 218} \times 1.233$$

$$= 0.376 \text{ kg/m}^3$$

1.6 Summary

The pressure exerted by a gas on the walls of its container arises from the collisions of the individual atoms or molecules with the walls, and the force exerted on the wall is equal to the rate of change of momentum of the atoms or molecules due to the collisions. The pressure at any point in a fluid is the same in all directions. The influence of gravity is such that the pressure at any plane in a static fluid, normal to the gravitational field, arises from the weight of the fluid above the plane, and thus the pressure at a point in a static fluid varies with position in the direction of the gravitational field. The variation of atmospheric pressure with altitude is given by the barometric formula.

In incompressible fluids the pressure is a linear function of position in the direction of the gravitational field, and thus in liquids, the pressure increases linearly with depth below the free surface of the liquid. The average pressure exerted by a liquid on a plate that is submerged vertically in the liquid is equal to the local value exerted at the centroid of the plate.

The phenomenon of buoyancy is explained by Archimedes' principle, which states that a body immersed in or floating in a fluid is acted upon by an upward force equal to the weight of the fluid displaced by the body, which acts through the center of gravity of the displaced volume.

Problems

PROBLEM 1.1

A 2 m × 3 m rectangular floodgate is placed vertically in water with the 2-m side at the free surface of the water. Calculate the force exerted by the water on one side of the floodgate. The density of the water is 997 kg/m^3.

PROBLEM 1.2

The center of a circular floodgate of radius 0.5 m is located in a dam at a depth of 2 m beneath the free surface of the water contained by the dam. Calculate the force exerted by the water on the floodgate. The density of water is 997 kg/m^3.

PROBLEM 1.3

A hemispherical bowl of radius 0.5 m is filled with water of density 997 kg/m^3. Calculate the force exerted on the bowl.

PROBLEM 1.4

A submariner is escaping from a damaged submarine at a depth of 50 m beneath the surface of the ocean. To prevent experiencing the "bends," the rate of decrease in pressure exerted on his body must not exceed 3400 Pa/s. What is the maximum rate at which he can ascend without experiencing the bends? The density of salt water is 1030 kg/m^3.

PROBLEM 1.5

When immersed in water of density 997 kg/m^3 a solid sphere is balanced with a mass of 10 g, and when immersed in an oil of density 800 kg/m^3 it balances with a mass of 11 g. Calculate (a) the volume and (b) the density of the sphere.

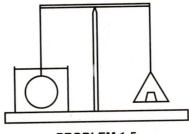

PROBLEM 1-5

PROBLEM 1.6

(a) When placed in water of density 997 kg/m³, a hollow sphere of inner radius 10 cm and outer radius 11 cm floats with half of its volume below the level of the free surface of the water. Calculate the density of the material from which the sphere is made.

(b) When placed in an oil the sphere floats with 60% of its volume below the level of the free surface of the oil. Calculate the density of the oil.

PROBLEM 1.7

A U-tube manometer contains mercury of density 13,600 kg/m³, and oil.

(a) When both ends of the tube are open to the atmosphere, h_1 = 0.17 m and Δh = 1 cm. Calculate the density of the oil.

(b) The open end of the left leg is sealed and is pressurized to a gauge pressure P, which causes the level of the free surface of the oil in the left leg to fall 5 mm. Calculate the value of P.

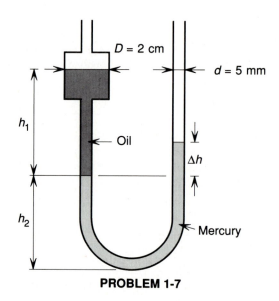

PROBLEM 1-7

PROBLEM 1.8

The altimeter in an airplane records a pressure of 70 kPa when the absolute pressure at sea level is 101 kPa and the air temperature is 288 K. Calculate the altitude of the airplane, assuming that the air temperature is not a function of altitude.

PROBLEM 1.9

In the hydraulic press shown, what mass can be raised by the ram when a mass of 1 kg is placed on the plunger?

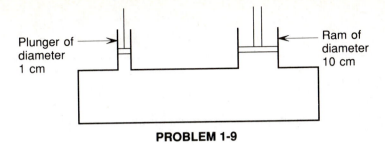

PROBLEM 1-9

PROBLEM 1.10

Calculate the gauge pressure in the oil at point D. The manometric fluids are mercury (of density 13600 kg/m^3) and water (of density 997 kg/m^3). The density of the oil is 800 kg/m^3.

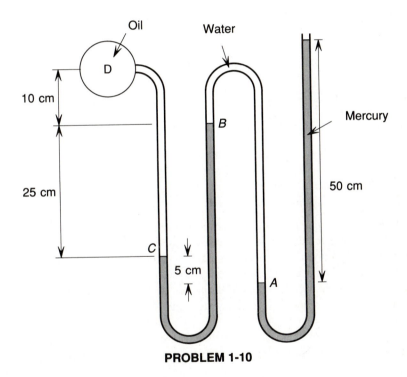

PROBLEM 1-10

2

Momentum Transport and Laminar Flow of Newtonian Fluids

2.1 Introduction

Treatment of the flow of fluid in a duct or conduit requires consideration of:

1. The geometry of the duct or conduit containing the fluid (e.g., the length, diameter, and inclination of the pipe or tube).
2. The magnitude of the force causing the flow.
3. The physical properties of the fluid.

In fluid flow, the important fluid properties are density and viscosity. It is common experience that greater suction is required to drink molasses through a drinking straw than is required to drink water through a straw at the same rate. This is so because molasses is more dense and more viscous than water. It is also common experience that more pressure is required to expel molasses from a horizontal drinking straw at a given rate than is required to expel water, and this is because molasses is more viscous than water.

The nature of fluid flow is also critically dependent on the magnitude of the applied force responsible for the flow, which in turn has a direct influence on the velocity of the flow. For the flow of a fluid of given physical properties in a conduit of given geometry, a critical velocity exists below which the flow is *laminar* and above which the flow is *turbulent*. The two regimes are illustrated in Fig. 2.1, which shows a thin stream of colored dye being introduced into a liquid flowing in a glass tube. At low flow rates, shown in Fig. 2.1(a), the dye flows with the fluid in a stable line parallel to the axis of the tube. The dye follows a streamline in the fluid and indicates that at low velocity, flow can be regarded as being the unidirectional movement of lamellae of fluid sliding over one another, with no macroscopic mixing or

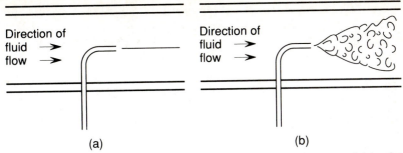

FIGURE 2.1.   Experimental demonstration of the difference between (a) laminar flow and (b) turbulent flow in a fluid.

intermingling of the fluid in the radial direction. This type of flow is known as *laminar flow*. If the flow velocity is increased to some value greater than a critical value, then, as shown in Fig. 2.1(b), the stream of dye emerging into the fluid is rapidly broken up and mixed with the fluid. In this situation macroscopic mixing is caused by turbulence in the fluid, and this type of flow is known as *turbulent flow*.

As a result of experiments similar to those illustrated in Fig. 2.1, and conducted in 1883, Osborne Reynolds [O. Reynolds, *Trans. R. Soc. London* (1883), vol. A174, p. 935] established the criterion for the transition from laminar to turbulent flow in terms of the dimensionless quantity:

$$\frac{\text{(characteristic length)} \times \text{(average velocity of the fluid)} \times \text{(density of the fluid)}}{\text{viscosity of the fluid}}$$

$$(2.1)$$

For flow in a tube of circular cross section, the characteristic length is the diameter of the tube, D. The average velocity of the fluid, v, is measured as the volume flow rate divided by the cross-sectional area of the pipe, and ρ and η are, respectively, the density and viscosity of the fluid. The dimensionless quantity given by Eq. (2.1) is called the *Reynolds number*, Re,

$$\text{Re} = \frac{Dv\rho}{\eta} \qquad (2.2)$$

and for fluid flow in a smooth pipe, the transition from laminar to turbulent flow begins at a Reynolds number of approximately 2100. From Eq. (2.2) it is seen that (1) increasing the diameter of the tube decreases the average velocity at which the transition occurs, and (2) increasing the flow velocity decreases the maximum pipe diameter in which laminar flow occurs. For flow of a fluid at a given velocity in a pipe of given diameter, the velocity at which the transition occurs is proportional to the viscosity of the fluid and is inversely proportional to the density of the fluid. The Reynolds number is the first of many dimensionless quantities that will be introduced in the treatment of transport phenomena, and as will be discussed in Chapter 4, rather than comprising an arbitrary selection of parameters, the Reynolds number is the ratio of two types of force acting on the fluid.

Laminar flow follows fundamental physical laws, and hence quantities such as the average flow velocity can be calculated from first principles. In contrast, determination of the average velocity in turbulent flow requires the use of correlations obtained from experimental observation, and thus before any fluid flow calculation can be made, it is necessary to determine whether the flow is laminar or turbulent. The value of the Reynolds number at which the transition from laminar to turbulent flow occurs in any flow geometry has to be determined by experiment. In the remainder of this chapter, the discussion will be restricted to laminar flow of incompressible fluids (i.e., fluids of constant density).

2.2 Newton's Law of Viscosity

Consider the situation shown in Fig. 2.2, in which a fluid is contained between two horizontal flat plates separated by the vertical distance Y. If the lower plate is held stationary and a force of constant magnitude F is applied to the upper plate in the x-direction, the latter begins to accelerate in the x-direction. Movement of the upper plate sets up a shear stress between the plate and the fluid, which opposes motion of the plate, and a *steady state* is reached when the applied force F is balanced by the shear force, in which state the upper plate has a constant velocity V. At steady state, Newton found that the velocity V is proportional to the applied force F and to the spacing Y, and is inversely proportional to the surface area of the plate, A:

$$V \propto \frac{FY}{A}$$

or

$$\frac{F}{A} \propto \frac{V}{Y} \qquad (2.3)$$

No slippage occurs at the interface between a fluid and the surface of a solid, and hence when the upper plate is moving with the velocity V, the layer of fluid in contact with it is also moving with velocity V. Similarly, the layer of fluid in contact with the lower plate is stationary. The phenomenon of no slippage causes the development of a velocity gradient in the fluid in the y-direction. In Eq. (2.3), V/Y, the velocity gradient in the y-direction, can be expressed in differential form as dv_x/dy, where

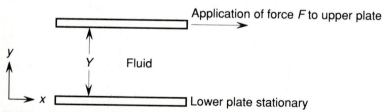

FIGURE 2.2. Flow in a fluid contained between horizontal parallel plates that are moving at differing velocities.

v_x is the velocity of a layer of liquid in the x-direction. Also, in Eq. (2.3), F/A is the shear stress at the interface between the upper plate and the fluid. Designating the shear stress as τ, Eq. (2.3) can be written as

$$\tau \propto \frac{dv_x}{dy}$$

that is, the shear stress is proportional to the velocity gradient in the fluid. The proportionality constant, η, defined as

$$\tau = -\eta \frac{dv_x}{dy} \tag{2.4}$$

is the viscosity of the fluid, and Eq. (2.4) is *Newton's law of viscosity*. Fluids that obey Newton's law are called *Newtonian fluids*, and in contrast, fluids that exhibit a nonlinear relationship between the shear stress τ and the velocity gradient dv_x/dy are called *non-Newtonian fluids*. Gases, simple organic and aqueous liquids, and liquid metals exhibit Newtonian behavior, and many complex liquids such as polymeric solutions and pastes are non-Newtonian in their behavior. The following discussion is restricted to consideration of incompressible Newtonian liquids in laminar flow.

In Fig. 2.3(a) the constant-velocity gradient across the fluid, V/Y, is indicated by arrows, the lengths of which represent the velocity of the layer of fluid located at y. With a constant velocity gradient, the shear stress, given by Eq. (2.4), is also constant. The shear stress, τ, can be considered by reference to Fig. 2.3(b) as follows. Because of the requirement of no slippage, the "layer" of fluid, of thickness δ, in contact with the lower plate is stationary. However, the next layer of fluid, also of thickness δ, is moving with a velocity $V\delta/Y$, and this causes a shear stress between the two layers of fluid. Similarly, the next layer of fluid, because it is moving with the velocity $2V\delta/Y$, gives rise to a shear stress between it and the slower-moving layer below it. The shear stress between two layers of fluid manifests a tendency to decrease the velocity of the faster-moving layer and increase the velocity of the slower-moving layer. The difference in velocity between the two layers (the velocity gra-

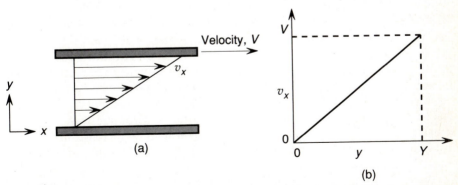

FIGURE 2.3. Variation of local flow velocity with position in the fluid between the plates shown in Fig. 2.2.

dient) is only sustained by the force applied to the upper plate; if this force were removed, the shear stress between the layers of fluid would cause cessation of the movement of the upper plate. The existence of a difference in velocity between the two layers (a velocity gradient in the fluid) gives rise to a transport of momentum from the faster-moving layer to the slower-moving layer, caused by the transport of slower-moving atoms or molecules in the slower-moving layer to the faster-moving layer, and vice versa. This effect can be illustrated by considering two passenger trains traveling with velocities v_1 and v_2 in the same direction on parallel tracks. If a person of mass m jumps from one train to the other, and another person of mass m jumps in the opposite direction, an amount of momentum equal to $m(v_2 - v_1)$ is transported from the faster-moving to the slower-moving train. Viewed from afar, the trains represent two adjacent layers of fluid, the jumping passengers represent atoms or molecules, and the transport of momentum is equivalent to a friction or shear stress between the two trains, which if not compensated for by a sustained difference between the driving forces of the trains, would eventually lead to both trains moving with the same velocity. As the thickness of the layers in Fig. 2.3(b) is made vanishingly small, the velocity gradient across the fluid becomes uniform and constant, as shown in Fig. 2.3(a).

The equivalence of shear stress and rate of momentum transport is seen by considering their units:

$$\text{shear stress} = \frac{\text{force}}{\text{area}} = \text{mass} \times \frac{\text{length}}{\text{time}^2} \times \frac{1}{\text{area}}$$

$$= \text{mass} \times \frac{\text{length}}{\text{time}} \times \frac{1}{\text{time}} \times \frac{1}{\text{area}}$$

$$= \frac{\text{mass} \times \text{velocity}}{\text{time} \times \text{area}}$$

$$= \frac{\text{momentum}}{\text{time} \times \text{area}}$$

$$= \text{transport of momentum per unit time per unit area}$$

$$= \text{rate of transport of momentum per unit area}$$

In Fig. 2.3(b) the velocity gradient, dv_x/dy, is positive, and as the direction of momentum transport is from the region of higher v_x to the region of lower v_x, the rate of momentum transport is a negative quantity (i.e., *momentum is transported down the velocity gradient*). The velocity gradient can thus be thought of as being the driving force for the transport of momentum. The difference between the signs of the velocity gradient and the direction of transport of momentum gives rise to the negative sign in Eq. (2.4). As the rate of transport of momentum in the *y-direction* due to fluid motion in the *x-direction* is equal to the shear stress, the latter, by convention, is denoted τ_{yx} (conversely, if momentum were transported in the *x*-direction due to motion in the *y*-direction, the shear stress would be designated τ_{xy}).

Therefore, for fluid flow in the *x*-direction, Newton's law of viscosity is written as

$$\tau_{yx} = -\eta \frac{dv_x}{dy} \tag{2.5}$$

In the cgs system, τ_{yx} has the units $dyn/cm^2 = g/cm\cdot s^2$. The velocity gradient dv_x/dy has the units $(cm/s)(1/cm) = 1/s$, and therefore η has the units $g/cm\cdot s$.
The traditional unit of viscosity is the poise, where

$$1 \text{ poise} = 1 \text{ P} = 1 \text{ g/cm}\cdot\text{s} \tag{2.6}$$

For simple liquids and gases a more convenient unit is the centipoise (cP), where $1 \text{ cP} = 10^{-2} \text{ P}$. As examples of magnitude, the viscosity of water at 20.22°C is 1 cP and the viscosity of a typical motor oil at 20°C is 8 P. In SI units

$$1 \text{ P} = 1 \frac{g}{cm\cdot s} \times 10^{-3} \frac{kg}{g} \times 10^2 \frac{cm}{m}$$

$$= 0.1 \frac{kg}{m\cdot s}$$

The units $kg/m\cdot s$ can be manipulated as

$$\frac{kg}{m\cdot s} = \frac{kg}{m^2}\left(\frac{m}{s^2}\right) s = \frac{N}{m^2}\cdot s = Pa\cdot s \text{ (pascal–second)}$$

Thus, in SI units, the viscosity of the typical motor oil at 20°C is 0.8 kg/m·s. or 0.8 Pa·s.

EXAMPLE 2.1

Two flat plates spaced 0.005 cm apart are separated by a lubricating oil of viscosity 0.2 Pa·s. If the lower plate is stationary and the upper plate moves with a velocity of 0.5 m/s, calculate the shear stress required to keep the upper plate in motion.

$$\frac{dv_x}{dy} = 0.5 \frac{m}{s} \times \frac{1}{5\times 10^{-5}}\left(\frac{1}{cm}\right) = 10^4 \text{ s}^{-1}$$

$$\eta = 0.2 \text{ Pa·s} = 0.2 \frac{kg}{m\cdot s}$$

Therefore,

$$\tau_{yx} = -\eta \frac{dv_x}{dy} = -0.2 \frac{kg}{m\cdot s} \times 10^4 \left(\frac{1}{s}\right)$$

$$= -2000 \frac{kg\cdot m}{s^2}\left(\frac{1}{m^2}\right)$$

$$= -2000 \frac{N}{m^2}$$

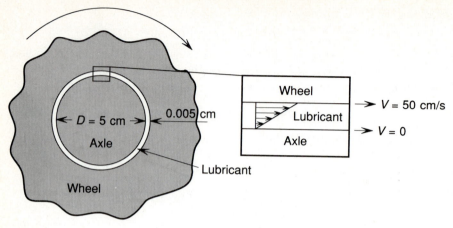

FIGURE 2.4. Couette flow.

The type of flow discussed in Example 2.1 is known as *Couette flow* and occurs in a friction bearing such as that shown in Fig. 2.4. A difference in velocity of 0.5 m/s between the upper plate (the rotating wheel) and the stationary lower plate (the axle of diameter 5 cm) corresponds to a rate of rotation of 190.9 rpm in Fig. 2.4. If the length L of the bearing is 5 cm, the area over which the shear stress acts is $\pi DL = \pi \times 5 \times 5 = 78.5 \text{ cm}^2 = 7.85 \times 10^{-3} \text{ m}^2$, and thus the shear force F_s on the bearing is

$$F_s = \tau_{yx}(\text{area})$$

$$= -2000 \frac{\text{N}}{\text{m}^2} \times 7.85 \times 10^{-3}(\text{m}^2)$$

$$= 15.7 \text{ N}$$

The rate at which mechanical energy is degraded to thermal energy by friction during rotation at 190.6 rpm is thus

$$\text{power} = \text{watts (W)} = \frac{\text{joules}}{\text{second}} \left(\frac{\text{J}}{\text{s}}\right)$$

$$= \frac{\text{N·m}}{\text{s}} = \text{N} \cdot \frac{\text{m}}{\text{s}}$$

$$= \text{force} \times \text{velocity}$$

$$= 15.7 \text{ N} \times 0.5 \frac{\text{m}}{\text{s}}$$

$$= 7.85 \text{ W}$$

2.3 Conservation of Momentum in Steady-State Flow

The nature of fluid flow is determined by considering the momentum balance in a control volume through which the fluid flows, and in such a calculation, two types of momentum transport are considered:

1. Momentum transport due to the existence of a velocity gradient in a direction normal to the direction of fluid flow (termed *viscous momentum transport*).
2. Momentum transport due to the motion of the fluid itself in the flow direction (termed *convective momentum transport*).

For consideration of Couette flow, the control volume, of dimensions Δx, Δy, and Δz, is fixed in space relative to the stationary lower plate, as shown in Fig. 2.5. As rate of momentum transport and force are equivalent, a momentum (or force) balance in the control volume, through which steady-state flow of the fluid occurs, is

(rate at which momentum enters the control volume)
− (rate at which momentum leaves the control volume)
+ (sum of the forces acting on the fluid in the control volume) = 0

As shown in Fig. 2.6, convective momentum due to the fluid flow in the x-direction enters the control volume through the yz face at x and leaves through the yz face at $x + \Delta x$. As viscous momentum is transported down the velocity gradient it enters the control volume through the xz face at $y + \Delta y$ and leaves through the xz face at y. The only force acting on the fluid in the control volume is the gravitational force on the mass of the fluid. However, as this force is acting in the y-direction, which is normal to the direction of fluid flow, it does not influence the fluid flow.

The rate of convective momentum transport has the units

$$\frac{\text{mass} \times \text{velocity}}{\text{time}} = \text{kg} \times \frac{\text{m}}{\text{s}} \times \frac{1}{\text{s}}$$

$$= \frac{\text{kg}}{\text{m}^3} \times \frac{\text{m}}{\text{s}} \times \text{m}^2 \times \frac{\text{m}}{\text{s}}$$

$$= \rho(\text{density}) \times v(\text{velocity}) \times A(\text{area}) \times v(\text{velocity})$$

$$= \rho v^2 A$$

where A is the cross-sectional area through which the fluid of density ρ flows.

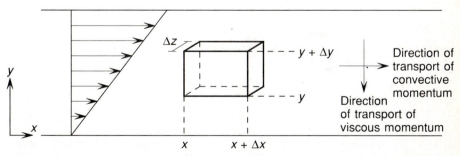

FIGURE 2.5. Control volume in Couette flow and directions of the transport of convective and viscous momentum.

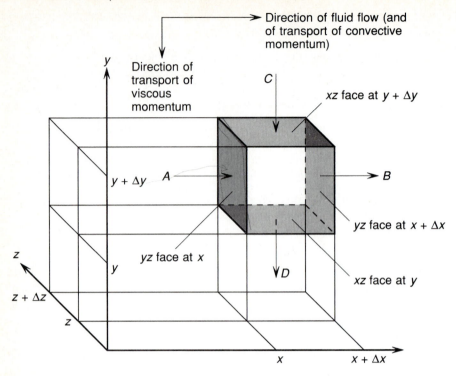

A. Convective momentum entering yz face
   at x at the rate $\rho v_x^2 \big|_x \, \Delta y \, \Delta z$
B. Convective momentum leaving through yz face
   at $x + \Delta x$ at the rate $\rho v_x^2 \big|_{x + \Delta x} \Delta y \, \Delta z$
C. Viscous momentum entering through xz face
   at $x + \Delta y$ at the rate $\tau_{yx} \big|_{y + \Delta y} \Delta x \, \Delta z$
D. Viscous momentum leaving through xz face
   at y at the rate $\tau_{yx} \big|_y \Delta x \, \Delta z$

FIGURE 2.6. Control volume in a fluid showing the directions of transport of convective and viscous momentum.

Therefore, the rate at which convective momentum enters the control volume through the yz face at x is $\rho v_x^2 \big|_x \, \Delta y \, \Delta z$ (where $v_x \big|_x$ is the value of v_x at x), and the rate at which convective momentum leaves the control volume through the yz face at $x + \Delta x$ is $\rho v_x^2 \big|_{x+\Delta x} \Delta y \, \Delta z$ (where $v_x \big|_{x+\Delta x}$ is the value of v_x at $x + \Delta x$).

The rate of viscous momentum transport has the units (shear stress) · (area). Therefore, the rate at which viscous momentum transport enters the control volume through the xz face at $y + \Delta y$ is $\tau_{yx} \big|_{y + \Delta y} \Delta x \, \Delta z$ (where $\tau_{yx} \big|_{y + \Delta y}$ is the shear stress at $y + \Delta y$), and the rate at which viscous momentum leaves the control volume through the xz face at y is $\tau_{yx} \big|_y \Delta x \, \Delta z$ (where $\tau_{yx} \big|_y$ is the shear stress at y).

The momentum balance is thus

> (rate at which convective momentum enters)
> − (rate at which convective momentum leaves)
> + (rate at which viscous momentum enters)
> − (rate at which viscous momentum leaves) = 0

that is,

$$(\rho v_x^2|_x - \rho v_x^2|_{x+\Delta x})\,\Delta y\,\Delta z + (\tau_{yx}|_{y+\Delta y} - \tau_{yx}|_y)\,\Delta x\,\Delta z = 0 \tag{i}$$

In steady-state flow, v_x is a function only of y (i.e., it does not vary with either x or time), hence $v_x|_x = v_x|_{x+\Delta x}$. Thus convective momentum enters and leaves the control volume at the same rate, so the first term in parentheses in Eq. (i) is zero. Dividing Eq. (i) by $\Delta x\,\Delta y\,\Delta z$ gives

$$\frac{\tau_{yx}|_{y+\Delta y} - \tau_{yx}|_y}{\Delta y} = 0$$

which in the limit of $\Delta y \to 0$ gives

$$\frac{d\tau_{yx}}{dy} = 0$$

or

$$\tau_{yx} = \text{a constant} = C_1 \tag{ii}$$

From Newton's law of viscosity, Eq. (2.3),

$$\tau_{yx} = -\eta\,\frac{dv_x}{dy} = C_1$$

Therefore,

$$\frac{dv_x}{dy} = -\frac{C_1}{\eta}$$

integration of which gives

$$v_x = -\frac{C_1}{\eta}\,y + C_2 \tag{iii}$$

The constants C_1 and C_2 are evaluated from the boundary conditions:

$$v_x = \begin{cases} 0 & \text{at } y = 0 \\ V & \text{at } y = Y \end{cases}$$

Application of the boundary conditions to Eq. (iii) gives

$$0 = -\frac{C_1}{\eta}\,0 + C_2 \qquad \text{therefore,} \quad C_2 = 0$$

and

$$V = -\frac{C_1}{\eta} Y \qquad \text{therefore,} \quad C_1 = -\eta \frac{V}{Y}$$

Substitution into Eq. (iii) gives

$$v_x = \frac{V}{Y} y \tag{2.7}$$

that is, v_x is proportional to the velocity gradient, is independent of the viscosity of the fluid, and is a linear function of y. Substitution into Eq. (ii) gives

$$\tau_{yx} = -\frac{V}{Y} \eta \tag{2.8}$$

that is, τ_{yx} is independent of y but is proportional to the viscosity of the fluid. Equation (2.7) is shown graphically in Fig. 2.3.

The procedure described above is standard for calculation of the shear stress and velocity distributions from a momentum balance. The steps in the procedure are as follows:

1. Write the momentum balance for the control volume and reduce it to a differential equation containing τ.
2. Integrate the differential equation to obtain τ.
3. Substitute the expression for τ into Newton's law of viscosity and integrate a second time to obtain the equation for v.
4. Evaluate the two integration constants from two boundary conditions.

Other types of fluid flow include (1) flow between two parallel plates, (2) free flow down an inclined plane, and (3) flow in a cylindrical pipe or tube. Laminar, Newtonian flow in these cases will be considered individually.

2.4 Fluid Flow Between Two Flat Parallel Plates

Couette flow is caused by the movement of one of the plates relative to the other, and the fact that no slippage occurs between the plates and the layers of fluid in contact with them gives rise to a linear variation of v_x with y. In contrast, fluid flow between two stationary horizontal flat plates requires that a force be exerted on the fluid in the direction of the flow. The occurrence of no slippage of the fluid in contact with the plates means that the velocity of the fluid in contact with both plates is zero, and hence the velocity, v_x, increases with vertical distance away from each plate into the body of the fluid. Symmetry then dictates that the maximum value of v_x occurs at the plane midway between the plates, and as $dv_x/dy = 0$ at this plane, from Eq. (2.5) the shear stress, τ_{yx}, is also zero at this plane. This will be used as a boundary condition in consideration of the momentum balance in the control volume.

Because of the symmetry, it is convenient to locate $y = 0$ at the plane midway between the plates and consider that the plates are a distance 2δ apart. The control volume can then be located on either side of the $y = 0$ plane at a distance sufficiently far removed from the entrance to the plates that the fluid flow through the control volume is fully developed and occurs at steady state. In fully developed flow the local velocities in the flow are independent of position in the direction of the flow, and at steady state they are independent of time. As viscous momentum is transported down the velocity gradient, it is transported from the $y = 0$ plane toward both plates. Consider the control volume of dimensions Δx, Δy, and Δz shown in Fig. 2.7. Convective momentum, due to fluid flow, enters through the yz face at x and leaves through the yz face at $x + \Delta x$, and viscous momentum, arising from the existence of a velocity gradient in the fluid, enters through the xz face at y and leaves through the xz face at $y + \Delta y$. Under conditions of fully developed steady-state flow resulting from the application of an external force, the static pressure in the fluid decreases linearly in the direction of flow. If $P|_x$ is the static pressure at x and $P|_{x+\Delta x}$ is the static pressure at $x + \Delta x$, the *pressure drop* across the control volume in the direction of flow is $P|_x - P|_{x+\Delta x}$. Thus the force due to the static pressure acting on the yz face of the control volume at x is $P|_x \, \Delta y \, \Delta z$, and the force acting on the yz face at $x + \Delta x$ is $P|_{x+\Delta x} \, \Delta y \, \Delta z$. As the fluid flow is horizontal, it is not influenced by gravity.

The components of the momentum balance are thus

rate at which convective momentum enters the control volume	$= \rho v_x^2\|_x \, \Delta y \, \Delta z$
rate at which convective momentum leaves the control volume	$= \rho v_x^2\|_{x+\Delta x} \, \Delta y \, \Delta z$
rate at which viscous momentum enters the control volume	$= \tau_{yx}\|_y \, \Delta x \, \Delta z$
rate at which viscous momentum leaves the control volume	$= \tau_{yx}\|_{y+\Delta y} \, \Delta x \, \Delta z$
sum of forces acting on the control volume	$= (P\|_x - P\|_{x+\Delta x}) \, \Delta y \, \Delta z$

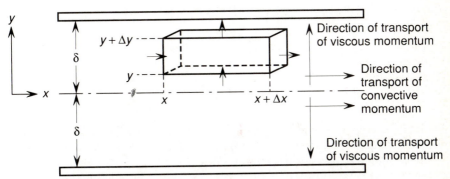

FIGURE 2.7. Control volume and the directions of transport of viscous and convective momentum in fluid flow between two parallel plates.

and the momentum balance, from Eq. (2.6), is

$$(\rho v_x^2|_x - \rho v_x^2|_{x+\Delta x}) \, \Delta y \, \Delta z + (\tau_{yx}|_y - \tau_{yx}|_{y+\Delta y}) \, \Delta x \, \Delta z$$
$$+ (P|_x - P|_{x+\Delta x}) \, \Delta y \, \Delta z = 0$$

Again, as the flow is fully developed, v_x is a function only of y, and hence the first term in the equation above is zero. Therefore,

$$(\tau_{yx}|_y - \tau_{yx}|_{y+\Delta y}) \, \Delta x \, \Delta z + (P|_x - P|_{x+\Delta x}) \, \Delta y \, \Delta z = 0$$

Dividing by $\Delta x \, \Delta y \, \Delta z$ gives

$$\frac{\tau_{yx}|_y - \tau_{yx}|_{y+\Delta y}}{\Delta y} + \frac{P|_x - P|_{x+\Delta x}}{\Delta x} = 0$$

In the limit as Δx and Δy approach zero,

$$\lim_{\Delta y \to 0} \frac{\tau_{yx}|_y - \tau_{yx}|_{y+\Delta y}}{\Delta y} = -\frac{d\tau_{yx}}{dy}$$

and

$$\lim_{\Delta x \to 0} \frac{P|_x - P|_{x+\Delta x}}{\Delta x} = -\frac{dP_x}{dx}$$

As the static pressure P decreases linearly with x, dP/dx is a negative quantity and hence $\Delta P/L = -dP/dx$, where ΔP is the magnitude of the decrease is pressure, or the *pressure drop*, occurring over the distance L. Therefore,

$$\frac{d\tau_{yx}}{dy} = \frac{\Delta P}{L}$$

The first integration gives

$$\tau_{yx} = \frac{\Delta P}{L} y + C_1$$

and the boundary condition that $\tau_{yx} = 0$ at $y = 0$ gives

$$C_1 = 0$$

Thus

$$\tau_{yx} = \frac{\Delta P}{L} y \qquad (2.9)$$

that is, τ_{yx} is proportional to the pressure drop per unit length through the fluid and is a linear function of y, with a maximum value of $(\Delta P \delta/L)$ at the plates.

Consideration of Newton's law of viscosity, Eq. (2.5),

$$\tau_{yx} = -\eta \frac{dv_x}{dy}$$

gives

$$-\eta \frac{dv_x}{dy} = \frac{\Delta P}{L} y$$

which, upon integration, gives

$$v_x = -\frac{\Delta P}{2L\eta} y^2 + C_2$$

Application of the boundary condition, $v_x = 0$ at $y = \pm\delta$, gives

$$C_2 = \frac{\Delta P}{2L\eta} \delta^2$$

and hence

$$v_x = \frac{\Delta P}{2L\eta} (\delta^2 - y^2) \tag{2.10}$$

that is, v_x is proportional to $\Delta P/L$, is inversely proportional to η, and is a parabolic function of y, with a maximum value of

$$v_{x,\text{max}} = \frac{\Delta P}{2L\eta} \delta^2 \tag{2.11}$$

at $y = 0$. The velocity and shear stress distributions are shown in Fig. 2.8. Thus, in contrast wtih Couette flow, in which (1) τ_{yx} is independent of y and (2) v_x is a linear function of y, in flow between parallel plates, (1) τ_{yx} is a linear function of y and (2) v_x is a parabolic function of y.

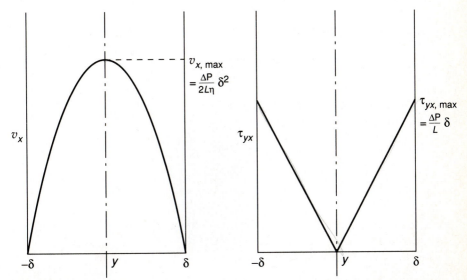

FIGURE 2.8. Distributions of local velocity and shear stress in fluid flow between two parallel plates.

The shear stresses acting on the plates are caused by the flow of a viscous fluid, which in turn is caused by the application of a normal pressure to the fluid in the direction of the flow. The shear stress and pressure drop through the fluid are related as follows. For plates of width W and length L, the shear force acting on the plates is

$$(\text{shear stress}) \cdot (\text{area}) = (\tau_{yx}|_{y=\delta})(2LW)$$

$$= \frac{\Delta P}{L} \delta \, (2LW)$$

$$= 2 \, \Delta P \, \delta W$$

which is equal to the decrease in the normal force over the length L, $= (\Delta P)(2\delta W)$.

Calculation of the mass flow rate and the volume flow rate requires knowledge of the average velocity of the fluid, $\bar{v}_x$. Calculation of the average velocity is illustrated in Fig. 2.9. The average velocity is such that the area $\bar{v}_x(2\delta)$ equals the area under the curve for v_x between $y = \delta$ and $-\delta$, or, more simply, the area $\bar{v}_x\delta$ equals the area under the curve for v_x between $y = 0$ and $y = \delta$:

$$\bar{v}_x\delta = \int_0^\delta v_x \, dy$$

$$= \int_0^\delta \frac{\Delta P}{2L\eta} (\delta^2 - y^2) \, dy$$

$$= \frac{\Delta P}{2L\eta} \left[\delta^2 y - \frac{y^3}{3} \right]_0^\delta$$

$$= \frac{\Delta P}{3L\eta} \delta^3$$

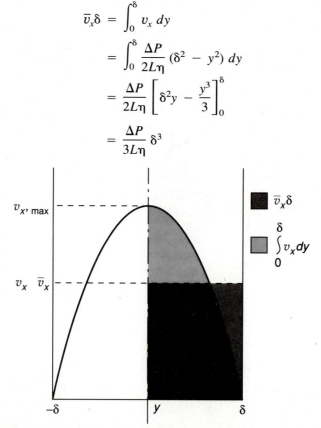

FIGURE 2.9. Calculation of the average flow velocity in fluid flow between parallel plates.

Therefore,

$$\bar{v}_x \text{ (average velocity of the fluid)} = \frac{\Delta P}{3L\eta} \delta^2 \qquad (2.12)$$

which is two-thirds of the value of $v_{x,max}$. The volume flow rate, $\dot{V}$, is then

$$\dot{V} = \text{(average velocity)} \times \text{(cross-sectional area)}$$
$$= \bar{v}_x \cdot 2\delta W$$
$$= \frac{2\,\Delta P}{3L\eta} \delta^3 W \qquad (2.13)$$

and the mass flow rate, $\dot{M}$, is

$$\dot{M} = \text{(volume flow rate)} \times \text{(density of the fluid)}$$
$$= \frac{2\,\Delta P}{3L\eta} \delta^3 W\rho \qquad (2.14)$$

EXAMPLE 2.2

An electrostatic dust precipitator consists of two electrically charged horizontal plates placed a distance 2δ apart, between which the electrical field strength is E. Gas containing dust particles of mass m and electrical charge e enters the space between the plates at the pressure P_0 and exits at the pressure P_L. Calculate the minimum length L of the plates which is required to guarantee that all of the dust particles are collected at the lower plate before the gas leaves the precipitator.

Consider a particle entering the precipitator which is virtually in contact with the upper plate at the location $y = \delta$ and $x = 0$. On entering the electrical field it experiences a downward force that causes it to accelerate in the $-y$-direction. If the influences of gravity and the viscous drag exerted by the gas on the particle are small enough to be ignored, the acceleration a of the particle is given by

$$F = -eE = ma = m\frac{dv_y}{dt}$$

$$\frac{dv_y}{dt} = -\frac{eE}{m}$$

Thus

$$v_y = -\frac{eE}{m}t + C_1$$

At $t = 0$, the time of entrance into the precipitator, $v_y = 0$ and hence $C_1 = 0$. Therefore,

$$v_y = \frac{dy}{dt} = -\frac{eE}{m}t$$

and hence

$$y = -\frac{eE}{2m} t^2 + C_2$$

At $t = 0$, $y = \delta$ and hence $C_2 = \delta$, such that

$$y = \delta - \frac{eE}{2m} t^2 \tag{i}$$

and thus the time t_f required for the particle to reach the lower plate, at $y = -\delta$, is

$$t_f = \left(\frac{4\delta m}{eE}\right)^{1/2} \tag{ii}$$

The x-component of the velocity of the particle is the local flow velocity in the gas, given by Eq. (2.10) as

$$v_x = \frac{P_0 - P_L}{2L\eta} (\delta^2 - y^2) \tag{2.10}$$

Substitution of Eq. (i) into Eq. (2.10) gives

$$v_x = \frac{P_0 - P_L}{2L\eta} \left[\delta^2 - \left(\delta - \frac{eE}{2m} t^2 \right) \right]$$

$$= \frac{P_0 - P_L}{2L\eta} \left[\frac{eE}{m} \delta t^2 - \left(\frac{eE}{2m} \right)^2 t^4 \right]$$

$$= \frac{dx}{dt}$$

and therefore

$$x = \frac{P_0 - P_L}{2L\eta} \left[\frac{eE}{m} \frac{\delta t^3}{3} - \left(\frac{eE}{2m} \right)^2 \frac{t^5}{5} \right] + C_3 \tag{iii}$$

At $t = 0$, $x = 0$ and hence $C_3 = 0$. Substitution of Eq. (ii) into Eq. (iii) gives the horizontal distance traveled by the particle before it reaches the lower plate as

$$x = \frac{P_0 - P_L}{2L\eta} \left[\frac{eE}{3m} \delta \left(\frac{4\delta m}{eE} \right)^{3/2} - \left(\frac{eE}{2m} \right)^2 \frac{1}{5} \left(\frac{4\delta m}{eE} \right)^{5/2} \right]$$

$$= \frac{P_0 - P_L}{2L\eta} \left(\frac{m}{eE} \right)^{1/2} \delta^{5/2} \left(\frac{16}{15} \right) \tag{iv}$$

This is the required minimum length L and hence

$$L^2 = \frac{P_0 - P_L}{\eta} \left(\frac{m}{eE} \right)^{1/2} \delta^{5/2} \left(\frac{8}{15} \right)$$

or

$$L = 0.730 \left(\frac{P_0 - P_L}{\eta}\right)^{1/2} \left(\frac{m}{eE}\right)^{1/4} \delta^{5/4} \qquad \text{(v)}$$

The trajectory of the path taken by the particle is given by Eqs. (i) and (iii).

Consider the situation in which $\delta = 0.05$ m, the average flow velocity is 1 m/s, and $m/eE = 20$ s^2/m. From Eq. (2.12),

$$\bar{v}_x = \frac{\Delta P}{3L\eta} \delta^2$$

Therefore,

$$\frac{\Delta P}{L\eta} = \frac{3\bar{v}_x}{\delta^2} = \frac{3 \times 1}{0.05^2} = 1200$$

and thus Eq. (iii) gives

$$x = \frac{1200}{2} \left[\frac{0.05}{3 \times 20} t^3 - \frac{1}{4}\left(\frac{1}{20}\right)^2 \frac{t^5}{5}\right]$$

$$= 0.5t^3 - 0.075t^5 \qquad \text{(vi)}$$

Equation (i) gives

$$y = 0.05 - \frac{1}{2 \times 20} t^2$$

$$= 0.05 - 0.025t^2 \qquad \text{(vii)}$$

The trajectory of the path given by Eqs. (vi) and (vii) is shown in Fig. 2.10. From Eq. (ii)

$$t_f = (4 \times 0.05 \times 20)^{1/2} = 2 \text{ s}$$

which on substitution into Eq. (vi) gives

$$L = x = 1.6 \text{ m}$$

Alternatively, Eq. (iv) gives

$$x = \frac{1200}{2} (20)^{1/2}(0.05)^{5/2} \times \frac{16}{15}$$

$$= 1.6 \text{ m}$$

With a working precipitator of a given length, an increase in the gas flow rate (which requires an increase in the rate of pressure drop $\Delta P/L$) requires an increase in the field strength E, which, from Eq. (iv), is such that

$$\frac{\Delta P}{LE^{1/2}} = \text{constant}$$

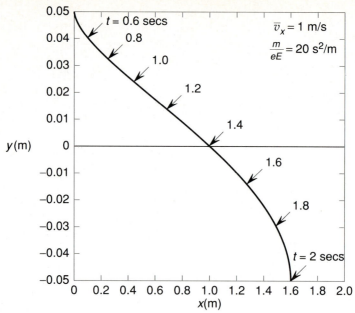

FIGURE 2.10.   Trajectory of an electrically charged particle of dust in an electrostatic dust precipitator.

2.5 Fluid Flow down an Inclined Plane

The free flow of fluid down an inclined plane is caused by the weight of the fluid (i.e., by the influence of gravity on the fluid). If the plane is inclined at an angle θ to the perpendicular, the component of gravity acting in the direction of flow down the plane is $g \cos \theta$. Consider the flowing film of fluid of thickness δ shown in Fig. 2.11. The fluid in contact with the inclined plane is motionless, and although there is no slippage at the interface between the fluid and the gas phase above it, the gas offers a negligible resistance to fluid flow at the surface provided that the viscosity of the gas is significantly less than that of the fluid. The film of gas in contact with the free surface of the liquid moves with the same velocity as the free surface, and hence the shear stress at the free surface of the fluid is zero. Thus v_x increases from zero at $y = \delta$ to a maximum value at the free surface, and the direction of transport of viscous momentum is from the free surface toward the inclined plane.

In considering the momentum balance, convective momentum enters the control volume through the yz face at x and leaves through the yz face at $x + \Delta x$. Again, as fully developed steady-state flow occurs, the rates at which convective momentum enters and leaves the control volume are equal. Viscous momentum enters the control volume through the xz face at y at the rate $\tau_{yx|y} \Delta x \Delta z$ and leaves through the xz face at $y + \Delta y$ at the rate $\tau_{yx|y+\Delta y} \Delta x \Delta z$. The force F acting on the fluid in the control volume in the direction of flow is

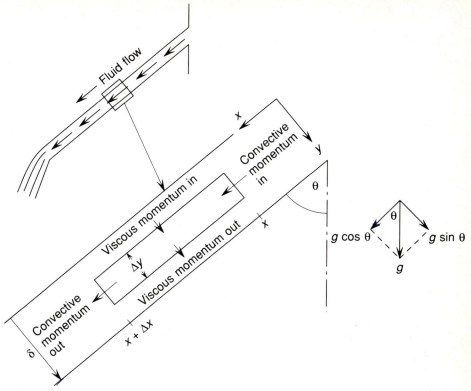

FIGURE 2.11. Control volume for consideration of fluid flow down an inclined plane.

$$F = \text{(mass of fluid)} g \cos \theta$$
$$= \text{(density)(volume)}(g \cos \theta)$$
$$= \rho g \cos \theta \, \Delta x \, \Delta y \, \Delta z$$

As v_x is a function only of y, the rate at which convective momentum enters the control volume is equal to the rate at which it leaves, and hence the momentum balance in the control volume is

$$(\tau_{yx|y} - \tau_{yx|y+\Delta y}) \, \Delta z \, \Delta x + \rho g \cos \theta \, \Delta x \, \Delta y \, \Delta z = 0$$

Dividing by $\Delta x \, \Delta y \, \Delta z$ and allowing Δy to approach zero gives

$$\frac{d\tau_{yx}}{dy} = \rho g \cos \theta$$

The first integration gives

$$\tau_{yx} = \rho g \cos \theta \, y + C_1$$

and as the shear stress at the free surface is zero (i.e., $\tau_{yx|y=0} = 0$), the integration constant C_1 is zero. Then, as

$$\tau_{yx} = \rho g \cos \theta \; y = -\eta \frac{dv_x}{dy}$$

the second integration gives

$$v_x = -\frac{\rho g \cos \theta}{\eta} \frac{y^2}{2} + C_2$$

The second boundary condition, $v_x = 0$ at $y = \delta$, allows C_2 to be evaluated as

$$C_2 = \frac{\rho g \cos \theta}{2\eta} \delta^2$$

and hence

$$v_x = \frac{\rho g \cos \theta}{2\eta} (\delta^2 - y^2) \tag{2.15}$$

v_x thus has a maximum value of

$$v_{x,\text{max}} = \frac{\rho g \cos \theta}{2\eta} \delta^2 \tag{2.16}$$

at $y = 0$.

The velocity and shear stress gradients are shown in Fig. 2.12. Thus:

1. The shear stress, τ_{yx}, is proportional to the force causing the flow (which is determined by the density of the fluid and the angle of inclination of the plate) and is a linear function of y.
2. The velocity, v_x, is proportional to the force causing the flow, is inversely proportional to the viscosity, and is a parabolic function of y.

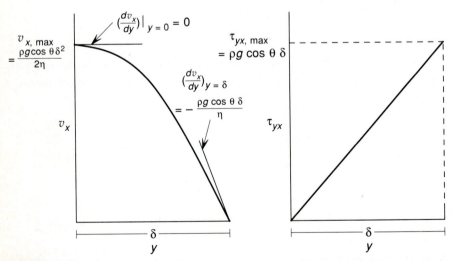

FIGURE 2.12.   Distributions of local velocity and shear stress in fluid flow down an inclined plane.

As before, the average velocity, $\bar{v}_x$, is such that the area $\bar{v}_x\delta$ equals the area $\int_0^\delta v_x\,dy$:

$$\bar{v}_x = \frac{1}{\delta}\int_0^\delta v_x\,dy$$

$$= \frac{1}{\delta}\left(\frac{\rho g \cos\theta}{2\eta}\right)\left[\delta^2 y - \frac{y^3}{3}\right]_0^\delta$$

$$= \frac{\rho g \cos\theta}{3\eta}\,\delta^2$$

which as in the case of flow between parallel plates, is two-thirds of v_{max}. The volume flow rate is

$$\dot{V} = \frac{\rho g \cos\theta}{3\eta}\,\delta^2(\delta W) \tag{2.17}$$

and the mass flow rate is

$$\dot{M} = \frac{\rho^2 g \cos\theta\delta^3 W}{3\eta} \tag{2.18}$$

From experimental observation, laminar flow down a vertical wall in the z-direction requires that

$$\text{Re} \le 25$$

where the Reynolds number is evaluated as

$$\text{Re} = \frac{4\bar{v}_z\delta\rho}{\eta}$$

EXAMPLE 2.3

Slag is continuously transferred, by means of a launder (an inclined plane) from a reverberatory furnace to a settling furnace. Calculate the maximum mass flow rate of the slag in the 1-m-wide launder at which the flow is still laminar, and investigate the relationships between $\bar{v}_x$, δ, and $\cos\theta$. The data are:

density of the slag, $\rho = 2627\ \text{kg/m}^3$
viscosity of the slag, $\eta = 0.31\ \text{Pa·s}$
$W = 1\ \text{m}$

For laminar flow

$$\text{Re} = \frac{4\bar{v}_x\delta\rho}{\eta} \le 25$$

From Eq. (2.18),

$$\dot{M} = \bar{v}_x\delta W\rho \qquad \text{therefore} \qquad \bar{v}_x\delta = \frac{\dot{M}}{W\rho} \tag{i}$$

Thus

$$\mathrm{Re} = \frac{4\dot{M}}{W\eta} \le 25$$

$$\dot{M} \le \frac{25W\eta}{4} = 25 \times 1 \text{ m} \times 0.31 \frac{\text{kg}}{\text{m·s}} \times \frac{1}{4}$$

$$= 1.94 \text{ kg/s}$$

Consider a mass flow rate of 1.94 kg/s. From Eq. (i),

$$\bar{v}_x \delta = 1.94 \frac{\text{kg}}{\text{s}} \times \frac{1}{1} \frac{1}{m} \times \frac{1}{2627} \frac{\text{m}^3}{\text{kg}}$$

$$= 7.38 \times 10^{-4} \text{ m}^2/\text{s} \tag{ii}$$

and from Eq. (2.18),

$$\dot{M} = \frac{\rho^2 g \cos \theta \, \delta^3 W}{3\eta}$$

Thus

$$\cos \theta \, \delta^3 = \frac{3\dot{M}\eta}{\rho^2 gW}$$

$$= 3 \times 1.94 \frac{\text{kg}}{\text{s}} \times 0.31 \frac{\text{kg}}{\text{m·s}} \times \frac{1}{(2627)^2} \frac{\text{m}^6}{\text{kg}^2}$$

$$\times \frac{1}{9.81} \frac{\text{s}^2}{\text{m}} \times \frac{1}{1} \frac{1}{m}$$

$$= 2.67 \times 10^{-8} \text{ m}^3 \tag{iii}$$

Equations (ii) and (iii) contain the three unknowns v_x, δ, and θ, and the value of one of the unknowns can be arbitrarily selected (subject to the condition that $0° \le \theta < 90°$). Select $\theta = 85°$; from Eq. (iii),

$$\delta^3 = \frac{2.67 \times 10^{-8}}{\cos 85°} = 3.06 \times 10^{-7} \text{ m}^3$$

Therefore, $\delta = 0.00674$ m and from Eq. (ii),

$$\bar{v}_x = \frac{7.38 \times 10^{-4}}{0.00674} = 0.109 \text{ m/s}$$

Select $\theta = 45°$; from Eq. (iii),

$$\delta^3 = \frac{2.67 \times 10^{-8}}{\cos 45°} = 3.77 \times 10^{-8} \text{ m}^3$$

Thus $\delta = 0.00335$ m and

$$\bar{v}_x = \frac{7.38 \times 10^{-4}}{0.00335} = 0.22 \text{ m/s}$$

Select $\theta = 0°$ (which corresponds to the slag flowing down the vertical wall of the reverberatory furnace); from Eq. (iii),

$$\delta^3 = 2.67 \times 10^{-8} \text{ m}^3$$

Thus $\delta = 0.003$ m and

$$\bar{v}_x = \frac{7.38 \times 10^{-4}}{0.003} = 0.247 \text{ m/s}$$

Therefore, for a given mass flow rate in laminar flow, $\bar{v}_x$ increases and δ decreases with decreasing θ. These variations, for the present example, are shown in Fig. 2.13.

2.6 Fluid Flow in a Vertical Cylindrical Tube

The most common type of fluid flow is that which occurs in a cylindrical tube or pipe. In Fig. 2.14 fluid flow occurs in the z-direction and hence the fluid velocity is denoted v_z. The velocity of the fluid in contact with the inner wall of the tube of radius R is zero and hence, by symmetry, v_z has its maximum value at the axis of the tube (i.e., at $r = 0$). As v_z increases with distance from the wall to the centerline, viscous momentum, arising from the shear stress τ_{rz}, is transported radially from the centerline toward the wall. The subscripts on τ_{rz} indicate that viscous momentum is transported in the r-direction due to fluid flow in the z-direction.

The consideration of fluid flow in a cylindrical tube is facilitated by the use of the cylindrical coordinates z, r and θ, illustrated in Fig. 2.14, and the control volume,

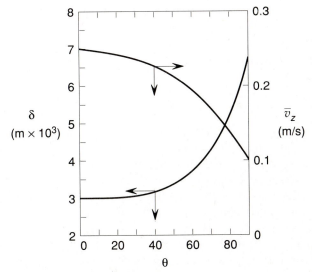

FIGURE 2.13. Variation, with angle of inclination, of the depth and average velocity in fluid flow down an inclined plane.

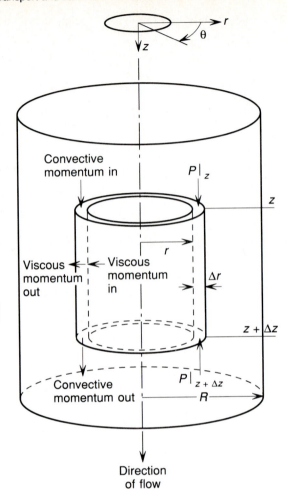

FIGURE 2.14.   Control volume for consideration of fluid flow in a vertical cylindrical pipe.

which is located far enough from the entrance of the tube that fully developed steady-state flow occurs through it, is a cylindrical shell of length Δz, inner radius r, outer radius $r + \Delta r$, and volume $2\pi r\, \Delta r\, \Delta z$. With fluid flow in the z-direction, convective momentum enters through the upper surface at the rate $\rho v_z^2|_z \cdot 2\pi r\, \Delta r$ and leaves through the bottom face at the rate $\rho v_z^2|_{z+\Delta z} \cdot 2\pi r\, \Delta r$. Viscous momentum enters the control volume through the inner surface at the rate $(\tau_{rz} \cdot 2\pi r\, \Delta z)|_r$ and leaves through the outer surface at the rate $(\tau_{rz} \cdot 2\pi r\, \Delta z)|_{r+\Delta r}$. The influence of gravity on the vertical fluid flow is such that the fluid in the control volume is subjected to the gravitational force of its own weight, $\rho g \cdot 2\pi r\, \Delta r\, \Delta z$, and if the flow is also influenced by an applied pressure, this contribution to the force balance is $P|_z \cdot 2\pi r\, \Delta r$ in the direction of flow at z and $P|_{z+\Delta z} \cdot 2\pi r\, \Delta r$ opposing the flow at $z + \Delta z$.

The momentum balance in the control volume is thus

$$(\rho v_z^2|_z - \rho v_z^2|_{z+\Delta z})2\pi r\, \Delta r + (\tau_{rz} \cdot 2\pi r\, \Delta z)|_r - (\rho_{rz}\cdot 2\pi r\, \Delta z)|_{r+\Delta r}$$
$$+ (P|_z - P|_{z+\Delta z})2\pi r\, \Delta r + \rho g \cdot 2\pi r\, \Delta r\, \Delta z = 0$$

As steady-state flow occurs, v_z is a function only of r and hence $v_z|_z = v_z|_{z+\Delta z}$, and the first term is zero.

Dividing by Δr and Δz and canceling the 2π terms gives

$$\frac{(r\tau_{rz})|_r - (r\tau_{rz})|_{r+\Delta r}}{\Delta r} + \frac{(P|_z - P|_{z+\Delta z})r}{\Delta z} + \rho gr = 0$$

In the limit Δr and $\Delta z \to 0$ this becomes

$$\frac{d(r\tau_{rz})}{dr} - \frac{dP}{dz}r + \rho gr = 0$$

Again the term $-\,dP/dz = \Delta P/L$, the *decrease* in the static pressure that occurs over the length L, and thus

$$\frac{d(r\tau_{rz})}{dr} = \left(\frac{\Delta P}{L} + \rho g\right)r$$

The first integration gives

$$r\tau_{rz} = \left(\frac{\Delta P}{L} + \rho g\right)\frac{r^2}{2} + C_1$$

As $\tau_{rz} = 0$ at $r = 0$ ($dv_z/dr = 0$ at $r = 0$), $C_1 = 0$ and thus

$$\tau_{rz} = \left(\frac{\Delta P}{L} + \rho g\right)\frac{r}{2}$$

$$= -\eta\frac{dv_z}{dr} \tag{2.19}$$

The second integration then gives

$$v_z = -\left(\frac{\Delta P}{L} + \rho g\right)\frac{r^2}{4\eta} + C_2$$

As $v_z = 0$ at $r = R$,

$$C_2 = \left(\frac{\Delta P}{L} + \rho g\right)\frac{R^2}{4\eta}$$

and thus

$$v_z = \left(\frac{\Delta P}{L} + \rho g\right)\frac{R^2 - r^2}{4\eta} \tag{2.20}$$

v_z is thus a paraboloid, as shown in Fig. 2.15(a), and Fig. 2.15(b) shows the variation of τ_{rz} across a diameter of the tube. The maximum value of v_z, which occurs at $r = 0$, is

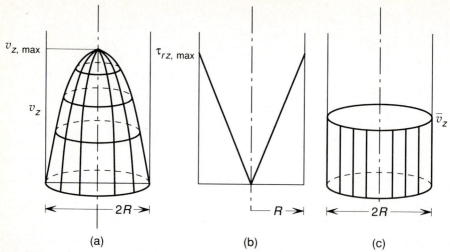

(a)   (b)   (c)

FIGURE 2.15. (a) Variation of local flow velocity with radial position in fluid flow in a cylindrical pipe; (b) variation of shear stress with radial position in fluid flow in a cylindrical pipe; (c) calculation of the average velocity in fluid flow in a cylindrical pipe.

$$v_{z,max} = \left(\frac{\Delta P}{L} + \rho g\right) \frac{R^2}{4\eta} \tag{2.21}$$

As illustrated in Fig. 2.15(a) and (c), the average value of the flow velocity $\bar{v}_z$ is such that the volume of the cylinder $\bar{v}_z \pi R^2$ in Fig. 2.15(c) is equal to the volume of the paraboloid in Fig. 2.15(a):

$$
\begin{aligned}
\bar{v}_z \pi R^2 &= \int_0^{2\pi} \int_0^R v_z r\, dr\, d\theta \\
&= 2\pi \int_0^R \left(\frac{\Delta P}{L} + \rho g\right) \frac{R^2 - r^2}{4\eta}\, r\, dr \\
&= 2\pi \frac{\Delta P/L + \rho g}{4\eta} \left[\frac{R^2 r^2}{2} - \frac{r^4}{4}\right]_0^R \\
&= 2\pi \left(\frac{\Delta P}{L} + \rho g\right) \frac{R^4}{16\eta}
\end{aligned}
$$

or

$$\bar{v} = \left(\frac{\Delta P}{L} + \rho g\right) \frac{R^2}{8\eta} \tag{2.22}$$

which is half the value of v_{max}. The volume flow rate $\dot{V}$ is then

$$\dot{V} = \bar{v}_z \pi R^2 = \pi \left(\frac{\Delta P}{L} + \rho g\right) \frac{R^4}{8\eta} \tag{2.23}$$

and Eq. (2.23) is known as the *Hagen–Poiseuille equation*. The mass flow rate $\dot{M}$ is

$$\dot{M} = \dot{V}\rho = \pi\rho\left(\frac{\Delta P}{L} + \rho g\right)\frac{R^4}{8\eta} \qquad (2.24)$$

EXAMPLE 2.4

For the purpose of examining the relative influences of density and viscosity, consider, separately, the flow of oil, water and air in a cylindrical tube of diameter 0.08 m at 25°C.

Data	viscosity (Pa·s)	density (kg/m³)
oil	0.8	888
water	8.57×10^{-4}	997
air	1.85×10^{-5}	1.177

Note that, while the densities of oil and water are comparable, the viscosity of the oil is almost three orders of magnitude greater than the viscosity of water. And, although the viscosity of water is only 46 times that of air, the density of water is almost three orders of magnitude greater than that of air.

Consider, first, the maximum average velocity of each fluid in a horizontal tube, at which the flow is still laminar, i.e., for which Re = 2100.

$$\text{Re} = 2100 = \frac{\bar{v}_x D\rho}{\eta}$$

or

$$\bar{v}_x = \frac{2100\eta}{D\rho} \qquad (i)$$

For oil;

$$\bar{v}_x = \frac{2100 \times 0.8}{0.08 \times 888} = 23.65 \text{ m/s}$$

for water;

$$\bar{v}_x = \frac{2100 \times 8.57 \times 10^{-4}}{0.08 \times 997} = 0.0226 \text{ m/s}$$

and for air;

$$\bar{v}_x = \frac{2100 \times 1.85 \times 10^{-5}}{0.08 \times 1.177} = 0.412 \text{ m/s}$$

Equation (i) shows that for flow at a given Reynolds number in a tube of given diameter, the average velocity is proportional to ratio η/ρ. This ratio, which itself is an important property of the fluid, is called the *kinematic viscosity*, v:

$$v \text{ (kinematic viscosity)} = \frac{\eta}{\rho}$$

and in SI, v has the units m^2/s. The traditional unit of kinematic viscosity is the stoke, where 1 stoke $= 1 \text{ cm}^2/\text{s}$.

The kinematic viscosities of oil, water, and air in m^2/s at 25°C are for oil, 9.01×10^{-4} for water, 8.60×10^{-7}, and for air, 1.57×10^{-5}. Therefore, for flow at a given Reynolds number in a tube of given diameter, the average velocities decrease in the order oil, air, water.

Consider, now, the pressure drops per unit length required in the 0.08-m-diameter horizontal pipe to obtain an average flow velocity of 0.0226 m/s (the maximum value of $\bar{v}_x$ for water at which the flow is still laminar in a 0.08-m-diameter tube). Rearrangement of Eq. (2.22), and noting that the gravitational force does not influence horizontal flow, gives

$$\frac{\Delta P}{L} = \frac{8\eta\bar{v}_x}{R^2} = \frac{8 \times 0.0226}{(0.04)^2}\eta = 113\eta \tag{ii}$$

Therefore, for oil,

$$\frac{\Delta P}{L} = 113 \times 0.8 = 90.4 \ \frac{\text{N}}{\text{m}^2\cdot\text{m}}$$

for water,

$$\frac{\Delta P}{L} = 113 \times 8.57 \times 10^{-4} = 0.0968 \ \frac{\text{N}}{\text{m}^2\cdot\text{m}}$$

and for air,

$$\frac{\Delta P}{L} = 113 \times 1.85 \times 10^{-5} = 0.0021 \ \frac{\text{N}}{\text{m}^2\cdot\text{m}}$$

Equation (ii) indicates that for laminar flow at a given average flow rate, in a tube of given diameter, the required pressure drop per unit length is proportional to the viscosity of the fluid. Therefore, in the present example, the value of $\Delta P/L$ decreases in the order oil, water, air.

Consider next the pressure drops per unit length required for flow upward in a vertical pipe of 0.08 m diameter at an average velocity of 0.0226 m/s. As the gravitational force in flow upward *opposes* the flow, the gravitational contribution is negative and hence Eq. (2.22) becomes

$$\bar{v}_z = \left(\frac{\Delta P}{L} - \rho g\right)\frac{R^2}{8\eta}$$

rearrangement of which gives

$$\frac{\Delta P}{L} = \frac{8\eta\bar{v}_z}{R^2} + \rho g$$

$$= 113\eta + 9.81\rho \qquad (2.25)$$

For oil,

$$\frac{\Delta P}{L} = 90.4 + 9.81 \times 888 = 90.4 + 8711 = 8802 \, \frac{N}{m^2 \cdot m}$$

for water,

$$\frac{\Delta P}{L} = 0.0968 + 9.81 \times 997 = 0.0968 + 9780 = 9780 \, \frac{N}{m^2 \cdot m}$$

and for air,

$$\frac{\Delta P}{L} = 0.0021 + 9.81 \times 1.177 = 0.0021 + 11.5 = 11.5 \, \frac{N}{m^2 \cdot m}$$

Examination shows that in each case the increase in $\Delta P/L$ is determined essentially by the density of the fluid, such that in contrast with horizontal flow, the vertical flow of water requires a greater $\Delta P/L$ value than does that of the more viscous but less dense oil.

As has been stated, the discussion thus far has been confined to consideration of the laminar flow of incompressible fluids. Thus strictly, the equations derived are not applicable to gases. For example, with respect to the Hagen–Poiseuille equation, Eq. (2.23), in flow through a horizontal pipe of length L in which the absolute pressure decreases from P_0 at $x = 0$ to P_L at $x = L$, the specific volume of an ideal gas entering the pipe at the higher pressure P_0 is smaller than that of the gas exiting at the pressure P_L by the factor P_L/P_0. As the pressure drop through the pipe is linear, the average pressure of the gas is $(P_0 + P_L)/2$, and this is the pressure of the gas at which the volume flow rate is calculated from Eq. (2.23). However, if the volume of the gas is measured at some other pressure, P_m, then from the ideal gas law for constant temperature,

$$(\text{volume measured at } P_m) \times P_m = \left(\text{volume calculated at } \frac{P_0 + P_L}{2}\right)$$

$$\times \frac{P_0 + P_L}{2}$$

and hence the

$$\text{actual flow rate} = \left(\text{volume calculated at } \frac{P_0 + P_L}{2}\right) \times \frac{P_0 + P_L}{2P_m}$$

$$= \pi \frac{P_0 - P_L}{L} \frac{R^4}{8\eta} \frac{P_0 + P_L}{2P_m} \qquad (2.26)$$

The significance of the correction depends on the magnitude of the pressure drop in the pipe. For example, in Example 2.4 it was calculated that the maximum average velocity of air in a horizontal pipe of diameter 0.08 m at which the flow is still laminar is 0.412 m/s. Equation (ii) gives the required pressure drop as 0.0381 N/m²·m. Thus the pressure drop across a 100-m length of pipe would be 3.81 N/m². At 1 atm absolute pressure $= 101,325$ N/m², this pressure drop is 3.76×10^{-5} atm, such that if $P_L = 1$ atm and the flow rate is measured at 1 atm pressure, the correction factor in Eq. (2.26) is $(2 + 3.76 \times 10^{-5})/2 = 1.00002$. For a correction factor of 1.1, a pressure drop of 0.2 atm would be required, and with $\Delta P/L = 0.0381$ N/m²·m, this would occur over a pipe length of 5.32×10^5 m (330 miles).

For flow in a horizontal pipe the mass flow rate is given by

$$\dot{M} = \bar{v}_x \pi R^2 \rho$$

Thus for flow of a compressible fluid in a horizontal tube of constant diameter, conservation of mass requires that the product $\bar{v}_x \rho$ be a constant, and consequently, as P, and hence ρ, decrease in the flow direction, the average linear flow rate and the volume flow rate increase.

EXAMPLE 2.5

Air at 298 K is flowing in a horizontal pipe of diameter 0.1 m. At one location, a, in the pipe the pressure is 700 kPa and the average flow velocity is 15 m/s, and at another location farther downstream, b, the pressure is 400 kPa. What is the mass flow rate in the pipe, and what is the average flow velocity at location b?

Taking air to be an ideal gas of molecular weight 0.02884 kg/g mol, the density of the air at location a is

$$\rho = \frac{1}{V} = \frac{MP}{RT} = \frac{0.02884 \times 700,000}{8.3144 \times 298} = 8.15 \text{ kg/m}^3$$

and thus the mass flow rate is

$$\dot{M} = (\bar{v}_x \pi R^2 \rho)_a = 15 \times \pi \times (0.05)^2 \times 8.15 = 0.961 \text{ kg/s}$$

As

$$(\bar{v}_x \rho)_a = (\bar{v}_x \rho)_b$$

and

$$\rho = \frac{1}{V} = \frac{MP}{RT}$$

then for isothermal flow,

$$(\bar{v}_x P)_a = (\bar{v}_x P)_b$$

and hence at location b,

$$(\bar{v}_x)_b = \frac{P_a}{P_b}(\bar{v}_x)_a = \frac{7}{4} \times 15 = 26.25 \text{ m/s}$$

In this example, with $\eta = 1.85 \times 10^{-5}$ Pa·s, the Reynolds number is 6.61×10^5 and hence the flow is not laminar, which means that Eq. (2.23) cannot be used to relate the volume flow rate and the rate of pressure drop in the flow direction.

Equation (2.25) shows that the pressure drop, ΔP, from $z = 0$ to $z = L$ in a motionless vertical column L of incompressible fluid of density ρ is

$$\Delta P = \rho g L \tag{iii}$$

For a compressible fluid (i.e., a gas in which ρ is a function of P),

$$-dP = \rho g \, dz$$

which has been derived previously as Eq. (1.7). For an ideal gas, $\rho = PM/RT$, where M is the molecular weight, and hence

$$-dP = \frac{PMg}{RT} \, dz$$

or

$$-\frac{dP}{P} = \frac{Mg}{RT} \, dz$$

Integration between $z = 0$ and $z = L$ gives

$$\ln \frac{P_L}{P_0} = -\frac{MgL}{RT}$$

or

$$P_L = P_0 \exp\left(-\frac{MgL}{RT}\right)$$

or

$$\Delta P = (P_0 - P_L) = P_0 \left[1 - \exp\left(-\frac{MgL}{RT}\right)\right]$$

If $MgL \ll RT$,

$$\exp\left(-\frac{MgL}{RT}\right) \approx 1 - \frac{MgL}{RT}$$

and hence

$$\Delta P = P_0 - P_L \approx P_0 \frac{MgL}{RT}$$

$$\approx P_0 \left(\frac{\rho_0}{P_0} gL\right)$$

$$\approx \rho_0 g L$$

as given by Eq. (iii).

Consideration of the variation of the density of air at 288 K with L gave Eq. (1.11) as

$$P_L = P_0 \exp(-1.181 \times 10^{-4} L) \qquad (1.11)$$

At $T = 288$ K and $P_0 = 101325$ Pa, the density of ideal gas air is

$$\rho = \frac{1}{V} = \frac{PM}{RT} = \frac{101,325 \times 0.02884}{8.3144 \times 288} = 1.22 \text{ kg/m}^3$$

and hence, assuming a constant density, Eq. (iii) gives

$$P_0 - P_L = 1.22 \times 9.81 \times L = 11.97L \qquad (iv)$$

Equation (iv) overestimates the value of P_L given by Eq. (1.11), but the overestimation is less than 1% for $L < 1135$ m, and hence for normal engineering applications involving vertical columns of static air, Eq. (iv) is adequate.

Pressure Drop per Unit Length, $\Delta P/L$

In the preceding examples $\Delta P/L$ is the decrease in the static pressure per unit length in the flow direction. A negative value of $\Delta P/L$ thus indicates that the pressure is increasing in the flow direction. Consider again the flow of oil in a vertical pipe of diameter 0.08 m at an average velocity of 1 m/s (which is less than the maximum value at which the flow is still laminar). From Eq. (2.22),

$$\bar{v}_z = \left(\frac{\Delta P}{L} \pm \rho g\right) \frac{R^2}{8\eta}$$

in which the positive sign pertains to flow downward and the negative sign pertains to flow upward. Thus for flow upward at 1 m/s,

$$\frac{\Delta P}{L} = \frac{\bar{v}8\eta}{R^2} + \rho g$$

$$= \frac{1 \times 8 \times 0.8}{(0.04)^2} + 888 \times 9.81$$

$$= 4000 + 8711 = 12,711 \frac{N}{m^2 \cdot m}$$

that is, the decrease in the static pressure in the upward-flowing oil is 12,711 N/m² per meter in the flow direction. For zero flow velocity (i.e., a static column of oil in the pipe), Eq. (iii) gives

$$\frac{\Delta P}{L} = \rho g = 8711 \frac{N}{m^2 \cdot m}$$

that is, the decrease in the static pressure per meter increase in the height of the static column is 8711 N/m². For vertical flow downward at 1 m/s,

$$\frac{\Delta P}{L} = \frac{\bar{v}8\eta}{R^2} - \rho g$$

$$= 4000 - 8711 = -4711 \text{ N/m}^2 \cdot \text{m}$$

and the negative sign indicates that the pressure drop is a negative quantity (i.e., that the static pressure increases by 4711 N/m² per meter in the flow direction).

EXAMPLE 2.6 ——————————————————————————————

Zinc is removed from liquid lead by reaction with chlorine gas to form $ZnCl_2$. Figure 2.16 shows chlorine gas being bubbled gently through a bath of liquid lead at 400°C. The graphite bubbling tube has an internal diameter of 1.5 mm and is immersed to a depth of 1 m in the liquid lead. Calculate the static pressure in the tube at the level of the surface of the bath required to give a bubbling rate of 0.2 kg of chlorine per hour. The atmospheric pressure at the surface of the bath is 101.3 kPa.

At 673 K,

$$\rho_{Pb} = 10{,}560 \text{ kg/m}^3$$

$$\eta_{Cl_2} = 2.87 \times 10^{-5} \text{ Pa·s}$$

The static pressure in the liquid lead at depth of 1 m is

$$P = P_{atmospheric} + \rho_{Pb}gh$$

$$= 101{,}300 + 10{,}560 \times 9.807 \times 1$$

$$= 204{,}900 \text{ Pa}$$

and this is thus the pressure in a static column of chlorine gas in the bubbling tube required to extend the static column to the lower end of the tube. Considering

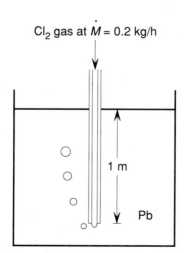

Cl_2 gas at $\dot{M} = 0.2$ kg/h

1 m

Pb

FIGURE 2.16.   Chlorine gas being bubbled through a bath of liquid lead.

chlorine gas to behave as an ideal gas of molecular weight 0.07091 kg/g mol, the density of the gas is

$$\rho = \frac{1}{V} = \frac{PM}{RT} = \frac{204{,}900 \times 0.07091}{8.3144 \times 673} = 2.596 \text{ kg/m}^3$$

Rearrangement of Eq. (2.24) gives the variation of the rate of pressure drop in a fluid flowing downward in a vertical tube with mass flow rate M as

$$\frac{\Delta P}{L} = \frac{8\dot{M}\eta}{\pi\rho R^4} - \rho g$$

$$= \frac{8 \times 0.2 \times 2.87 \times 10^{-5}}{3.142 \times 2.596 \times 3600 \times (0.75 \times 10^{-3})^4} - 2.596 \times 9.807$$

$$= 4936 - 25$$

$$= 4911 \text{ Pa/m}$$

Thus with a pressure of 204,900 Pa at its point of exit from the bubbling tube, the static pressure in the flowing chlorine 1 m above the point of exit is

$$204{,}900 + 4911 = 209{,}800 \text{ Pa}$$

The Reynolds number is

$$\frac{\bar{v}D\rho}{\eta} = \frac{\dot{M}}{\pi R^2 \rho} \times \frac{2R\rho}{\eta} = \frac{2\dot{M}}{\pi R\eta}$$

$$= \frac{2 \times 0.2}{3.142 \times 0.75 \times 10^{-3} \times 2.87 \times 10^{-5} \times 3600}$$

$$= 1642$$

which indicates that the flow is laminar and hence Eq. (2.24) can be used.

Summary of the Fluid Flow Equations

	Couette	Parallel Plates	Inclined Plane	Cylindrical Tube
τ	$\tau_{yx} = -\dfrac{V}{Y}\eta$	$\tau_{yx} = \dfrac{\Delta P}{L}y$	$\tau_{yx} = \rho g \cos\theta y$	$\tau_{rz} = \left(\dfrac{\Delta P}{L} \pm \rho g\right)\dfrac{r}{2}$
v	$v_x = \dfrac{V}{Y}y$	$v_x = \dfrac{\Delta P}{2L\eta}(\delta^2 - y^2)$	$v_x = \dfrac{\rho g \cos\theta}{2\eta}(\delta^2 - y^2)$	$v_z = \left(\dfrac{\Delta P}{L} \pm \rho g\right)\dfrac{R^2 - r^2}{4\eta}$
		$\bar{v}_x = \dfrac{\Delta P}{3L\eta}\delta^3$	$\bar{v}_x = \dfrac{\rho g \cos\theta}{3\eta}\delta^2$	$\bar{v}_z = \left(\dfrac{\Delta P}{L} \pm \rho g\right)\dfrac{R^2}{8\eta}$
		$= \tfrac{2}{3}v_{x,\max}$	$= \tfrac{2}{3}v_{x,\max}$	$= \tfrac{1}{2}v_{z,\max}$

2.7 Capillary Flowmeter

In the preceding derivations fluid flow was considered between *horizontal* plates, down an *inclined plane*, and in a *vertical* cylindrical tube. In all three cases both the shear stress τ and the velocity of the fluid are proportional to $\Delta P/L + \rho g \cos \theta$. In the case of horizontal flow $\theta = 90°$, and hence $\cos \theta = 0$, and thus only $\Delta P/L$ appears in the expressions for τ and v, and in the case of vertical flow $\theta = 0°$ and hence $\cos \theta = 1$, such that the expressions for τ and v contain $\Delta P/L + \rho g$. However, in the general case of inclined flow occurring under the influence of an externally imposed pressure, the terms $\Delta P/L$ and $\rho g \cos \theta$ appear in the expressions for shear stress and velocity.

The significance of the terms $\Delta P/L + \rho g \cos \theta$ can be appreciated from an examination of the capillary flowmeter shown in Fig. 2.17. This simple device involves a glass tube containing a length L of glass capillary tubing of inside diameter d inclined at an angle θ to the vertical, with a U-tube manometer, containing a manometer fluid of density ρ_m, placed across the length of capillary. When a second fluid of density ρ_f, which is immiscible with the manometer fluid, flows through the capillary tube, a difference in height h occurs between the levels of the manometer fluid in the two arms of the manometer. If P_0 is the static pressure in the flowing

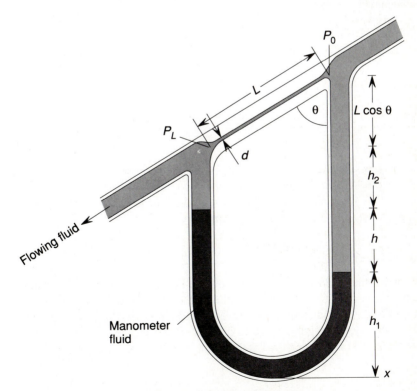

FIGURE 2.17.   Capillary flowmeter.

fluid at the entrance to the capillary and P_L is the static pressure in the fluid at the point of exit from the capillary, the pressure at the level X in the manometer, P_X, is, from consideration of the left arm of the manometer,

P_X = pressure due to the weight of the column of
manometer fluid of height $h_1 + h$

+ pressure due to the weight of the column of
second fluid of height h_2

+ P_L

= $(h_1 + h)\rho_m g + h_2 \rho_f g + P_L$

Also from consideration of the right arm of the manometer,

P_X = pressure due to the weight of the column of
manometer fluid of height h_1

+ pressure due to the weight of the column of
second fluid of height $h + h_2 + L \cos \theta$

+ P_0

= $h_1 \rho_m g + (h + h_2 + L \cos \theta)\rho_f g + P_0$

Rearrangement gives

$$P_0 - P_L = h(\rho_m - \rho_f)g - L \cos \theta \, \rho_f g$$

and hence the pressure difference measured at the manometer, $h(\rho_m - \rho_f)g$, is

$$h(\rho_m - \rho_f)g = (P_0 - P_L) + L \cos \theta \, \rho_f g$$

or

$$\frac{h(\rho_m - \rho_f)g}{L} = \frac{\Delta P}{L} + \rho_f \cos \theta \, g$$

Thus the pressure difference measured at the manometer includes the decrease in the static pressure across the capillary and the contribution due to the influence of gravity on the fluid, and consequently, use of the flowmeter does not require knowledge of the angle of inclination of the capillary.

EXAMPLE 2.7

Consider the use of the flowmeter shown in Fig. 2.17 to measure the volume flow rate of water at 20°C. The manometer fluid is an organic liquid, of density 1154 kg/m³, which is immiscible with water, and when the water is flowing through the capillary, of length 0.1 m and diameter 0.005 m, the difference between the levels of the manometer fluid is 0.030 m. The data are

$$L = 0.1 \text{ m}$$

$$D = 0.005 \text{ m}$$

$$\rho_m = 1154 \text{ kg/m}^3$$

$$\rho_f = 998 \text{ kg/m}^3$$

$$\eta_f = 9.44 \times 10^{-4} \text{ Pa·s}$$

$$h = 0.030 \text{ m}$$

Then the

$$\text{pressure difference measured at the manometer} = (\rho_m - \rho_f)gh$$

$$= (1154 - 998) \frac{\text{kg}}{\text{m}^3} \times 9.81 \frac{\text{m}}{\text{s}^2} \times 0.03 \text{ m}$$

$$= 45.9 \frac{\text{kg}}{\text{m·s}^2}$$

and

$$\frac{\text{measured pressure difference}}{L} = 45.9 \frac{\text{kg}}{\text{m·s}^2} \times \frac{1}{0.1} \frac{1}{\text{m}}$$

$$= 459 \text{ kg/m}^2\text{·s}^2$$

$$= \frac{\Delta P}{L} + \rho_f g \cos \theta$$

From Eq. (2.23),

$$\dot{V} = \pi \left(\frac{\Delta P}{L} + \rho_f g \cos \theta \right) \frac{R^4}{8\eta}$$

$$= 3.142 \times 459 \frac{\text{kg}}{\text{m}^2\text{·s}^2} \times (0.0025)^4 \text{ m}^4 \times \frac{1}{8 \times 9.44 \times 10^{-4}} \frac{\text{m·s}}{\text{kg}}$$

$$= 7.46 \times 10^{-6} \text{ m}^3/\text{s}$$

The average velocity of the flow is

$$\bar{v} = \frac{\dot{V}}{\pi R^2} = \frac{7.46 \times 10^{-6}}{\pi (0.0025)^2} = 0.38 \text{ m/s}$$

and thus the Reynolds number is

$$\text{Re} = \frac{\bar{v} D \rho_f}{\eta_f} = \frac{0.38 \times 0.005 \times 998}{9.44 \times 10^{-4}}$$

$$= 2008$$

which, being less than 2100, indicates that the flow is laminar.

Similarly, if an independent means is available for measuring the volume flow rate, the manometer can be used to measure the viscosity of the fluid.

EXAMPLE 2.8

When the volume flow rate of air at 20°C, through a capillary of 0.0025 m diameter and 0.1 m in length, is 4.45×10^{-5} m^3/s, the pressure difference measured at a dibutyl phthalate manometer is 9.6×10^{-3} m. Calculate the viscosity of air at 20°C. The data are

$$\rho_{air} = 1.21 \text{ kg/m}^3$$

$$\rho_m = 901.7 \text{ kg/m}^3$$

$$D = 0.0025 \text{ m}$$

$$L = 0.1 \text{ m}$$

$$h = 9.6 \times 10^{-3} \text{ m}$$

$$\dot{V} = 4.45 \times 10^{-5} \text{ m}^3/\text{s}$$

Then the

measured pressure drop $= (\rho_m - \rho_f)hg$

$$= (901.7 - 1.21) \frac{\text{kg}}{\text{m}^3} \times 9.6 \times 10^{-3} \text{ m} \times 9.81 \frac{\text{m}}{\text{s}^2}$$

$$= 84.8 \text{ kg/m·s}^2$$

and

$$\frac{\text{measured pressure drop}}{L} = 84.8 \frac{\text{kg}}{\text{m·s}^2} \times \frac{1}{0.1} \frac{1}{\text{m}}$$

$$= 848 \text{ kg/m}^2\text{·s}^2$$

$$= \Delta P/L + \rho_f g \cos \theta$$

Then, from rearrangement of Eq. (2.23),

$$\eta = \pi \left(\frac{\Delta P}{L} + \rho_f g \cos \theta \right) \frac{R^4}{8\dot{V}}$$

$$= 3.142 \times 848 \frac{\text{kg}}{\text{m}^2\text{·s}^2} \times (0.00125)^4 \text{ m}^4 \times \frac{1}{8 \times 4.45 \times 10^{-5}} \frac{\text{s}}{\text{m}^3}$$

$$= 1.83 \times 10^{-5} \text{ kg/m·s}$$

The average velocity is

$$\bar{v} = \frac{\dot{V}}{A} = \frac{4.45 \times 10^{-5}}{\pi \times (0.00125)^2} = 9.07 \text{ m/s}$$

and hence the Reynolds number is

$$\text{Re} = \frac{\bar{v} D \rho}{\eta} = \frac{9.07 \times 0.0025 \times 1.21}{1.83 \times 10^{-5}} = 1500$$

which indicates that the flow is laminar.

2.8 Fluid Flow in an Annulus

In Fig. 2.18 a fluid flows vertically upward through the annulus between two concentric cylindrical tubes, overflows at the top, and flows down the outside. Consider flow in the annulus and flow in the free-falling film on the outer surface.

Flow in the Annulus

The fluid in contact with the walls of the annulus is immobile (i.e., $v_z = 0$ at $r = aR$ and $r = R$) and hence $v_{z,max}$ occurs at some value of r between aR and R, at which point $\tau_{rz} = 0$. The directions of viscous momentum transport are from the value of r at which $v_{z,max}$ occurs toward both walls, and the gravitational force due to the weight of the fluid acts in a direction opposite to that of the flow.

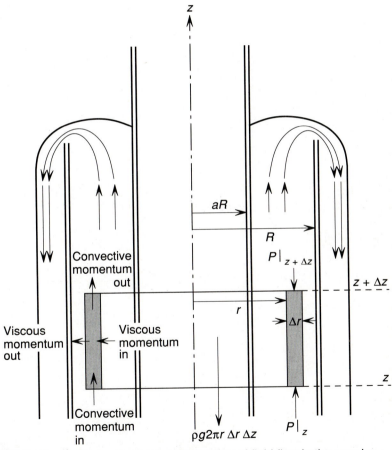

FIGURE 2.18. Control volume for consideration of fluid flow in the annulus between two concentric vertical cylinders.

Consider the momentum balance in the control volume of inner diameter r, outer diameter $r + \Delta r$, and length Δz. As fully developed steady-state flow is being considered, the rate of transport of convective momentum into the control volume through its lower surface equals the rate of transport of convective momentum out of the volume through its upper surface, and hence convective momentum does not contribute to the momentum balance. If the inner surface of the control volume is located at a radius greater than that at which $dv_z/dr = 0$, viscous momentum is transported down the velocity gradient into the control volume through the inner wall at r and out of the volume through the outer wall at $r + \Delta r$.

The momentum balance is then

$$(\tau_{rz} 2\pi r\, \Delta z)|_r - (\tau_{rz} 2\pi r\, \Delta z)|_{r+\Delta r} + (P|_z - P|_{z+\Delta z}) 2\pi r\, \Delta r - \rho g 2\pi r\, \Delta r\, \Delta z = 0$$

$$(\text{i})$$

which leads to

$$-\frac{d(r\tau_{rz})}{dr} + \frac{\Delta P}{L} r - \rho g r = 0$$

or

$$\frac{d(r\tau_{rz})}{dr} = \left(\frac{\Delta P}{L} - \rho g \right) r$$

Conversely, if the outer surface of the control volume is located at a radius which is less that at which $dv_z/dr = 0$, viscous momentum is transported down the velocity gradient in the $-r$ direction into the control volume through the outer wall at $r + \Delta r$ and out of the volume through the inner wall at r. The momentum balance is thus

$$(-\tau_{rz} 2\pi r\, \Delta z)|_{r+\Delta r} - (-\tau_{rz} 2\pi r\, \Delta z)|_r$$
$$+ (P|_z - P|_{z+\Delta z}) 2\pi r\, \Delta r - \rho g 2\pi r\, \Delta r\, \Delta z = 0$$

which again leads to

$$-\frac{d(r\tau_{rz})}{dr} + \frac{\Delta P}{L} r - \rho g r = 0$$

or

$$\frac{d(r\tau_{rz})}{dr} = \left(\frac{\Delta P}{L} - \rho g \right) r$$

The first integration gives

$$r\tau_{rz} = \left(\frac{\Delta P}{L} - \rho g \right) \frac{r^2}{2} + C_1$$

or

$$\tau_{rz} = \left(\frac{\Delta P}{L} - \rho g \right) \frac{r}{2} + \frac{C_1}{r} = -\eta \frac{dv_z}{dr} \qquad (\text{ii})$$

and the second integration gives

$$v_z = -\left(\frac{\Delta P}{L} - \rho g\right)\frac{r^2}{4\eta} - \frac{C_1 \ln r}{\eta} + C_2 \tag{iii}$$

The integration constants C_1 and C_2 are evaluated using the boundary conditions of

$$v_z \begin{cases} 0 & \text{at } r = R \\ 0 & \text{at } r = aR \end{cases}$$

as

$$C_1 = \left(\frac{\Delta P}{L} - \rho g\right)\frac{R^2}{4}\frac{1 - a^2}{\ln a}$$

and

$$C_2 = \left(\frac{\Delta P}{L} - \rho g\right)\frac{R^2}{4\eta}\left[1 + (1 - a^2)\frac{\ln R}{\ln a}\right]$$

which on substitution into Eq. (iii) gives

$$v_z = \left(\frac{\Delta P}{L} - \rho g\right)\frac{R^2}{4\eta}\left[1 - \left(\frac{r}{R}\right)^2 + \frac{1 - a^2}{\ln a}\ln\frac{R}{r}\right] \tag{iv}$$

and on substitution into Eq. (ii) gives

$$\tau_{rz} = \left(\frac{\Delta P}{L} - \rho g\right)\left(\frac{r}{2} + \frac{R^2}{4r}\frac{1 - a^2}{\ln a}\right) \tag{v}$$

$v_{z,\max}$ occurs where $\tau_{rz} = 0$, which from Eq. (v) is where

$$\frac{r}{2} = \frac{R^2}{4r}\frac{a^2 - 1}{\ln a}$$

or

$$r = R\left(\frac{a^2 - 1}{2\ln a}\right)^{1/2}$$

Substitution into Eq. (iv) gives

$$v_{z,\max} = \left(\frac{\Delta P}{L} - \rho g\right)\frac{R^2}{4\eta}\left[1 + \frac{1 - a^2}{2\ln a}\left(1 + \ln\frac{a^2 - 1}{2\ln a}\right)\right] \tag{vi}$$

At $r = R$, Eq. (v) gives

$$\tau_{rz|R} = \left(\frac{\Delta P}{L} - \rho g\right)\frac{R}{2}\left(1 + \frac{1 - a^2}{2\ln a}\right)$$

which is a positive quantity in accordance with $dv_z/dr|_R < 0$ and at $r = aR$, Eq. (v) gives

$$\tau_{rz|aR} = \left(\frac{\Delta P}{L} - \rho g\right)\left(\frac{aR}{2} + \frac{R}{4a}\frac{1 - a^2}{\ln a}\right)$$

which is a negative quantity in accordance with $dv_z/dr|_{aR} > 0$. The velocity and shear stress distributions are shown in Fig. 2.19. The average velocity, $\bar{v}_z$, is obtained from

$$\bar{v}_z \int_0^{2\pi} \int_{aR}^{R} r \, dr \, d\theta = \int_0^{2\pi} \int_{aR}^{R} v_z r \, dr \, d\theta$$

as

$$\bar{v}_z = \left(\frac{\Delta P}{L} - \rho g \right) \frac{R^2}{8\eta} \left(\frac{1 - a^4}{1 - a^2} + \frac{1 - a^2}{\ln a} \right) \tag{vii}$$

and the volume flow rate $\dot{V}$ and mass flow rate $\dot{M}$ are

$$\dot{V} = \bar{v}_z \pi R^2 (1 - a^2) \tag{viii}$$

and

$$\dot{M} = \dot{V} \rho \tag{ix}$$

Free Flow down the Outside of the Tube

The falling film shown in Fig. 2.20 is such that its free surface is at a radius of kR and its inner surface is at R. Thus $v_z|_R = 0$ and $v_z|_{kR} = v_{z,\text{max}}$. Also at the free surface $\tau_{rz} = 0$, and as the film is free falling, it is only subject to the force of its own weight. The control volume is a cylindrical shell of inner diameter r, outer diameter $r + \Delta r$, and length Δz, and as the direction of the transfer of viscous momentum is in the $-r$ direction from the free surface toward the wall (down the velocity gradient), viscous momentum enters through the outer face and leaves through the inner face of the control volume. Again, as steady-state flow occurs through the control volume, the transfer of convective momentum does not appear in the momentum balance.

The momentum balance is thus

$$(-\tau_{rz} \cdot 2\pi r \, \Delta z)|_{r+\Delta r} - (-\tau_{rz} \cdot 2\pi r \, \Delta z)|_r + \rho g \cdot 2\pi r \, \Delta r \, \Delta z = 0$$

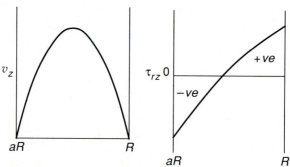

FIGURE 2.19.   Variations of local flow velocity and shear stress with radial position in fluid flow in an annulus.

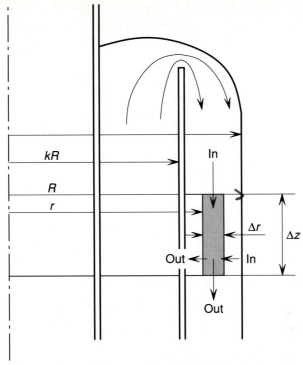

FIGURE 2.20.   Control volume for consideration of fluid flow down the outer wall of a vertical cylindrical pipe.

which leads to

$$-\frac{d(r\tau_{rz})}{dr} + \rho gr = 0$$

and hence

$$r\tau_{rz} = \rho g \frac{r^2}{2} + C_1$$

From the boundary condition that $\tau_{rz} = 0$ at $r = kR$,

$$C_1 = -\rho g \frac{(kR)^2}{2}$$

and hence

$$\tau_{rz} = \frac{\rho gr}{2} - \frac{\rho g(kR)^2}{2r} = -\eta \frac{dv_z}{dz}$$

A second integration gives

$$v_z = \frac{\rho g}{\eta} \frac{(kR)^2}{2} \ln r - \frac{\rho g}{\eta} \frac{r^2}{4} + C_2$$

and from the boundary condition that $v_{rz} = 0$ at $r = R$,

$$C_2 = \frac{\rho g}{\eta} \left[\frac{R^2}{4} - \frac{(kR)^2 \ln R}{2} \right]$$

Therefore,

$$v_z = \frac{\rho g}{4\eta} \left[R^2 - r^2 + 2(kR)^2 \ln \frac{r}{R} \right]$$

The average velocity $\bar{v}_z$ is obtained from

$$\bar{v}_z \int_0^{2\pi} \int_R^{kR} r \, dr \, d\theta = \int_0^{2\pi} \int_R^{kR} v_z r \, dr \, d\theta$$

as

$$\bar{v}_z = \frac{\rho g R^2}{8\eta} \left(1 - 3k^2 + \frac{4k^4}{k^2 - 1} \ln k \right) \tag{x}$$

and the volume flow rate $\dot{V}$ and mass flow rate $\dot{M}$ are

$$\dot{V} = \bar{v}_z \pi[(kR)^2 - R^2] = \bar{v}_z \pi R^2 (k^2 - 1) \tag{xi}$$

and

$$\dot{M} = \dot{V}\rho = \rho \bar{v}_z \pi R^2 (k^2 - 1)$$

EXAMPLE 2.9 _____

Calculate the maximum value of the average velocity of water flowing upward through an annulus formed between concentric tubes of radii 0.01 m and 0.02 m at which the film formed by overflow is still laminar.

For water at 25°C,

$$\rho = 997 \text{ kg/m}^3$$

$$\eta = 8.57 \times 10^{-4} \text{ Pa·s}$$

From Eq. (x) the average velocity in the falling film is

$$\bar{v}_z = \frac{\rho g}{8\eta} R^2 \left(1 - 3k^2 + \frac{4k^4}{k^2 - 1} \ln k \right)$$

$$= \frac{997 \times 9.81 \times (0.02)^2}{8 \times 8.57 \times 10^{-4}} \left(1 - 3k^2 + \frac{4k^4}{k^2 - 1} \ln k \right)$$

$$= 571 \left(1 - 3k^2 + \frac{4k^4}{k^2 - 1} \ln k \right) \qquad \text{m/s} \tag{xii}$$

For laminar flow in the film $25 \le \text{Re} = \bar{v}_z 4\delta\rho/\eta$, where δ (the film thickness)

$= kR - R = R(k - 1) = 0.02(k - 1)$ meters. Therefore, at the limit of the regime of laminar flow.

$$25 = \frac{4\bar{v}_z \times 0.02(k - 1)\rho}{\eta}$$

$$= \frac{\bar{v}_z(k - 1) \times 4 \times 0.02 \times 997}{8.57 \times 10^{-4}}$$

or

$$\bar{v}_z(k - 1) = 2.686 \times 10^{-4} \tag{xiii}$$

Combination of Eqs. (xii) and (xiii) gives

$$\bar{v}_z = \frac{2.686 \times 10^{-4}}{k - 1} = 571 \left(1 - 3k^2 + \frac{4k^4}{k^2 - 1} \ln k \right)$$

or

$$(k - 1) \left(1 - 3k^2 + \frac{4k^4}{k^2 - 1} \ln k \right) = 4.71 \times 10^{-7}$$

which has the solution $k = 1.006$. Therefore, from Eq. (xiii),

$$\bar{v}_z \text{ (in falling film)} = \frac{2.686 \times 10^{-4}}{k - 1} = \frac{2.686 \times 10^{-4}}{0.006} = 0.0448 \text{ m/s}$$

and

$$\delta \text{ (film thickness)} = 0.02 \ (k - 1) = 0.02 \times 0.006 = 1.2 \times 10^{-4} \text{ m}$$

Therefore, from Eq. (xi), the volume flow rate is

$$\dot{V} = \bar{v}_x \pi R^2 (k^2 - 1)$$
$$= 0.0448 \times \pi \times (0.02)^2 \times (1.006^2 - 1)$$
$$= 6.776 \times 10^{-7} \text{ m}^3/\text{s}$$

From Eq. (viii), the volume flow rate in the annulus is

$$\dot{V} = \bar{v}_z \pi [R^2 - (aR)^2]$$
$$= 6.776 \times 10^{-7} \text{ m}^3/\text{s}$$

and hence the average velocity in the annulus, $\bar{v}_z$, is

$$\bar{v}_z = \frac{6.776 \times 10^{-7}}{\pi(0.02^2 - 0.01^2)} = 7.19 \times 10^{-4} \text{ m/s}$$

For laminar flow is the annulus, $Re \leq 2000$, where the Reynolds number is given as

$$Re = \frac{2R(1 - a)\bar{v}_z \rho}{\eta}$$

With $\bar{v}_z = 7.19 \times 10^{-4}$ m/s,

$$Re = \frac{2 \times 0.02 \times (1 - 0.5) \times 7.19 \times 10^{-4} \times 997}{8.57 \times 10^{-4}}$$

$$= 17$$

and hence flow in the annulus is laminar.

2.9 Mean Residence Time

The occurrence of velocity gradients in a fluid flowing in a conduit means that the time during which an element of fluid in the flow remains in the conduit varies with the position of the element in the flow. The mean residence time, Θ, of a fluid flowing in the x-direction in a conduit of length L is given by

$$\Theta = \frac{L}{\bar{v}_x} \tag{2.27}$$

Consider a fluid flowing down a 30-m length of inclined plane of width 10 m, and let the angle of inclination of the plane and the physical properties of the fluid be such that the depth of the fluid is 0.5 m and the volume flow rate is 10^{-3} m^3/s. Then, from Eq. (2.17),

$$\dot{V} = \frac{\rho g \cos \theta}{3 \eta} \delta^2 (\delta W) = 10^{-3} \text{ m}^3/\text{s}$$

which, with $\delta = 0.5$ m and $W = 10$ m, gives

$$\frac{\rho g \cos \theta}{\eta} = \frac{3 \dot{V}}{\delta^3 W} = \frac{3 \times 10^{-3}}{0.5^3 \times 10} = 0.0024 \ s^{-1} \cdot m^{-1}$$

and hence

$$\bar{v}_x = \frac{\rho g \cos \theta}{3 \eta} \delta^2 = \frac{0.0024}{3} \times 0.5^2 = 0.0002 \text{ m/s}$$

Thus, from Eq. (2.27), the mean residence time of the fluid is

$$\Theta = \frac{L}{\bar{v}_x} = \frac{30}{0.0002} = 1.5 \times 10^5 \text{ s}$$

and from Eq. (2.15), the variation of the local flow velocity with depth y is

$$v_x = \frac{\rho g \cos \theta}{2 \eta} (\delta^2 - y^2) = \frac{0.0024}{2} (0.5^2 - y^2)$$

$$= 0.0003 - 0.0012 y^2 \text{ m/s}$$

This variation of v_x with y is shown in Fig 2.21. The layer of fluid in the flow that has the mean residence time is located at the value of y at which $v_x = \bar{v}_x$;

$$0.0003 - 0.0012 y^2 = 0.0002$$

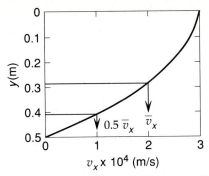

FIGURE 2.21. Variation, with depth, of the local flow velocity in fluid flow on an inclined plane.

or

$$y = 0.289 \text{ m}$$

and thus the fluid flowing at levels between $y = 0.289$ m and $y = 0.5$ m remains in the 30-m length of inclined plane for times greater or equal to the mean residence time of 1.5×10^5 s. Similarly, the layer of fluid that has twice the mean residence time is located at the value of y at which $v_x = 0.5\bar{v}_x$ (i.e., at $y = 0.408$ m) and thus the fluid at levels between $y = 0.408$ m and $y = 0.5$ m remains in the length of inclined plane for times greater or equal to twice the mean residence time.

The time required for the layer of fluid at level y to flow through the length L is given by

$$t = \frac{L}{v_x|_y}$$

or

$$t = \frac{30}{0.0003 - 0.0012y^2}$$

This variation of t with y is shown in Fig 2.22. The fraction of fluid that remains in the length of inclined plane for a period of time equal or greater than $n\Theta$ occurs between the value of y at which $v_x = \bar{v}_x/n$ and $y = \delta$, that is, between

$$0.0003 - 0.0012y^2 = \frac{0.0002}{n}$$

or

$$y_1 = \left(\frac{0.0003 - 0.0002/n}{0.0012} \right)^{1/2}$$

and

$$y_2 = 0.5$$

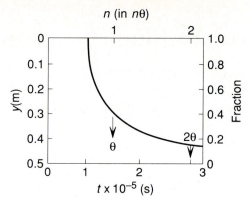

FIGURE 2.22. Variation, with depth, of the residence time of fluid flowing on an inclined plane.

and is thus obtained as

$$\text{fraction} = \frac{y_2 - y_1}{y_2}$$

$$= 1 - \left(\frac{0.0003 - 0.0002/n}{0.0012 \times 0.25} \right)^{1/2}$$

$$= 1 - \left(1 - \frac{2}{3n} \right)^{1/2}$$

This variation, which is also shown in Fig 2.22, shows that 42.2% of the fluid remains in the 30-m length of inclined plane for times greater than the mean residence time and 18.4% remains for times greater than twice the mean residence time.

2.10 Calculation of Viscosity from the Kinetic Theory of Gases

A monatomic gas is a population of atoms that move randomly in all directions within the vessel containing the gas. The atoms are constantly colliding with one another, and hence exchanging energy and momentum with one another, and at thermal equilibrium the distribution of the velocities of the atoms is determined only by the temperature of the gas and by the masses of the atoms. Integration of the distribution of velocities from infinity to zero gives the average speed $\bar{c}$ of the atoms in m/s as

$$\bar{c} = \left(\frac{8 \hbar T}{\pi m} \right)^{1/2}$$

where $\hbar$ is Boltzmann's constant (1.38054×10^{-23} J/K), T the absolute temperature of the gas in kelvin, and m the mass of an atom in kilograms.

The pressure that the gas exerts on the wall of its container is caused by collisions of the atoms with the wall, and the force exerted by the gas is simply the rate of change of momentum of the atoms due to their change in direction of movement after impace with the wall.

Calculation of the viscosity of a gas from kinetic theory is facilitated by making the following simplifying assumptions:

1. The atoms are rigid, nonattracting spheres of diameter d which undergo perfectly elastic collisions with one another.
2. All of the atoms travel with the same speed: the average speed $\bar{c}$.
3. All of the atoms are moving in directions parallel to the x-, y-, or z-axis, which means that one-sixth of them are moving in each of the $+x$, $-x$, $+y$, $-y$, $+z$, and $-z$ directions.

In considering the transport properties of a gas, knowledge is required of (1) the rate at which atoms pass through unit cross-sectional area in the volume containing the gas, and (2) the average distance traveled by the atoms between collisions with one another.

Figure 2.23 shows a rectangular volume of unit cross-sectional area and length $\bar{c}$. If the density of the gas is n atoms per unit volume, the rectangular volume contains $n\bar{c}$ atoms, one-sixth of which are traveling in the $-y$-direction. Consequently, the number of atoms passing through the base of the volume in unit time (i.e., through unit cross-sectional area) is $n\bar{c}/6$. A collision between two atoms occurs when the distance between their centers decreases to d, and therefore every atom excludes a volume of $\frac{4}{3}\pi d^3$ to every other atom. As shown in Fig. 2.24, an atom traveling with the velocity $\bar{c}$ sweeps out a volume $\pi d^2\bar{c}$ in unit time, and as this volume contains $n\pi d^2\bar{c}$ atoms, the traveling atom experiences $n\pi d^2\bar{c}$ collisions with other atoms in unit time. This argument is overly simple, in that the assumption is made that all of

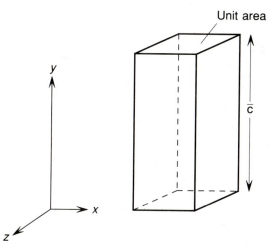

FIGURE 2.23. Volume element for considering the rate at which atoms in a gas phase pass through unit area.

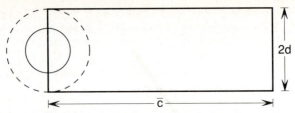

FIGURE 2.24.   Volume in a gas phase for consideration of the number of collisions that an atom experiences with other atoms in unit time.

the n atoms in the swept volume are stationary. If the relative velocities of the atoms are taken into consideration, the number of collisions per second is correctly calculated as $\sqrt{2}\, n\pi d^2 \bar{c}$, and hence the average distance traveled between collisions, termed the *mean free path*, λ, is

$$\lambda = \frac{\text{distance traveled in unit time}}{\text{number of collisions in unit time}} = \frac{\bar{c}}{\sqrt{2}\, n\pi d^2 \bar{c}} = \frac{1}{\sqrt{2}\, n\pi d^2}$$

Consider a gas undergoing laminar flow in the x-direction, with a linear velocity gradient in the y-direction. In Fig. 2.25 the flow is represented by layers or sheets of atoms, separated from one another by the mean free path, λ. Although the atoms in the sheets are moving in the x, y, and z directions with the velocity $\bar{c}$ relative to the sheets, the sheets are moving in the x-direction with a velocity v_x, the magnitude of which is characteristic of the position of the sheet on the y-axis. As the sheets are separated by the distance λ, atoms moving up or down on the y-axis undergo collisions only in the sheets, and hence on jumping vertically from one sheet to another, an atom transports momentum characteristic of the sheet from which the jump started. The number of atoms per unit area moving vertically from sheet A to sheet O in unit time is $n\bar{c}/6$, and as each molecule has the x-momentum characteristic of sheet A (i.e., $mv_x|_{-\lambda}$), the rate of transport of momentum per unit area toward O in the y-direction is $\frac{1}{6}n\bar{c}mv_x|_{-\lambda}$. Similarly, the rate of transport of momentum per unit

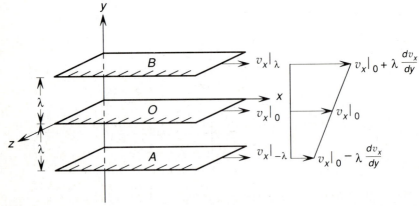

FIGURE 2.25.   Construction for theoretical calculations of the viscosity of a gas.

area toward the layer at O in the $-y$-direction due to atoms jumping from sheet B is $\frac{1}{6}n\bar{c}mv_x|_\lambda$. Thus the momentum flux in the y-direction across the layer at O is

$$\frac{1}{6}n\bar{c}mv_x|_{-\lambda} - \frac{1}{6}n\bar{c}mv_x|_\lambda = \frac{1}{6}n\bar{c}m\left(v_x|_0 - \lambda\,\frac{dv_x}{dy}\right) - \frac{1}{6}n\bar{c}m\left(v_x|_0 + \lambda\,\frac{dv_x}{dy}\right)$$

$$= -\frac{1}{3}n\bar{c}m\lambda\,\frac{dv_x}{dy}$$

As the momentum flux is equal to the shear stress, τ_{yx},

$$\tau_{yx} = -\frac{1}{3}n\bar{c}m\lambda\,\frac{dv_x}{dy}$$

which in comparison with Newton's law of viscosity, Eq. (2.4), gives the viscosity in terms of the properties of the gas as

$$\eta = \tfrac{1}{3}n\bar{c}m\lambda \tag{2.28}$$

Substituting for $\bar{c}$ and λ gives

$$\eta = \frac{2}{3}\frac{(m k T)^{1/2}}{\pi^{3/2}\,d^2} \tag{2.29}$$

Rigorous theory of rigid-sphere molecules, which does not require the simplifying assumptions made above and which takes into account the fact that a flowing gas is not in a state of complete equilibrium, gives the accurate result

$$\eta = \frac{5}{16}\frac{(m k T)^{1/2}}{\pi^{1/2}\,d^2} \tag{2.30}$$

With $k = 1.38054 \times 10^{-16}$ erg/K, T in kelvin, m in grams, and d in centimeters, Eq. (2.30) gives the viscosity of the gas in poise (g/cm·s), and the conversion of Eq. (2.30) to practical units gives

$$\eta = \frac{2.6693 \times 10^{-5}\sqrt{MT}}{d^2} \tag{2.31}$$

where M = atomic weight ($m \times$ Avogadro's number)

   T is in kelvin

   d is in angstroms, Å (1 Å $= 10^{-10}$ m)

   Avogadro's number $= 6.0232 \times 10^{23}$ per g mol

Equation (2.31) shows that the viscosity of a gas is independent of the pressure of the gas and is a linear function of $T^{1/2}$.

   The theoretical prediction that the viscosity of a gas is independent of pressure might, at first sight, seem surprising. It might seem that with increasing pressure, there would be more "jostling" of the atoms, such that for a given velocity gradient, there would be a greater shear stress and hence greater viscosity in the gas. The independence of viscosity on density n, and hence on pressure, is indicated by Eq. (2.28), which shows η to be proportional to both n and λ. However, as λ is inversely proportional to n, the influence of n on η is nullified. This canceling effect can be seen with reference to Fig. 2.25 as follows. With decreasing density, n, fewer atoms are traveling up and down vertically, but as λ increases with decreasing density, the smaller number of atoms are traveling a proportionately greater distance between the sheets. The greater distance traveled exactly compensates for the smaller number of atoms jumping, and hence the momentum flux is independent of the density. The theoretical calculation of the viscosity of a gas was first made by Maxwell in 1860, and the subsequent experimental observation that the viscosity of a gas is independent of pressure (at least at reasonably low pressures) was a distinct triumph for the kinetic theory.

Calculation of η from Eq. (2.31) requires knowledge of d, which in the absence of any other information, requires that the viscosity be known at some temperature. The viscosity of helium at 0°C is 1.86×10^{-4} P and hence, at 0°C, from Eq. (2.31), with an atomic weight of 4.003 g/g mol,

$$d = \left(\frac{2.6693 \times 10^{-5}\sqrt{MT}}{\eta} \right)^{1/2}$$

$$= \left[\frac{2.6693 \times 10^{-5} \times (4.003 \times 273)^{1/2}}{1.86 \times 10^{-4}} \right]^{1/2}$$

$$= 2.18 \text{ Å}$$

Therefore, Eq. (2.31) gives

$$\eta_{He} = \frac{2.6693 \times 10^{-5} \times \sqrt{4.003T}}{(2.18)^2}$$

$$= 1.123 \times 10^{-5}T^{1/2}$$

or

$$\log \eta_{He} = \tfrac{1}{2} \log T - 4.949 \tag{2.32}$$

A similar calculation for Ne, which has a viscosity of 2.973×10^{-4} P at 0°C and an atomic weight of 20.183 g/g mol, gives $d_{Ne} = 2.58$ Å. Thus

$$\eta_{Ne} = 1.799 \times 10^{-5}T^{1/2}$$

or

$$\log \eta_{Ne} = \tfrac{1}{2} \log T - 4.745 \tag{2.33}$$

Equations (2.32) and (2.33) are shown in comparison with the measured viscosities of He and Ne in Fig. 2.26. The slopes of the experimental lines for Ne and He are, respectively, 0.65 and 0.67:

$$\eta_{He} \propto T^{0.65}$$

and

$$\eta_{Ne} \propto T^{0.67}$$

instead of $\eta \propto T^{0.5}$, as required by Eq. (2.31). The dependence on temperature of the viscosities of all real gases is $\eta \propto T^n$, with n in the range 0.6 to 1.0. Agreement

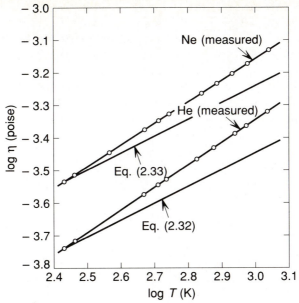

FIGURE 2.26. Variations, with temperature, of the measured and calculated viscosities of Ne and He.

between Eq. (2.31) and the actual viscosities of real gases would thus require that the diameters of the atoms decrease with increasing temperature; for example, the diameter of He would have to decrease from 2.18 Å at 273 K to 1.94 Å at 1000 K, and the diameter of Ne would have to decrease from 2.58 Å at 273 K to 2.35 Å at 1000 K.

This discrepancy occurs because the atoms of a gas are not nonattracting hard spheres; they are ''soft'' spheres surrounded by fields of force that attract at large distances of separation and repel at short interatomic distances. Atoms colliding in a high-temperature gas penetrate more deeply into the force fields of one another than do colliding atoms in a low-temperature gas, and hence the apparent diameters of the atoms do decrease with increasing temperature. The force field comprises a short-range repulsive force which, for mathematical convenience, is assumed to be proportional to r^{-13}, and a longer-range attractive force, which is proportional to r^{-7}. The calculated force between two Ar atoms, and the attractive (negative) and repulsive (positive) components of the force, are shown in Fig. 2.27(a). The relationship between the net force, f, between two atoms and the potential energy, ϕ, of the two-atom system is

$$f = -\frac{d\phi}{dr}$$

and hence the potential energy at any interatomic distance is obtained as

$$\phi = -\int_{\infty}^{r} f \, dr$$

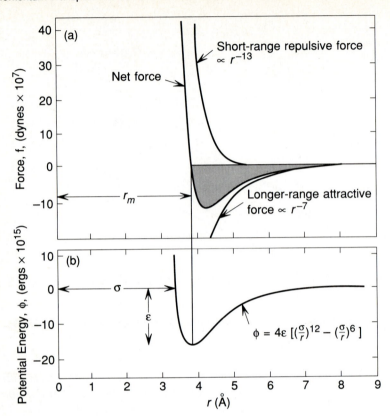

FIGURE 2.27. Variation, with interatomic distance, of (a) the force acting between two atoms, and (b) the potential energy of two atoms.

The variation of ϕ with interatomic distance r, for Ar, obtained by integrating the force from infinity to r, is shown in Fig. 2.27(b). At the value of r at which $f = 0$ in Fig. 2.27(a), the potential energy, obtained as the shaded area in Fig. 2.27(a), has a maximum negative value ϵ, and this interatomic distance is designated r_m. The collision diameter of the atom, σ, is taken as being the value of r at which $\phi = 0$.

The potential energy shown in Fig. 2.27(b) is the *Lennard-Jones 6-12 potential,* given by

$$\phi = 4\epsilon \left[\left(\frac{\sigma}{r} \right)^{12} - \left(\frac{\sigma}{r} \right)^{6} \right] \tag{2.34}$$

in which σ, the *collision diameter*, is the interatomic distance at which $\phi = 0$, and ϵ, the *characteristic energy of interaction between the atoms*, is the minimum value of ϕ, which occurs at $r_m = 2^{1/6}\sigma$. The r^{-12} and r^{-6} terms arise from integration of the r^{-13} and r^{-7} terms in the expression for the force. The 6-12 potentials in Ne, Ar, and CO_2 are shown in Fig. 2.28.

The "softness" of the atoms, which is determined by the magnitudes of the attractive and repulsive forces, can be quantified in terms of the compressibility of the gas, which, in turn, is obtained from the equation of state for the gas relating

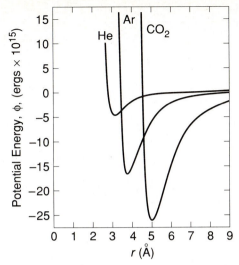

FIGURE 2.28. Variation, with interatomic distance, of the potential energies of two atoms of He, Ar, and CO_2.

pressure, temperature, and volume. Analysis of measured $P-V-T$ relationships thus allows determination of the values of σ and ϵ in Eq. (2.34).

The effect of the softness of atoms on the viscosity calculated from kinetic theory has been dealt with independently by Chapman and Enskog, who introduced the *collision integral*, Ω, into Eq. (2.31) to give the *Chapman–Enskog formula*,

$$\eta = 2.6693 \times 10^{-5} \frac{\sqrt{MT}}{\sigma^2 \Omega} \tag{2.35}$$

The collision integral, Ω, which is a measure of the softness of the atom, is a function of the dimensionless quantity kT/ϵ.

Although, strictly, Chapman–Enskog theory applies only to monatomic gases, it has been applied successfully to calculation of the viscosity of reasonably spherical polyatomic molecules. The values of ϵ/k and σ for several common gases are listed in Table 2.1, and the variation of Ω with kT/ϵ is listed in Table 2.2.

Comparison between the Chapman–Enskog equation and measured data at 273 K for several gases is shown in Fig. 2.29 as a plot of η versus $\sqrt{M}/\sigma^2\Omega$. The line, with a slope of $2.6693 \times 10^{-5} \sqrt{273}$, is from Eq. (2.35) and the data in Tables 2.1 and 2.2, and the open circles are experimentally measured viscosities.

EXAMPLE 2.10 _____

Calculate the viscosities of O_2 and N_2 at 100°C.
   From Table 2.1:

	M	σ (Å)	ϵ/kT (K)
O_2	32.00	3.433	113
N_2	28.02	3.681	91.5

TABLE 2.1

Gas	Molecular Weight, M	Lennard-Jones Parameters	
		σ (Å)	$\epsilon/\bar{k}$
H_2	2.016	2.915	38.0
He	4.003	2.576	10.2
Ne	20.183	2.789	35.7
Ar	39.944	3.418	124
N_2	28.02	3.681	91.5
O_2	32.00	3.433	113
CO	28.01	3.590	110
CO_2	44.01	3.996	190
SO_2	64.07	4.290	252
F_2	38.00	3.653	112
Cl_2	70.91	4.115	357
CH_4	16.04	3.822	137
CS_2	76.14	4.438	488
Air	28.97	3.617	97.0

Source: J. O. Hirschfelder, C. F. Curtiss, and R. B. Bird, *Molecular Theory of Gases and Liquids*, Wiley, New York, 1954, pp. 1110–1112.

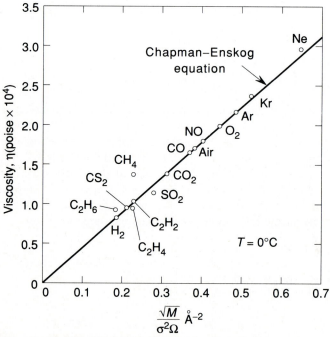

FIGURE 2.29. Comparison between the measured and calculated viscosities of several gases.

TABLE 2.2

$\dfrac{kT}{\epsilon}$	Ω	$\dfrac{kT}{\epsilon}$	Ω	$\dfrac{kT}{\epsilon}$	Ω	$\dfrac{kT}{\epsilon}$	Ω
0.30	2.785	1.30	1.399	2.60	1.081	4.60	0.9422
0.35	2.628	1.35	1.375	2.70	1.069	4.70	0.9382
0.40	2.492	1.40	1.353	2.80	1.058	4.80	0.9343
0.45	2.368	1.45	1.333	2.90	1.048	4.90	0.9305
0.50	2.257	1.50	1.314	3.00	1.039	5.0	0.9269
0.55	2.156	1.55	1.296	3.10	1.030	6.0	0.8963
0.60	2.065	1.60	1.279	3.20	1.022	7.0	0.8727
0.65	1.982	1.65	1.264	3.30	1.014	8.0	0.8538
0.70	1.908	1.70	1.248	3.40	1.007	9.0	0.8379
0.75	1.841	1.75	1.234	3.50	0.9999	10.0	0.8242
0.80	1.780	1.80	1.221	3.60	0.9932	20.0	0.7432
0.85	1.725	1.85	1.209	3.70	0.9870	30.0	0.7005
0.90	1.675	1.90	1.197	3.80	0.9811	40.0	0.6718
0.95	1.629	1.95	1.186	3.90	0.9755	50.0	0.6504
1.00	1.587	2.00	1.175	4.00	0.9700	60.0	0.6335
1.05	1.549	2.10	1.156	4.10	0.9649	70.0	0.6194
1.10	1.514	2.20	1.138	4.20	0.9600	80.0	0.6076
1.15	1.482	2.30	1.122	4.30	0.9553	90.0	0.5973
1.20	1.452	2.40	1.107	4.40	0.9507	100.0	0.5882
1.25	1.424	2.50	1.093	4.50	0.9464		

Source: J. O. Hirschfelder, R. B. Bird, and E. L. Spotz, *Chem. Rev.* (1949), vol. 44, p. 205.

For oxygen at 373 K, $kT/\epsilon = 373/113 = 3.30$, which from Table 2.2 gives $\Omega = 1.014$. Therefore,

$$\eta_{O_2,373K} = \frac{2.6693 \times 10^{-5}\sqrt{MT}}{\sigma^2\,\Omega} = \frac{2.6693 \times 10^{-5} \times \sqrt{32 \times 373}}{(3.433)^2 \times 1.014}$$
$$= 2.44 \times 10^{-4}\ P$$

For nitrogen at 373 K, $kT/\epsilon = 373/91.5 = 4.07$, which from interpolation of the data in Table 2.2 gives $\Omega = 0.9664$. Thus

$$\eta_{N_2,373K} = \frac{2.6693 \times 10^{-5}\sqrt{28.02 \times 373}}{(3.681)^2 \times 0.9664} = 2.08 \times 10^{-4}\ P$$

The viscosities of gas mixtures can be calculated from Wilke's formula [C. R. Wilke, *J. Chem. Phys.* (1950), vol. 18, pp. 517–519]

$$\eta_{mix} = \sum_{i=1}^{n} \frac{X_i\eta_i}{\sum_{j=1}^{n} X_j\Phi_{ij}} \tag{2.36}$$

where X_i is the mole fraction of component i in the n-component mixture, η_i is the viscosity of pure i, and

$$\Phi_{ij} = \frac{1}{\sqrt{8}} \left(1 + \frac{M_i}{M_j} \right)^{-1/2} \left[1 + \left(\frac{\eta_i}{\eta_j} \right)^{1/2} \left(\frac{M_j}{M_i} \right)^{1/4} \right]^2 \qquad (2.37)$$

where M_i is the molecular weight of component i.

EXAMPLE 2.11

Calculate the viscosities of mixtures of H_2 and CO_2 at 300 K.
At 300 K,

	M	η (P)
H_2	2.016	0.89×10^{-4}
CO_2	44.01	1.5×10^{-4}

From Eq. (2.37), with $H_2 \equiv i$ and $CO_2 \equiv j$,

$$\Phi_{H_2-CO_2} = \frac{1}{\sqrt{8}} \left(1 + \frac{2.016}{44.01} \right)^{-1/2} \left[1 + \left(\frac{0.89}{1.5} \right)^{1/2} \left(\frac{44.01}{2.016} \right)^{1/4} \right]^2$$

$$= 2.455$$

and with $CO_2 \equiv i$ and $H_2 \equiv j$,

$$\Phi_{CO_2-H_2} = \frac{1}{\sqrt{8}} \left(1 + \frac{44.01}{2.016} \right)^{-1/2} \left[1 + \left(\frac{1.5}{0.89} \right)^{1/2} \left(\frac{2.016}{44.01} \right)^{1/4} \right]^2 = 0.190$$

Note that

$$\Phi_{ii} = \frac{1}{\sqrt{8}} (2)^{-1/2} [2]^2 = 1$$

Then, from Eq. (2.36),

$$\eta_{mix} = \frac{X_{H_2} H_{H_2}}{X_{H_2} \Phi_{H_2-H_2} + X_{CO_2} \Phi_{H_2-CO_2}} + \frac{X_{CO_2} \eta_{CO_2}}{X_{CO_2} \Phi_{CO_2-CO_2} + X_{H_2} \Phi_{CO_2-H_2}}$$

$$= \frac{0.89 \times 10^{-4} X_{H_2}}{X_{H_2} + 2.455 X_{CO_2}} + \frac{1.5 \times 10^{-4} X_{CO_2}}{X_{CO_2} + 0.190 X_{H_2}}$$

which as $X_{H_2} + X_{CO_2} = 1$ reduces to

$$\eta_{mix} = \frac{0.89 \times 10^{-4} X_{H_2}}{2.455 - 1.455 X_{H_2}} + \frac{1.5 \times 10^{-4} (1 - X_{H_2})}{1 - 0.81 X_{H_2}}$$

This variation is shown as the solid line in Fig. 2.30.

In an n-component mixture, n^2 values of Φ are required, but as n of these Φ equal unity, only $n(n - 1)$ need be calculated.

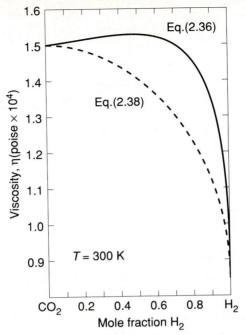

FIGURE 2.30. Variation, with composition, of calculated viscosities of mixtures in the system CO_2–H_2.

EXAMPLE 2.12

Calculate the viscosity of a mixture (on a molar basis) of 50% CO, 30% CO_2, and 20% water vapor at 400 K.

At 400 K

	M	η (P)	X_i
CO	28.01	2.21×10^{-4}	0.5
CO_2	44.01	1.94×10^{-4}	0.3
H_2O	34.02	1.32×10^{-4}	0.2

From Eq. (2.37),

$$\Phi_{CO-CO} = \Phi_{CO_2-CO_2} = \Phi_{H_2O-H_2O} = 1$$

$$\Phi_{CO-CO_2} = 1.332$$

$$\Phi_{CO_2-CO} = 0.744$$

$$\Phi_{CO-H_2O} = 1.456$$

$$\Phi_{H_2O-CO} = 0.716$$

$$\Phi_{CO_2-H_2O} = 1.066$$

$$\Phi_{H_2O-CO_2} = 0.938$$

Therefore,

$$\eta_{mix} = \frac{X_{CO}\eta_{CO}}{X_{CO}\Phi_{CO-CO} + X_{CO_2}\Phi_{CO-CO_2} + X_{H_2O}\Phi_{CO-H_2O}}$$

$$+ \frac{X_{CO_2}\eta_{CO_2}}{X_{CO_2}\Phi_{CO_2-CO_2} + X_{CO}\Phi_{CO_2-CO} + X_{H_2O}\Phi_{CO_2-H_2O}}$$

$$+ \frac{X_{H_2O}\eta_{H_2O}}{X_{H_2O}\Phi_{H_2O-H_2O} + X_{CO_2}\Phi_{H_2O-CO_2} + X_{CO}\Phi_{H_2O-CO}}$$

$$= \frac{0.5 \times 2.21 \times 10^{-4}}{0.5 + (0.3 \times 1.332) + (0.2 \times 1.456)}$$

$$+ \frac{0.3 \times 1.94 \times 10^{-4}}{0.3 + (0.5 \times 0.744) + (0.2 \times 1.066)}$$

$$+ \frac{0.2 \times 1.32 \times 10^{-4}}{0.2 + (0.3 \times 0.938) + (0.5 \times 0.716)}$$

$$= 1.90 \times 10^{-4} \text{ P}$$

If the molecular weights of the components of the gas mixture are similar, the viscosity of the gas mixture can be calculated from

$$\eta_{mix} = \frac{\Sigma X_i \eta_i (M_i)^{1/2}}{\Sigma X_i (M_i)^{1/2}} \tag{2.38}$$

Figure 2.31 shows the viscosities of O_2-N_2 mixtures at 300 K calculated from Eqs. (2.36) and (2.38), and the dashed line in Fig. 2.30 shows the viscosities of H_2-CO_2 mixtures at 300 K calculated from Eq. (2.38). The viscosities and molecular weights of N_2 and O_2 give $\Phi_{O_2-N_2} = 1.006$ and $\Phi_{N_2-O_2} = 0.993$, and thus variation of viscosity calculated from Eq. (2.36) is insignificantly different from that given by the simpler Eq. (2.38). The calculated value of the composition of air, (79% $N_2-21\%O_2$), of 1.85×10^{-4} P is in exact agreement with the value measured. On the other hand, the differences between the viscosities and molecular weights of CO_2 and H_2 are sufficiently large that Eq. (2.38) gives a poor estimation of the viscosities of H_2-CO_2 mixtures.

2.11 Viscosities of Liquid Metals

In contrast with the case of the gaseous state, there is no reasonable kinetic theory of liquids, and most theories of liquids are based on models which employ parameters that lack fundamental significance and which require a priori assumptions as to the structures, interactions, or mechanisms of transport in the fluid. Although the exact nature of the forces between atoms in a liquid metal has not been determined, Chap-

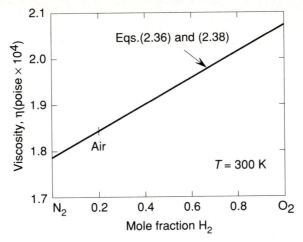

FIGURE 2.31. Variation, with composition, of calculated viscosities of mixtures in the system N_2 and O_2.

man [T. Chapman, *AIChE J.* (1966), vol. 12, p. 395] assumed that they can be described to a fair approximation by a potential function of the form given by Eq. (2.34), which contains only two parameters: an energy parameter ϵ and a distance parameter δ. Under this assumption, the reduced viscosity of a liquid metal, η^*, is a universal function of a reduced temperature T^* and a reduced volume V^*:

$$\eta^* = f(T^*, V^*) \tag{2.39}$$

where

$$\eta^* = \frac{\eta \delta^2 \mathring{N}_O}{(MRT)^{1/2}} \tag{2.40}$$

$$T^* = \frac{\mathring{k}T}{\epsilon} \tag{2.41}$$

$$V^* = \frac{1}{n\delta^3} \tag{2.42}$$

The characteristic distance parameter, δ, is taken as being Goldschmidt's atomic diameter for the metal (which is the atomic diameter calculated assuming a coordination number of 12) and n is number of atoms per unit volume. In Eq. (2.40), with η expressed in $kg \cdot m^{-1} \cdot s^{-1}$ and δ in meters, R is 8.3144 $J \cdot g\ mol^{-1} \cdot K^{-1}$ and the molecular weight M is in kg/g mol.

The form of Eq. (2.39) has been determined for sodium and potassium from calculation of the Lennard-Jones parameters from measured x-ray scattering curves. This variation is shown in Fig. 2.32. Chapman then took the experimentally measured values of the viscosities of 19 liquid metals, shown in Fig. 2.33, and determined the corresponding values of ϵ required to make the reduced viscosity data fall on the line given in Fig. 2.32. For example, from Fig. 2.33, the viscosity

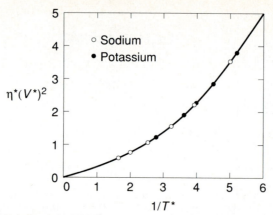

FIGURE 2.32. Variation of $\eta^*(V^*)^2$ with $1/T^*$ for sodium and potassium as determined from calculations of the Lennard-Jones parameters from measured x-ray scattering curves.

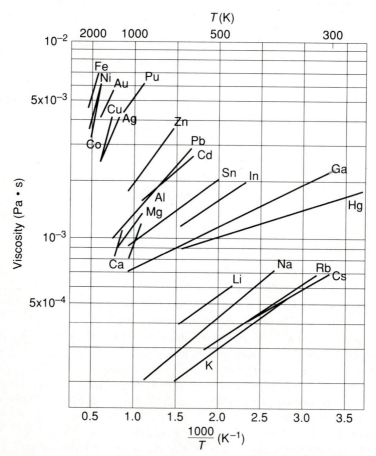

FIGURE 2.33. Variations, with temperature, of the viscosities of several liquid metals.

of liquid Sn is 2×10^{-3} kg·m^{-1}·s^{-1} at $1/T = 0.002$ K^{-1} (500 K). At this temperature the density of Sn is 6930 kg/m^3 and, with $\delta = 3.16$ Å and $M = 0.1187$ kg/g mol, Eq. (2.40) gives

$$\eta^* = \frac{2 \times 10^{-3} \times (3.16 \times 10^{-10})^2 \times 6.023 \times 10^{23}}{(0.1187 \times 8.3144 \times 500)^{1/2}}$$

$$= 5.41$$

The value of n in Eq. (2.42) is obtained as

$$n = 6930 \, \frac{\text{kg}}{\text{m}^3} \times \frac{1}{0.1187} \, \frac{\text{g mol}}{\text{kg}} \times 6.023 \times 10^{23} \, \frac{1}{\text{g mol}} \times \frac{1}{(10^{10})^3} \, \frac{\text{m}^3}{\text{Å}^3}$$

$$= 3.52 \times 10^{-2} \, \text{Å}^{-3}$$

and hence

$$V^* = \frac{1}{3.52 \times 10^{-2} \times (3.16)^3} = 0.901$$

Therefore, $\eta^*(V^*)^2 = 4.39$, and, from Fig. 2.32, $1/T^* = 5.3$. Thus, from Eq. (2.41),

$$\frac{\epsilon}{kT} = \frac{T}{T^*} = 500 \times 5.3 = 2650 \text{ K}$$

The values of ϵ/k determined by Chapman for the 19 metals shown in Fig. 2.33 and the corresponding values of δ are listed in Table 2.3, and the fit of the calculated values of $\eta^*(V^*)^2$ with $1/T^*$ is shown in Fig. 2.34. The derived value of ϵ/k is a linear function of the melting temperature of the metal, T_m, as shown in Fig. 2.35, and the correlation

$$\frac{\epsilon}{k} = 5.20 T_m \qquad (2.43)$$

TABLE 2.3 Goldschmidt Atomic Diameters and the Energy Parameters of Metals

Metal	δ (Å)	ϵ/k (K)	Metal	δ (Å)	ϵ/k
Na	3.84	1,970[a]	Rb	5.04	1,600
K	4.76	1,760[a]	Ag	2.88	6,400
Li	3.14	2,350	Cd	3.04	3,300
Mg	3.20	4,300	In	3.14	2,500
Al	2.86	4,250	Sn	3.16	2,650
Ca	4.02	5,250	Cs	5.40	1,550
Fe	2.52	10,900	Au	2.88	6,750
Co	2.32	9,550	Hg	3.10	1,250
Ni	2.50	9,750	Pb	3.50	2,800
Cu	2.56	6,600	Pu	3.1	5,550
Zn	2.74	4,700			

From T. W. Chapman, *AIChE J.* (1966), vol. 12, pp. 395–400.

[a]Determined from x-ray scattering data.

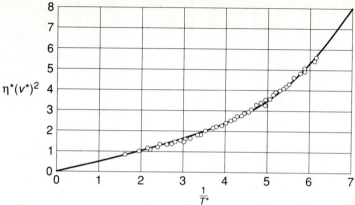

FIGURE 2.34. Variation of $\eta^*(V^*)^2$ with $1/T^*$ for the liquid metals shown in Fig. 2.33.

allows the viscosity of any liquid metal be estimated from knowledge of its melting temperature, density, and value of δ.

EXAMPLE 2.13 _____

Estimate the viscosity of liquid chromium at its melting temperature of 2171 K. The data are

$$\text{density at 2171 K} = 6.4 \text{ g/cm}^3$$

$$M = 52 \text{ g/g mol}$$

$$\delta = 2.57 \text{ Å}$$

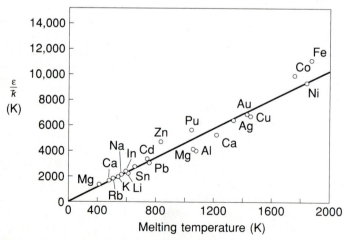

FIGURE 2.35. Variation of ϵ/k with melting temperature for the metals shown in Fig. 2.33.

From Eq. (2.43),

$$\frac{\epsilon}{k} = 5.20 \times 2171 = 11,290$$

$$T^* = \frac{2171}{11,290} = 0.192$$

From Fig. 2.34, $1/T^* = 5.2$ gives $\eta^*(V^*)^2 = 3.9$:

$$V^* = \frac{1}{6.4 \times \frac{1}{52} \times 6.023 \times 10^{23} \times (10^{-8})^3 \times (.257)^3}$$

$$= 0.795$$

$$\eta^* = \frac{3.9}{(0.795)^2} = 6.17$$

$$\eta = \frac{6.17 \times (0.052 \times 8.3144 \times 2171)^{1/2}}{(2.57 \times 10^{-10})^2 \times 6.023 \times 10^{23}}$$

$$= 4.75 \times 10^{-3} \text{ kg·m}^{-1}\text{·s}^{-1}$$

Figure 2.33 shows the temperature dependence of the viscosities of liquid metals to be of the form

$$\eta = A \exp\left(\frac{\Delta H^*}{RT}\right) \tag{2.44}$$

in which expression ΔH^* is the activation enthalpy for the thermally activated process of viscous flow. Comparison of Eq. (2.44) with Eq. (2.31) shows that although the hydrodynamic theories of the flow of liquids and gases are similar, the kinetic-molecular mechanisms are very different; the viscosities of gases increase with increasing temperature and the viscosities of liquids decrease with increasing temperature. The elementary flow process in a liquid involves either the squeezing of flow units (atoms of molecules) between pairs of other flow units or the creation of small holes in the liquid into which a flow unit can move, and ΔH^* is the energy barrier that a flow unit must overcome to squeeze successfully between its neighbors or is the energy required to form a hole. The fraction of flow units that are sufficiently energetic to overcome the energy barrier to flow increases exponentially with increasing temperature and hence the viscosity of the liquid decreases exponentially with increasing temperature. The values of ΔH^* for a number of liquids are listed in Table 2.4 in comparison with the corresponding molar enthalpies of evaporation. The energy required to create a hole of molecular size in a liquid is $\Delta H_{evap}/N_O$ and the observation, in Table 2.4, that the ratio of ΔH^* to ΔH_{evap} is between 0.25 and 0.33 suggests that the flow process requires the creation of holes in the liquid which are between one-third and one-fourth of the volume of a molecule. In contrast with molecular liquids, the ratios of ΔH^* to ΔH^*_{evap} for liquid metals vary from $\frac{1}{8}$ to $\frac{1}{20}$, which suggests that the flow unit in a liquid metal is an ion embedded in the free electron cloud, and that the unit which evaporates is the larger neutral atom.

TABLE 2.4

Liquid	ΔH^* (kJ/g mol)	ΔH_{evap} (kJ/g mol)	$\dfrac{\Delta H^*}{\Delta H_{evap}}$
CCl_4	10.5	27.6	0.379
C_6H_6	10.6	27.9	0.381
CH_4	3.01	7.61	0.395
A	2.16	5.94	0.363
N_2	1.88	5.06	0.371
O_2	1.67	6.15	0.271
CS_2	5.36	24.8	0.216
Na	6.07	97.9	0.0620
K	4.73	79.5	0.0595
Ag	20.2	254.0	0.0794
Hg	2.72	56.9	0.0478

The viscosity of a liquid varies with pressure as

$$\eta = Be^{P\Delta V^*/RT} \tag{2.45}$$

in which ΔV^* is the volume of the hole in the liquid that must be created for the flow process to occur. The variations, with temperature at atmospheric pressure, of the viscosities of several common fluids are shown in Fig. 2.36.

2.12 Summary

The characteristics of fluid flow are determined by the geometry of the containing duct, the magnitude of the force causing the flow, and the density and viscosity of the fluid. At low enough flow velocities laminar flow occurs, which is characterized by lamellae of fluid sliding over one another, with no macroscopic mixing of the fluid in directions normal to the direction of flow. The viscosity of the fluid causes the lamellae of fluid in contact with the walls of the containing duct to be stationary, and velocity gradients are established in the fluid in directions normal to the flow. The shear stresses occurring between the lamellae of fluid are proportional to the local velocity gradients, and the proportionality constant between the two is the viscosity of the fluid. The shear stress is formally equivalent to a rate of transport of momentum down the velocity gradient.

The shear stresses, velocity distributions, and hence average linear velocities and volume and mass flow rates are determined by momentum balances on control volumes in the fluid. The momentum balance requires that the difference between the rates at which convective and viscous momentum leave and enter the control volume plus the sum of the forces acting on the fluid in the control volume be zero. The balance leads to a differential equation containing the shear stress, which when integrated twice yields the variations of shear stress and local flow velocity with

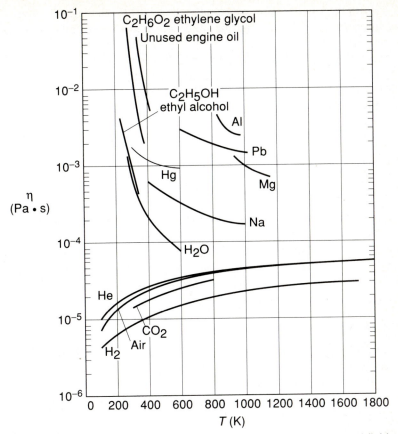

FIGURE 2.36. Variations, with temperature, of the viscosities of several fluids.

position in the flow. The two integration constants are obtained from boundary conditions for the particular type of flow.

The limits of laminar flow in all flow geometries are identified by values of the Reynolds number for the flow. The upper limit for laminar flow in a cylindrical pipe occurs at a Reynolds number of 2100. The average flow velocity in a horizontal pipe at a given Reynolds number is determined by the kinematic viscosity of the fluid, and the rate of decrease in the static pressure of the fluid in the flow direction is determined by the viscosity of the fluid. In flow that is not horizontal the flow is influenced by gravity, and hence the density of the fluid is important in determining the flow characteristics.

Application of the kinetic theory of gases to a gas comprising rigid noninteracting spherical atoms that undergo elastic collisions with one another shows that the viscosity of the gas is proportional to the mass and the diameter of the atom and to the square root of the temperature of the gas and is independent of the pressure of the gas. The experimental observation that the viscosity of a gas is proportional to T^n, where n varies in the range 0.6 to 1.0, is accounted for by a more sophisticated application of the kinetic theory which takes into consideration the existence of force fields surrounding real atoms and the influence of these force fields on collisions between atoms.

In contrast with gases, viscous flow in liquids is a thermally activated process, and hence the viscosity of a liquid decreases exponentially with increasing temperature. The rate of decrease is determined by the magnitude of the activation energy for the flow process.

Problems

PROBLEM 2.1

A shaft of outer diameter D meters is rotating at r revolutions per minute in a journal with a thin film of lubricant of viscosity η and thickness δ between the shaft and the journal. Derive an expression that relates the torque, T, required to overcome the friction in the fluid to D, r, η, and δ. Calculate the torque and the rate of degradation of mechanical energy to thermal energy when $D = 2.5$ cm, $\delta = 0.1$ mm, $r = 1000$ rpm, and $\eta = 3 \times 10^{-3}$ Pa·s.

PROBLEM 2.2

A cylindrical shaft of diameter 5 cm slides through a cylindrical hole in a 0.5-m-thick block. The clearance is 0.15 mm and the annulus between the shaft and the block is filled with a lubricant of viscosity 5×10^{-2} Pa·s. Calculate the force required to move the shaft through the hole at a velocity of 0.15 m/s. Ignore the influence of any end effects.

PROBLEM 2.3

The shaft in Problem 2.2 is rotated in the hole and it is found that a torque of 1 N·m is required to sustain a speed of rotation of 60 rpm. Calculate the viscosity of the lubricant in the annulus between the shaft and the block. Assume a linear velocity gradient through the lubricant and ignore any end effects.

PROBLEM 2.4

A 5-mm-thick film of oil of density 888 kg/m³ and viscosity 0.8 Pa·s flows down a vertical wall. (a) Calculate the mass flow rate. If the wall were inclined at 45° to the vertical and the mass flow rate were the same, what would (b) the film thickness and (c) the average flow velocity be?

PROBLEM 2.5

A layer of molten slag of density 2700 kg/m³ and viscosity 0.3 Pa·s is being transferred from one reverberatory furnace to another by flow down a plane between the two furnaces which is inclined at an angle of 80° to the vertical. The inclined plane is 5 m in width and 5 m in length and the mass flow rate of the slag is 7.5 kg/s.

Calculate:

(a) The thickness of the layer
(b) The average linear flow velocity of the slag
(c) The mean residence time of the slag on the plane
(d) The fraction of the slag that remains on the plane for times equal to or greater than the mean residence time

Neglect end effects.

PROBLEM 2.6

A capillary flowmeter is being designed to measure the volumetric flow rate of air. The diameter of the capillary is 0.5 mm and the maximum value of h that can be read on the manometer is 1 m. Calculate the minimum length of capillary at which, with a reading of $h = 1$ m, the flow through the capillary is still laminar. The manometric fluid is carbon tetrachloride of density 1594 kg/m^3 and the density and viscosity of air at 298 are, respectively, 1.19 kg/m^3 and 1.86×10^{-5} Pa·s.

PROBLEM 2.7

When air at 300 K is flowing through a horizontal tube of inner diameter 1 cm, the local flow velocity measured at a radius of 3 mm for the centerline of the tube is found to be 1 m/s. Calculate:

(a) The rate of pressure drop in the flowing air
(b) The average linear velocity of the flow
(c) The Reynolds number
(d) The shear stress exerted on the tube by the air

At 300 K the density and viscosity of air are, respectively, 1.177 kg/m^3 and 1.85×10^{-5} Pa·s.

PROBLEM 2.8

Show that in laminar fluid flow in a tube of circular cross section, half of the fluid remains in the tube for times equal to or greater than the mean residence time, and half remains in the tube for times equal to or less than the mean residence time.

PROBLEM 2.9

For laminar flow in a tube of circular cross section of radius R, by what factor would R have to be increased to:

(a) Double the average linear velocity?
(b) Double the volume flow rate?

PROBLEM 2.10

A horizontal pipeline 20 miles long delivers petroleum at a certain flow rate when the pressure drop over the line is 3000 kPa. If the pressure drop over the 20 miles remains at 3000 kPa, which of the three following procedures provides the largest increase in the volume flow rate of the oil? Assume that the flow is laminar.

(1) Doubling of the last 10 miles of the pipe with an identical pipe.
(2) Replacement of the last 10 miles of the pipe with a pipe that has twice the diameter of the original.
(3) Replacement of the last 10 miles of pipe with a pipe that has twice the flow area of the original.

PROBLEM 2.11

Calculate the viscosity of an equimolar mixture of CO_2 and CO at 1000°C.

PROBLEM 2.12

Calculate the viscosity of an equimolar mixture of CH_4 and H_2 at 1000°C.

PROBLEM 2.13

Estimate the viscosity of liquid iron at 1600°C. The density of liquid iron at 1600°C is 7160 kg/m^3.

PROBLEM 2.14

The surface of a plane, which is inclined to the vertical at 60°, is coated with a film of lubricating oil of thickness 1×10^{-4} m and viscosity 1.96×10^{-2} Pa·s. Calculate the terminal velocity reached by a cubical box of mass 20 kg and dimensions 1 m $\times$ 1 m $\times$ 1 m when it is placed on the inclined plane and allowed to slide on the film of oil.

PROBLEM 2.15

Oil of density 888 kg/m^3 and viscosity 0.8 Pa·s is flowing in a horizontal pipe of diameter 0.15 m. If the local flow velocity at the centerline is 10 m/s, what is:

(a) The shear stress exerted by the oil on the wall of the pipe?
(b) The rate of decrease of the static pressure in the oil in the flow direction?
(c) Write an equation for the variation of the local flow velocity with radial position in the pipe.

PROBLEM 2.16

A cylindrical wire of radius b is pulled at the velocity V through a cylindrical tube of radius a. The wire is concentric with the tube and the annular space between the two is filled with an oil. Determine:

(a) The variation of local flow velocity of the oil with V, a, b, and radial position
(b) The expression for the volume flow rate of oil through the tube

The tube connects two large reservoirs of oil.

PROBLEM 2.17

Two immiscible incompressible fluids are flowing between two horizontal flat plates of length L, width W, and vertical separation 2δ. The flow rates are adjusted such that the thickness of each layer is δ. The fluid in the lower layer is more dense and more viscous than the fluid in the upper layer. Calculate:

(a) The variation of v_x with y in both layers of fluid
(b) The variation of τ_{yx} with y in both layers
(c) The value of y at which v_x has its maximum value in both layers

The value of $\Delta P/L$ is the same in both layers.

PROBLEM 2.18

The flow network in the figure is made of capillary tubing. Lengths 1, 2, 3, and 5 have internal diameters of 2 mm and their lengths are as follows: 1, 282 m; 2, 6.32 m; 3, 5.83 m; 4, 4.24 m; and 5, 5.83 m. Laminar flow of a fluid is in the directions shown by the arrows. Calculate the internal diameter of length, 4, which eliminates fluid flow in length 5.

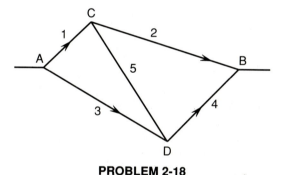

PROBLEM 2-18

3

Equations of Continuity and Conservation of Momentum and Fluid Flow Past Submerged Objects

3.1 Introduction

In Chapter 2 the shear stress and velocity distributions in several fluid flow situations were developed by constructing a momentum balance on a control volume of fluid in the particular situations. In this chapter general equations are developed which can, in principle, be applied to any flow situation. These are the *equation of continuity* and the *equation of conservation of momentum*, from which the velocity distribution can be derived.

3.2 Equation of Continuity

The *equation of continuity* is simply a mass balance which requires that the rate of accumulation of mass within a control volume in the flowing fluid be the difference between the rates at which mass enters and leaves the control volume:

rate of accumulation of mass =

(rate at which mass enters) − (rate at which mass leaves)   (3.1)

Consider the control volume in Cartesian coordinates shown in Fig. 3.1(a). The velocity of the fluid, $\mathbf{v}$, which, being a vector quantity, has both magnitude and direction, can be broken down into its components v_x in the x-direction, v_y in the y-direction, and v_z in the z-direction, as shown in Fig. 3.1(b), where

$$\mathbf{v}^2 = v_x^2 + v_y^2 + v_z^2$$

The rate at which mass enters the control volume through the x-face at x is obtained as

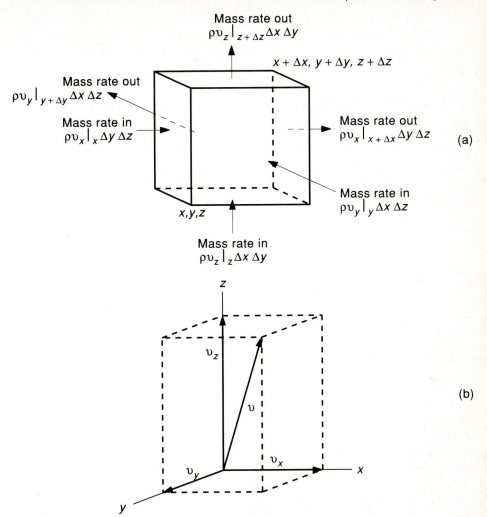

FIGURE 3.1. (a) Control volume for consideration of fluid flow in cartesian coordinates; (b) x, y, and z components of velocity v.

$$\text{density} \left(\frac{\text{kg}}{\text{m}^3}\right) \times x\text{-component of the velocity} \left(\frac{\text{m}}{\text{s}}\right) \times \text{area (m}^3)$$

that is, as

$$\rho v_x|_x \, \Delta y \, \Delta z$$

and the rate at which mass leaves the control volume through the x-face at $x + \Delta x$ is

$$\rho v_x|_{x+\Delta x} \, \Delta y \, \Delta z$$

Similarly, mass enters through the y-face at y at a rate of $\rho v_y|_y \, \Delta x \, \Delta z$ and leaves though the y-face at $y + \Delta y$ at a rate of $\rho v_y|_{y+\Delta y} \, \Delta x \, \Delta z$, and mass enters through the z-face at z at a rate of $\rho v_z|_z \, \Delta x \, \Delta y$ and leaves through the z-face at $z + \Delta z$ at a rate of $\rho v_z|_{z+\Delta z} \, \Delta x \, \Delta y$.

The difference between the rates at which mass enters and leaves the control volume is thus

$$\rho v_x|_x \, \Delta y \, \Delta z - \rho v_x|_{x+\Delta x} \, \Delta y \, \Delta z + \rho v_y|_y \, \Delta x \, \Delta z - \rho v_y|_{y+\Delta y} \Delta x \, \Delta z$$
$$+ \rho v_z|_z \, \Delta x \, \Delta y - \rho v_z|_{z+\Delta z} \, \Delta x \, \Delta y$$

which, from the mass balance, is the rate of accumulation of mass in the control volume. For a control volume of fixed volume, the rate of accumulation of mass is the rate of increase of the density of the fluid times the volume,

$$\frac{\partial \rho}{\partial t} \, \Delta x \, \Delta y \, \Delta z$$

Making the equation and dividing by $\Delta x \, \Delta y \, \Delta z$ gives

$$\frac{\partial \rho}{\partial t} = -\frac{\rho v_x|_{x+\Delta x} - \rho v_x|_x}{\Delta x} - \frac{\rho v_y|_{y+\Delta y} - \rho v_y|_y}{\Delta y} - \frac{\rho v_z|_{z+\Delta z} - \rho v_z|_z}{\Delta z}$$

which in the limits of $\Delta x \to 0$, $\Delta y \to 0$, and $\Delta z \to 0$ gives

$$\frac{\partial \rho}{\partial t} = -\left[\frac{\partial}{\partial x} (\rho v_x) + \frac{\partial}{\partial y} (\rho v_y) + \frac{\partial}{\partial z} (\rho v_z) \right] \tag{3.2}$$

Equation (3.2) is the equation of continuity. If the density of the fluid is constant, Eq. (3.2) becomes

$$\frac{\partial v_x}{\partial x} + \frac{\partial v_y}{\partial y} + \frac{\partial v_z}{\partial z} = 0 \tag{3.3}$$

which in shorthand notation can be written as

$$\nabla \cdot \mathbf{v} = 0 \tag{3.4}$$

3.3 Conservation of Momentum

Conservation of momentum requires that the differences between the rates at which momentum enters and leaves the control volume equals the rate of accumulation of momentum in the control volume. The fluid flows considered in Chapter 2 were all fully developed and at steady state, in which case there was no accumulation of momentum within the control volume. As before, the momentum balance requires consideration of viscous momentum transport, convective momentum transport, and the "pressure" and gravitational forces influencing the fluid flow through the control volume, and momentum transport must be considered in each of the x, y, and z directions.

Consider the shear stresses arising from the x-component of the flow. As shown in Fig. 3.2, these are the shear stresses acting on the y-faces, which cause the transport of viscous momentum in the y-direction into the y-face at y and out of the y-face at $y + \Delta y$, and the shear stresses acting on the z-faces, which cause the transport of viscous momentum into the z-face at z and out of the z-face at $z + \Delta z$. The difference between the rates at which viscous momentum enters and leaves through the y and z faces is thus

$$\tau_{yx}|_y \, \Delta x \, \Delta z \; - \; \tau_{yx}|_{y+\Delta y} \, \Delta x \, \Delta z \; + \; \tau_{zx}|_z \, \Delta x \, \Delta y \; - \; \tau_{zx}|_{z+\Delta z} \, \Delta x \, \Delta y \qquad \text{(i)}$$

In addition, a provision must be made for the possibility that v_x is varying with x (i.e., that the fluid is accelerating or decelerating in the x-direction). Such an acceleration would result from the application of a normal stress to the x-face of the control volume, which by virtue of causing the transport of viscous momentum in the x-direction by fluid flow in the x-direction, is designated τ_{xx}. The difference

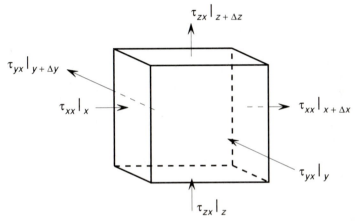

(a) Directions of transport of momentum due to x-component of velocity

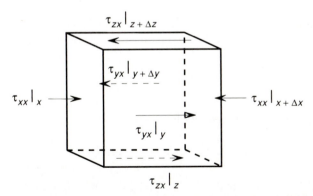

(b) Directions of viscous forces due to momentum transport

FIGURE 3.2.   (a) Directions of transport of momentum arising from the x-component of velocity; (b) directions of the viscous forces arising from momentum transport.

between the rates at which this viscous momentum enters the x-face at x and leaves through the x-face at $x + \Delta x$ is thus

$$\tau_{xx}|_x \, \Delta y \, \Delta z - \tau_{xx}|_{x+\Delta x} \, \Delta y \, \Delta z \tag{ii}$$

Mass enters the x-face at x at the rate $\rho v_x|_x \, \Delta y \, \Delta z$ and leaves through the x-face at $x + \Delta x$ at the rate $\rho v_x|_{x+\Delta x} \, \Delta y \, \Delta z$, and as momentum is the product of mass and velocity, the difference between the rates at which x-directed momentum enters and leaves the control volume is

$$\rho v_x v_x|_x \, \Delta y \, \Delta z - \rho v_x v_x|_{x+\Delta x} \, \Delta y \, \Delta z \tag{iii}$$

Mass enters through the y-face at y at the rate $\rho v_y|_y \, \Delta x \, \Delta z$ and leaves through the y-face at $y + \Delta y$ at the rate $\rho v_y|_{y+\Delta y} \, \Delta x \, \Delta z$. The rate at which this mass transfer causes x-directed convective momentum to enter the control volume is thus the product of the mass transfer rate $\rho v_y|_y \, \Delta x \, \Delta z$ and the x-component of the velocity v_x (i.e., $\rho v_x v_y|_y \, \Delta x \, \Delta z$), and the rate at which the x-directed momentum leaves the control volume is $\rho v_x v_y|_{y+\Delta y} \, \Delta x \, \Delta z$. The difference between the rates at which x-directed momentum enters and leaves through the y-faces is thus

$$\rho v_x v_y|_y \, \Delta x \, \Delta z - \rho v_x v_y|_{y+\Delta y} \, \Delta x \, \Delta z \tag{iv}$$

and similarly, the difference between the rates at which x-directed momentum enters and leaves through the z-faces is

$$\rho v_x v_z|_z \, \Delta x \, \Delta y - \rho v_z v_z|_{z+\Delta z} \, \Delta x \, \Delta y \tag{v}$$

If the static pressure in the fluid at the x-face at x is $P|_x$ and at the x-face at $x + \Delta x$ is $P|_{x+\Delta x}$, the "pressure" force contribution to the x-directed momentum balance is

$$P|_x \, \Delta y \, \Delta z - P|_{x+\Delta x} \, \Delta y \, \Delta z \tag{vi}$$

and the gravitational contribution is

$$\rho g_x \, \Delta x \, \Delta y \, \Delta z \tag{vii}$$

For the momentum balance, the sum of the terms given by Eqs. (i) to (vii) equals the rate of accumulation of x-directed momentum in the control volume, which is given by

$$\frac{\partial(\rho v_x)}{\partial t} \, \Delta x \, \Delta y \, \Delta x$$

Dividing both sides of this equality by $\Delta x \, \Delta y \, \Delta z$ gives

$$\frac{\partial(\rho v_x)}{\partial t} = -\left(\frac{\tau_{xx}|_{x+\Delta x} - \tau_{xx}|_x}{\Delta x} + \frac{\tau_{yx}|_{y+\Delta y} - \tau_{yx}|_y}{\Delta y} + \frac{\tau_{zx}|_{z+\Delta z} - \tau_{zx}|_z}{\Delta z} \right)$$

$$- \left(\frac{\rho v_x v_x|_{x+\Delta x} - \rho v_x v_x|_x}{\Delta x} + \frac{\rho v_x v_y|_{y+\Delta y} - \rho v_x v_y|_y}{\Delta y} + \frac{\rho v_x v_z|_{z+\Delta z} - \rho v_x v_z|_z}{\Delta z} \right)$$

$$- \frac{P|_{x+\Delta x} - P|_x}{\Delta x} + \rho g_x$$

which, in the limit of $\Delta x \to 0$, $\Delta y \to 0$, and $\Delta z \to 0$ gives

$$\frac{\partial(\rho v_x)}{\partial t} = -\left(\frac{\partial \tau_{xx}}{\partial x} + \frac{\partial \tau_{yx}}{\partial y} + \frac{\partial \tau_{zx}}{\partial z}\right)$$

$$-\left[\frac{\partial(\rho v_x v_x)}{\partial x} + \frac{\partial(\rho v_x v_y)}{\partial y} + \frac{\partial(\rho v_x v_z)}{\partial z}\right] - \frac{\partial P}{\partial x} + \rho g_x \qquad (3.5)$$

Similarly, the y-directed momentum balance gives

$$\frac{\partial(\rho v_y)}{\partial t} = -\left(\frac{\partial \tau_{xy}}{\partial x} + \frac{\partial \tau_{yy}}{\partial y} + \frac{\partial \tau_{zy}}{\partial z}\right)$$

$$-\left[\frac{\partial(\rho v_y v_x)}{\partial x} + \frac{\partial(\rho v_y v_y)}{\partial y} + \frac{\partial(\rho v_y v_z)}{\partial z}\right] - \frac{\partial P}{\partial y} + \rho g_y \qquad (3.6)$$

and the z-directed momentum balance gives

$$\frac{\partial(\rho v_z)}{\partial t} = -\left(\frac{\partial \tau_{xz}}{\partial x} + \frac{\partial \tau_{yz}}{\partial y} + \frac{\partial \tau_{zz}}{\partial z}\right)$$

$$-\left[\frac{\partial(\rho v_z v_x)}{\partial x} + \frac{\partial(\rho v_z v_y)}{\partial y} + \frac{\partial(\rho v_z v_z)}{\partial z}\right] - \frac{\partial P}{\partial z} + \rho g_z \qquad (3.7)$$

Combination of Eqs. (3.5), (3.6), and (3.7) gives the equation for conservation of momentum in shorthand notation as

$$\frac{\partial}{\partial t}\rho v = -[\nabla \cdot \tau] - [\nabla \cdot \rho vv] - \nabla P + \rho g \qquad (3.8)$$

In Eq. (3.8), v is the velocity vector and the term on the left-hand side represents the sum of the three velocity components

$$\frac{\partial(\rho v_x)}{\partial t} + \frac{\partial(\rho v_y)}{\partial t} + \frac{\partial(\rho v_z)}{\partial t}$$

The term $\nabla \cdot \tau$ is called the *stress tensor* and contains the nine components of τ,

$$\tau_{xx}, \quad \tau_{xy}, \quad \tau_{xz}, \quad \tau_{yx}, \quad \tau_{yy}, \quad \tau_{yz}, \quad \tau_{zx}, \quad \tau_{zy}, \text{ and } \tau_{zz}$$

The term $\nabla \cdot \rho vv$ is the dyadic product of ρv and v and contains the nine components of the momentum flux. The term ∇P is shorthand for

$$\frac{\partial P}{\partial x} + \frac{\partial P}{\partial y} + \frac{\partial P}{\partial z}$$

and ρg represents the components of the gravitational force

$$\rho g_x + \rho g_y + \rho g_z$$

To obtain the velocity distributions from Eq. (3.8), the relationships between the stresses and the velocity gradients must be known, which are given by Newton's general laws of viscosity as follows:

$$\text{\textit{Normal stresses:}} \quad \tau_{xx} = -2\eta \frac{\partial v_x}{\partial x} + \frac{2}{3}\eta(\nabla \cdot \mathbf{v}) \tag{3.9}$$

$$\tau_{yy} = -2\eta \frac{\partial v_y}{\partial y} + \frac{2}{3}\eta(\nabla \cdot \mathbf{v}) \tag{3.10}$$

$$\tau_{zz} = -2\eta \frac{\partial v_z}{\partial z} + \frac{2}{3}\eta(\nabla \cdot \mathbf{v}) \tag{3.11}$$

$$\text{\textit{Shear stresses:}} \quad \tau_{xy} = \tau_{yx} = -\eta\left(\frac{\partial v_y}{\partial x} + \frac{\partial v_x}{\partial y}\right) \tag{3.12}$$

$$\tau_{xz} = \tau_{zx} = -\eta\left(\frac{\partial v_z}{\partial x} + \frac{\partial v_x}{\partial z}\right) \tag{3.13}$$

$$\tau_{yz} = \tau_{zy} = -\eta\left(\frac{\partial v_z}{\partial y} + \frac{\partial v_y}{\partial z}\right) \tag{3.14}$$

All of the flow situations discussed in Chapter 2 were for unidirectional fully developed steady-state flow of an incompressible fluid (i.e., a fluid of constant density). Thus for unidirectional steady flow in the x-direction, constant density gives $\nabla \cdot \mathbf{v} = 0$ from Eq. (3.3), $v_x = $ constant, $v_z = v_y = 0$ gives $\tau_{xx} = \tau_{yy} = \tau_{zz} = 0$, $\tau_{xz} = \tau_{zx} = \tau_{yz} = \tau_{zy} = 0$, and

$$\tau_{yz} = -\eta \frac{\partial v_x}{\partial y}$$

as given by Eq. (2.5).

3.4 Navier–Stokes Equation for Fluids of Constant Density and Viscosity

For a fluid of constant density, the equation of continuity, Eq. (3.4), gives

$$\nabla \cdot \mathbf{v} = \frac{\partial v_x}{\partial x} + \frac{\partial v_y}{\partial y} + \frac{\partial v_z}{\partial z} = 0$$

and the x-component of the $\nabla \cdot \rho\mathbf{v}\mathbf{v}$ term in Eq. (3.8) can be written as

$$\rho\left(v_x\frac{\partial v_x}{\partial x} + v_x\frac{\partial v_x}{\partial x} + v_x\frac{\partial v_y}{\partial y} + v_y\frac{\partial v_x}{\partial y} + v_x\frac{\partial v_z}{\partial z} + v_z\frac{\partial v_x}{\partial z}\right)$$

$$= \rho\left[v_x\frac{\partial v_x}{\partial x} + v_y\frac{\partial v_x}{\partial y} + v_z\frac{\partial v_x}{\partial z} + v_x\left(\frac{\partial v_x}{\partial x} + \frac{\partial v_y}{\partial y} + \frac{\partial v_z}{\partial z}\right)\right]$$

From Eq. (3.3), the term in the inner parentheses is zero. Substituting from Eqs. (3.9), (3.12), and (3.13), and noting that $\nabla \cdot \mathbf{v} = 0$, the x-component of the $\nabla \cdot \tau$ term in Eq. (3.8) can be written as

$$\frac{\partial}{\partial x}\left(-2\eta\,\frac{\partial v_x}{\partial x}\right) - \frac{\partial}{\partial y}\left(\eta\,\frac{\partial v_y}{\partial x} + \eta\,\frac{\partial v_x}{\partial y}\right) - \frac{\partial}{\partial z}\left(\eta\,\frac{\partial v_z}{\partial x} + \eta\,\frac{\partial v_x}{\partial z}\right)$$

$$= -\eta\left(2\,\frac{\partial^2 v_x}{\partial x^2} + \frac{\partial^2 v_y}{\partial y\partial x} + \frac{\partial^2 v_x}{\partial y^2} + \frac{\partial^2 v_z}{\partial z\partial x} + \frac{\partial^2 v_x}{\partial z^2}\right)$$

$$= -\eta\left[\frac{\partial^2 v_x}{\partial x^2} + \frac{\partial^2 v_x}{\partial y^2} + \frac{\partial^2 v_x}{\partial z^2} + \frac{\partial}{\partial x}\left(\frac{\partial v_x}{\partial x} + \frac{\partial v_y}{\partial y} + \frac{\partial v_z}{\partial z}\right)\right]$$

Again, the term in the inner parentheses is zero. The x-directed momentum balance for a fluid of constant ρ and η is thus

$$\rho\,\frac{\partial v_x}{\partial t} = -\rho\left(v_x\,\frac{\partial v_x}{\partial x} + v_y\,\frac{\partial v_x}{\partial y} + v_z\,\frac{\partial v_x}{\partial z}\right)$$

$$+ \eta\left(\frac{\partial^2 v_x}{\partial x^2} + \frac{\partial^2 v_x}{\partial y^2} + \frac{\partial^2 v_x}{\partial z^2}\right) - \frac{\partial P}{\partial x} + \rho g_x$$

or

$$\rho\left(\frac{\partial v_x}{\partial t} + v_x\,\frac{\partial v_x}{\partial x} + v_y\,\frac{\partial v_x}{\partial y} + v_z\,\frac{\partial v_x}{\partial z}\right)$$

$$= \eta\left(\frac{\partial^2 v_x}{\partial x^2} + \frac{\partial^2 v_x}{\partial y^2} + \frac{\partial^2 v_x}{\partial z^2}\right) - \frac{\partial P}{\partial x} + \rho g_x \qquad \text{(i)}$$

The left-hand side of Eq. (i) can be considered as follows. The x-component of the velocity depends on both position in the flow and time:

$$v_x = v_x(t,x,y,z)$$

and hence

$$dv_x = \frac{\partial v_x}{\partial t}\,dt + \frac{\partial v_x}{\partial x}\,dx + \frac{\partial v_x}{\partial y}\,dy + \frac{\partial v_x}{\partial z}\,dz$$

and the total differential of v_x with respect to time is

$$\frac{dv_x}{dt} = \frac{\partial v_x}{\partial t} + \frac{\partial v_x}{\partial x}\frac{dx}{dt} + \frac{\partial v_x}{\partial y}\frac{dy}{dt} + \frac{\partial v_x}{\partial z}\frac{dz}{dt} \qquad \text{(ii)}$$

Consider an observer moving about independently in the flowing fluid. Equation (ii) is then the total rate of change of v_x with time observed by the observer as the point of observation is moved an infinitesimal distance in the fluid when the independent motion of the observer is described by the derivatives, dx/dt, dy/dt, and dz/dt. If the motion of the observer is the same as that of the fluid, then $dx/dt = v_x$, $dy/dt = v_y$, and $dz/dt = v_z$, and in this special case the total derivative given by Eq. (ii) is called the *substantial derivative of* v_x, Dv_x/Dt, given by

$$\frac{Dv_x}{Dt} = \frac{\partial v_x}{\partial t} + v_x\,\frac{\partial v_x}{\partial x} + v_y\,\frac{\partial v_x}{\partial y} + v_z\,\frac{\partial v_x}{\partial z} \qquad \text{(iii)}$$

The general form of the substantial derivative is

$$\frac{D}{Dt} = \frac{\partial}{\partial t} + v_x \frac{\partial}{\partial x} + v_y \frac{\partial}{\partial y} + v_z \frac{\partial}{\partial z} \tag{iv}$$

and it can be applied to any property of a fluid, the magnitude of which varies with time and position. In considering the pressure in a fluid, Eq. (iv) gives the total differential of P with respect to time as

$$\frac{dP}{dt} = \frac{\partial P}{\partial t} + \frac{\partial P}{\partial x}\frac{dx}{dt} + \frac{\partial P}{\partial y}\frac{dy}{dt} + \frac{\partial P}{\partial z}\frac{dz}{dt} \tag{v}$$

For unidirectional flow in the x-direction in a fully developed steady-state flow, the partial derivatives $\partial P/\partial t$, $\partial P/\partial y$, and $\partial P/\partial z$ are zero and hence Eq. (v) becomes

$$\frac{dP}{dt} = \frac{dP}{dx}\frac{dx}{dt}$$

In such a flow the static pressure in the fluid normally decreases in the direction of the flow, (i.e., dP/dx is negative). Thus an observer moving in the fluid in the flow direction would experience a decrease in pressure with time and an observer moving in the fluid in the $-x$ direction would experience an increase in pressure with time. If the observer were moving with the flow at the velocity v_x, the observed rate of change of pressure with time would be

$$\frac{DP}{Dt} = v_x \frac{dP}{dx}$$

Further, if the pressures at all locations in the fluid were varying with time and the rate of change at the location of the observer were $\partial P/\partial t$, the rate of change of pressure experienced by the observer moving with the flow at the velocity v_x would be

$$\frac{DP}{Dt} = \frac{\partial P}{\partial t} + v_x \frac{\partial P}{\partial x}$$

The term in parentheses on the left side of Eq. (i) is thus recognized as being the substantial derivative of v_x, and hence Eq. (i) can be written as

$$\rho \frac{Dv_x}{Dt} = \eta \left(\frac{\partial^2 v_x}{\partial x^2} + \frac{\partial^2 v_x}{\partial y^2} + \frac{\partial^2 v_x}{\partial z^2} \right) - \frac{\partial P}{\partial x} + \rho g_x$$

Similarly, the y and z components can be written, respectively, as

$$\rho \frac{Dv_y}{Dt} = \eta \left(\frac{\partial^2 v_y}{\partial x^2} + \frac{\partial^2 v_y}{\partial y^2} + \frac{\partial^2 v_y}{\partial z^2} \right) - \frac{\partial P}{\partial y} + \rho g_y$$

and

$$\rho \frac{Dv_z}{dt} = \eta \left(\frac{\partial^2 v_z}{\partial x^2} + \frac{\partial^2 v_z}{\partial y^2} + \frac{\partial^2 v_z}{\partial z^2} \right) - \frac{\partial P}{\partial z} + \rho g_z$$

and the complete equation is then

$$\rho \frac{D\mathbf{v}}{Dt} = \eta \nabla^2 \mathbf{v} - \nabla P + \rho \mathbf{g} \tag{3.15}$$

which is the Navier–Stokes equation. Equation (3.15) states that force equals mass times acceleration. The left-hand side is (mass) × (acceleration), which is thus equal to the sum of the viscous forces ($\eta \nabla^2 \mathbf{v}$), the "pressure" forces (∇P), and the gravitational force, ρg.

The equations of continuity and conservation of momentum are expressed in cartesian coordinates, cylindrical coordinates, and spherical coordinates in Tables 3.1 to 3.4.

TABLE 3.1   Equation of Continuity in Several Coordinate Systems

Rectangular coordinates (x, y, z):

$$\frac{\partial \rho}{\partial t} + \frac{\partial}{\partial x} (\rho v_x) + \frac{\partial}{\partial y} (\rho v_y) + \frac{\partial}{\partial z} (\rho v_z) = 0 \tag{A}$$

Cylindrical coordinates (r, θ, z):

$$\frac{\partial \rho}{\partial t} + \frac{1}{r} \frac{\partial}{\partial r} (\rho r v_r) + \frac{1}{r} \frac{\partial}{\partial \theta} (\rho v_\theta) + \frac{\partial}{\partial z} (\rho v_z) = 0 \tag{B}$$

Spherical coordinates (r, θ, ϕ):

$$\frac{\partial \rho}{\partial t} + \frac{1}{r^2} \frac{\partial}{\partial r} (\rho r^2 v_r) + \frac{1}{r \sin \theta} \frac{\partial}{\partial \theta} (\rho v_\theta \sin \theta) + \frac{1}{r \sin \theta} \frac{\partial}{\partial \phi} (\rho v_\phi) = 0 \tag{C}$$

TABLE 3.2   Equation of Motion in Rectangular Coordinates (x, y, z)

In terms of τ:

x-component: $\rho \left(\dfrac{\partial v_x}{\partial t} + v_x \dfrac{\partial v_x}{\partial x} + v_y \dfrac{\partial v_x}{\partial y} + v_z \dfrac{\partial v_x}{\partial z} \right)$

$$= -\frac{\partial P}{\partial x} - \left(\frac{\partial \tau_{xx}}{\partial x} + \frac{\partial \tau_{yx}}{\partial y} + \frac{\partial \tau_{zx}}{\partial z} \right) + \rho g_x \tag{A}$$

y-component: $\rho \left(\dfrac{\partial v_y}{\partial t} + v_x \dfrac{\partial v_y}{\partial x} + v_y \dfrac{\partial v_y}{\partial y} + v_z \dfrac{\partial v_y}{\partial z} \right)$

$$= -\frac{\partial P}{\partial y} - \left(\frac{\partial \tau_{xy}}{\partial x} + \frac{\partial \tau_{yy}}{\partial y} + \frac{\partial \tau_{zy}}{\partial z} \right) + \rho g_y \tag{B}$$

z-component: $\rho \left(\dfrac{\partial v_z}{\partial t} + v_x \dfrac{\partial v_z}{\partial x} + v_y \dfrac{\partial v_z}{\partial y} + v_z \dfrac{\partial v_z}{\partial z} \right)$

$$= -\frac{\partial P}{\partial z} - \left(\frac{\partial \tau_{xz}}{\partial x} + \frac{\partial \tau_{yz}}{\partial y} + \frac{\partial \tau_{zz}}{\partial z} \right) + \rho g_z \tag{C}$$

TABLE 3.2   (*continued*)

In terms of velocity gradients for a Newtonian fluid with constant ρ and η:

$$x\text{-component:} \quad \rho\left(\frac{\partial v_x}{\partial t} + v_x\frac{\partial v_x}{\partial x} + v_y\frac{\partial v_x}{\partial y} + v_z\frac{\partial v_x}{\partial z}\right)$$

$$= -\frac{\partial P}{\partial x} + \eta\left(\frac{\partial^2 v_x}{\partial x^2} + \frac{\partial^2 v_x}{\partial y^2} + \frac{\partial^2 v_x}{\partial z^2}\right) + \rho g_x \tag{D}$$

$$y\text{-component:} \quad \rho\left(\frac{\partial v_y}{\partial t} + v_x\frac{\partial v_y}{\partial x} + v_y\frac{\partial v_y}{\partial y} + v_z\frac{\partial v_y}{\partial z}\right)$$

$$= -\frac{\partial P}{\partial y} + \eta\left(\frac{\partial^2 v_y}{\partial x^2} + \frac{\partial^2 v_y}{\partial y^2} + \frac{\partial^2 v_y}{\partial z^2}\right) + \rho g_y \tag{E}$$

$$z\text{-component:} \quad \rho\left(\frac{\partial v_z}{\partial t} + v_x\frac{\partial v_z}{\partial x} + v_y\frac{\partial v_z}{\partial y} + v_z\frac{\partial v_z}{\partial z}\right)$$

$$= -\frac{\partial P}{\partial z} + \eta\left(\frac{\partial^2 v_z}{\partial x^2} + \frac{\partial^2 v_z}{\partial y^2} + \frac{\partial^2 v_z}{\partial z^2}\right) + \rho g_z \tag{F}$$

TABLE 3.3   Equation of Motion in Cylindrical Coordinates (r, θ, z)

In terms of τ:

$$r\text{-component:} \quad \rho\left(\frac{\partial v_r}{\partial t} + v_r\frac{\partial v_r}{\partial r} + \frac{v_\theta}{r}\frac{\partial v_r}{\partial \theta} - \frac{v_\theta^2}{r} + v_z\frac{\partial v_r}{\partial z}\right)$$

$$= -\frac{\partial P}{\partial r} - \left[\frac{1}{r}\frac{\partial}{\partial r}(r\tau_{rr}) + \frac{1}{r}\frac{\partial \tau_{r\theta}}{\partial \theta} - \frac{\tau_{\theta\theta}}{r} + \frac{\partial \tau_{rz}}{\partial z}\right] + \rho g_r \tag{A}$$

$$\theta\text{-component:} \quad \rho\left(\frac{\partial v_\theta}{\partial t} + v_r\frac{\partial v_\theta}{\partial r} + \frac{v_\theta}{r}\frac{\partial v_\theta}{\partial \theta} + \frac{v_r v_\theta}{r} + v_z\frac{\partial v_\theta}{\partial z}\right)$$

$$= -\frac{1}{r}\frac{\partial P}{\partial \theta} - \left[\frac{1}{r^2}\frac{\partial}{\partial r}(r^2\tau_{r\theta}) + \frac{1}{r}\frac{\partial \tau_{\theta\theta}}{\partial \theta} + \frac{\partial \tau_{\theta z}}{\partial z}\right] + \rho g_\theta \tag{B}$$

$$z\text{-component:} \quad \rho\left(\frac{\partial v_z}{\partial t} + v_r\frac{\partial v_z}{\partial r} + \frac{v_\theta}{r}\frac{\partial v_z}{\partial \theta} + v_z\frac{\partial v_z}{\partial z}\right)$$

$$= -\frac{\partial P}{\partial z} - \left[\frac{1}{r}\frac{\partial}{\partial r}(r\tau_{rz}) + \frac{1}{r}\frac{\partial \tau_{\theta z}}{\partial \theta} + \frac{\partial \tau_{zz}}{\partial z}\right] + \rho g_z \tag{C}$$

In terms of velocity gradients for a Newtonian fluid with constant ρ and η:

$$r\text{-component:} \quad \rho\left(\frac{\partial v_r}{\partial t} + v_r\frac{\partial v_r}{\partial r} + \frac{v_\theta}{r}\frac{\partial v_r}{\partial \theta} - \frac{v_\theta^2}{r} + v_z\frac{\partial v_r}{\partial z}\right)$$

$$= -\frac{\partial P}{\partial r} + \eta\left[\frac{\partial}{\partial r}\left(\frac{1}{r}\frac{\partial}{\partial r}(rv_r)\right) + \frac{1}{r^2}\frac{\partial^2 v_r}{\partial \theta^2} - \frac{2}{r^2}\frac{\partial v_\theta}{\partial \theta} + \frac{\partial^2 v_r}{\partial z^2}\right] + \rho g_r \tag{D}$$

$$\theta\text{-component:} \quad \rho\left(\frac{\partial v_\theta}{\partial t} + v_r\frac{\partial v_\theta}{\partial r} + \frac{v_\theta}{r}\frac{\partial v_\theta}{\partial \theta} + \frac{v_r v_\theta}{r} + v_z\frac{\partial v_\theta}{\partial z}\right)$$

$$= -\frac{1}{r}\frac{\partial P}{\partial \theta} + \eta \left[\frac{\partial}{\partial r}\left(\frac{1}{r}\frac{\partial}{\partial r}(rv_\theta)\right) + \frac{1}{r^2}\frac{\partial^2 v_\theta}{\partial \theta^2} + \frac{2}{r^2}\frac{\partial v_r}{\partial \theta} + \frac{\partial^2 v_\theta}{\partial z^2} \right] + \rho g_\theta \qquad (E)$$

$$z\text{-component: } \rho \left(\frac{\partial v_z}{\partial t} + v_r \frac{\partial v_z}{\partial r} + \frac{v_\theta}{r}\frac{\partial v_z}{\partial \theta} + v_z \frac{\partial v_z}{\partial z} \right)$$

$$= -\frac{\partial P}{\partial z} + \eta \left[\frac{1}{r}\frac{\partial}{\partial r}\left(r\frac{\partial v_z}{\partial r}\right) + \frac{1}{r^2}\frac{\partial^2 v_z}{\partial \theta^2} + \frac{\partial^2 v_z}{\partial z^2} \right] + \rho g_z \qquad (F)$$

Application of the Navier–Stokes Equation to Simple Flow Systems

In the case of fully developed steady-state unidirectional flow of fluid of constant density and viscosity between parallel plates, shown in Fig. 2.7, v_y and v_z are both zero. Thus in Table 3.1, Eq. (A),

$$\frac{\partial \rho}{\partial t} = v_y = v_z = 0 \qquad \text{and hence} \qquad \frac{\partial v_x}{\partial x} = 0$$

From Table 3.2, Eq. (A),

$$\frac{\partial \tau_{xx}}{\partial x}, \quad \frac{\partial \tau_{zx}}{\partial y}, \quad \text{and} \quad g_x$$

are zero and hence

$$0 = \frac{\partial P}{\partial x} - \frac{\partial \tau_{yx}}{\partial y}$$

From Table 3.2, Eq. (D),

$$0 = -\frac{\partial P}{\partial x} + \eta \frac{\partial^2 v_x}{\partial y^2}$$

integration of which leads to Eq. (2.10).

   For fully developed steady-state unidirectional free flow of fluid of constant density down an inclined plane, shown in Fig. 2.10, v_y and v_z are zero. Thus, from Table 3.1, Eq. (A),

$$\frac{\partial \rho}{\partial t} = v_y = v_z = 0 \qquad \text{and hence} \qquad \frac{\partial v_x}{\partial x} = 0$$

From Table 3.2 Eq. (A),

$$\frac{\partial \rho}{\partial x}, \quad \frac{\partial \tau_{zx}}{\partial y}, \quad \text{and} \quad \frac{\partial \tau_{xx}}{\partial x}$$

are zero and hence

$$0 = -\frac{\partial \tau_{yx}}{\partial y} + \rho g_x$$

TABLE 3.4 Equation of Motion in Spherical Coordinates (r, θ, ϕ)

In terms of τ:

r-component: $\rho \left(\dfrac{\partial v_r}{\partial t} + v_r \dfrac{\partial v_r}{\partial r} + \dfrac{v_\theta}{r} \dfrac{\partial v_r}{\partial \theta} + \dfrac{v_\phi}{r \sin \theta} \dfrac{\partial v_r}{\partial \phi} - \dfrac{v_\theta^2 + v_\phi^2}{r} \right)$

$$= -\frac{\partial P}{\partial r} - \left[\frac{1}{r^2} \frac{\partial}{\partial r} (r^2 \tau_{rr}) + \frac{1}{r \sin \theta} \frac{\partial}{\partial \theta} (\tau_{r\theta} \sin \theta) + \frac{1}{r \sin \theta} \frac{\partial \tau_{r\phi}}{\partial \phi} - \frac{\tau_{\theta\theta} + \tau_{\phi\phi}}{r} \right] + \rho g_r$$

θ-component: $\rho \left(\dfrac{\partial v_\theta}{\partial t} + v_r \dfrac{\partial v_\theta}{\partial r} + \dfrac{v_\theta}{r} \dfrac{\partial v_\theta}{\partial \theta} + \dfrac{v\phi}{r \sin \theta} \dfrac{\partial v_\theta}{\partial \phi} + \dfrac{v_r v_\theta}{r} - \dfrac{v_\phi^2 \cot \theta}{r} \right)$

$$= -\frac{1}{r} \frac{\partial P}{\partial \theta} - \left[\frac{1}{r^2} \frac{\partial}{\partial r} (r^2 \tau_{r\theta}) + \frac{1}{r \sin \theta} \frac{\partial}{\partial \theta} (\tau_{\theta\theta} \sin \theta) + \frac{1}{r \sin \theta} \frac{\partial \tau_{\theta\phi}}{\partial \phi} + \frac{\tau_{r\theta}}{r} - \frac{\cot \theta}{r} \tau_{\phi\phi} \right] + \rho g_\theta$$

ϕ-component: $\rho \left(\dfrac{\partial v_\phi}{\partial t} + v_r \dfrac{\partial v_\phi}{\partial r} + \dfrac{v_\theta}{r} \dfrac{\partial v_\phi}{\partial \theta} + \dfrac{v_\phi}{r \sin \theta} \dfrac{\partial v_\phi}{\partial \phi} + \dfrac{v_\phi v_r}{r} + \dfrac{v_\theta v_\phi}{r} \cot \theta \right)$

$$= -\frac{1}{r \sin \theta} \frac{\partial P}{\partial \phi} - \left[\frac{1}{r^2} \frac{\partial}{\partial r} (r^2 \tau_{r\phi}) + \frac{1}{r} \frac{\partial \tau_{\theta\phi}}{\partial \theta} + \frac{1}{r \sin \theta} \frac{\partial \tau_{\phi\phi}}{\partial \phi} + \frac{\tau_{r\phi}}{r} + \frac{2 \cot \theta}{r} \tau_{\theta\phi} \right] + \rho g_\phi$$

In terms of velocity gradients for a Newtonian fluid with constant ρ and η[a]:

r-component: $\rho \left(\dfrac{\partial v_r}{\partial t} + v_r \dfrac{\partial v_r}{\partial r} + \dfrac{v_\theta}{r} \dfrac{\partial v_r}{\partial \theta} + \dfrac{v_\phi}{r \sin \theta} \dfrac{\partial v_r}{\partial \phi} - \dfrac{v_\theta^2 + v_\phi^2}{r} \right)$

$$= -\frac{\partial P}{\partial r} + \eta \left(\nabla^2 v_r - \frac{2}{r^2} v_r - \frac{2}{r^2} \frac{\partial v_\theta}{\partial \theta} - \frac{2}{r^2} v_\theta \cot \theta - \frac{2}{r^2 \sin \theta} \frac{\partial v_\phi}{\partial \phi} \right) + \rho g_r$$

θ-component: $\rho \left(\dfrac{\partial v_\theta}{\partial t} + v_r \dfrac{\partial v_\theta}{\partial r} + \dfrac{v_\theta}{r} \dfrac{\partial v_\theta}{\partial \theta} + \dfrac{v\phi}{r \sin \theta} \dfrac{\partial v_\theta}{\partial \phi} + \dfrac{v_r v_\theta}{r} - \dfrac{v_\phi^2 \cot \theta}{r} \right)$

$$= -\frac{1}{r} \frac{\partial P}{\partial \theta} + \eta \left(\nabla^2 v_\theta + \frac{2}{r^2} \frac{\partial v_r}{\partial \theta} - \frac{v_\theta}{r^2 \sin^2 \theta} - \frac{2 \cos \theta}{r^2 \sin^2 \theta} \frac{\partial v_\phi}{\partial \phi} \right) + \rho g_\theta$$

ϕ-component: $\rho \left(\dfrac{\partial v_\phi}{\partial t} + v_r \dfrac{\partial v_\phi}{\partial r} + \dfrac{v_\theta}{r} \dfrac{\partial v_\phi}{\partial \theta} + \dfrac{v_\phi}{r \sin \theta} \dfrac{\partial v_\phi}{\partial \phi} + \dfrac{v_\phi v_r}{r} + \dfrac{v_\theta v_\phi}{r} \cot \theta \right)$

$$= -\frac{1}{r \sin \theta} \frac{\partial P}{\partial \phi} + \eta \left(\nabla^2 v_\phi - \frac{v_\phi}{r^2 \sin^2 \theta} + \frac{2}{r^2 \sin \theta} \frac{\partial v_r}{\partial \phi} + \frac{2 \cos \theta}{r^2 \sin^2 \theta} \frac{\partial v_\theta}{\partial \phi} \right) + \rho g_\phi$$

Source: R. B. Bird, W. E. Stewart, and E. N. Lightfoot, *Transport Phenomena*, Wiley, New York, 1960, pp. 83–91.
[a]In these equations

$$\nabla^2 = \frac{1}{r^2} \frac{\partial}{\partial r} \left(r^2 \frac{\partial}{\partial r} \right) + \frac{1}{r^2 \sin \theta} \frac{\partial}{\partial \theta} \left(\sin \theta \frac{\partial}{\partial \theta} \right) + \frac{1}{r^2 \sin^2 \theta} \left(\frac{\partial^2}{\partial \phi^2} \right)$$

From Table 3.2, Eq. (D),

$$0 = \eta \frac{\partial^2 v_x}{\partial y^2} + \rho g_x$$

integration of which leads to Eq. (2.15). For fully developed steady-state flow in a vertical cylindrical tube, shown in Fig. 2.13, v_r and v_θ are zero. Thus from Table 3.1, Eq. (B),

$$\frac{\partial \rho}{\partial t} = v_r = v_\theta = 0 \qquad \text{and} \qquad \text{hence} \qquad \frac{\partial}{\partial r}(r v_r) = 0$$

From Table 3.3, Eq. (C),

$$0 = -\frac{\partial P}{\partial z} - \frac{1}{r}\frac{\partial}{\partial r}(r\tau_{rz}) + \rho g_z$$

and from Table 3.3, Eq. (F),

$$0 = -\frac{\partial P}{\partial z} + \eta \left[\frac{1}{r}\frac{\partial}{\partial r}\left(r\frac{\partial v_z}{\partial r} \right) \right] + \rho g_z$$

integration of which leads to Eq. (2.20).

3.5 Fluid Flow over a Horizontal Flat Plate

Consider the flow of fluid over a horizontal flat plate as shown in Fig. 3.3. Far away from the plate the fluid is flowing in the x-direction with the constant free-stream velocity v_∞ (which is independent of x, y, z, and time). The absence of a viscosity-induced velocity profile requires that the fluid have zero viscosity, and such a fluid is said to be *inviscid* or *ideal*. The assumption of a flat free-stream velocity profile is reasonable for fluids of low viscosity such as air or water at locations far from the bounding surface, except for flow at very low Reynolds numbers. When the fluid flows over the flat plate it experiences the influence of the viscous drag caused by the requirement of no slippage of the fluid at the surface of the plate. The influence of this drag is such that at any distance along the plate, the velocity of the fluid in the x-direction, v_x, increases with increasing y from a value of zero at $y = 0$ to v_∞ at some distance into the fluid. The viscous drag and the consequent velocity gradient in the y-direction give rise to what is referred to as a *boundary layer* of fluid over the plate, the upper surface of which is located at the value of y at which $v_x/v_\infty = 0.99$.

Alternatively, the boundary layer could be envisaged as being on a flat plate moving with constant velocity v_∞ through a stagnant fluid. The fluid in contact with the plate at $y = 0$ has the velocity v_∞ and v_x decreases from v_∞ to zero with distance away from the plate. Measurement of v_x relative to the velocity of the plate gives a situation that is identical to consideration of fluid flowing over a stationary plate.

A packet of fluid flowing in the x-direction at a constant value of y moves with the velocity v_∞ until it passes through the upper surface of the boundary layer.

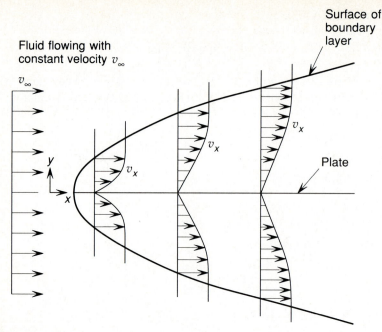

FIGURE 3.3. Growth of the momentum boundary layer on a horizontal flat plate immersed in a flowing fluid.

Thereafter, v_x (at the constant value of y) is continuously decreased by the viscous drag in the boundary layer, (i.e., $\partial v_x/\partial x$ in the boundary layer is a negative quantity). Consequently, from the equation of continuity, Eq. (3.3) or Table 3.1, Eq. (A), there is a component of flow in the y-direction within the boundary layer in compliance with

$$\frac{\partial v_x}{\partial x} = -\frac{\partial v_y}{\partial y} \tag{i}$$

The questions arising are thus how thick is the boundary layer, how does the thickness vary with the properties of the fluid and distance down the plate, and what is the magnitude of the viscous drag force exerted by the fluid on the flat plate? The equations of conservation of momentum in Table 3.2, Eqs. (D) and (E), give for the x-component

$$\rho\left(v_x \frac{\partial v_x}{\partial x} + v_y \frac{\partial v_x}{\partial y}\right) = -\frac{\partial P}{\partial x} + \eta\left(\frac{\partial^2 v_x}{\partial x^2} + \frac{\partial^2 v_x}{\partial y^2}\right) \tag{ii}$$

and for the y-component

$$\rho\left(v_x \frac{\partial v_y}{\partial x} + v_y \frac{\partial v_y}{\partial y}\right) = -\frac{\partial P}{\partial y} + \eta\left(\frac{\partial^2 v_y}{\partial x^2} + \frac{\partial^2 v_y}{\partial y^2}\right) \tag{iii}$$

The properties of the boundary layer can be obtained from Eqs. (i) to (iii) only if some assumptions are made as to the relative magnitudes, and hence the relative importance, of the individual terms in the equations, and the necessary assumptions require physical intuition and experience. Prandtl suggested that since the boundary layer is presumed to be very thin and lies on a solid surface, v_y is very small in comparison with v_x and $\partial v_x/\partial y$ is large in comparison with $\partial v_x/\partial x$. Also, all of the terms containing v_y and its derivatives are small, which reduces the problem to one of simultaneous solution of Eqs. (i) and (ii). Furthermore, in a fluid flowing with a flat velocity profile, the absence of the effects of viscosity means that sustained flow does not require a pressure gradient. That is, $\partial P/\partial x = 0$, and if there is no pressure gradient in the x-direction in the bulk flow, there is no pressure gradient within the boundary layer, and thus $\partial P/\partial x$ in Eq. (ii) is zero.

The solution arrived at with these assumptions gives the expression for the thickness of the boundary layer, δ, as

$$\delta = 5.0 \left(\frac{\nu}{v_\infty} \right)^{1/2} x^{1/2} \tag{3.16}$$

or

$$\frac{\delta}{x} = 5.0 (\mathrm{Re}_x)^{-1/2} \tag{3.17}$$

that is, δ increases as $x^{1/2}$, and at any value of x, increases with increasing kinematic viscosity ν and decreases with increasing v_∞. Equation (3.17) is referred to as the "exact solution" to calculation of the thickness of the boundary layer.

3.6 Approximate Integral Method of Obtaining Boundary Layer Thickness

The thickness of the boundary layer can be obtained from a momentum balance on a control volume in the boundary layer and an assumption as to the form of the velocity profile in the boundary layer. The control volume of length Δx and unit width in the z-direction is shown in Fig. 3.4. The upper surface of the boundary layer is the upper surface of the control volume and a mass balance on the control volume gives

> (rate at which mass enters through front face at x, $\dot{M}_1$)
> \+ (rate at which mass enters through upper surface, $\dot{M}_2$)
> \- (rate at which mass leaves through rear face at $x + \Delta x$, $\dot{M}_3$)
> $= 0$

The mass entering through the front face per unit time per unit width is

$$\dot{M}_1 = \int_0^\delta \rho v_x \, dy$$

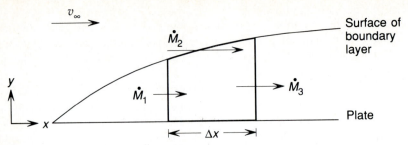

FIGURE 3.4. Control volume for consideration of fluid flow in the momentum boundary layer on a horizontal flat plate.

the mass leaving through the rear face per unit time per unit width is

$$\dot{M}_3 = \int_0^\delta \rho v_x \, dy + \frac{d}{dx} \int_0^\delta \rho v_x \, dy \, \Delta x$$

and hence the mass entering through the upper surface per unit time and unit width is

$$\dot{M}_2 = \dot{M}_3 - \dot{M}_1 = \frac{d}{dx} \int_0^\delta \rho v_x \, dy \, \Delta x$$

The momentum balance on the control volume is then

    (rate at which convective momentum enters through front face)
    + (rate at which convective momentum enters through upper surface)
    − (rate at which convective momentum leaves through rear face)
    − (rate at which viscous momentum leaves through bottom surface)
    = 0

As $\partial P / \partial x = 0$, the momentum balance does not contain a contribution from "pressure" forces.

The rate at which convective momentum enters through the front face is

$$\int_0^\delta \rho v_x^2 \, dy$$

The rate at which convective momentum enters through the upper surface is

$$v_\infty \frac{d}{dx} \int_0^\delta \rho v_x \, dy \, \Delta x$$

the rate at which convective momentum leaves through the rear face is

$$\int_0^\delta \rho v_x^2 \, dy + \frac{d}{dx} \int_0^\delta \rho v_x^2 \, dy \, \Delta x$$

and the rate at which viscous momentum leaves through the bottom surface is

$$- \tau_{yx}|_{y=0} \, \Delta x$$

The momentum balance is thus

$$\frac{d}{dx} \int_0^\delta \rho v_\infty v_x \, dy \, \Delta x - \frac{d}{dx} \int_0^\delta \rho v_x^2 \, dy \, \Delta x + \tau_{yx}|_{y=0} \, \Delta x = 0$$

Canceling the Δx terms and noting that $\tau_{yx}|_{y=0} = -\eta(\partial v_x/\partial y)|_{y=0}$ and $\eta/\rho = v$ gives

$$\frac{d}{dx} \int_0^\delta (v_\infty v_x - v_x^2) \, dy = v \left(\frac{\partial v_x}{\partial y}\right)_{y=0} \tag{3.18}$$

Equation (3.18) is known as the von Kármán integral relation for zero pressure gradient and its solution requires knowledge of the velocity gradient in the boundary layer. In view of the fact that flow in the boundary layer is laminar and a parabolic velocity gradient has been found in all of the laminar flow situations considered previously, it might be assumed that this velocity profile is also parabolic, being given by

$$v_x = ay^{1/2} \tag{i}$$

However, as seen in Fig. 3.3, $\partial v_x/\partial y$ has a finite value at the plate and is zero at the surface of the boundary layer. Inspection of Eq. (i) shows that $\partial v_x/\partial y$ is infinite at the plate, which implies an infinite shear stress at the plate, and $\partial v_x/\partial y$ is never zero. The expected parabolic velocity profile in the boundary layer is thus approximated by the expression

$$v_x = ay + by^3$$

The thickness of the boundary layer, δ, is now defined as the value of y at which $\partial v_x/\partial y$ becomes zero (i.e., the value of y at which v_x has increased from zero at $y = 0$ to v_∞). The boundary conditions

$$v_x = v_\infty \qquad \text{at } y = \delta$$

and

$$\frac{\partial v_x}{\partial y} = 0 \qquad \text{at } y = \delta$$

give

$$a = \frac{3}{2}\frac{v_\infty}{\delta} \qquad \text{and} \qquad b = -\frac{1}{2}\frac{v_\infty}{\delta^3}$$

such that

$$v_x = v_\infty \left[\frac{3}{2}\left(\frac{y}{\delta}\right) - \frac{1}{2}\left(\frac{y}{\delta}\right)^3\right] \tag{3.19}$$

and

$$\frac{\partial v_x}{\partial y}\bigg|_{y=0} = \frac{3}{2}\frac{v_\infty}{\delta} \tag{3.20}$$

The term $(v_\infty v_x - v_x^2)$ in Eq. (3.18) is thus evaluated as

$$v_\infty^2 \left[\frac{3}{2} \left(\frac{y}{\delta}\right) - \frac{1}{2} \left(\frac{y}{\delta}\right)^3 - \frac{9}{4} \left(\frac{y}{\delta}\right)^2 + \frac{3}{2} \left(\frac{y}{\delta}\right)^4 - \frac{1}{4} \left(\frac{y}{\delta}\right)^6 \right]$$

which on integration and evaluation between $y = \delta$ and $y = 0$ gives

$$0.1393 v_\infty^2 \delta$$

Thus Eq. (3.18) reduces to

$$\frac{d}{dx} (0.1393 v_\infty^2 \delta) = \frac{3 v v_\infty}{2\delta}$$

or

$$\delta \, d\delta = \frac{10.769 v}{v_\infty} dx$$

Integrating between $\delta = \delta$ at $x = x$ and $\delta = 0$ at $x = 0$ gives

$$\frac{\delta^2}{2} = \frac{10.769 v}{v_\infty} x$$

or

$$\delta = 4.64 \left(\frac{v}{v_\infty}\right)^{1/2} x^{1/2} \tag{3.21}$$

or

$$\frac{\delta}{x} = 4.64 (\text{Re}_x)^{-1/2} \tag{3.22}$$

which are very close to the solutions given by Eqs. (3.16) and (3.17).

The value of the constant in Eqs. (3.21) and (3.22) is not very sensitive to the form assumed for the velocity gradient in the boundary layer. For example, the assumption of a linear gradient

$$v_x = v_\infty \frac{y}{\delta}$$

gives

$$\frac{\delta}{x} = \frac{3.46}{(\text{Re}_x)^{1/2}}$$

and the assumption of a gradient given by

$$v_x = v_\infty \sin \frac{\pi y}{2\delta}$$

gives

$$\frac{\delta}{x} = \frac{4.79}{(Re_x)^{1/2}}$$

The three velocity gradients are shown in comparison with one another in Fig. 3.5 for conditions of $v_\infty = 1$ m/s and values of x at which $\delta = 0.1$ m. The equations for the velocity gradients are applied only in the range $0 \le y \le \delta$.

The shear stress exerted by the plate on the fluid at $y = 0$ is τ_0 and hence the shear stress exerted by the fluid on the plate is $-\tau_0$. Thus the drag force, F_K, exerted by the fluid on the plate of width W and length L is

$$F_K = \int_0^W \int_0^L (-\tau_0) \, dx \, dz$$

$$= \int_0^W \int_0^L \left(\eta \left. \frac{\partial v_x}{\partial y} \right|_{y=0} \right) dx \, dz$$

which on substitution from Eq. (3.20) gives

$$\int_0^W \int_0^L \frac{3}{2} \frac{v_\infty}{\delta} \eta \, dx \, dz$$

Substituting from Eq. (3.21) and integrating gives

$$F_K = 0.646\sqrt{v_\infty^3 \eta \rho L W^2} \tag{3.23}$$

The "exact" solution gives

$$F_K = 0.664\sqrt{v_\infty^3 \eta \rho L W^2} \tag{3.24}$$

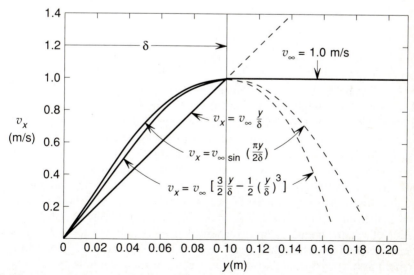

FIGURE 3.5. Possible variations of local flow velocity in the momentum boundary layer on a horizontal flat plate with distance from the plate.

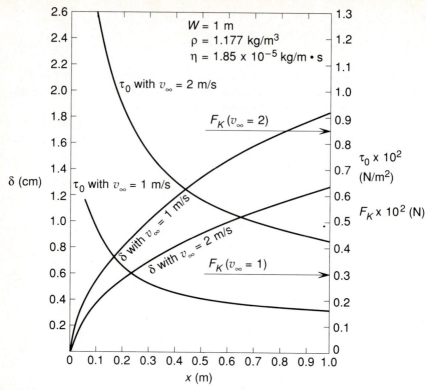

FIGURE 3.6. Variations of the momentum boundary thickness and the local shear stress acting on the plate with distance along the surface of the plate from its leading edge.

It is to be noted that Eqs. (3.23) and (3.24) give the drag force on one side of the plate only.

Figure 3.6 shows the boundary layers in air ($\nu = 1.57 \times 10^{-5}$ m²/s at 300 K) flowing over a 1-m length of flat plate at $v_\infty = 1$ m/s and $v_\infty = 2$ m/s. At $v_\infty = 1$ m/s δ has grown to 1.84 cm at the trailing edge of the plate and at $v_\infty = 2$ m/s δ is 1.3 cm at the trailing edge. Figure 3.6 also shows the variation of the local shear stress τ_0 with distance along the plate. As, from Eq. (3.20), $\partial v_x/\partial y|_{y=0}$ is inversely proportional to δ and δ increases as $x^{1/2}$, τ_0 is inversely proportional to $x^{1/2}$ and hence approaches infinity as $x \to 0$. The local shear stress is also proportional to $v_\infty^{3/2}$. The drag forces on the plate for $v_\infty = 1$ m/s and $v_\infty = 2$ m/s, obtained for $W = 1$ m from Eq. (3.23) are also shown in Fig. 3.6. Figure 3.7 shows v_x as a function of x in the boundary layer for three values of y. When $v_\infty = 2$ m/s at $y = 0.1$ cm, the fluid enters the boundary layer at $x = 0.01$ m. At $x = 0.1$ m v_x has decreased to 0.71 m/s and at $x = 1$m v_x has been decreased to 0.23 m/s. At $y = 0.25$ cm the fluid enters the boundary layer at $x = 0.06$ m and v_x is decreased to 0.57 m/s at $x = 1.0$ m. Figure 3.7 shows that the severity of retardation of v_x decreases with increasing y.

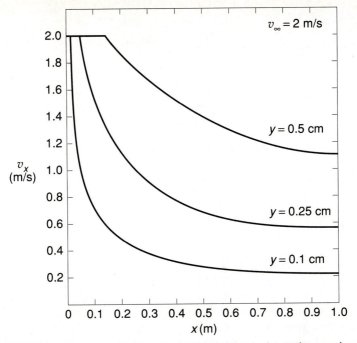

FIGURE 3.7. Variations, with x, of the local flow velocity at three values of y.

EXAMPLE 3.1

Compare the flow of air with that of water over a flat plate 0.5 m in length and 0.25 m wide at $v_\infty = 0.5$ m/s and 300 K.

At 300 K for air,

$$\rho = 1.177 \text{ kg/m}^3$$

$$\eta = 1.85 \times 10^{-5} \text{ Pa·s}$$

$$v = 1.57 \times 10^{-5} \text{ m}^2/\text{s}$$

For water,

$$\rho = 997 \text{ kg/m}^3$$

$$\eta = 8.57 \times 10^{-4} \text{ Pa·s}$$

$$v = 8.6 \times 10^{-7} \text{ m}^2/\text{s}$$

At the trailing edge of the plate (i.e., $x = 0.5$ m), the Reynolds number is

$$\text{Re}_L = \frac{v_\infty L}{v}$$

Therefore, for water,

$$Re_L = \frac{0.5 \times 0.5}{8.6 \times 10^{-7}} = 291,000$$

and for air,

$$Re_L = \frac{0.5 \times 0.5}{1.57 \times 10^{-5}} = 15,900$$

Therefore, at the trailing edge, Eq. (3.17) gives

$$\delta = \frac{5.0 \times 0.5}{(291,000)^{1/2}} = 4.64 \times 10^{-3} \text{ m} \qquad \text{for water}$$

and

$$\delta = \frac{5.0 \times 0.5}{(15,900)^{1/2}} = 1.98 \times 10^{-2} \text{ m} \qquad \text{for air}$$

Thus, for a given geometry and flow velocity, the larger kinematic viscosity of air causes the boundary layer in air to grow more rapidly than that in water, and hence at any given distance from the leading edge of the plate, the air boundary layer thickness is larger than that of water.

The drag force on one side of the plate is given by Eq. (3.24) as

$$F_K = 0.664\sqrt{v_\infty^3 \eta\rho LW^2}$$

Therefore, for water,

$$F_K = 0.664\sqrt{0.5^3 \times 8.57 \times 10^{-4} \times 997 \times 0.5 \times 0.25^2}$$
$$= 3.84 \times 10^{-2} \text{ N}$$

and for air,

$$F_K = 0.664\sqrt{0.5^3 \times 1.85 \times 10^{-5} \times 1.177 \times 0.5 \times 0.25^2}$$
$$= 1.94 \times 10^{-4} \text{ N}$$

Thus the larger density and viscosity of air cause the drag force exerted by the water boundary layer to be greater than that exerted by the air boundary layer. Considering both sides of the plate, the rate at which mechanical energy is degraded to thermal energy by friction in the boundary layer is $2F_K v_\infty$.

In order that the thickness of the air boundary layer at the trailing edge be decreased to that of the water boundary layer, v_∞ would have to be increased to the value at which $Re_L = 291,000$:

$$v_\infty = \frac{291,000 \times v}{L} = \frac{291,000 \times 1.85 \times 10^{-5}}{0.5}$$
$$= 10.8 \text{ m/s}$$

and with $v_\infty = 10.8$ m/s, the drag force exerted on the plate by air would be the same as that exerted by water at $v_\infty = 0.5$ m/s.

Entry Length at Entrance to a Pipe

The growth of a boundary layer on a solid surface gives rise to the concept of an entry length, L_E, at the entrance to a pipe, which is required before fully developed hydrodynamic flow is developed in the pipe. Figure 3.8 shows the entry length when fluid enters the space between flat plates, which is defined as the length required for the two boundary layers to make contact with one another. From Eq. (3.17), if the distance between the plates is Δ, the value of x at which $\delta = \Delta/2$ (i.e., the entry length, L_E) is

$$\frac{\Delta}{2} = 5.0 \left(\frac{\nu}{v_\infty}\right)^{1/2} L_E^{1/2}$$

or

$$L_E = 0.01\Delta^2 \frac{v_\infty}{\nu} \tag{3.25}$$

However, from a mass balance in the entry length, if v_x at every value of y in the boundary layers is decreasing with increasing x, the velocity of the inviscid core between the two boundary layers must be increasing with increasing x, and hence L_E is not given correctly by Eq. (3.25).

For laminar flow into a circular pipe, Langhaar derived

$$\frac{L_E}{D} \sim 0.05\text{Re} = 0.05 \frac{\bar{v}D}{\nu} \tag{3.26}$$

where $\bar{v}$ is the average flow velocity in the fully developed hydrodynamic flow beyond the entry length.

3.7 Creeping Flow Past a Sphere

If a sphere of radius R and diameter D made of a solid material of density ρ_s is released in a fluid of density ρ, and $\rho_s > \rho$, the sphere begins to sink in the fluid. This is because the gravitational force, $\frac{4}{3}\pi R^3 \rho_s g$, acting downward on the sphere, is

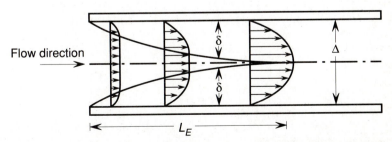

FIGURE 3.8. Illustration of the merging of the momentum boundary layers on the surfaces of parallel flat plates, which defines the entry length required for development of fully developed hydrodynamic flow in the fluid between the plates.

greater than the buoyancy force, $\frac{4}{3}\pi R^3 \rho g$, acting upward. The net downward force causes the sphere to accelerate in the downward direction, but the relative motion of the fluid past the sphere causes the fluid to exert a retarding drag force on the sphere, which increases in magnitude with increasing velocity of the sphere. Eventually, the sphere attains a velocity at which the retarding drag force equals $\frac{4}{3}\pi R^3 (\rho_s - \rho) g$, and thereafter the sphere sinks at this constant velocity, which is termed the terminal velocity, v_t. If the density and viscosity of the fluid and the terminal velocity of the sphere are such that the Reynolds number, defined as

$$\text{Re} = \frac{D v_t \rho}{\eta}$$

is less than 0.1, the consequent *creeping flow* of fluid past the sphere can be analyzed as follows. Consider the sphere to be stationary in a fluid that is flowing with the constant velocity v_∞ in the positive z-direction (which is equivalent to the sphere falling with the velocity v_∞ in the negative z-direction through a stationary fluid). The sphere experiences two types of force, one acting in directions normal to the surface of the sphere, as shown in Fig. 3.9(a), and another acting tangentially at the surface, as shown in Fig. 3.9(b).

In Fig. 3.9(b), the normal pressure at the surface of the sphere is given by the expression

$$P = P_0 - \rho g z - \frac{3}{2}\frac{\eta v_\infty}{R}\cos\theta \qquad (3.27)$$

where θ is a spherical coordinate of the position on the surface of the sphere as shown in Fig. 3.10. In Eq. (3.27), P_0 is the pressure at the plane of the equator of

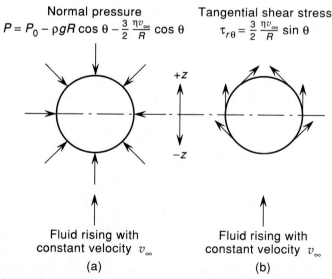

FIGURE 3.9.   (a) Normal pressures exerted on a sphere immersed in a flowing fluid; (b) tangential shear stresses acting on a sphere immersed in a flowing fluid.

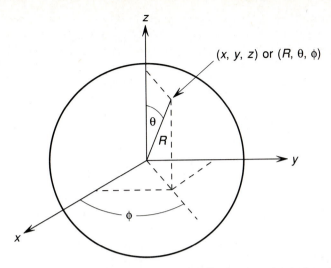

FIGURE 3.10.   Conversion of Cartesian coordinates to spherical coordinates.

the sphere (at $z = 0$) in the fluid far away from the sphere, $-\rho g z$ is the contribution due to the weight of the fluid, and the third term arises from the motion of the fluid past the sphere. The values of z and θ for any point on the surface of the sphere are related by $z = R \cos \theta$ and thus the second term in Eq. (3.27) can be written as $-\rho g R \cos \theta$. Thus if both the sphere and the fluid were motionless (i.e., $v_\infty = 0$), Eq. (3.27) gives the normal pressure acting downward on the north pole of the sphere as $P_0 - \rho g R$ and the normal pressure acting upward on the south pole of the sphere as $P_0 + \rho g R$. As shown in Fig. 3.11, the z-component of the normal force acting on the sphere per unit area is $-P \cos \theta$, and hence the total force acting in the z-direction on the sphere is

$$F_n = \int_{area} (-P \cos \theta) \, dA \tag{3.28}$$

where dA is an element of the surface area expressed in spherical coordinates. As illustrated in Fig. 3.12,

$$dA = R^2 \sin \theta \, d\theta \, d\phi \tag{3.29}$$

which, on integration over the surface of the sphere (i.e., from $\theta = 0$ to $\theta = \pi$ and from $\phi = 0$ to $\phi = 2\pi$), gives

$$-R^2 (\cos \pi - \cos 0)(2\pi - 0) = -R^2[(-1) - (1)]2\pi$$
$$= 4\pi R^2$$

which is the surface area of a sphere of radius R. Substitution of Eq. (3.29) into Eq. (3.28) gives

$$F_n = \int_0^{2\pi} \int_0^{\pi} (-P \cos \theta) R^2 \sin \theta \, d\theta \, d\phi$$

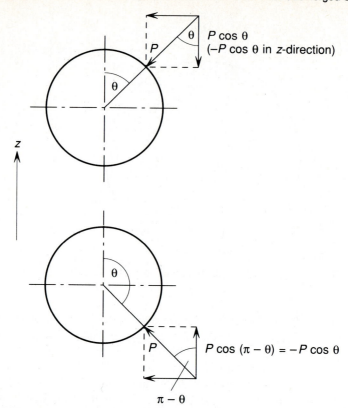

FIGURE 3.11. Determination of the vertical components of the normal forces acting on a sphere immersed in a flowing fluid.

and substitution of Eq. (3.27) gives

$$F_n = \int_0^{2\pi} \int_0^{\pi} \left(-P_0 + \rho g R \cos\theta + \frac{3}{2}\frac{\eta v_\infty}{R}\cos\theta \right) \cos\theta \; R^2 \sin\theta \; d\theta \; d\phi$$

$$(3.30)$$

From the identities

$$\int_0^{\pi} \cos\theta \sin\theta \; d\theta = \tfrac{1}{2}\sin^2\theta \big|_0^{\pi} = 0$$

and

$$\int_0^{\pi} \cos^2\theta \sin\theta \; d\theta = -\tfrac{1}{3}\cos^3\theta \big|_0^{\pi} = -\tfrac{1}{3}(-1-1) = \tfrac{2}{3}$$

Equation (3.30) is evaluated as

$$F_n = \tfrac{4}{3}\pi R^3 \rho g + 2\pi\eta R v_\infty \qquad (3.31)$$

In Eq. (3.31), the first term on the right can be recognized as being the buoyancy

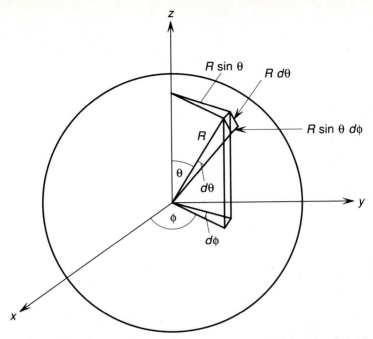

FIGURE 3.12.   The differential element of area on the surface of a sphere in spherical coordinates.

force exerted on the sphere, and the second term is a drag force called the *form drag*. Thus if the fluid is stationary with respect to the sphere (i.e., if $v_\infty = 0$), Eq. (3.32) shows that the z-component of the normal force exerted by the fluid on the sphere is simply the buoyancy force.

In Fig. 3.9(b) the tangential pressure, or shear stress, acting on the fluid at the surface of the sphere is given by the expression

$$\tau_{r\theta} = \tfrac{3}{2}\frac{\eta v_\infty}{R}\sin\theta \tag{3.32}$$

Thus the shear force acting tangentially on the surface of the sphere in the θ-direction per unit area is $-\tau_{r\theta}$ and the z-component of this force is $(-\tau_{r\theta})(-\sin\theta)$. The total force acting on the sphere is thus

$$F_t = \int_0^{2\pi}\int_0^{\pi}\left(\frac{3}{2}\frac{\eta v_\infty}{R}\sin\theta\right)\sin\theta\,R^2\sin\theta\,d\theta\,d\phi$$

which from the identity

$$\int \sin^3\theta = (\tfrac{1}{3}\cos^3\theta - \cos\theta)\big|_0^{\pi} = \tfrac{4}{3}$$

gives

$$F_t = 4\pi\eta R v_\infty \tag{3.33}$$

F_t is the *friction* drag force acting on the sphere.

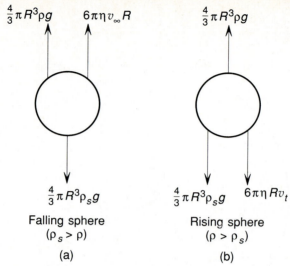

FIGURE 3.13. Force balances on spheres (a) falling ($\rho_s > \rho$) and (b) rising ($\rho > \rho_s$) through a fluid at their terminal velocities.

Addition of Eqs. (3.31) and (3.32) gives the total force acting on the sphere as

$$F = \tfrac{4}{3}\pi R^3 \rho g + 6\pi\eta R v_\infty \tag{3.34}$$

Thus, when a sphere of density ρ_s and radius R is dropped into a fluid of density ρ, it accelerates to a constant terminal velocity v_t, at which the point the downward force due to the influence of gravity on the sphere, $\tfrac{4}{3}\pi R^3 \rho_s g$, is equal to the upward buoyancy and drag forces, $\tfrac{4}{3}\pi R^3 \rho g + 6\pi\eta R v_t$,

$$\tfrac{4}{3}\rho R^3(\rho_s - \rho)g = 6\pi\eta R v_t \tag{3.35}$$

provided that Re < 0.1. Equation (3.35) is known as *Stokes' law*. If $\rho_s < \rho$, the sphere rises in fluid of the terminal velocity v_t. The directions of the forces are shown in Fig. 3.13.

EXAMPLE 3.2

When a hollow sphere of diameter 0.005 m and composite density 1500 kg/m³ is dropped into a column of oil of density 888 kg/m³, it attains a terminal velocity of 0.01 m/s. Calculate the viscosity of the oil.

First assume that Re < 0.1, in which case Eq. (3.35) can be used and rearranged as

$$\eta = \tfrac{2}{9}R^2 \frac{\rho_s - \rho}{v_t} g$$

$$= \frac{2 \times 0.0025^2 \times (1500 - 888) \times 9.81}{9 \times 0.01}$$

$$= 0.834 \text{ kg/m·s}$$

The Reynolds number is

$$\text{Re} = \frac{Dv_t\rho}{\eta}$$

$$= \frac{0.005 \times 0.01 \times 888}{0.834}$$

$$= 0.053$$

which, being less than 0.1, indicates that the calculation is valid.

EXAMPLE 3.3

The product of the deoxidation of liquid steel by the addition of aluminum is solid alumina. On the assumption that the solid alumina forms as small spheres, calculate the size of the smallest sphere that can float to the surface of a 1.5-m depth of quiescent liquid steel in 20 min given

$$\rho(\text{steel}) = 7160 \text{ kg/m}^3$$

$$\rho(\text{Al}_2\text{O}_3) = 3980 \text{ kg/m}^3$$

$$\eta(\text{steel}) = 0.0061 \text{ kg/m·s}$$

On the assumption that the sphere is formed at the bottom of the liquid bath, the required terminal velocity is $1.5/(20 \times 60) = 1.25 \times 10^{-3}$ m/s. Thus on the assumption that the Reynolds number is less than 0.1, the minimum radius is obtained from Eq. (3.35) as

$$R^2 = \frac{9\eta v_t}{2(\rho - \rho_s)g} = \frac{9 \times 0.0061 \times 1.25 \times 10^{-3}}{2 \times (7160 - 3980) \times 9.81}$$

$$= 1.1 \times 10^{-9} \text{ m}^2$$

Therefore,

$$R = 3.32 \times 10^{-5} \text{ m } (= 33.2\mu\text{m})$$

The Reynolds number is then

$$\text{Re} = \frac{2Rv_t\rho}{\eta} = \frac{2 \times 3.32 \times 10^{-5} \times 1.25 \times 10^{-3} \times 7160}{0.0061}$$

$$= 0.1$$

and hence the calculation is valid.

Suppose now that it was required to know what time a particle of radius 1×10^{-3} m would take to rise through the 1.5-m depth of liquid steel. Application of Eq. (3.35) would give

$$v_t = \frac{2R^2(\rho - \rho_s)g}{9\eta} = \frac{2 \times (1 \times 10^{-3})^2 \times (7160 - 3980) \times 9.81}{9 \times 0.0061}$$

$$= 1.136 \text{ m/s}$$

which means that $1.5/1.136 = 1.32$ s would be required. However, calculation of the Reynolds number as

$$\mathrm{Re} = \frac{2Rv_t\rho}{\eta} = \frac{2 \times 1 \times 10^{-3} \times 1.136 \times 7160}{0.0061}$$

$$= 2667$$

shows the calculation to be invalid, as $\mathrm{Re} > 0.1$. This problem will be considered again in Chapter 5.

3.8 Summary

Mass and momentum balances on a control volume in a flowing fluid produce universal equations named, respectively, the equation of continuity and the equation of conservation of momentum, which can be applied to all flow situations. For fluids of constant density and viscosity these equations lead to the Navier–Stokes equation, which shows that the sum of the viscous, "pressure," and gravitational forces acting in or on a flowing fluid is equal to mass times acceleration.

When a fluid flows over the surface of a submerged plate the requirement of no slippage between the plate and the film of fluid with which it is in contact sets up a velocity gradient in the fluid in a direction normal to the plate in which the local flow velocity varies from zero at the plate to the free-stream velocity of the fluid. This phenomenon causes the formation of a momentum boundary layer on the plate, and the upper surface of the boundary layer is the locus of where, on moving from the plate into the fluid, the local flow velocity reaches the free-stream value. The thickness of the boundary layer is proportional to the square root of the distance along the plate from the leading edge and is increased with increasing kinematic viscosity and decreasing free-stream velocity of the fluid. The viscous retardation of the fluid when it enters the momentum boundary layer causes a viscous drag between the fluid and the plate, and the local shear stress exerted at any position on the plate increases with increasing viscosity and free-stream velocity and decreases with increasing boundary layer thickness. Consequently, the local shear stress on the plate is proportional to the inverse of the square root of distance from the leading edge. The total shear force exerted on the plate is obtained by integrating the local shear stress over the area of the plate.

When a solid sphere is released in a motionless fluid it experiences the influence of its weight and its buoyancy. If its weight is greater than its buoyancy, the sphere accelerates in the downward-vertical direction, and if its buoyancy is greater than its weight, it accelerates in the upward-vertical direction. The motion of the sphere through the fluid causes the fluid to exert a viscous drag on the sphere, the magnitude of which increases with increasing velocity of the sphere. The accelerating sphere thus attains a constant terminal velocity when the difference between the weight of the sphere and its buoyancy is exactly balanced by the drag force exerted by the fluid. For creeping flow of a fluid past a submerged sphere, which occurs at Reynolds numbers less than 0.1, a force balance, which considers the normal and shear forces acting on the sphere, leads to Stokes' law.

Problems

PROBLEM 3.1 ──────────────────────────────

The viscosities of experimental molten glasses are being determined by measuring the terminal velocity of a platinum sphere falling through a column of molten glass. When dropped into a column of a standard glass of viscosity 10 Pa·s and density 2500 kg/m³, the measured terminal velocity is 0.0258 m/s, and when dropped into an experimental glass of density 3000 kg/m³, the measured velocity is 0.0168 m/s. Calculate the viscosity of the experimental molten glass. The density of platinum is 21,450 kg/m³.

PROBLEM 3.2 ──────────────────────────────

Water flows over a flat plate of length 1 m. At the location $x = x_1$ from the leading edge of the plate the velocity profile in the boundary layer is

$$v_x = 0.46y - 0.36y^3$$

in which the units of v are m/s and y is in cm. Calculate:

(a) The thickness of the boundary layer at $x = x_1$
(b) The velocity v_∞ of the water
(c) The value of x_1
(d) The velocity profile in the boundary layer at $x = 0.9$ m

The density and viscosity of the water are, respectively, 995 kg/m³ and 6.92 × 10^{-4} Pa·s.

PROBLEM 3.3 ──────────────────────────────

Water flows over a submerged square flat plate of area 1 m² at a velocity of 0.3 m/s. Calculate:

(a) The local shear stress on the plate at $x = 0.5$ m
(b) The total shear force on one side of the plate
(c) The rate of transport of viscous momentum at $x = 0.5$ m and $y =$ half the momentum boundary thickness at $x = 0.5$ m.
(d) The rate of transport of convective momentum at $x = 0.5$ m and $y =$ half the momentum boundary thickness at $x = 0.5$ m.

For water $\rho = 995$ kg/m³ and $\eta = 6.92 \times 10^{-4}$ Pa·s.

PROBLEM 3.4 ──────────────────────────────

Small glass spheres of density 2620 kg/m³ are allowed to fall through CCl_4 of density 1590 kg/m³ and viscosity 9.58 × 10^{-4} Pa·s. Calculate:

(a) The maximum diameter of sphere for which the flow obeys Stokes' law
(b) The terminal velocity that a sphere of this diameter attains

PROBLEM 3.5

A steel ball of radius 0.01 m is dropped through a molten glass to determine the viscosity of the glass. The density of the glass is half the density of the steel ball and the ball attains a terminal velocity of 0.03 m/s. Calculate:

(a) The kinematic viscosity of the glass.

(b) The radius of the largest steel ball that can be used along with Stokes' law to determine the kinematic viscosity of the glass.

(c) What is the terminal velocity of this largest steel ball?

PROBLEM 3.6

A flat plate of dimensions 1.2 m × 0.5 m is towed, parallel to its long side, through engine oil of viscosity 0.8 Pa·s and kinematic viscosity $9.2 \times 10^{-4} \text{m}^2/\text{s}$. Calculate the towing power required.

4

Turbulent Flow

4.1 Introduction

The particles in a flowing fluid experience two types of force: inertial and viscous. The *inertial force* acting on an element of fluid arises from the weight of fluid in the element and is equal to the mass of the element times its acceleration, and the *viscous force*, which acts between adjacent layers of fluid flowing at differing velocities, is equal to the product of the viscosity-induced shear stress between the layers and the area on which it acts. The influence of these forces on the characteristics of the fluid flow differ from one another in that the viscous forces tend to stabilize the unidirectional motion of the flow, and the inertia forces tend to disrupt the unidirectional motion, and consequently, the nature of the flow is determined by the relative magnitudes of the two types of force. At low enough flow velocities the viscous forces dominate and their stabilizing influence causes laminar flow. However, with increasing velocity the tendency toward disruption of laminar flow increases. The inertial force acting on an element of fluid is its mass ρL^3 (where L is a characteristic length) times its acceleration, given by velocity/time or v/t (where time is expressed as L/v). Thus the inertial force is

$$\rho L^3 \times \frac{v^2}{L} = \rho L^2 v^2$$

The viscous force is the product of the shear stress, which, from Eq. (2.4), is $\eta(dv_x/dy)$, or $\eta(v/L)$, times the area L^2. Thus the viscous force is

$$\left(\eta \frac{v}{L}\right) L^2 = \eta v L$$

and the ratio of the inertia force to the viscous force is

$$\frac{\rho L^2 v^2}{\eta v L} = \frac{v L \rho}{\eta}$$

which is the Reynolds number defined in Eq. (2.2).

In 1883, by means of experiments similar to those illustrated in Fig. 2.1, Reynolds showed that when the average velocity of a fluid in a circular tube, and hence the Reynolds number, exceed some critical value, the inertial forces become more influential than the viscous forces and the flow becomes turbulent. The actual transition from laminar to completely turbulent flow occurs over the range $2100 < \text{Re} < 3000$, and in this range experimental measurement of the flow properties does not always give reproducible results. For example, laminar flow in a very smooth-walled vibration-free tube can be achieved at Reynolds numbers greater than 2100, but if the tube is subjected to an impact or a vibration, the smooth streamlines in the laminar flow can be replaced by turbulent eddies and the entire nature of the flow changes. With increasing flow rate, the transition range ends at $\text{Re} \sim 3000$, and at $\text{Re} > 3000$, experimental measurement of the flow properties again gives reproducible results.

Although the notion of turbulence is familiar to anyone who has flown in an airplane, a comprehensive definition of turbulence as it pertains to fluid flow is difficult. Turbulent flow is chaotic and irregular in that the local velocities and the local pressure in the fluid vary with time and position in the flow, and thus time-averaged properties must be considered. The variation of the local velocity v_x at a fixed location in turbulent flow is illustrated in Fig. 4.1. The time-averaged value of v_x is defined as

$$\tilde{v}_x = \frac{1}{t} \int_0^t v_x \, dt$$

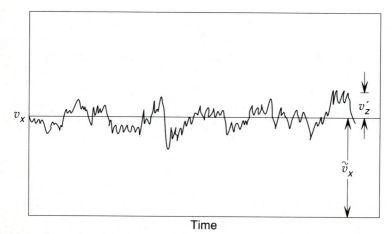

FIGURE 4.1 Variation, with time, of the *x*-component of the local flow velocity at a fixed location in turbulent flow.

and the value of v_x at any instant in time is thus

$$v_x = \bar{v}_x + v'_x$$

where v'_x is the magnitude of the fluctuation in the x-component of the local velocity. By definition of $\bar{v}_x$ the time-averaged value of v'_x is zero. In unidirectional turbulent flow fluctuations in the velocity also occur in the y and z directions, although the time-averaged values of v_y and v_z are zero. The intensity of the turbulence is determined by the magnitudes of the fluctuations in the velocities, and the time-averaged magnitude of v'_x is a measure of the amplitudes of the fluctuations. The *intensity I* of the turbulence is thus defined as

$$I = \frac{\sqrt{\frac{1}{3}(\overline{v'^2_x} + \overline{v'^2_y} + \overline{v'^2_z})}}{\bar{v}_x}$$

in which $\overline{v'_y}$ and $\overline{v'_z}$ are, respectively, the time-averaged fluctuations of the velocity in the y and z directions.

In turbulent flow the chaos and the consequent macroscopic mixing in the radial direction of the tube cause the velocity profile to be much flatter than the parabolic variation characteristic of laminar flow. Figure 4.2 shows an experimentally measured velocity profile for turbulent flow of water in a cylindrical tube of diameter 0.01 m at a Reynolds number of 4000 in comparison with what the profile would be like if the flow were laminar with the same average velocity. The variation of $\bar{v}/v_{max}$ with Reynolds number is shown in Fig. 4.3, which shows that as the transition range between laminar and completely turbulent flow is traversed, the value of $\bar{v}/v_{max}$ increases rapidly from the value of 0.5, which is the value characteristic of laminar flow, to 0.75, and thereafter, it increases more slowly to a more or less constant value of about 0.8.

As a result of the fact that no slippage occurs at the wall of the tube, a thin layer of fluid undergoing laminar flow exists next to the wall, and this layer is separated from the turbulent core of the flow by a transition region. A semiempirical derivation gives the velocity distribution in the turbulent core for turbulent flow up to $Re = 10^5$ as

$$\frac{v}{v_{max}} = \left(\frac{R - r}{R}\right)^{1/7} \tag{4.1}$$

and Eq. (4.1) is known as *the Blasius one-seventh power law*. Equation (4.1), with $v_{max} = 0.668$ m/s, is shown as the dashed line in Fig. 4.2. The one-seventh power law is only an approximation; it does not give the physical requirement of $dv/dr = 0$ at $r = 0$, but the integral

$$\bar{v} = \frac{v_{max}}{\pi R^2} \int_0^R 2\pi \left(\frac{R - r}{R}\right)^{1/7} r \, dr$$

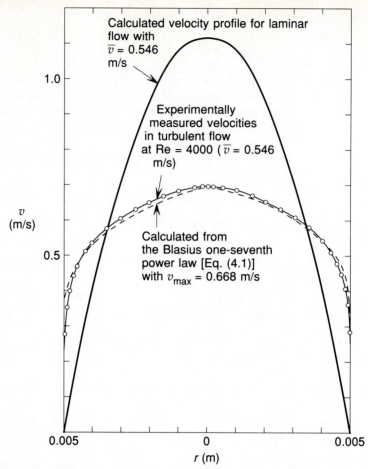

FIGURE 4.2 Experimentally measured local flow velocities in turbulent flow at a Reynolds number of 4000 in a cylindrical pipe in comparison with velocity profiles calculated assuming laminar flow at a Reynolds number of 4000 and calculated from the Blasius one-seventh power law [Eq. 4.1] with $v_{max} = 0.668$ m/s.

does give

$$\frac{\bar{v}}{v_{max}} = 0.817 \tag{4.2}$$

which is in close agreement with the experimental measurements of turbulent flow shown in Fig. 4.3. The variation of $\bar{v}/v_{max}$ with Re in the range $10^4 \leq \mathrm{Re} \leq 10^7$ is

$$\frac{\bar{v}}{v_{max}} = 0.62 + 0.04 \log \frac{\mathrm{Re}\, v_{max}}{\bar{v}}$$

Because of the existence of a thin layer of laminar flow adjacent to the tube wall, the one-seventh power law cannot be extended to $r = R$.

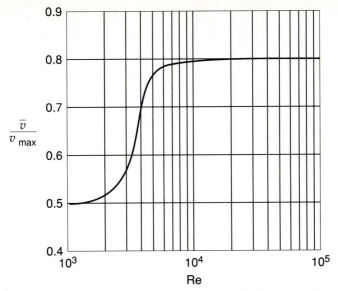

FIGURE 4.3   Variation of $\bar{v}/v_{max}$ with Reynolds number in the range of transition from laminar flow to turbulent flow in a cylindrical pipe.

4.2   Graphical Representation of Fluid Flow

The variations of $\Delta P/L$ with $\bar{v}$ for the flow of water at 25°C in horizontal pipes of diameter 0.01, 0.05, and 0.1 m are shown in Fig. 4.4. As both $\Delta P/L$ and $\bar{v}$ vary over several orders of magnitude, the variations are presented as log-log plots. For laminar flow the variations are given by the Hagen–Poiseuille equation, Eq. (2.22), rearranged as

$$\frac{\Delta P}{L} = \frac{8\eta}{R^2}\,\bar{v} \tag{4.3}$$

or

$$\log\frac{\Delta P}{L} = \log\bar{v} + \log\frac{8\eta}{R^2} \tag{4.4}$$

which are lines with slopes of 1 in Fig. 4.4. For completely turbulent flow the variations on a log-log plot up to Re $= 10^5$ are also straight lines with slopes of 1.75:

$$\frac{\Delta P}{L} \propto (\bar{v})^{1.75} \tag{4.5}$$

The increase in the value of $\Delta P/L$ required for a given average velocity in turbulent flow, over the value that would be required for laminar flow (if laminar flow could be achieved at the same velocity in the same pipe) is caused by turbulence-induced friction in the flow.

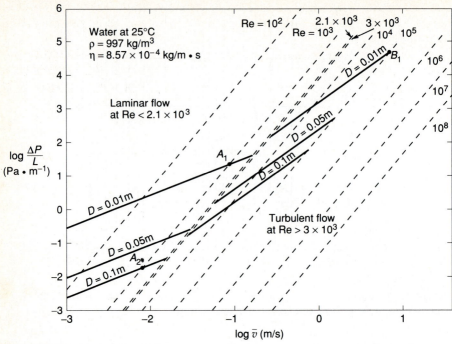

FIGURE 4.4 Variation of $\Delta P/L$ with average flow velocity for water at 25°C flowing in a cylindrical pipe.

In Fig. 4.4, lines of constant Re are linear with a slope of 3. This relationship can be seen in laminar flow by substituting R, expressed in terms of the Reynolds number:

$$\text{Re} = \frac{D\bar{v}\rho}{\eta}$$

or

$$R = \frac{\text{Re}\,\eta}{2\bar{v}\rho}$$

into Eq. (4.3) to give

$$\frac{\Delta P}{L} = 8\eta \left(\frac{2\bar{v}\rho}{\text{Re}\,\eta}\right)^2 \bar{v} = \frac{32\rho^2}{\text{Re}^2\eta}\bar{v}^3$$

or

$$\log \frac{\Delta P}{L} = 3 \log \bar{v} + \log \frac{32\rho^2}{\text{Re}^2\eta} \tag{4.6}$$

The lines of Re = 2100 and Re = 3000 for a fluid of given density and viscosity divide Fig. 4.4 into a region of laminar flow, the transition region, and a region of turbulent flow.

Although Fig. 4.4 is an adequate representation of the flow of water at 25°C (i.e., a fluid of density 997 kg/m^3 and viscosity 8.57×10^{-4} Pa·s), it is obvious that a separate graph would be needed for every combination of density and viscosity. It is thus desirable to develop a single graph, similar in nature to Fig. 4.4, which presents the characteristics of fluid flow for all combinations of the important properties D, $\bar{v}$, ρ, and η for both laminar and turbulent flow. Such a graphical representation is made possible by introduction of the friction factor, f.

4.3 Friction Factor and Turbulent Flow in Cylindrical Pipes

An expression for the shear force, F_s, exerted by a flowing fluid on the walls of its conduit can be formulated as follows:

$$F_s = \text{force}$$

$$= \frac{\text{mass} \cdot \text{length}}{\text{time}^2}$$

$$= \text{area} \times \frac{\text{mass} \cdot \text{length}}{\text{area} \cdot \text{time}^2}$$

$$= \text{area} \times \text{mass} \times \left(\frac{\text{length}}{\text{time}}\right)^2 \times \frac{1}{\text{volume}}$$

$$= \text{area} \times \frac{\text{mass} \cdot \text{velocity}^2}{\text{volume}}$$

The force thus has the units of the product of an area and the kinetic energy of the fluid per unit volume of the fluid. Selecting the area as some area characteristic of the system, A, and designating the average kinetic energy of the fluid per unit volume as $K = \frac{1}{2}\rho\bar{v}^2$, the formulation for the force can be written as

$$F_K = fAK \tag{4.7}$$

In Eq. (4.7) all of our present ignorance of the nature of the flow has been placed in the factor f, and Eq. (4.7) is the formal definition of the *friction factor*, f.

For flow in a cylindrical pipe of radius R the characteristic area is the wall area wetted by the fluid, $2\pi RL$, and hence Eq. (4.7) becomes

$$F_K = f(2\pi RL)(\tfrac{1}{2}\rho\bar{v}^2) \tag{4.8}$$

A force balance between the shear force F_s exerted by the flowing fluid on a length L of the wall of a tube of radius R, and the decrease in the normal force exerted on the fluid over the length L, gives

$$F_s = \tau_0(2\pi RL) = \Delta P(\pi R^2) \tag{4.9}$$

where τ_0 is the shear stress exerted by the fluid on the wall. Combinations of Eq. (4.8) with Eq. (4.9) give alternative definitions of f as

$$f = \frac{\tau_0}{\frac{1}{2}\rho\bar{v}^2} \tag{4.10}$$

and

$$f = \frac{1}{4}\left(\frac{D}{L}\right)\frac{\Delta P}{\frac{1}{2}\rho\bar{v}^2} \tag{4.11}$$

In laminar flow the relationship between $\Delta P/L$ and $\bar{v}$ is given by Eq. (2.22), and hence f for laminar flow can be calculated from first principles. However, for turbulent flow, the determination of f from Eq. (4.11) requires experimental measurement of the relationship between $\Delta P/L$ and $\bar{v}$.

The analysis of fluid flow is greatly simplified by the fact that in long, smooth tubes, the friction factor is a function only of the Reynolds number, a relationship that can be demonstrated by application of the *pi theorem* as follows. $\Delta P/L$ must be some function of the flow properties D, $\bar{v}$, ρ, and η; that is,

$$\frac{\Delta P}{L} = \phi(D, \bar{v}, \rho, \eta) \tag{4.12}$$

where ϕ is any function equal to

$$\alpha(D^a\bar{v}^b\rho^c\eta^e) \tag{4.13}$$

in which α is dimensionless.

The units of the various terms involved are

$$\frac{\Delta P}{L} = \frac{\text{mass}}{\text{length}^2 \cdot \text{time}^2} = \frac{M}{L^2 t^2}$$

$$D = \text{length} = L$$

$$\bar{v} = \frac{\text{length}}{\text{time}} = \frac{L}{t}$$

$$\rho = \frac{\text{mass}}{\text{length}^3} = \frac{M}{L^3}$$

$$\eta = \frac{\text{mass}}{\text{length} \cdot \text{time}} = \frac{M}{Lt}$$

and the equality of dimensions in Eq. (4.13) thus requires that

$$\frac{M}{L^2 t^2} = \alpha\left[L^a\left(\frac{L}{t}\right)^b\left(\frac{M}{L^3}\right)^c\left(\frac{M}{Lt}\right)^e\right]$$

That is, for M,

$$1 = c + e \tag{i}$$

for L,

$$-2 = a + b - 3c - e \qquad \text{(ii)}$$

and for t,

$$-2 = -b - e \qquad \text{(iii)}$$

Equations (i) to (iii) are three equations containing four unknowns and hence any one of the unknowns can be specified arbitrarily. Choosing e gives

$$c = 1 - e$$
$$b = 2 - e$$

and

$$\begin{aligned} a &= -2 - 2 + e + 3 - 3e + e \\ &= -1 - e \end{aligned}$$

and substitution into Eq. (4.13) then gives

$$\begin{aligned} \frac{\Delta P}{L} &= \alpha \left(\frac{1}{D D^e} \frac{\bar{v}^2}{\bar{v}^e} \frac{\rho}{\rho^e} \eta^e \right) \\ &= \alpha \left[\left(\frac{\eta}{D \bar{v} \rho} \right)^e \frac{\rho \bar{v}^2}{D} \right] \end{aligned}$$

or

$$\frac{\Delta P}{L} \frac{D}{\rho \bar{v}^2} = \alpha \left(\frac{D \bar{v} \rho}{\eta} \right)^{-e}$$

or

$$\frac{\Delta P}{L} \frac{D}{\rho \bar{v}^2} = \phi(\text{Re}) \qquad \text{(4.14)}$$

Comparison with Eq. (4.11) shows the left-hand side of Eq. (4.14) to be twice the friction factor, and hence Eq. (4.14) shows that f is a function only of Re.

In laminar flow the relationship between $\Delta P/L$ and $\bar{v}$ is given by the Hagen–Poiseuille equation, substitution of which into Eq. (4.11) to eliminate $\Delta P/L$ gives

$$\begin{aligned} f &= \frac{1}{4} D \frac{\bar{v} 8 \eta}{R^2} \frac{2}{\rho \bar{v}^2} \\ &= 16 \frac{\eta}{\bar{v} D \rho} \\ &= \frac{16}{\text{Re}} \qquad \text{(4.15)} \end{aligned}$$

Thus laminar fluid flow in long, smooth-walled tubes is represented by the variation of f with Re, shown as a log-log plot in Fig. 4.5.

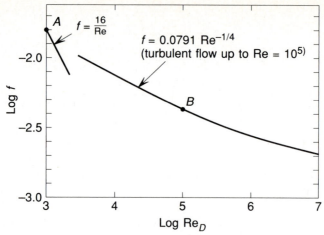

FIGURE 4.5 Variation of the friction factor with Reynolds number for fluid flow in a smooth-walled cylindrical pipe.

A comparison of Figs. 4.4 and 4.5 can be made as follows. Point A in Fig. 4.5 is the point $\log f = -1.796$, or $f = 0.016$, Re $= 10^3$; therefore,

$$f = 0.016 = \frac{1}{4}\left(\frac{\Delta P}{L}\right)\frac{D}{\frac{1}{2}\rho\bar{v}^2} \qquad \text{at} \quad \text{Re} = 10^3 = \frac{\bar{v}D\rho}{\eta}$$

For water at 20°C, with $\rho = 997$ kg/m^3 and $\eta = 8.57 \times 10^{-4}$ Pa·s,

$$f = 0.016 = \frac{1}{4}\left(\frac{\Delta P}{L}\right)\frac{D}{\frac{1}{2} \times 997\bar{v}^2} \qquad \text{or} \qquad \frac{\Delta P}{L} = 31.90\,\frac{\bar{v}^2}{D} \qquad \text{(i)}$$

when

$$\text{Re} = 10^3 = \frac{\bar{v}D \times 997}{8.57 \times 10^{-4}} \qquad \text{or} \qquad \bar{v} = \frac{8.596 \times 10^{-4}}{D} \qquad \text{(ii)}$$

Equation (ii) gives $\bar{v} = 0.086$ m/s ($\log \bar{v} = -1.07$), and Eq. (i) gives $\Delta P/L = 23.6$ N/m^2·m ($\log \Delta P/L = 1.37$).

The point $\log \Delta P/L = 1.37$, $\log \bar{v} = -1.07$ is the intersection of the lines for Re $= 10^3$ and $D = 0.10$ m (indicated as point A_1) in Fig. 4.4. With $D = 0.1$ m, Eq. (ii) gives $\bar{v} = 0.0086$ m/s ($\log \bar{v} = -2.07$), and Eq. (i) gives $\Delta P/L = 0.0257$ N/m^2·m ($\log \Delta P/L = -1.63$). The point $\log \Delta P/L = -1.63$, $\log \bar{v} = 2.07$, is the intersection of the lines for Re $= 10^3$ and $D = 0.1$ m (indicated as point A_2) in Fig. 4.4.

For turbulent flow up to Re $= 10^5$, the friction factor is given by the Blasius friction equation,

$$f = 0.0791\text{Re}^{-1/4} \qquad (4.16)$$

which is a correlation with experimental observation and is consistent with the Blasius one-seventh power law given by Eq. (4.1). Combination of Eqs. (4.16) and (4.11) gives

$$\log \frac{\Delta P}{L} = \log \bar{v} + \log \frac{0.1582\eta^{0.25}\rho^{0.75}}{D^{1.25}}$$

for turbulent flow at Re $\leq 10^5$ in a pipe of given diameter, and

$$\log \frac{\Delta P}{L} = 3 \log \bar{v} + \log \frac{0.1582\rho^2}{\eta \mathrm{Re}^{1.25}}$$

for turbulent flow at a constant Reynolds number up to a value of 10^5.

Point B on the Blasius line in Fig. 4.5 is the point $\log f = -2.35$, or $f = 0.00445$, Re $= 10^5$; therefore,

$$f = 0.00445 = \frac{1}{4}\left(\frac{\Delta P}{L}\right)\frac{D}{\frac{1}{2} \times 997\bar{v}^2} \quad \text{or} \quad \frac{\Delta P}{L} = 8.870 \frac{\bar{v}^2}{D} \qquad \text{(iii)}$$

when

$$\mathrm{Re} = 10^5 = \frac{\bar{v}D \times 997}{8.57 \times 10^{-4}} \quad \text{or} \quad \bar{v} = \frac{0.08596}{D} \qquad \text{(iv)}$$

With $D = 0.01$ m, Eq. (iv) gives $\bar{v} = 8.60$ m/s ($\log \bar{v} = 0.934$), and Eq. (iii) gives $\Delta P/L = 65{,}530$ N/m²·m ($\log \Delta P/L = 4.82$). The point $\log \Delta P/L = 4.82$, $\log \bar{v} = 0.934$ is the intersection of the lines for Re $= 10^5$ and $D = 0.1$ m (indicated as point B_1) in Fig. 4.4.

Thus all of the information contained by lines of constant Re in Fig. 4.4 is represented by a single point in Fig. 4.5, and Fig. 4.5 contains all of the required information on the steady-state flow of a fluid of any density, and any viscosity in a long smooth-walled cylindrical pipe of any diameter (i.e., any given flow). For laminar flow,

$$\mathrm{Re} \leq 2100, \qquad f = \frac{16}{\mathrm{Re}} \qquad (4.17)$$

and for turbulent flow up to Re $= 10^5$, the friction factor is given by

$$f = 0.0791\mathrm{Re}^{-1/4}$$

EXAMPLE 4.1

Calculate the pressure drop required to pass water at 300 K through a 300-m length of smooth 0.05-m-inside diameter (I.D.) pipe at the rate of 1.5×10^{-3} m³/s:

$$\bar{v} = \frac{1.5 \times 10^{-3}}{\pi(0.025)^2} = 0.764 \text{ m/s}$$

At 300 K,

$$\rho = 997 \text{ kg/m}^3$$

$$\eta = 8.57 \times 10^{-4} \text{ Pa·s}$$

$$\text{Re} = \frac{D\bar{v}\rho}{\eta} = \frac{0.05 \times 0.764 \times 997}{8.57 \times 10^{-4}} = 4.444 \times 10^4$$

As $3000 < \text{Re} < 10^5$, the friction factor can be calculated from the Blasius friction equation [Eq. (4.17)] as

$$f = 0.0791\text{Re}^{-1/4} = \frac{0.0791}{(4.444 \times 10^4)^{0.25}} = 5.448 \times 10^{-3}$$

and rearrangement of Eq. (4.11) gives

$$\Delta P = \frac{2fL\rho\bar{v}^2}{D} = \frac{2 \times 5.448 \times 10^{-3} \times 300 \times 997 \times (0.764)^2}{0.05}$$

$$= 3.80 \times 10^4 \text{ N/m}^2$$

Example 4.1 was straightforward because enough information was available to calculate Re. If Re cannot be calculated from the information available, the point on one of the lines in Fig. 4.5 which represents the flow must be found by trial and error.

EXAMPLE 4.2

Calculate the flow rate of water at 80°F in a smooth 0.07-m-I.D. pipe when $\Delta P/L$ is 125 N/m²·m.

Assumption 1: The flow is laminar (i.e., Re < 2100):

$$f = \frac{16}{\text{Re}} = \frac{16\eta}{\bar{v}D\rho} \tag{i}$$

and from Eq. (4.11),

$$f = \frac{1}{4}\left(\frac{D}{L}\right)\frac{\Delta P}{\frac{1}{2}\rho\bar{v}^2} \tag{ii}$$

Combination of Eqs. (i) and (ii) to eliminate f gives

$$\bar{v} = \frac{1}{32}\left(\frac{\Delta P}{L}\right)\frac{D^2}{\eta}$$

$$= \frac{1}{32} \times 125 \times \frac{0.07^2}{8.57 \times 10^{-4}}$$

$$= 22.33 \text{ m/s}$$

A check must now be made to see if this value of $\bar{v}$ gives a Reynolds number of less than 2100.

$$\mathrm{Re} = \frac{\bar{v}D\rho}{\eta} = \frac{22.33 \times 0.07 \times 997}{8.57 \times 10^{-4}}$$

$$= 1.82 \times 10^{6}$$

As Re > 2100, the assumption of laminar flow is thus incorrect.

Assumption 2: $3000 < \mathrm{Re} < 10^5$, in which case the Blasius expression for the friction factor can be used. From Eq. (4.17),

$$f = 0.0791\mathrm{Re}^{-0.25} = \frac{0.0791\eta^{0.25}}{\bar{v}^{0.25}D^{0.25}\rho^{0.25}} \qquad \text{(iii)}$$

and from Eq. (4.11),

$$f = \frac{1}{4}\left(\frac{D}{L}\right)\frac{\Delta P}{\frac{1}{2}\rho\bar{v}^2} \qquad \text{(iv)}$$

Combination of Eqs. (iii) and (iv) to eliminate f gives

$$\bar{v} = \left(6.321\frac{\Delta P}{L}\frac{D^{1.25}}{\rho^{0.75}\eta^{0.25}}\right)^{4/7}$$

$$= \left[6.321 \times 125 \times \frac{0.07^{1.25}}{997^{0.75}(8.57 \times 10^{-4})^{0.25}}\right]^{4/7}$$

$$= 0.964 \text{ m/s}$$

Substitution of $\bar{v}$ into Eqs. (iii) and (iv) gives

$$f = 0.0791 \times \left(\frac{8.57 \times 10^{-4}}{0.964 \times 0.07 \times 997}\right)^{0.25} = 4.73 \times 10^{-3}$$

and

$$f = \frac{1}{4} \times 125 \times \frac{0.07}{0.5 \times 997 \times (0.964)^2} = 4.73 \times 10^{-3}$$

Furthermore,

$$\mathrm{Re} = \frac{\bar{v}D\rho}{\eta} = \frac{0.964 \times 0.07 \times 997}{8.57 \times 10^{-4}} = 7.85 \times 10^4$$

which, as required, lies between 3000 and 10^5. Thus assumption 2 is correct and $\bar{v} = 0.964$ m/s. Therefore,

$$\bar{v} = 0.964 \text{ m/s}$$

$$\dot{V} = \bar{v}A = 0.964 \times \pi \times (0.035)^2 = 3.71 \times 10^{-3} \text{ m}^3/\text{s}$$

$$\dot{M} = \dot{V}\rho = 3.71 \times 10^{-3} \times 997 = 3.70 \text{ kg/s}$$

The foregoing discussion has been limited to consideration of fluid flow in smooth-walled tubes. Any roughness of the tube wall increases the friction factor and hence decreases the flow rate obtained for any given value of $\Delta P/L$. The influ-

TABLE 4.1

Material	Absolute Roughness, ϵ (mm)
Drawn tubing	0.0015
Commercial steel or wrought iron	0.046
Welded-steel pipe	0.046
Galvanized iron	0.15
Cast iron, average	0.259
Concrete	0.3–3

ence of surface roughness on the flow is determined by the average height of the protuberances on the surface, ϵ, and on the diameter of the tube, D, and the *relative roughness* is given by the ratio ϵ/D. Values of ϵ for typical surfaces are listed in Table 4.1.

The influence of relative roughness on the variation of f with Re for flow in tubes is shown in Fig. 4.6, which is known as *the Moody diagram* [L. F. Moody, *ASME Trans.* (1944), vol. 66, p. 671]. In Fig. 4.6 it can be seen that for any relative roughness, a critical Reynolds number occurs above which the friction factor is independent of the Reynolds number, and the value of this critical Reynolds number decreases with increasing relative roughness. The dashed line in Fig. 4.6 is the locus of the critical Reynolds numbers and defines the limit of fully rough flow.

The variations of f, Re, and ϵ/D for turbulent flow are given by Haaland's correlation [S. E. Haaland, *J. Fluids Eng.* (March 1983)]:

$$f^{-1/2} = -3.6 \log_{10} \left[\left(\frac{\epsilon}{3.7D} \right)^{1.11} + \frac{6.9}{\text{Re}} \right] \tag{4.18}$$

EXAMPLE 4.3

In Example 4.1 it was calculated that a pressure drop of 3.80×10^4 N/m^2 is required to pass water at 300 K through a 300-m length of smooth 0.05-m-I.D. pipe at the rate of 1.5×10^{-3} m^3/s. Now calculate the pressure drop required if the relative roughness of the pipe is 0.002. The data given are

$$\bar{v} = 0.764 \text{ m/s}$$

$$\rho = 997 \text{ kg/m}^3$$

$$\eta = 8.57 \times 10^{-4} \text{ Pa·s}$$

$$\text{Re} = 4.444 \times 10^4$$

$$L = 300 \text{ m}$$

$$D = 0.05 \text{ m}$$

The friction factor is obtained from Eq. (4.18) as

$$f^{-1/2} = -3.6 \log_{10} \left[\left(\frac{0.002}{3.7} \right)^{1.11} + \frac{6.9}{4.444 \times 10^4} \right]$$

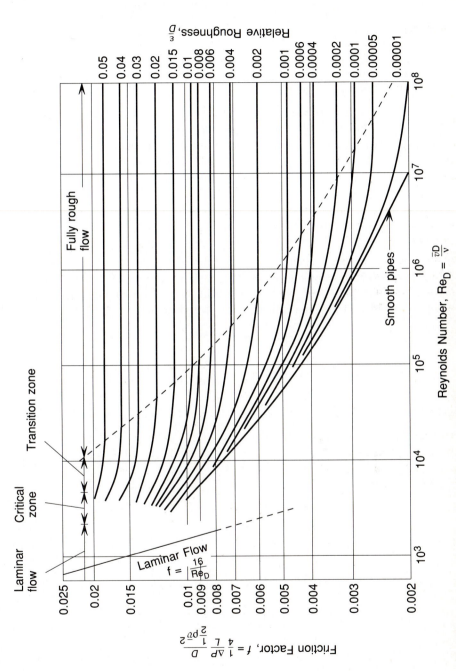

FIGURE 4.6 Moody diagram showing the variation of the friction factor with Reynolds number and relative roughness of the wall for fluid flow in a cylindrical pipe.

149

or

$$f = 6.647 \times 10^{-3}$$

Then rearrangement of Eq. (4.11) gives

$$\Delta P = \frac{2fL\rho\bar{v}^2}{D} = \frac{2 \times 6.647 \times 10^{-3} \times 300 \times 997 \times (0.764)^2}{0.05}$$

$$= 4.64 \times 10^4 \text{ N/m}^2$$

In this example, a relative roughness of 0.002 increases the friction factor from 5.448×10^{-3} (the value for a smooth pipe) to 6.647×10^{-3}, with the consequence that the pressure drop required for the same flow rate is increased from $3.80 \times 10^4 \text{ N/m}^2$ (for a smooth pipe) to $4.64 \times 10^4 \text{ N/m}^2$.

EXAMPLE 4.4

In Example 4.2 it was calculated that an average velocity of 0.964 m/s is obtained for water at 300 K flowing in a smooth 0.07-m-I.D. pipe when the pressure drop per unit length is 125 N/m². Now calculate the average velocity obtained if the relative roughness of the pipe is 0.002. The data given are

$$\Delta P/L = 125 \text{ N/m}^2 \cdot \text{m}$$

$$\rho = 997 \text{ kg/m}^3$$

$$\eta = 8.57 \times 10^{-4} \text{ Pa·s}$$

$$D = 0.07 \text{ m}$$

Solving gives

$$\text{Re} = \frac{\bar{v}D\rho}{\eta} \quad \text{or} \quad \bar{v} = \frac{\eta \text{ Re}}{D\rho} = \frac{8.57 \times 10^{-4}}{0.07 \times 997} \text{ Re}$$

or

$$\bar{v} = 1.228 \times 10^{-5} \text{Re} \text{ m/s} \tag{i}$$

From Eq. (4.11),

$$\bar{v} = \left[\frac{1}{4} \left(\frac{D}{L} \right) \frac{\Delta P}{\frac{1}{2}\rho} \right]^{1/2} f^{-1/2}$$

$$= \left(\frac{1}{4} \times 0.07 \times 125 \times \frac{1}{\frac{1}{2} \times 997} \right)^{1/2} f^{-1/2}$$

or

$$\bar{v} = 0.06624 f^{-1/2} \tag{ii}$$

and from Eq. (4.18),

$$f^{-1/2} = -3.6 \log_{10}\left[\left(\frac{0.002}{3.7}\right)^{1.11} + \frac{6.9}{Re}\right]$$

$$= -3.6 \log_{10}\left(2.363 \times 10^{-4} + \frac{6.9}{Re}\right) \qquad \text{(iii)}$$

Substitution of Eq. (iii) into Eq. (ii) gives

$$\bar{v} = -0.06624 \times 3.6 \log_{10}\left(2.363 \times 10^{-4} + \frac{6.9}{Re}\right)$$

and combination of this with Eq. (i) to eliminate $\bar{v}$ gives

$$1.228 \times 10^{-5} Re = -0.2385 \log_{10}\left(2.363 \times 10^{-4} + \frac{6.9}{Re}\right) \qquad \text{(iv)}$$

Equation (iv) has the solutions

$$Re = 6.91$$

and

$$Re = 6.733 \times 10^4$$

The solution of $Re = 6.91$ is out with the range of applicability of Eq. (4.18) and hence the physical solution is $Re = 6.733 \times 10^4$. From Eqs. (i) and (iii),

$$\bar{v} = 0.827 \text{ m/s}$$

$$f = 0.00641$$

Thus, in this example, a relative roughness of 0.002 increases the friction factor from 0.00473 (for a smooth pipe) to 0.00641, with the consequence that with the same value of $\Delta P/L$, the average flow velocity decreases from 0.964 m/s (for a smooth pipe) to 0.827 m/s.

Fluid Flow in Noncircular Ducts

The equations developed for turbulent flow in circular ducts can be applied to flow in noncircular ducts by replacing the diameter D with the *equivalent diameter*, D_e, defined as

$$D_e = \frac{4 \times \text{cross-sectional flow area}}{\text{wetted perimeter}} = \frac{4A}{P_w} \qquad \text{(4.19)}$$

For a rectangular duct of dimensions $z_1 \times z_2$,

$$D_e = \frac{4z_1 z_2}{2(z_1 + z_2)} = \frac{2z_1 z_2}{z_1 + z_2}$$

and the transition from laminar to turbulent flow in noncircular ducts also begins at $Re = 2100$. Simple substitution of D_e for D cannot be made when the flow is laminar, as can be illustrated by considering laminar flow between two flat parallel plates of width W separated by a distance of 2δ. In this case

$$D_e = \frac{4(2\delta W)}{2(W + 2\delta)}$$

and if $W \gg \delta$, this can be simplified as

$$D_e = 4\delta \quad \text{or} \quad \delta = \frac{D_e}{4}$$

Substitution into Eq. (2.12) gives

$$\bar{v} = \frac{\Delta P}{3L\eta}\delta^2 = \frac{\Delta P}{3L\eta}\frac{D_e^2}{16}$$

or

$$\frac{\Delta P}{L} = \frac{48\bar{v}\eta}{D_e^2}$$

The friction factor is then given as

$$f = \frac{1}{4}\left(\frac{\Delta P}{L}\right)\frac{D_e}{\frac{1}{2}\rho\bar{v}^2} = \frac{1}{4}\left(\frac{48\bar{v}\eta}{D_e^2}\right)\frac{D_e}{\frac{1}{2}\rho\bar{v}^2}$$

$$= \frac{24\eta}{D_e\bar{v}\rho}$$

$$= \frac{24}{Re}$$

This expression differs from that given by Eq. (4.15) for laminar flow in circular ducts.

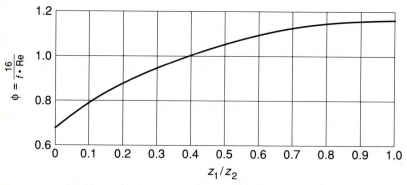

FIGURE 4.7 Correction factor required in Eq. (4.21).

For laminar flow in noncircular ducts the friction factor is defined as

$$f = \frac{16}{\phi \, \text{Re}} \tag{4.21}$$

where ϕ varies with the ratio z_1/z_2 as shown in Fig. 4.7. As z_1/z_2 approaches zero (i.e., as the geometry approaches that of flow between flat parallel plates in which $W \gg \delta$), ϕ approaches $\frac{2}{3}$ and the friction factor given by Eq. (4.21) approaches that given by Eq. (4.20).

4.4 Flow Over a Flat Plate

In Chapter 3 the velocity gradient at $y = 0$ in the laminar boundary layer on the surface of a flat plate was given by Eq. (3.20) as

$$\left.\frac{\partial v_x}{\partial y}\right|_{y=0} = \frac{3}{2}\left(\frac{v_\infty}{\delta}\right) \tag{3.20}$$

which, from

$$\tau_0 = -\eta \left.\frac{\partial v_x}{\partial y}\right|_{y=0}$$

and Eq. (3.21), gives the local shear stress on the surface of the plate at a distance x from the leading edge as

$$\tau_0 = 0.323 \left(\frac{\rho\eta v_\infty^3}{x}\right)^{1/2} \tag{4.22}$$

The local friction factor at x, f_x, is thus obtained from Eqs. (4.22) and (4.10) as

$$f_x = \frac{\tau_0}{\frac{1}{2}\rho v_\infty^2} = 0.646 \left(\frac{\eta}{\rho x v_\infty}\right)^{1/2} = 0.646(\text{Re}_x)^{-1/2} \tag{4.23a}$$

where Re_x is the local Reynolds number at x. The "exact" solution gives

$$f_x = 0.664(\text{Re}_x)^{-1/2} \tag{4.23b}$$

As the drag force, F_K, exerted by the laminar boundary layer on one side of a flat plate of width W and length L is given by Eq. (3.24) as

$$F_K = 0.664 \, (v_\infty^3 \eta \rho L W^2)^{1/2}$$

the average friction factor, $\bar{f}_L$, is obtained from Eq. (4.7) as

$$\bar{f}_L = \frac{F_K}{AK} = \frac{0.664(v_\infty^3 \eta \rho L W^2)^{1/2}}{WL\frac{1}{2}\rho v_\infty^2}$$

$$= 1.328(\text{Re}_L)^{-1/2} \tag{4.24}$$

The local Reynolds number for flow over a flat plate, given by

$$\text{Re}_x = \frac{v_\infty x \rho}{\eta}$$

increases linearly with x, and at a value of Re_x of approximately 3×10^5, a transition from laminar boundary flow to turbulent boundary flow begins. The transition is completed in the range $3 \times 10^5 \leq \text{Re}_x \leq 3 \times 10^6$ and the boundary layer is as shown in Fig. 4.8. The thickness of the boundary layer increases significantly and the turbulence causes a flattening of the velocity profile and increases the shear stress exerted on the plate.

For Reynolds numbers up to 10^5 in turbulent pipe flow, the Blasius one-seventh power law, Eq. (4.1), gives the velocity distribution as

$$\frac{v_x}{v_{x,\text{max}}} = \left(\frac{y}{R}\right)^{1/7}$$

where y is the distance measured from the pipe wall toward the center line. By analogy, the velocity distribution in the turbulent layer on a flat plate is given by

$$\frac{v_x}{v_\infty} = \left(\frac{y}{\delta}\right)^{1/7} \tag{4.25}$$

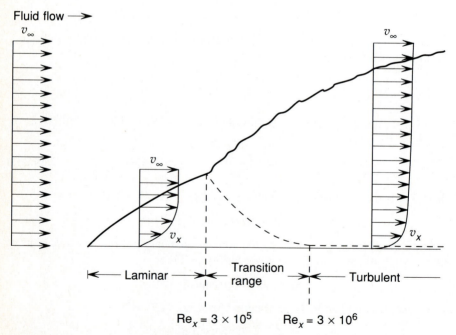

FIGURE 4.8 Transition from laminar flow to turbulent flow in the momentum boundary layer on the surface of a flat plate. The transition begins at a local Reynolds number of 3×10^5 and is complete at a local Reynolds number of 3×10^6.

which differs significantly from the parabolic velocity distribution in a laminar boundary layer given by Eq. (3.19). The use of Eq. (4.25) as the velocity profile in the development of the von Kármán integral expression gives

$$\frac{d\delta}{dx} = 10.286 \left(\frac{\tau_0}{\rho v_\infty^2}\right) \qquad (4.26)$$

where τ_0 is the shear stress exerted on the plate. Although the expression $\tau_0 = -\eta(\partial v_x/\partial y)|_{y=0}$ is valid at the surface of the plate, the existence of a thin laminar layer in contact with plate means that the one-seventh power law does not hold as y approaches zero, and thus the one-seventh power law velocity distribution cannot be used to obtain τ_0 in Eq. (4.26). Therefore, the Blasius equation for turbulent flow in a pipe, which is given by Eq. (4.16) and which is consistent with the one-seventh power law, is used as follows. Combination of Eq.s (4.10) and (4.16) gives

$$f = 0.0791 \left(\frac{D\bar{v}\rho}{\eta}\right)^{-1/4} = \frac{\tau_0}{\frac{1}{2}\rho\bar{v}^2}$$

or

$$\frac{\tau_0}{\rho\bar{v}^2} = 0.0396 \left(\frac{D\bar{v}\rho}{\eta}\right)^{-1/4} \qquad (4.27)$$

Equation (4.27), which is for turbulent flow in a pipe, is adapted to boundary layer flow over a flat plate by substituting $D = 2\delta$, and from Eq. (4.2), $\bar{v} = 0.817 v_{max} = 0.817 v_\infty$;

$$\frac{\tau_0}{\rho v_\infty^2} = 0.0396 \times (0.817)^2 \times (2 \times 0.817)^{-1/4} \left(\frac{\delta v_\infty \rho}{\eta}\right)^{-1/4}$$

$$= 0.0233 \left(\frac{\delta v_\infty}{\nu}\right)^{-1/4} \qquad (4.28)$$

which, in combination with Eq. (4.26), gives

$$\frac{\tau_0}{\rho v_\infty^2} = \frac{1}{10.286}\left(\frac{d\delta}{dx}\right)$$

or

$$\frac{d\delta}{dx} = 0.240 \left(\frac{\delta v_\infty}{\nu}\right)^{-1/4}$$

or

$$\delta^{1/4}\, d\delta = 0.240 \left(\frac{v_\infty}{\nu}\right)^{-1/4} dx \qquad (4.29)$$

If the laminar boundary layer at the front end of the plate is ignored, integration of Eq. (4.26) from x and δ to $x = \delta = 0$ gives

$$\frac{4}{5} \delta^{5/4} = 0.240 \left(\frac{v_\infty}{\nu}\right)^{-1/4} x$$

which, on rearrangement, becomes

$$\delta = 0.381 \left(\frac{v_\infty x}{\nu}\right)^{-1/5} x \tag{4.30}$$

$$= 0.381 \, (\mathrm{Re}_x)^{-1/5} x \tag{4.31}$$

Thus, in a turbulent boundary layer δ increases as $x^{4/5}$, in contrast with the behavior of a laminar boundary layer in which, from Eq. (3.16), δ increases as $x^{1/2}$.

From Eqs. (4.28) and (4.30) the local friction factor, f_x is obtained as

$$f_x = \frac{\tau_0}{\frac{1}{2}\rho v_\infty^2} = 0.0466 \left(\frac{\delta v_\infty}{\nu}\right)^{-1/4}$$

$$= 0.0466 \left[0.382 \left(\frac{v_\infty x}{\nu}\right)^{-1/5} x \right]^{-1/4} \left(\frac{v_\infty}{\nu}\right)^{-1/4}$$

$$= 0.0593 \left(\frac{v_\infty x}{\nu}\right)^{-1/5}$$

$$= 0.0593 (\mathrm{Re}_x)^{-1/5} \tag{4.32}$$

The drag force on the plate over length L and width W is

$$F_K = \int_0^W \int_0^L \tau_0 \, dx \, dz \tag{4.33}$$

which, from Eq. (4.32), is

$$F_K = \frac{1}{2} \rho v_\infty^2 \int_0^W \int_0^L 0.0593 \left(\frac{v_\infty x}{\nu}\right)^{-1/5} dx \, dz$$

$$= \frac{1}{2} \rho v_\infty^2 (0.0741) \left(\frac{v_\infty}{\nu}\right)^{-1/5} W L^{4/5} \tag{4.34}$$

and hence the average friction factor for the area $W \times L$ is

$$\bar{f}_L = \frac{F_K}{AK} = \frac{F_K}{W L \frac{1}{2} \rho v_\infty^2}$$

$$= 0.0741 \left(\frac{v_\infty L}{\eta}\right)^{-1/5}$$

$$= 0.0741 (\mathrm{Re}_L)^{-1/5} \tag{4.35}$$

Equation (4.35) is valid in the range $5 \times 10^5 < \mathrm{Re}_L < 10^7$. A more complex analysis

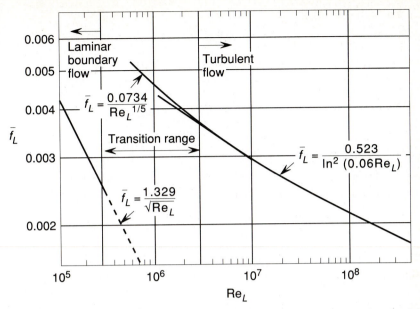

FIGURE 4.9 Variation of the average friction factor with Reynolds number for fluid flow over a flat plate.

(F. M. White, *Viscous Fluid Flow*, McGraw-Hill, New York, 1974, pp. 498 and 500) gives

$$f_x = \frac{0.455}{\ln^2(0.06\text{Re}_x)} \tag{4.36}$$

and

$$\bar{f}_L = \frac{0.523}{\ln^2(0.06\text{Re}_L)} \tag{4.37}$$

as the local and average friction factor, respectively, for a smooth flat plate at any turbulent Reynolds number. The variation of $\bar{f}_L$ with Re_L is shown in Fig. 4.9.

EXAMPLE 4.5

A flat rectangular plate of dimensions 1 m × 0.5 m is towed at a velocity of 10 m/s through water at 300 K. Which of the two configurations, (1) being towed in the direction parallel to the longer side of 1 m and (2) being towed in the direction parallel to the shorter side of 0.5 m, causes the greater dissipation of power?

For water at 300 K:

$$\rho = 997 \ kg/m^3$$

$$\eta = 8.57 \times 10^{-4} \ Pa\cdot s$$

$$\nu = 8.6 \times 10^{-7} \ m^2/s$$

(1) $L = 1$ m *and* $W = 0.5$ m: At the trailing edge,

$$Re_L = \frac{v_\infty L}{\nu} = \frac{10 \times 1}{8.6 \times 10^{-7}} = 1.16 \times 10^7$$

and hence a transition from laminar to turbulent boundary flow begins at that location on the plate at which $Re_x = 3 \times 10^5$:

$$x = \frac{3 \times 10^5 \times 8.6 \times 10^{-7}}{10} = 0.026 \ m$$

The boundary layer is thus laminar over the first 2.6% of the plate, which is a small enough percentage that it can be assumed that the layer is turbulent over the entire plate. Thus, from Eq. (4.37), the average friction fraction for the entire plate is

$$\bar{f}_L = \frac{0.523}{\ln^2(0.06 Re_L)} = \frac{0.523}{\ln^2(0.06 \times 1.16 \times 10^7)}$$
$$= 0.00289$$

and hence, from Eq. (4.7), the drag force on one side of the plate is

$$F_K = fAK = 0.00289 \times (1 \times 0.5) \times (\tfrac{1}{2} \times 997 \times 10^2)$$
$$= 72 \ N$$

The power dissipated by friction on both sides of the plate is thus

$$power = 2F_K v_\infty = 2 \times 72 \times 10 = 1440 \ W$$

(2) $L = 0.5$ m *and* $W = 1$ m: In this configuration, at the trailing edge

$$Re_L = 0.5 \times 1.16 \times 10^7 = 5.8 \times 10^6$$

and the boundary layer is laminar over the first 5.2% of the plate, which is still small enough to be ignored. The average friction factor for the entire plate is now

$$\bar{f}_L = \frac{0.523}{\ln^2(0.06 \times 5.8 \times 10^6)} = 0.00321$$

and hence

$$F_K = 0.00321 \times (0.5 \times 1) \times (\tfrac{1}{2} \times 997 \times 10^2)$$
$$= 80 \ N$$

and power is dissipated at the rate $2 \times 80 \times 10 = 1600$ W. Thus case (2) causes the dissipation of more power than does case (1). This result can be understood

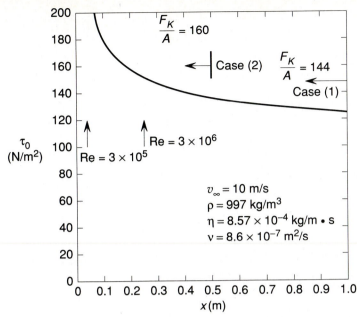

FIGURE 4.10   Variation, with distance from the leading edge, of the local shear stress exerted on a flat plate by water flowing at $v_\infty = 10$ m/s.

in terms of the variation of the local shear stress, τ_0, with distance x along the plate. Equation (4.36) gives, indirectly, the variation of the local friction factor with x, and hence combination of Eqs. (4.36) and (4.10) gives the variation of the local shear stress with x as

$$\tau_0 = \frac{\frac{1}{2}\rho v_\infty^2 0.455}{\ln^2(0.06 v_\infty x/v)} = \frac{22{,}682}{\ln^2(6.98 \times 10^5 x)}$$

$$= \frac{2.27 \times 10^4}{181 + 26.9 \ln x + \ln^2 x}$$

which is shown in Fig. 4.10. The average shear stress over the length of plate from $x = x_1$ to $x = x_2$ is obtained from

$$\bar{\tau}_0 = \frac{1}{x_2 - x_1} \int_{x_1}^{x_2} \tau_0 \, dx$$

which, with $x_1 = 0$ and $x_2 = 1$ m, gives $\bar{\tau}_0 = 144$ N/m², in agreement with

$$\frac{F_K}{A} = \frac{72}{0.5} = 144 \text{ N/m}^2$$

obtained in case (1). Similarly, with $x_1 = 0$ and $x_2 = 0.5$, $\bar{\tau}_0 = 160$ N/m², in agreement with

$$\frac{F_K}{A} = \frac{80}{0.5} = 160 \text{ N/m}^2$$

obtained in case (2).

Thus, because $\bar{\tau}_0$ for the length from $x = 0$ to x decreases with increasing x, the overall drag force acting on a plate of constant area given by $A = WL$ decreases as L increases (and hence W decreases).

4.5 Flow Past a Submerged Sphere

The force balance, illustrated in Fig. 3.12, shows that for a sphere falling at its terminal velocity through a fluid, the drag force exerted on the sphere by the fluid equals the difference between the weight of the sphere and its buoyancy in the fluid:

$$F_K = \tfrac{4}{3}\pi R^3 (\rho_s - \rho)g \qquad (4.38)$$

For the case of creeping flow, Stokes' law gives the drag force as the sum of the form drag and the friction drag,

$$F_K = 6\pi\eta R v_t \qquad (4.39)$$

In general, however,

$$F_K = fAK = f(\pi R^2)(\tfrac{1}{2}\rho v_t^2) \qquad (4.40)$$

and hence from Eqs. (4.38) and (4.40),

$$f(\rho R^2)(\tfrac{1}{2}\rho v_t^2) = \tfrac{4}{3}\rho R^3 (\rho_s - \rho)g \qquad (4.41)$$

For creeping flow, combination of Eqs. (4.39) and (4.40) gives

$$f = 12 \frac{\eta}{v_t R \rho} = \frac{24}{\text{Re}} \qquad (4.42)$$

Equation (4.42), which is derived from Stokes' law, is valid when $\text{Re} < 0.1$. In the range $2 < \text{Re} < 5 \times 10^2$,

$$f \approx \frac{18.5}{\text{Re}^{3/5}} \qquad (4.43)$$

and in the range $5 \times 10^2 < \text{Re} < 2 \times 10^5$,

$$f \approx 0.44 \qquad (4.44)$$

Equation (4.44) is known as Newton's law. The variation of f with Re for flow past a submerged sphere is shown in Fig. 4.11.

EXAMPLE 4.6 _____

In Example 3.2 an attempt was made to calculate the time that would be required for an alumina particle of radius 10^{-3}m to rise through a 1.5-m depth of liquid steel. Reconsideration of that example indicates that the friction factor is given

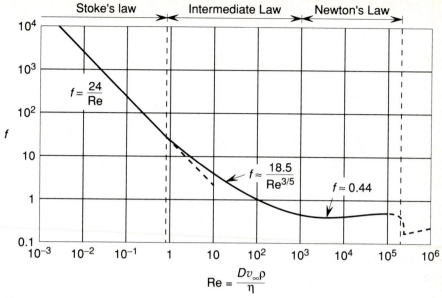

FIGURE 4.11 Variation of the friction factor with Reynolds number for fluid flow past sphere.

by Eq. (4.43), and not, as orginally tacitly assumed, by Eq. (4.42). Thus for

$$\rho(\text{steel}) = 7160 \text{ kg/m}^3$$

$$\rho(\text{Al}_2\text{O}_3) = 3980 \text{ kg/m}^3$$

$$\eta(\text{steel}) = 0.0061 \text{ Pa·s}$$

and the assumption that $2 < \text{Re} < 5 \times 10^2$,

$$\text{Re} = \frac{D\rho v_t}{\eta} = \frac{2 \times 10^{-3} \times 7160}{0.0061} v_t$$

$$= 2348 v_t$$

and from Eq. (4.43),

$$f = \frac{18.5}{(D\rho v_t/\eta)^{3/5}} = \frac{18.5}{(2348)^{3/5} v_t^{3/5}} = \frac{0.1757}{v_t^{3/5}} \tag{i}$$

Rearrangement of Eq. (4.41) gives

$$v_t = \left[\frac{8R(\rho - \rho_s)g}{3\rho f} \right]^{1/2}$$

$$= \left[\frac{8 \times 10^{-3} \times (7160 - 3980) \times 9.81)}{3 \times 7160 f} \right]^{1/2}$$

$$= \frac{0.1078}{f^{1/2}} \tag{ii}$$

Combining Eqs. (i) and (ii) to eliminate f gives

$$v_t = 0.144 \text{ m/s}$$

and

$$\text{Re} = \frac{2 \times 10^{-3} \times 0.144 \times 7160}{0.0061} = 337$$

which is in the range of applicability of Eq. (4.43).

In the original example, in which creeping flow was assumed, the terminal velocity was calculated as being 1.136 m/s.

EXAMPLE 4.7

Calculate the terminal velocity attained by a steel sphere of diameter 0.01 m when it is dropped in still air. The density of the steel ball is 7500 kg/m³ and the density and viscosity of air are, respectively, 1.177 kg/m³ and 1.85×10^{-5} Pa·s.

(1) Assume that Stoke's law is valid (i.e., Re $\leq$ 0.1). From Stokes' law, Eq. (3.35),

$$v_t = \frac{2R^2(\rho_s - \rho)g}{9\eta} = \frac{2 \times 0.005^2 \times (7500 - 1.177) \times 9.81}{9 \times 1.85 \times 10^{-5}}$$

$$= 2.21 \times 10^4 \text{ m/s}$$

for which

$$\text{Re} = \frac{v_t D\rho}{\eta} = \frac{2.21 \times 10^4 \times 0.01 \times 1.177}{1.85 \times 10^{-5}} = 1.41 \times 10^7$$

Therefore Stokes' law is not valid.

(2) Assume that the intermediate law is valid (i.e., $2 \leq$ Re $\leq$ 500):

$$F_K = \tfrac{4}{3}\pi R^3(\rho_s - \rho)g = \tfrac{4}{3}\pi \times (0.005)^3 \times (7500 - 1.177) \times 9.81$$

$$= 3.853 \times 10^{-2}$$

$$= fAK$$

$$= f(\pi R^2) \times (\tfrac{1}{2}\rho v_t^2)$$

$$= f \times (\pi \times 0.005^2 \times 0.5 \times 1.177)v_t^2$$

$$= 4.624 \times 10^{-5} f v_t^2$$

Therefore,

$$f v_t^2 = 833.3$$

From the intermediate law

$$f = \frac{18.5}{\text{Re}^{3/5}} = 18.5 \times \left(\frac{1.85 \times 10^{-5}}{0.01 \times 1.177}\right)^{3/5} \frac{1}{v_t^{3/5}} = \frac{0.3846}{v_t^{3/5}} \tag{i}$$

or

$$fv_t^{3/5} = 0.3846 \tag{ii}$$

and combination of Eqs. (i) and (ii) gives

$$v_t^{7/5} = 2167$$

or

$$v_t = 241 \text{ m/s}$$

Thus

$$\text{Re} = \frac{241 \times 0.01 \times 1.177}{1.85 \times 10^{-5}} = 1.54 \times 10^5$$

which, being greater than 500, indicates that the intermediate law is not valid.

(3) Assume that Newton's law is valid (i.e., $500 \leq \text{Re} \leq 2 \times 10^5$). Insertion of $f = 0.44$ into Eq. (i) gives

$$v_t = \left(\frac{833.3}{0.44}\right)^{1/2} = 43.5 \text{ m/s}$$

and

$$\text{Re} = \frac{43.5 \times 0.01 \times 1.177}{1.85 \times 10^{-5}} = 2.77 \times 10^4$$

Thus Newton's law is valid and the terminal velocity is 43.5 m/s.

4.6 Flow Past a Submerged Cylinder

Figure 4.12(a) shows a cylinder in fluid crossflow (i.e. the fluid flow is in a direction normal to the axis of the cylinder). As the fluid in the free stream approaches the cylinder its velocity decreases and falls to zero at the point of forward stagnation shown in Fig. 4.12(a) and (b). The consequent disappearance of the kinetic energy per unit volume causes an increase in the pressure of the fluid, and thus fluid flow around the cylinder, and the formation of a laminar momentum boundary layer on its surface, begin under conditions of a negative pressure gradient in the flow direction. This causes the velocity of the fluid to increase and, eventually, point B in Fig. 4.12(b) is reached, at which the flow velocity is a maximum and the pressure gradient in the flow direction is zero. At this point the velocity gradient in the momentum boundary layer is similar to that on a flat plate. Beyond point B the pressure gradient in the flow direction is positive, which tends to hinder the flow, and hence the flow velocity and the momentum of the flow decrease. The positive pressure gradient also has a significant influence on the shape of the velocity profile in the boundary layer, as shown at point D. Eventually, point E in Fig. 4.12(b) is reached, at which the momentum of flow in the boundary layer near the surface is insufficient to overcome

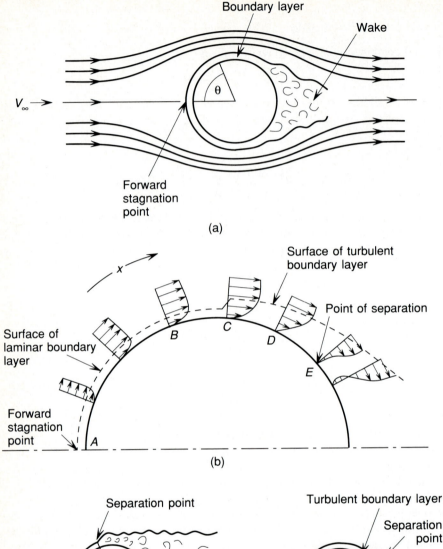

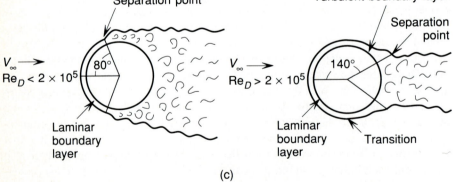

FIGURE 4.12 (a) Fluid flow past a cylinder; (b) variation, with position, of the velocity profile in the momentum boundary layer on a cylinder; (c) influence of Reynolds number on the separation point in the boundary layer on a cylinder.

the influence of the positive pressure gradient, and hence the velocity gradient at the surface of the cylinder falls to zero. As reversal of the direction of the flow is prevented by the oncoming flow, the boundary layer becomes detached from the surface. This phenomenon is known as *boundary separation*, and downstream from this point, a turbulent wake is formed that fills the void formed by the separated flow.

The transition from laminar to turbulent flow in the boundary layer occurs at a Reynolds number of 2×10^5, where $\mathrm{Re}_D = v_\infty D / v$ and the characteristic length is the diameter of the cylinder, D. Since the momentum of the fluid in a turbulent boundary layer is greater than that in a laminar boundary layer, separation of a turbulent layer is delayed and hence occurs at a greater value of θ than does separation of a laminar layer. As shown in Fig. 4.12(c), separation of a laminar layer occurs at $\theta \sim 80°$ and separation of a turbulent layer occurs at $\theta \sim 140°$. Figure 4.12(b) is drawn for $\mathrm{Re}_D > 2 \times 10^5$. As in the case for fluid flow past a sphere, the drag force exerted on a cylinder by the fluid has two components: the friction drag arising from the shear stress at the surface, and the form drag due to the pressure differential in the flow direction caused by the formation of the wake. The friction factor is defined as

$$f = \frac{F_D}{A \cdot \frac{1}{2} \rho v_\infty^2}$$

where F_D is the drag force on the cylinder and A is the projected area of the cylinder perpendicular to the direction of flow of the free stream. The variation of f with Re_D is shown in Fig. 4.13. With $\mathrm{Re}_D < 2 \times 10^5$, separation makes a negligible contribution to the drag force. However, with increasing Re_D the effect of separation and its contribution to the drag force become important and the significant decrease in the friction factor that occurs at $\mathrm{Re}_D > 2 \times 10^5$ is due to the boundary layer transition [at point C in Fig. 4.12(b)], which delays separation and, by decreasing the extent of the wake, decreases the magnitude of the form drag.

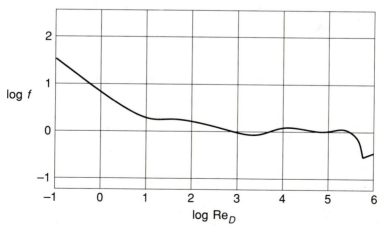

FIGURE 4.13 Variation of the friction factor with Reynolds number for fluid flow past a cylinder.

EXAMPLE 4.8

A vertical cylindrical flag pole of diameter 0.1 m and height 20 m is subjected to a 15-m/s horizontal wind. Calculate the drag force exerted on the pole by the wind and the bending moment about the base of the pole.

With air of density 1.117 kg/m³ and viscosity 1.85×10^{-5} Pa·s,

$$Re = \frac{15 \times 0.1 \times 1.117}{1.85 \times 10^{-5}}$$

$$= 90{,}600$$

Thus log Re = 4.96, and hence from Fig. 4.13, log f = 0.1 and f = 1.26. Thus the drag force is

$$F_D = fA \cdot \tfrac{1}{2}\rho v_\infty^2$$

$$= 1.26 \times 20 \times \pi \times 0.1 \times \tfrac{1}{2} \times 1.117 \times 15^2$$

$$= 269 \text{ N}$$

As the pressure on the pole is uniform over its length, the resultant force acts at the midpoint of the pole and the bending moment about the base of the pole is

$$269 \times 10 = 2690 \text{ N·m}$$

Summary of the Fluid Flow Equations

	Laminar	Turbulent
Pipe flow	$f = 16/Re_D$, Eq. (4.15)	$f = 0.0791Re_D^{-1/4}$, Eq. (4.16), in the range $3 \times 10^3 < Re_D < 10^5$
		$f^{-1/2} = 3.6 \log_{10}\left[\left(\dfrac{\epsilon}{3.7D}\right)^{1.11} + \dfrac{6.9}{Re_D}\right]$, Eq. (4.18), for all turbulent Re_D
Flow over a flat plate	$f_x = 0.664Re_x^{-1/2}$, Eq. (4.23b) $\overline{f}_L = 1.328Re_L^{-1/2}$, Eq. (4.24)	$f_x = 0.0593Re_x^{-1/5}$, Eq. (4.32) $\overline{f}_L = 0.0741Re_L^{-1/5}$, Eq. (4.34) in the range $5 \times 10^5 < Re_D < 10^7$
		$f_x = \dfrac{0.455}{\ln^2(0.06Re_x)}$, Eq. (4.36)
		$\overline{f}_L = \dfrac{0.523}{\ln^2(0.06Re_L)}$, Eq. (4.37), for all turbulent Re
Flow past a submerged sphere	$f = 24/Re_D$ for $Re_D < 0.1$, $f \approx 18.5/Re_D^{3/5}$ for $2 < Re_D < 5 \times 10^2$,	Eq. (4.42) Eq. (4.43)
	$f \approx 0.44$ for $5 \times 10^2 < Re_D < 2 \times 10^5$,	Eq. (4.44)

4.7 Flow Through Packed Beds

D'Arcy's Law

From experimental observation, D'Arcy determined that at sufficiently low rates of fluid flow through a packed bed, the flow rate is proportional to the pressure drop per unit length of bed:

$$\dot{V} = \frac{k_D A \; \Delta P}{L} \tag{4.45}$$

Equation (4.45) is *D'Arcy's law*, and with $\dot{V}$ in cm³/s, A in cm², and $\Delta P/L$ in dyn/cm³, the proportionality constant, k_D (which is called the permeability coefficient), has the units cm⁴/dyn·s. The value of k_D in any flow situation depends on the physical properties of the fluid and the packing characteristics of the bed. The specific permeability, P, defined as

$$P = k_D \left(\frac{cm^4}{dyn \cdot s}\right) \times \eta \left(\frac{dyn \cdot s}{cm^2}\right) \tag{4.46}$$

has the units of cm² and a d'Arcy (the unit of specific permeability) is 1×10^{-8} cm².

Tube Bundle Theory

Consider a packed bed made up of a number of twisted and tangled solid tubes, the diameters of which are very much less than the diameter of the bed. The porosity of the bed is continuous throughout its length and the tubes are evenly distributed across the cross section of the bed so that no channeling occurs when a fluid flows through the bed. A cross section of the bed is shown in Fig. 4.14. As shown, the cross-sectional area of the bed through which the fluid can flow (the cross-sectional area of the voids in the bed) is significantly less than the cross-sectional area of the bed, $A = \pi R^2$. The void fraction of the bed, ω, is defined as

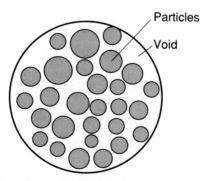

FIGURE 4.14 Illustration of the difference between the cross sectional area of a bed and the actual cross-sectional area available for fluid flow in the bed.

$$\omega = \frac{\text{cross-sectional area of the voids}}{A} \tag{4.47}$$

Measurement of the volume flow rate of fluid through the bed $\dot{V}$, allows the calculation of two average velocities: the superficial velocity, $\bar{v}_s$, calculated as

$$\bar{v}_s = \frac{\dot{V}}{A}$$

and the actual average velocity of the fluid through the voids, $\bar{v}$, calculated as

$$\bar{v} = \frac{\dot{V}}{\text{cross-sectional area of the voids}}$$

From Eq. (4.47) the two velocities are related as

$$\dot{V} = \bar{v}\omega A = \bar{v}_s A$$

or

$$\bar{v}_s = \bar{v}\omega \tag{4.48}$$

Tube bundle theory begins by adapting Eq. (2.22) for laminar flow through a vertical pipe,

$$\bar{v} = \left(\frac{\Delta P}{L} + \rho g\right)\frac{R^2}{8\eta}$$

to

$$\bar{v} = K_1 \frac{\Delta P}{L}\frac{R_h^2}{\eta}$$

or from Eq. (4.48),

$$\bar{v}_s = K_1 \frac{\Delta P}{L}\frac{R_h^2}{\eta}\omega \tag{4.49}$$

in which R_h is the hydraulic radius, defined as

$$R_h = \frac{\text{volume of the voids}}{\text{total surface area of the tubes wetted by the fluid}} \tag{4.50}$$

With a bed of volume V and porosity ω,

$$\text{volume of the voids} = \omega V$$

and the volume of the tubes (or particles) $= (1 - \omega)V$. Thus if S' is the total surface area of the tubes (or particles), S_0, the total surface area of the tubes (or particles) *per unit volume of particles*, is

$$S_0 = \frac{S'}{(1 - \omega)V} \tag{4.51}$$

and

$$R_h = \frac{\omega V}{S_0(1 - \omega)V} = \frac{\omega}{S_0(1 - \omega)} \tag{4.52}$$

Substitution of Eq. (4.52) into Eq. (4.49) gives

$$\bar{v}_s = K_1 \frac{\Delta P}{L\eta S_0^2} \frac{\omega^3}{(1 - \omega)^2} \tag{4.53}$$

and for laminar fluid flow through the packed bed, K_1 has been evaluated as $1/4.2$, so that

$$\bar{v}_s = \frac{1}{4.2} \left(\frac{\Delta P}{L\eta S_0^2} \right) \frac{\omega^3}{(1 - \omega)^2} \tag{4.54}$$

Equation (4.54), known as the *Blake–Kozeny equation*, provides a physical explanation for D'Arcy's law. Comparison with Eq. (4.45), written as

$$\bar{v}_s = \frac{\dot{V}}{A} = k_D \frac{\Delta P}{L}$$

shows that the permeability coefficient, k_D, is given by

$$k_D = \frac{1}{4.2} \frac{\omega^3}{\eta S_0^2(1 - \omega)^2}$$

and from comparison with Eq. (4.46), the specific permeability, P, is given by

$$P = k_D \times \eta = \frac{\omega^2}{4.2S_0^2(1 - \omega)^2}$$

The specific permeability is thus determined only by the porosity of the bed and the total surface area of the particles per unit volume of particles.

Again, transition from laminar to turbulent flow begins at a critical value of the Reynolds number, which is derived for application to packed beds as follows:

$$Re = \frac{\bar{v}D\rho}{\eta} = \frac{\bar{v}_s}{\omega} 2R_h \frac{\rho}{\eta} = \frac{\bar{v}_s}{\omega} 2 \frac{\omega}{S_0(1 - \omega)} \frac{\rho}{\eta}$$

$$= \frac{2\rho\bar{v}_s}{\eta(1 - \omega)S_0} \tag{4.55}$$

For packed beds, the factor of 2 in Eq. (4.55) is omitted and a new Reynolds number, Re_c, is defined as

$$Re_c = \frac{\rho\bar{v}_s}{\eta(1 - \omega)S_0} \tag{4.56}$$

Laminar flow through the packed bed occurs with $Re_c < 2$ and the transition from laminar to fully turbulent flow occurs in the range $2 < Re_c < 1000$. Turbulent flow requires the introduction of the friction factor, which is obtained from Eq. (4.11) as

$$f = \frac{1}{4}\frac{D}{L}\frac{\Delta P}{\frac{1}{2}\rho\bar{v}^2}$$

$$= \frac{1}{4}\frac{2R_h}{L}\frac{2}{\rho\bar{v}_s^2}\frac{\Delta P\omega^2}{\rho\bar{v}_s^2}$$

$$= \frac{\omega}{S_0(1-\omega)}\frac{\Delta P}{L}\frac{\omega^2}{\rho\bar{v}_s^2} \tag{4.57}$$

For fully turbulent flow the friction factor has the constant value of 0.292, and thus for $Re_c > 1000$,

$$\frac{\Delta P}{L} = \frac{0.292S_0(1-\omega)\rho\bar{v}_s^2}{\omega^3} \tag{4.58}$$

and Eq. (4.58) is known as the *Burke–Plummer equation*. The variation of $\bar{v}_s$ with $\Delta P/L$ over the entire range of Reynolds numbers is given by the sum of Eqs. (4.54) and (4.58) as

$$\frac{\Delta P}{L} = \frac{4.2\eta\bar{v}_sS_0^2(1-\omega)^2}{\omega^3} + \frac{0.292\rho\bar{v}_s^2(1-\omega)S_0}{\omega^3} \tag{4.59}$$

Equation (4.59) is simplified by dividing both sides by $\rho\bar{v}_s^2(1-\omega)S_0/\omega^3$, to get

$$\frac{\Delta P\omega^3}{L\rho\bar{v}_s^2(1-\omega)S_0} = \frac{4.2\eta S_0(1-\omega)}{\rho\bar{v}_s} + 0.292 \tag{4.60}$$

and noting, from Eq. (4.57), that the term on the left side of Eq. (4.60) is the friction factor, f, and from Eq. (4.56) that the first term on the right side of Eq. (4.60) is $4.2Re_c$,

$$f = \frac{4.2}{Re_c} + 0.292 \tag{4.61}$$

which shows again that the friction factor is a function only of the Reynolds number. Thus, for a required value of $\bar{v}_s$ through the packed bed, Re_c is obtained from Eq. (4.56), f is obtained from Eq. (4.61), and the required value of $\Delta P/L$ is obtained from Eq. (4.59). At very low Reynolds numbers the term $4.2/Re_c$, in Eq. (4.61) is larger than 0.292 and thus the Blake–Kozeny contribution to Eq. (4.59) predominates, and at very large Reynolds numbers, $4.2/Re_c < 0.292$, and thus the Burke–Plummer contribution predominates.

Spherical Particles and Ergun's Equation

For spherical particles of radius R and diameter d_p, the ratio of the surface area to the volume, S_0, is

$$S_0 = \frac{4\pi R^2}{\frac{4}{3}\pi R^3} = \frac{3}{R} = \frac{6}{d_p} \tag{4.62}$$

substitution of which into Eq. (4.59) gives

$$\frac{\Delta P}{L} = \frac{150\eta\bar{v}_s(1 - \omega)^2}{d_p^2\omega^3} + \frac{1.75\rho\bar{v}_s^2(1 - \omega)}{d_p\omega^3} \qquad (4.63)$$

Equation (4.63) is known as *Ergun's equation*. Substitution of Eq. (4.62) into Eqs. (4.56) and (4.57) gives, respectively,

$$Re_c = \frac{d_p\rho\bar{v}_s}{6\eta(1 - \omega)}$$

and

$$f = \frac{d_p\omega^3\Delta P}{6(1 - \omega)L\rho\bar{v}_s}$$

When working with packed beds of spherical particles, it is convenient to omit the factor 6 in the expressions for the Reynolds number and the friction factor by defining a new Reynolds number as

$$Re_E = 6Re_c = \frac{d_p\rho\bar{v}_s}{\eta(1 - \omega)} \qquad (4.64)$$

and a new friction factor as

$$f_E = 6f = \frac{d_p\Delta P\omega^3}{L\rho\bar{v}_s^2(1 - \omega)} \qquad (4.65)$$

Dividing Eq. (4.63) by $\rho\bar{v}_s^2(1 - \omega)/d_p\omega^3$ then gives

$$f_E = \frac{150}{Re_E} + 1.75 \qquad (4.66)$$

Equation (4.66) is shown in Fig. 4.15.

Consider a perfectly close-packed bed of spherical particles of diameter d_p. Recalling the face-centered cubic crystal structure, the atoms of diameter d are in contact with one another along the diagonal of a face of the unit cell. The length of this diagonal is thus $2d$ and the length of the side of a unit cell is $\sqrt{2}\, d$. The volume of the unit cell (the bed volume) is thus $2\sqrt{2}\, d^3$. The FCC unit cell contains four atoms (the particles in the bed) each of volume $\frac{4}{3}\pi(d/2)^3$, and thus the bed porosity, ω, is

$$\omega = \frac{2\sqrt{2}\, d^3 - \frac{16}{3}\pi(d/2)^3}{2\sqrt{2}\, d^3} = 0.26$$

which is the minimum possible value for packed beds of uniformly sized spheres. Less-well-packed beds have larger porosities.

Real packed beds are not made up of uniformly sized spheres; they contain a range of particles of different sizes which are not necessarily spherical. One of the more practical means of dealing with the variation in the size of the particles is to define the *specific area mean diameter*, $\bar{d}_p$, as

$$\bar{d}_p = \frac{1}{\Sigma(wt\ \%)_n/d_{p,n}}$$

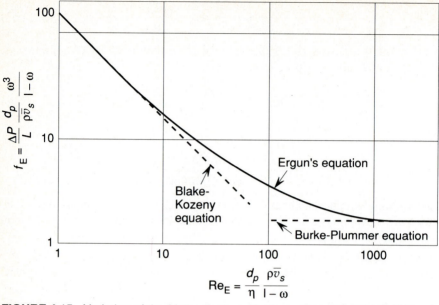

FIGURE 4.15 Variation of the friction factor with Reynolds number for fluid flow in a packed bed.

where $(\text{wt } \%)_n$ is the weight percentage of the population of particles that contains particles of diameter d_p. In practice, $(\text{wt } \%)_n$ is determined as the weight percentage of a sample of the bed that is retained between two sieves and the diameter is taken as being the average of the sieve sizes. Deviation from spherical shape is quantified by the shape factor λ, as follows. For a particle of volume V and surface area A, the diameter d_s of the sphere, which has the same volume, V, is

$$d_s = \left(\frac{6V}{\pi}\right)^{1/3}$$

The surface area of this equivalent-volume sphere, A_s, is thus

$$A_s = \pi d_s^2$$

and the shape factor of the particles of volume V and surface area A is defined as

$$\lambda = \frac{A}{A_s}$$

Thus for a cube of side a,

$$d_s = \left(\frac{6a^3}{\pi}\right)^{1/3}$$

$$A_s = \pi \left(\frac{6a^3}{\pi}\right)^{2/3}$$

and

$$\lambda = \frac{6a^2}{\pi(6a^3/\pi)^{2/3}} = 1.24$$

For nonuniformly sized particles,

$$S_0 = \frac{6\lambda}{d_p}$$

and Ergun's equation becomes

$$\frac{\Delta P}{L} = \frac{150\eta\bar{v}_s\lambda^2(1 - \omega)^2}{\bar{d}_p^2\omega^3} + \frac{1.75\rho\bar{v}_s^2\lambda(1 - \omega)}{\bar{d}_p\omega^3} \tag{4.67}$$

EXAMPLE 4.9

A packed bed of porosity $\omega = 0.4$ consists of uniformly sized spheres of diameter 0.05 m contained in a vertical cylindrical vessel of height 15 m and diameter 5 m. Air at 500 K, passing vertically downward through the bed, exits at a pressure of 1.2×10^5 Pa. Determine the relationship between the mass flow rate of the air, $\dot{M}$, and the pressure at the entrance to the bed, P_0. The data are

$$L = 15 \text{ m}$$

$$D = 5 \text{ m}$$

$$d_p = 0.05 \text{ m}$$

$$P_L = 1.2 \times 10^5 \text{ Pa}$$

$$\omega = 0.4$$

$$\eta \text{ (at 500 K)} = 2.68 \times 10^{-5} \text{ Pa·s}$$

$$\rho \text{ (at } T = 500 \text{ and } P = 1 \text{ atm)} = 0.706 \text{ kg/m}^3$$

Equation (4.63) gives the expression for the pressure drop ΔP over the length L of a packed bed in terms of the superficial velocity $\bar{v}_s$ and the density ρ. In the present example the density of the gas is a function of the pressure in the bed, and thus Eq. (7.19) should be written as a differential equation,

$$-\frac{dP}{dz} = \frac{150\eta\bar{v}_s(1 - \omega)^2}{d_p^2\omega^3} + \frac{1.75\rho\bar{v}_s^2(1 - \omega)}{d_p\omega^3}$$

$$= \frac{150 \times 2.68 \times 10^{-5} \times (1 - 0.4)^2}{(0.05)^2 \times (0.4)^3}\bar{v}_s + \frac{1.75 \times (1 - 0.4)}{0.05 \times (0.4)^3}\rho\bar{v}_s^2$$

$$= 9.05\bar{v}_s + 328.1\rho\bar{v}_s^2 \tag{i}$$

The mass flow rate is given by

$$\dot{M} = \bar{v}_s A\rho$$

and hence

$$\bar{v}_s\rho = \frac{\dot{M}}{A} = \frac{\dot{M}}{\pi \times (2.5)^2} = 0.0509\dot{M} \qquad \text{(ii)}$$

Assuming that air behaves as an ideal gas (i.e., at constant T),

$$P_1V_1 = P_2V_2$$

or

$$\frac{P_1}{\rho_1} = \frac{P_2}{\rho_2}$$

and given that $\rho = 0.706$ kg/m^3 at $P = 1$ atm $= 101{,}325$ Pa,

$$\frac{P}{\rho} = \frac{101{,}325}{0.706}$$

or

$$\rho = 6.968 \times 10^{-6}P \qquad \text{(iii)}$$

substitution of which into Eq. (ii) gives

$$\bar{v}_s = \frac{0.0509\dot{M}}{\rho} = \frac{7306\dot{M}}{P} \qquad \text{(iv)}$$

Substitution of Eqs. (ii) and (iv) into Eq. (i) gives

$$-\frac{dP}{dz} = \frac{9.05 \times 7306\dot{M}}{P} + \frac{328.1 \times 0.0509\dot{M} \times 7306\dot{M}}{P}$$

$$= \frac{66{,}119\dot{M} + 122{,}012\dot{M}^2}{P}$$

or

$$P\,dP = -(66{,}119\dot{M} + 122{,}012\dot{M}^2)\,dz$$

and integration gives

$$P^2 = -2(66{,}119\dot{M} + 122{,}012\dot{M}^2)z + \text{const}$$

At $z = 15$, $P = P_L = 1.2 \times 10^5$ Pa, and hence

$$\text{const} = (1.2 \times 10^5)^2 + 2(66{,}119\dot{M} + 122{,}012\dot{M}^2)15$$

and thus

$$P = [1.44 \times 10^{10} + 2(66{,}119\dot{M} + 122{,}012\dot{M}^2)(15 - z)]^{1/2} \qquad \text{(v)}$$

The variations in pressure through the bed for various values of $\dot{M}$ are shown in Fig. 4.16. The entry pressure P_0 at $z = 0$ is

$$P_0 = [1.44 \times 10^{10} + 30(66{,}119\dot{M} + 122{,}012\dot{M}^2)]^{1/2} \qquad \text{(vi)}$$

Alternatively, the problem could be approached using average values for ρ and $\bar{v}_s$ in Eq. (4.63). Taking the average pressure in the bed as $(P_0 + P_L)/2$, the

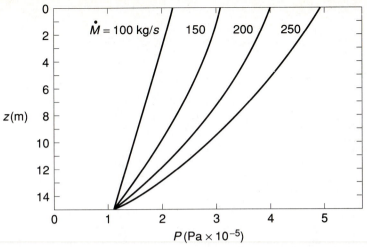

FIGURE 4.16 Variation of pressure with position in a packed bed for four mass flow rates.

average value of ρ is obtained from Eq. (iii) as

$$\rho = \frac{6.968 \times 10^{-6} \times (P_0 + P_L)}{2}$$

and the average value of $\bar{v}_s$ is obtained from Eq. (iv) as

$$\bar{v}_s = \frac{2 \times 7306\dot{M}}{P_0 + P_L}$$

Substitution into Eq. (4.63) then gives

$$\frac{P_0 - P_L}{15} = \frac{150 \times 2.68 \times 10^{-5} \times (1 - 0.4)^2 \times 2 \times 7306\dot{M}}{(0.05)^2 \times (0.4)^3 \times (P_0 + P_L)}$$

$$+ \frac{1.75 \times (1 - 0.4)}{0.05 \times (0.4)^3} \times 0.0509\dot{M} \times \frac{2 \times 7306\dot{M}}{P_0 + P_L}$$

$$= \frac{132,238\dot{M} + 244,024\dot{M}^2}{P_0 + P_L}$$

Therefore,

$$(P_0 - P_L)(P_0 + P_L) = P_0^2 - P_L^2 = 15(132,238\dot{M} + 244,024\dot{M}^2)$$

which is identical with Eq. (vi).

4.8 Fluidized Beds

If the flow velocity of fluid passing upward through the packed bed is increased (and hence the pressure drop across the bed is increased), a point is reached at which the pressure drop times the cross-sectional area of the bed equals the weight of the bed:

$$\frac{\Delta P}{L} = (\rho_s - \rho)g(1 - \omega) \qquad (4.68)$$

and at this point the packed bed has become a fluidized bed. The relationship between $\log \Delta P$ and $\log \bar{v}_s$ is shown in Fig. 4.17. AB is the variation of ΔP with $\bar{v}_s$ for flow through a packed bed. Between B and C the bed expands slightly and rearranges itself to minimize its resistance to flow, and at C it becomes fluidized. Further increase in $\bar{v}_s$ does not influence the pressure drop across the now-fluidized bed, but does cause an increase in the porosity of the bed, which means that the fluidized bed expands. Eventually, point D is reached, at which ω has increased to unity, and at this point $\bar{v}_s$ is the terminal free-fall velocity of each of the particles in the fluid. Further increase in $\bar{v}_s$ causes the particles to be elutriated (i.e., blown out of the container), and line DE is the variation of ΔP with $\bar{v}$ in an empty tube. Point C is the point of minimum fluidization and the porosity of the bed in this state, ω_{mf}, is, equivalently, both the maximum porosity of the fixed bed and the minimum porosity of the fluidized bed. At the point of minimum fluidization, combination of Eqs. (4.67) and (4.68) gives

$$\frac{\Delta P}{L} = \frac{150\eta\bar{v}_s\lambda^2(1 - \omega)^2}{d_p^2\omega^3} + \frac{1.75\rho\bar{v}_s^2\lambda(1 - \omega)}{d_p\omega^3} = (\rho_s - \rho)(1 - \omega)g \qquad (4.69)$$

and multiplication of Eq. (4.69) by $d_p^3\rho/(1 - \omega)\eta^2$ gives

$$\frac{150(1 - \omega)\lambda^2}{\omega^3}\left[\frac{\rho\bar{v}_s d_p}{\eta}\right] + \frac{1.75\lambda}{\omega^3}\left[\frac{\rho\bar{v}_s d_p}{\eta}\right]^2 = \frac{d_p^3(\rho_s - \rho)\rho g}{\eta^2} \qquad (4.70)$$

The term in brackets in Eq. (4.70) is the Reynolds number for fluid flow past a sphere of radius d_p,

$$\text{Re}_d = \frac{\rho\bar{v}_s d_p}{\eta}$$

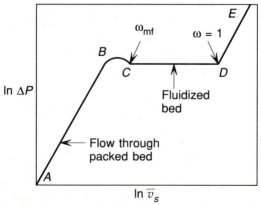

FIGURE 4.17 Variation of the pressure drop with superficial flow velocity for fluid flow in a packed bed and a fluidized bed.

and the term on the right-hand side of Eq. (4.70) is a dimensionless number, the value of which is determined by the size and density of the particles and by the physical properties of the fluid. This number is called the *Galileo number*, Ga:

$$Ga = \frac{d_p^3(\rho_s - \rho)\rho g}{\eta^2} \tag{4.71}$$

and thus Eq. (4.71) can be written as

$$\frac{150(1 - \omega)\lambda^2}{\omega^3} Re_d + \frac{1.75\lambda}{\omega^3} Re_d^2 = Ga \tag{4.72}$$

Wen and Yu (C. Y. Wen and Y. H. Yu, *Fluid Particle Technology*, Chem. Eng. Prog. Symp. Ser. 62 AIChE, New York, 1966) have determined that at the point of minimum fluidization

$$\frac{(1 - \omega_{mf})\lambda^2}{\omega_{mf}^3} \approx 11 \quad \text{and} \quad \frac{\lambda}{\omega_{mf}^3} \approx 14 \tag{4.73}$$

The approximate nature of Eq. (4.73) can be seen by considering spherical particles for which $\lambda = 1$. One expression gives

$$\omega = \left(\frac{1}{14}\right)^{1/3} = 0.41$$

and the other gives $\omega = 0.38$. Thus at minimum fluidization, ω_{mf} is approximately 0.4. Substitution of Eq. (4.73) into Eq. (4.72) gives

$$1650 Re_{d,mf} + 24.5 Re_{d,mf}^2 = Ga$$

or

$$Re_{d,mf} = (1134 + 0.0408\, Ga)^{1/2} - 33.7 \tag{4.74}$$

where $Re_{d,mf}$ is the Reynolds number at minimum fluidization. Wen and Yu found, empirically, that between points *C* and *D* in Fig. 4.17, the Reynolds number and the porosity of the bed are related as

$$\omega^{4.7} Ga = 18 Re_d + 2.70 Re_d^{1.687} \tag{4.75}$$

At point *D*, the flow velocity is the terminal velocity given by rearrangement of Eq. (4.41) as

$$v_t = \left[\frac{4 d_p(\rho_s - \rho)g}{3 f \rho}\right]^{1/2} \tag{4.76}$$

EXAMPLE 4.10

Consider the behavior of a bed of porosity $\omega = 0.35$ consisting of uniformly sized spherical particles of alumina of diameter 10^{-3} m. The fluid is air of density 1.177 kg/m^3 and viscosity 1.85×10^{-5} Pa·s, and the density of alumina is 3990 kg/m.^{3} From Eq. (4.71), the Galileo number is

$$Ga = \frac{(10^{-3})^3 \times (3990 - 1.177) \times 1.177 \times 9.81}{(1.85 \times 10^{-5})^2}$$

$$= 1.346 \times 10^5$$

and thus from Eq. (4.74) the Reynolds number at minimum fluidization is

$$Re_{d,mf} = (1134 + 0.0408 \times 1.346 \times 10^5)^{1/2} - 33.7$$

$$= 47.7$$

The superficial velocity of the fluid at minimum fluidization is obtained from the Reynolds number as

$$\bar{v}_{s,mf} = \frac{Re_d \eta}{\rho d_p}$$

$$= \frac{47.7 \times 1.85 \times 10^{-5}}{1.177 \times 10^{-3}}$$

$$= 0.75 \text{ m/s}$$

The bed porosity at minimum fluidization is obtained from Eq. (4.75) as

$$\omega^{4.7} \times 1.346 \times 10^5 = (18 \times 47.7) + (2.70 \times 47.7^{1.687})$$

(i.e., $\omega = 0.43$).

Consider now the variation of ω with $\bar{v}_s$ in the range of $\bar{v}_s$ over which the bed is fluidized. The upper limit of $\bar{v}_s$ is that at which elutriation occurs (i.e., is the terminal free-fall velocity of an alumina particle in air). This velocity is obtained from Eq. (4.76) as

$$v_t = \left[\frac{4 \times 10^{-3} \times (3990 - 1.177) \times 9.81}{3 \times 1.177f}\right]^{1/2}$$

$$= \frac{6.658}{f^{1/2}} \tag{i}$$

If, at this point, $Re_d < 0.1$, Re_d and f are related via Eq. (4.42) as

$$f = \frac{24}{Re_d} = \frac{24}{(10^{-3} \times v_t \times 1.177)/(1.85 \times 10^{-5})}$$

$$= \frac{0.377}{v_t} \tag{ii}$$

Eliminating f from Eqs. (i) and (ii) gives

$$f = \frac{(6.658)^2}{v_t^2} = \frac{0.377}{v_t}$$

or

$$v_t = 117.1 \text{ m/s}$$

and

$$Re_d = \frac{10^{-3} \times 117.1 \times 1.177}{1.85 \times 10^{-5}} = 7480$$

As the calculated Re_d is greater than 0.1, Eq. (4.42) is not applicable at the point of elutriation. If $2 < Re_d < 5 \times 10^2$, the friction factor and Re_d are related by Eq. (4.43) as

$$f \approx \frac{18.5}{Re_d^{3/5}} = \frac{18.5}{[(10^{-3} \times 1.177)/(1.85 \times 10^{-5})]^{3/5} \, v_t^{3/5}}$$

$$= \frac{1.531}{v_t^{3/5}} \tag{iii}$$

and elimination of f from Eqs. (i) and (iii) gives

$$f = \frac{(6.658)^2}{v_t^2} = \frac{1.531}{v^{3/5}}$$

or

$$v_t = 11.07 \text{ m/s}$$

and

$$Re_d = \frac{10^{-3} \times 11.07 \times 1.177}{1.85 \times 10^{-5}} = 704$$

which, being greater than 500, indicates that Eq. (4.43) is not valid. If $500 \le Re_d \le 2 \times 10^5$, the friction factor is obtained from Eq. (4.44) as $f = 0.44$, which, with Eq. (i), gives

$$v_t = \frac{6.658}{0.44^{1/2}} = 10.0 \text{ m/s}$$

and

$$Re_d = \frac{10^{-3} \times 10.0 \times 1.177}{1.85 \times 10^{-5}} = 639$$

which is within the applicable range. Thus, with $Re_d = 639$, the bed porosity at the point of elutriation is obtained from Eq. (4.75) as

$$\omega^{4.7} \times 1.346 \times 10^5 = (18 \times 639) + (2.70 \times 639^{1.687})$$

or $\omega = 1$. The variation of ω with $\bar{v}_s$ in the range $0.75 \text{ m/s} < \bar{v}_s < 10 \text{ m/s}$ is obtained from Eq. (4.75) and is as follows:

$\bar{v}_s$	Re_d	ω
0.75	47.7	0.43
2	127	0.60
4	255	0.75
6	382	0.87
8	509	0.96
10	639	1

4.9 Summary

A flowing fluid experiences viscous forces, which tend to stabilize the flow, and inertia forces, which tend to disrupt the flow. The inertia force is proportional to the square of the fluid velocity, and the viscous force is proportional to the velocity, and thus with increasing flow velocity in a fluid, a point is eventually reached at which the disruptive tendency of the inertia force is greater than the stabilizing tendency of the viscous force, and orderly laminar flow is replaced by chaotic turbulent flow. The transition from laminar to turbulent flow in any flow situation begins at a definite value of the Reynolds number. In turbulent flow the local flow velocities and the local static pressure fluctuate with time, and hence consideration of steady state in turbulent flow requires the consideration of time-averaged properties. The eddies occurring in turbulent flow cause the velocity profile in turbulent flow in cylindrical pipes to be much flatter than in laminar flow, and in contrast with laminar flow, the variations with position in the flow of shear stress and velocity cannot be derived from first principles.

Consideration of turbulent flow requires knowledge of the friction factor. The drag force exerted by a flowing fluid on the walls of its container has the units of area times kinetic energy per unit volume and the friction factor is the proportionality contant between these two terms. Alternatively, the friction factor is the ratio of the shear stress at the wall to the kinetic energy of the fluid per unit volume, and is a function only of the Reynolds number of the flow. Thus knowledge of the friction factor allows the determination of the friction force at any Reynolds number.

Any surface roughness on the containing wall increases the friction between the fluid and the wall, and hence for a given rate of pressure drop in the fluid, decreases the average flow velocity. The relative roughness of a pipe wall is quantified as the ratio of the average height of the protuberances on the wall to the diameter of the pipe, and with increasing relative roughness at any Reynolds number in turbulent flow a regime of fully turbulent flow is reached in which the friction factor is independent of the Reynolds number. In this regime the friction factor is a function only of the relative roughness.

In flow of a fluid over the surface of a submerged flat plate the Reynolds number increases with distance along the plate, and at a Reynolds number of 3×10^5 a transition from laminar flow in the momentum boundary layer to turbulent flow begins. With further increase in distance down the plate the laminar portion of the boundary layer shrinks to form a thin laminar layer in contact with the plate beneath the turbulent boundary layer and the transition to turbulence is complete at a Reynolds number of 3×10^6. The turbulent boundary layer is much thicker than the laminar layer and it exerts a much greater shear stress on the plate.

Three regimes occur in flow of fluid past a submerged sphere; a regime in which Stokes' law is obeyed, an intermediate regime, and a regime in which the friction factor is essentially independent of the Reynolds number. In flow past both a submerged sphere and a submerged cylinder, the phenomenon of boundary separation can occur. At the stagnation point of the fluid in contact with the sphere or the cylinder, the kinetic energy of the fluid appears as an increase in local pressure, and hence initially, the pressure in the boundary layer decreases in the flow direction. This causes an increase in the local flow velocity in the boundary layer to a maximum value at which point the pressure gradient in the layer has decreased to zero. Beyond this point the pressure gradient in the boundary is positive, which hinders the flow, and eventually a point is reached at which the momentum of the fluid near the surface is insufficient to overcome the positive pressure gradient, and hence the local velocity falls to zero. As reversal of the direction of the flow is prevented by the oncoming flow the boundary becomes detached from the surface and the void formed by the separated flow is filled with a turbulent wake.

Fluid flow through packed beds is considered in terms of tube bundle theory, which adopts an analogy of the Hagen–Poiseuille equation in which the pipe radius is replaced by a hydraulic radius and a measurable superficial flow velocity is defined in terms of the true linear velocity of the fluid through the voids in the packed bed. At a high enough flow velocity through a packed bed the pressure drop in the fluid across the bed times the cross-sectional area of the bed becomes equal to the weight of the bed, at which point the bed expands slightly to minimize its resistance to the fluid flow and becomes fluidized with a porosity of approximately 0.4. Further increase in the flow velocity causes the bed to expand, and when the expansion has increased the bed porosity to unity, the individual particles of the bed are elutriated. A fluidized bed thus exists between a minimum flow velocity determined by the minimum Reynolds number for fluidization of the bed, and a maximum flow velocity which is the terminal velocity attained by the individual particles during free fall through the fluid.

Problems

PROBLEM 4.1

Water at 300 K is flowing through a cylindrical pipe of diameter 0.05 m at an average flow velocity of 1.5 m/s. Calculate:

(a) The local flow velocity at the centerline
(b) The local flow velocity at a radial distance from the centerline of 0.005 m
(c) The radial position at which the local flow velocity equals the average value

PROBLEM 4.2

Water at 300 K is pumped at an average linear flow velocity of 2 m/s through a 30-m length of horizontal pipe of inside diameter 0.025 m and relative roughness 0.004.

(a) Calculate the pressure drop over the length of pipe.
(b) The rough pipe is replaced by a smooth-walled pipe of that diameter, which with the same pressure drop, gives the same average linear flow velocity. Calculate the required diameter of the smooth-walled pipe.

PROBLEM 4.3

Calculate the pressure drop required to give a flow rate of water of 40 kg/s through a 500-m length of 0.2-m-I.D. pipe of relative roughness 2.3×10^{-4}.

PROBLEM 4.4

The average flow velocity of water at 300 K in a smooth pipe of internal diameter 0.07 m is 1 m/s when the rate of pressure drop is 125 Pa/m. Calculate the average flow velocity if the relative roughness were 0.002.

PROBLEM 4.5

A lead shot of diameter 3 mm is fired upward into still air at 300 K over open water. Calculate:

- (a) The drag force that the shot experiences as it leaves the gun barrel at a velocity of 150 m/s
- (b) The terminal velocity that it attains when it falls through the still air
- (c) The terminal velocity that it attains when it falls through the still water

The density of lead is 11,340 kg/m^3.

PROBLEM 4.6

A balloon of mass 0.4 kg is inflated with helium gas to a pressure of 300 kPa, at which pressure the diameter of the balloon is 1.5 m. When tethered to the ground on a windy day the tether makes an angle of 13° to the vertical. If the wind is blowing horizontally, what is its velocity? The air temperature and pressure are, respectively, 300 K and 101.3 kPa, and the helium in the balloon behaves as an ideal gas. The drag force exerted by the wind on the tether can be ignored. The atomic weight of helium is 4 g/g mol.

PROBLEM 4.7

A small pufferfish (which can be approximated as a sphere of 2 cm diameter) can sustain a power output of 25 W. What is the maximum velocity that the fish can attain when swimming horizontally in quiescent water? Ignore the influence of gravity on the motion of the fish.

PROBLEM 4.8

Air at a temperature of 350 K and an average pressure of 101.3 kPa is flowing through a packed bed of spheres of diameter 0.01 m. The bed is 0.10 m in diameter and 0.2 m in height and has a porosity of 0.35. Calculate the pressure drop over the bed required to give a mass flow rate of air of 0.05 kg/s.

PROBLEM 4.9

Water flows, under the influence of gravity, through a packed bed of spheres of diameter 2 cm and porosity 0.35. The free surface of the water in the bed is at the same level as the upper surface of the bed, and the upper and lower surfaces of the bed are in contact with atmospheric air. Calculate the superficial velocity of the water flowing through the bed.

PROBLEM 4.10

A vertical cylindrical packed bed comprises a cylindrical central core of radius R_A packed with spherical particles of diameter 0.08 m surrounded concentrically with an annular region of outer radius R_B which is packed with spherical particles of diameter 0.02 m. The porosities of the central bed and the annular bed are, respectively, 0.4 and 0.25. Calculate the ratio R_B/R_A for which the volume flow rate of a fluid through the central core equals the volume flow rate through the annular space when the flow of fluid through the bed is turbulent.

PROBLEM 4.11

Water at 300 K flows over a flat plate of dimensions 1 m $\times$ 1 m in a direction parallel to one side. Derive expressions for the drag force exerted on one side of the plate as a function of v_∞, and plot this variation in the range 0.1 m/s $< v_\infty < 10$ m/s.

PROBLEM 4.12

Water at 300 K flows in a horizontal cast iron pipe (of absolute roughness 0.259 mm) 2 m in diameter. The rate of pressure drop in the pipe is 15 Pa/m. Calculate the volume flow rate of the water.

PROBLEM 4.13

In the looping pipe system shown in the figure the pressure drop between points A and B is 150 kPa. The lengths and diameters of the three pipes are

    (1) $L = 600$ m, $D = 0.3$ m
    (2) $L = 450$ m, $D = 0.25$ m
    (3) $L = 700$ m, $D = 0.35$ m

Calculate the average linear flow velocities in each of the three legs. The fluid is water at 300 K and the pipes are smooth.

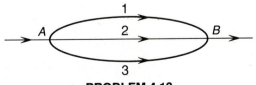

PROBLEM 4.13

PROBLEM 4.14

In the figure the volume flow rate of water in pipes A and B is 1 m³/s. The lengths and diameters of pipes 1 and 2 are

(1) $L = 2400$ m, $D = 0.6$ m
(2) $L = 3000$ m, $D = 0.4$ m

Calculate:

(a) The average linear flow velocities in pipes 1 and 2
(b) The pressure drop from A to B

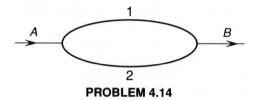

PROBLEM 4.14

PROBLEM 4.15

Calculate the drag force exerted on a fastball thrown at 95 mph through still air. A regulation baseball has a diameter of 7.3 cm.

PROBLEM 4.16

An attempt is to be made to reduce Fe_2O_3 with hydrogen gas by fluidizing a bed of the ore with hydrogen at 1200 K. If the particles of ore are spheres of diameter 10^{-3} m, calculate the range of flow rates that can be used to fluidize the bed. The density of Fe_2O_3 is 5300 kg/m^3.

5

Mechanical Energy Balance and Its Application to Fluid Flow

5.1 Introduction

A fluid flowing in a conduit has, by virtue of its mass and motion, a kinetic energy. If the fluid flow is not horizontal, the potential energy of the fluid varies in the flow direction and, in all cases of real fluids, the friction arising from the shear stress exerted on the fluid by the containing conduit causes a degradation of mechanical energy to thermal energy. All of the mechanical energy of a flowing fluid is obtained from the work done on the fluid by some external agency, and the rate at which this work is done is the power required to sustain the flow. The most common power supply is the pump, which effects an increase in the pressure of the fluid at some point in the pipeline. The question arises: What are the power requirements of a given fluid flowing in a given geometrical configuration at a given flow rate? The answer requires that an energy balance be made in the entire flow system.

5.2 Bernoulli's Equation

Figure 5.1 shows an incompressible, inviscid fluid flowing through a section of pipe between locations 1 and 2. The static pressures in the fluid at planes 1 and 2 are, respectively, P_1 and P_2, the fluid enters the section at 1 with the flow velocity v_1 and leaves at 2 with a flow velocity of v_2. As the fluid is inviscid (i.e., has no viscosity), the pipe wall does not exert a viscous drag on the fluid and hence there is no variation in the local flow velocity of the fluid across the cross-sectional area of the pipe.

The fluid entering the section at plane 1 is being pushed by the fluid behind it, and the work done on unit mass of the fluid entering is $P_1 V$, where $V = 1/\rho$ is the

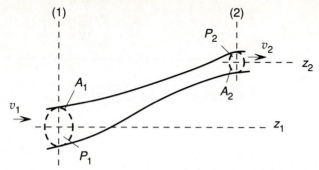

FIGURE 5.1 Incompressible fluid flowing through a section of pipe between locations 1 and 2.

volume per unit mass of fluid (i.e., the work done on the fluid $= P_1/\rho$). Similarly, the fluid leaving the section at plane 2 is pushing on the fluid in front of it and the work done by unit mass of the fluid leaving is $P_2 V$ (i.e., the work done by the fluid $= P_2/\rho$). The difference between the work done on the fluid and the work done by the fluid is called the *flow work done on the fluid*.

The kinetic energy of the fluid entering at plane 1 is

$$E_K = \tfrac{1}{2}mv_1^2$$

or the kinetic energy per unit mass of fluid entering is $\tfrac{1}{2}v_1^2$. Similarly, the kinetic energy per unit mass of fluid leaving at plane 2 is $\tfrac{1}{2}v_2^2$. The potential energy of the fluid entering at plane 1 is

$$E_p = mgz_1$$

or the potential energy per unit mass of fluid entering is gz_1. Similarly, the potential energy per unit mass of fluid leaving at plane 2 is gz_2.

Conservation of energy between locations 1 and 2 requires that

(flow work done on the fluid) = (change in kinetic energy)

+ (change in potential energy)

that is,

$$\frac{P_1}{\rho} - \frac{P_2}{\rho} = \frac{1}{2}(v_2^2 - v_1^2) + g(z_2 - z_1) \tag{5.1}$$

Equation (5.1) is *Bernoulli's equation*, and because the fluid is inviscid, the local flow velocities at planes 1 and 2 are independent of position in the cross-sectional area. The mass flow rate in the pipe, $\dot{M}$, is thus obtained as

$$\dot{M} = \rho A_1 v_1 = \rho A_2 v_2$$

However, if the fluid is not inviscid, the local flow velocity at any position along the pipe is a function of radial position, and this must be taken into account in calculating the mass flow rate and the rates at which kinetic energy enters and leaves the section of pipe. When the local flow velocity v is a function of position in the cross-sectional area of the flow, the mass flow rate $\dot{M}$ is obtained as

$$\dot{M} = \int_0^A \rho v \, dA \tag{5.2}$$

and the rate at which kinetic energy enters the section of pipe at location 1 is

$$\frac{\text{kinetic energy}}{\text{mass}} \times \frac{\text{mass}}{\text{unit time}} = \frac{1}{2} v_1^2 \times \dot{M}$$

which from Eq. (5.2) gives

$$\text{kinetic energy entering in unit time} = \tfrac{1}{2} \int_0^A \rho v_1^3 \, dA$$

or for an incompressible fluid flowing in a pipe of circular cross section,

$$\text{kinetic energy entering in unit time} = \tfrac{1}{2} \cdot 2\pi\rho \int_0^R v_1^3 r \, dr \tag{5.3}$$

For horizontal laminar flow in a circular tube of radius R, Eq. (2.20) gives

$$v_x = \frac{\Delta P}{L} \frac{R^2 - r^2}{4\eta}$$

which on substitution into Eq. (5.3) gives

$$\begin{aligned}
\text{kinetic energy entering in unit time} &= \pi\rho \left(\frac{\Delta P}{L}\right)^3 \left(\frac{1}{4\eta}\right)^3 \times \\
&\quad \cdot \int_0^R (R^6 - 3R^4 r^2 + 3R^2 r^4 - r^6) r \, dr \\
&= \pi\rho \left(\frac{\Delta P}{L}\right)^3 \left(\frac{1}{4\eta}\right)^3 \frac{R^8}{8} \\
&= \pi\rho \left(\frac{\Delta P}{L}\right)^3 \left(\frac{R^2}{8\eta}\right)^3 R^2 \tag{5.4}
\end{aligned}$$

From Eq. (2.22) the average flow velocity $\bar{v}_x$ is

$$\bar{v}_x = \frac{\Delta P}{L} \frac{R^2}{8\eta}$$

and hence Eq. (5.4) gives

$$\begin{aligned}
\text{kinetic energy entering in unit time} &= \pi\rho \bar{v}_1^3 R^2 \\
&= \rho A \bar{v}_1^3 \\
&= (\rho A \bar{v}_1)\bar{v}_1^2
\end{aligned}$$

and as the mass flow rate $\dot{M}$ is

$$\dot{M} = \int_0^A \rho v \, dA = \rho A \bar{v}$$

$$\text{kinetic energy entering in unit time} = \dot{M}\bar{v}_1^2$$

or

$$\text{kinetic energy entering per unit mass} = \bar{v}_1^2 \qquad (5.5)$$

Similarly, the kinetic energy leaving per unit mass at plane 2 is $\bar{v}_2^2$.

If the fluid flow had been fully turbulent, the kinetic energy per unit mass would have been obtained as approximately $\frac{1}{2}\bar{v}^2$. Both laminar and turbulent flow conditions are accommodated by the equation

$$\text{kinetic energy per unit mass} = \frac{\bar{v}^2}{2\beta}$$

where $\beta = 0.5$ for laminar flow and β can be taken as being unity for turbulent flow. Substitution into Eq. (5.1) thus gives

$$\frac{P_1}{\rho} - \frac{P_2}{\rho} = \frac{\bar{v}_2^2}{2\beta_2} - \frac{\bar{v}_1^2}{2\beta_1} + g(z_2 - z_1) \qquad (5.6)$$

in which the units are

$$\frac{m^2}{s^2} = \frac{kg}{kg} \cdot \frac{m^2}{s^2} = \left(kg \cdot \frac{m}{s^2} \cdot m \right) \frac{1}{kg}$$

$$= \frac{joules}{kg}$$

Equation (5.6) must be modified to account for the dissipation of energy caused by the viscous drag exerted on the flowing fluid by the tube wall. If between planes 1 and 2 this quantity, which is termed the *friction loss*, is E_f, the energy balance becomes

flow work done on the fluid = (increase in the kinetic energy)
+ (increase in the potential energy)
+ (friction loss)

Furthermore, if between locations 1 and 2, heat Q is added to unit mass of the fluid and work w is done on unit mass of the fluid, the energy balance becomes

$$\text{flow work} + Q + w = \Delta E_k + \Delta E_p + E_f$$

or

$$\left(\frac{P_2}{\rho} - \frac{P_1}{\rho} \right) + \left(\frac{\bar{v}_2^2}{2\beta_2} - \frac{\bar{v}_1^2}{2\beta_1} \right) + g(z_2 - z_1) - Q - w + E_f = 0 \qquad (5.7)$$

Equation (5.7) is called the *modified Bernoulli equation*.

5.3 Friction Loss, E_f

Consider a short length of horizontal pipe ($z_2 - z_1 = 0$) in which the flow velocity is constant ($\bar{v}_2 = \bar{v}_1$) and Q and w are zero. For this situation Eq. (5.7) gives

$$\frac{P_2 - P_1}{\rho} = -E_f \tag{5.8}$$

which illustrates that the pressure drop in the flowing fluid is caused by the viscous drag exerted on the fluid by the pipe wall. In Eq. (5.8), $P_2 < P_1$ and hence $E_f > 0$. The decrease in the normal force exerted on the fluid over the length of pipe, L, is equal to the shear force exerted on the fluid at the pipe wall:

$$(P_1 - P_2)\pi R^2 = \tau_0 2\pi RL$$

and hence Eq. (5.8) becomes

$$E_f = \frac{2\tau_0 L}{R\rho} \tag{5.9}$$

In Eq. (4.10) the friction factor is defined in terms of τ_0 as

$$f = \frac{\tau_0}{\frac{1}{2}\rho \bar{v}^2}$$

substitution of which into Eq. (5.9) gives

$$E_f = 2f \frac{L}{D} \bar{v}^2 \tag{5.10}$$

which is the general expression for calculating the friction loss. The dependence of E_f on the various flow parameters can be demonstrated by considering laminar flow, for which, from Eq. (4.15),

$$f = \frac{16}{\text{Re}} = \frac{16\eta}{D\bar{v}\rho}$$

Substitution into Eq. (5.10) gives

$$E_f = \frac{32\eta \bar{v} L}{D^2 \rho}$$

which shows that the friction loss per unit mass of fluid per unit length of flow is proportional to the viscosity and the velocity, and is inversely proportional to the cross-sectional area of flow and the density.

Equation (5.8) shows that the friction loss is equal to the pressure drop in the flow direction, $(P_1 - P_2)$, divided by the density of the fluid. Thus, for vertical flow in the upward direction over the distance $z_2 - z_1 = h$, Eq. (5.7) gives

$$(P_2 - P_1)_{\text{for upward vertical flow over the distance } h} = -\rho gh + (P_2 - P_1)_{\text{for horizontal flow over the distance } h}$$

and for vertical flow in the downward direction over the distance $z_2 - z_1 = -h$,

$$(P_2 - P_1)_{\text{for downward vertical flow over the distance } h} = \rho gh + (P_2 - P_1)_{\text{for horizontal flow over the distance } h}$$

5.4 Influence of Bends, Fittings, and Changes in the Pipe Radius

The occurrence of bends, valves, and abrupt change of radii in a pipe system increases the extent to which the energy of the fluid is dissipated and hence increases the magnitude of the friction loss. The increase in the friction loss caused by the presence of a bend in a pipe is calculated in terms of the increase in the length of a straight pipe which would be required to produce the same increase in friction loss. Thus any bend or fitting has an *equivalent length*, L_e, of straight pipe; for example, the presence of a 45° elbow in a cylindrical pipe is equivalent to increasing the length of straight pipe by 15 pipe diameters, and thus for a 45° elbow,

$$\frac{L_e}{D} = 15$$

The values of L_e/D for various fittings are listed in Table 5.1, and the friction loss for a pipe system containing fittings is calculated as

$$E_f = 2f \left[\left(\frac{L}{D} \right)_{pipe} + \sum \left(\frac{L_e}{D} \right)_{fittings} \right] \bar{v}^2 \qquad (5.11)$$

Sudden changes in the radius of pipe through which the fluid is flowing, such as sudden expansions or sudden contractions, are dealt with in terms of the *friction loss*

TABLE 5.1 The Values of L_e/D for Various Fittings.

Fitting	L_e/D
45° elbow	15
90° elbow, standard radius	31
90° elbow, medium radius	26
90° elbow, long sweep	20
90° square elbow	65
180° close return bend	75
Swing check-valve, open	77
"T" (as "L" entering run)	65
"T" (as "L" entering branch)	90
Gate valve, open	7
Gate valve, ¼ closed	40
Gate valve, ½ closed	190
Gate valve, ¾ closed	840
Globe valve, open	340
Angle valve, open	170

Source: W. M. Rohsenow and H. Y. Choi, "Heat, Mass and Momentum Transfer," Prentice-Hall, Englewood Cliffs, New Jersey, 1961, p. 64.

factor, e_f, and the consequent friction loss is calculated as

$$E_f = \tfrac{1}{2} e_f \bar{v}^2 \tag{5.12}$$

For a sudden contraction

$$e_f = 0.5 \left(1 - \frac{A_s}{A_L} \right) \tag{5.13}$$

and for a sudden expansion

$$e_f = \left(1 - \frac{A_s}{A_L} \right)^2 \tag{5.14}$$

where A_s is the cross-sectional area of the smaller diameter pipe and A_L is the cross-sectional area of the larger-diameter pipe. Equations (6.13) and (6.14) apply to turbulent flow, and the velocity used in Eq. (5.12) is the flow velocity in the smaller-diameter pipe.

EXAMPLE 5.1 ───

Calculate the power required to pump water at a volume flow rate of 6×10^{-3} m^3/s through a smooth pipe of diameter 0.1 m from the lower to the higher reservoir as shown in Fig. 5.2. The density and viscosity of water are, respectively, 997 kg/m^3 and 8.57×10^{-4} Pa·s.

From Eq. (5.7),

$$\frac{P^2}{\rho} - \frac{P_1}{\rho} + \frac{\bar{v}_2^2}{2\beta_2} - \frac{\bar{v}_1^2}{2\beta_1} + g(z_2 - z_1) - w + E_f = 0$$

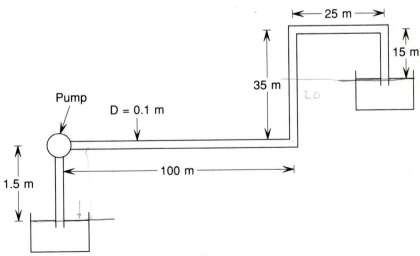

FIGURE 5.2   Flow arrangement considered in Example 5.1.

Levels 1 and 2 are taken as being the surfaces of the water in the lower and higher reservoirs, respectively. As both surfaces are in contact with the atmosphere, $P_1 = P_2$, and as both surfaces are virtually stationary, $\bar{v}_1 \sim \bar{v}_2 \sim 0$. Equation (5.7) thus reduces to

$$w = g(z_2 - z_1) + E_f \qquad \text{(i)}$$

The increase in the potential energy of the water per unit mass is

$$g(z_2 - z_1) = 9.81(1.5 + 35 - 15) = 211 \ \text{m}^2/\text{s}^2$$

and the friction loss in the pipe is obtained as

$$E_f = 2f \left[\left(\frac{L}{D}\right)_{\text{pipe}} + 3 \left(\frac{L_e}{D}\right)_{\text{elbow}} \right] \bar{v}^2$$

Evaluation of f requires evaluation of Re, which, in turn, requires knowledge of the average flow velocity in the pipe, $\bar{v}$.

$$\bar{v} = \frac{\dot{V}}{A} = \frac{6 \times 10^{-3}}{(\pi/4) \times D^2} = \frac{6 \times 10^{-3} \times 4}{\pi \times 0.1^2} = 0.764 \ \text{m/s}$$

Therefore,

$$\text{Re} = \frac{\bar{v} D \rho}{\eta} = \frac{0.764 \times 0.1 \times 997}{8.57 \times 10^{-4}} = 8.89 \times 10^4$$

and thus the flow is turbulent. Therefore,

$$f^{-1/2} = -3.6 \log \frac{6.9}{\text{Re}} \qquad \text{or} \qquad f = 4.57 \times 10^{-3}$$

Therefore, the friction loss

$$E_f = 2 \times 4.57 \times 10^{-3} \left[\frac{1.5 + 100 + 35 + 25 + 15}{0.1} + (3 \times 31) \right]$$
$$\times 0.764^2$$
$$= 9.91 \ \text{m}^2/\text{s}^2$$

At both the entrance and the exit $A_L \gg A_s$, and thus at the entrance $e_f = 0.5$ and the exit $e_f = 1$. Therefore,

$$\text{friction loss due to entry and exit} = \tfrac{1}{2} e_f \bar{v}^2$$
$$= \tfrac{1}{2}(1 + 0.5) \times 0.764^2$$
$$= 0.44 \ \text{m}^2/\text{s}^2$$

Substitution into Eq. (i) gives

$$w = 211 + 9.91 + 0.44$$
$$= 221 \ \text{m}^2/\text{s}^2$$

which, being a positive quantity, indicates that work is done on the fluid. The units

$$\frac{m^2}{s^2} = \frac{kg}{kg} \cdot \frac{m^2}{s^2} = \left(kg \frac{m}{s^2}\right) \cdot m \cdot \frac{1}{kg} = \frac{J}{kg}$$

The mass flow rate is $\dot{M} = \dot{V}\rho = 6 \times 10^{-3} \times 997 = 5.98$ kg/s, and thus the power required of the pump, W, is

$$W = w\dot{M} = 221 \frac{J}{kg} \times 5.98 \frac{kg}{s}$$
$$= 1322 \text{ W}$$

Of this power, 211/221, or 95.5%, is required to increase the potential energy of the water by the required amount at the required rate, and the remaining 4.5% is dissipated as friction loss.

The increase in pressure across the pump can also be calculated from Eq. (5.7). If plane 2 is located at the entrance to the pump and plane 1 is located at the surface of water in the lower reservoir, the application of Eq. (5.7) gives

$$\frac{P_2}{\rho} - \frac{P_1}{\rho} + \frac{\bar{v}_2^2}{2\beta_2} - \frac{\bar{v}_1^2}{2\beta_1} + g(z_2 - z_1) + E_f = 0$$

Within the pipe $\bar{v} = 0.764$ m/s $= \bar{v}_2$, $f = 4.57 \times 10^{-3}$, $P_1 =$ atmospheric pressure $= 101,325$ N/m^2, and $\bar{v}_1 \sim 0$. Thus

$$\frac{P_2}{997} - \frac{101,325}{997} + \frac{0.764^2}{2} + (9.81 \times 1.5)$$
$$+ \left(2 \times 4.57 \times 10^{-3} \times \frac{1.5}{0.1} \times 0.764^2\right) = 0$$

which gives

$$P_2 = 86,250 \text{ N/m}^2$$

Note that if the fluid were stationary, $E_f = 0$ and Eq. (5.7) collapses to the barometric formula

$$P_2 - P_1 = -\rho g(z_2 - z_1)$$
$$= -997 \times 9.81 \times 1.5$$
$$= -14,670 \text{ N/m}^2$$

in which case

$$P_2 = 101,325 - 14,670$$
$$= 86,660 \text{ N/m}^2$$

The pressure at the outlet side of the pump is obtained by application of Eq. (5.7) to the 175-m length of pipe on the delivery side of the pump. In this case plane 2 is located at the surface of the water in the higher reservoir, plane 1 is located at the outlet of the pump, and $P_2 = 101,325$ N/m^2. Thus

$$g(z_2 - z_1) = 9.81 \times (35 - 15) = 196.2 \text{ m}^2/\text{s}^2$$

and

$$E_f = 2 \times 4.57 \times 10^{-3}$$

$$\times \left[\frac{100 + 35 + 25 + 15}{0.1} + (3 \times 31) \right] \times 0.764^2$$

$$= 9.8 \text{ m}^2/\text{s}^2$$

Thus, with $\bar{v}_1 = 0.764$ m/s and $\bar{v}_2 \sim 0$,

$$\frac{P_2}{\rho} - \frac{P_1}{\rho} = -\frac{0.764^2}{2} - 9.8 - 196.2 = -206 \text{ m}^2/\text{s}^2$$

and

$$P_1 = (997 \times 206) + 101,325$$
$$= 306,700 \text{ N/m}^2$$

Thus the increase in pressure across the pump is $306,700 - 86,250 = 220,500$ N/m^2.

The application of Eq. (5.7) to the pump itself, with plane 2 at the outlet and plane 1 at the inlet, gives

$$\frac{P_2 - P_1}{\rho} = w$$

or

$$w = \frac{220,500}{997} = 221 \text{ m}^2/\text{s}^2$$

as was calculated in the first part of this example.

The rate at which the static pressure in the water decreases with increase in z places a limit on the vertical distance between the entrance to the pump and the level of the surface in the lower reservoir. If the static pressure falls to the value of the saturated vapor of water, the water boils (i.e., is converted to water vapor) and the bulk flow of liquid water ceases. At 300 K the saturated vapor pressure of water is 3570 Pa, and thus for a linear flow rate of 0.764 m/s, the height L at which the static pressure has decreased from 101,325 Pa to 3570 Pa is obtained from Eq. (5.7):

$$\frac{3570}{997} - \frac{101,325}{997} + \frac{0.764^2}{2} + (9.81 \times L)$$

$$+ \left(2 \times 4.57 \times 10^{-3} \times \frac{L}{0.1} \times 0.764^2 \right) = 0$$

as $L = 9.92$ m. A doubling of the linear flow rate to 1.53 m/s decreases the friction factor to 3.97×10^{-3} and decreases L to 9.69 m. Pressure-induced formation of a vapor phase is known as *cavitation*.

Just as work must be done on a fluid (by means of a pump) to increase its potential energy, useful work can be obtained from a fluid (from a turbine) that is flowing downward under the influence of gravity. In such a case w is a negative quantity.

EXAMPLE 5.2

Water is fed from a reservoir to a turbine through a vertical concrete pipe of diameter 0.5 m and absolute roughness 5×10^{-4} m at a mass flow rate of 1000 kg/s. The pressure of the water at a point 50 m above the entrance to the turbine is 6×10^5 Pa, and the pressure at the exit from the turbine is 1×10^5 Pa. Calculate the power generated by the turbine. The density and viscosity of the water are, respectively, 997 kg/m^3 and 8.57×10^{-4} Pa·s.

With $\dot{M} = 1000$ kg/s and $D = 0.5$ m, the linear flow velocity is

$$\bar{v} = \frac{\dot{M}}{\pi R^2 \rho} = \frac{1000}{\pi \times 0.25^2 \times 997} = 5.11 \text{ m/s}$$

and thus the Reynolds number is

$$\text{Re} = \frac{\bar{v} D \rho}{\eta} = \frac{5.11 \times 0.5 \times 1000}{8.57 \times 10^{-4}} = 2.98 \times 10^6$$

which indicates that the flow is turbulent. The friction factor is obtained from Eq. (4.18) as

$$f^{-1/2} = -3.6 \log_{10}\left[\left(\frac{\epsilon}{3.7D}\right)^{1.11} + \frac{6.9}{\text{Re}}\right]$$

$$= -3.6 \log_{10}\left[\left(\frac{0.0005}{3.7 \times 0.5}\right)^{1.11} + \frac{6.9}{2.98 \times 10^6}\right]$$

which gives $f = 0.00494$.

Locating plane 1 50 m above the turbine and plane 2 at the exit to the turbine gives

$$E_f = 2f \frac{L}{D} \bar{v}^2$$

$$= 2 \times 0.00494 \times \frac{50}{0.5} \times 5.11^2$$

$$= 25.81 \text{ m}^2/\text{s}^2$$

As $v_1 = v_2$, Eq. (5.7) is

$$\frac{P_2 - P_1}{\rho} + g(z_2 - z_1) - w + E_f = 0$$

$$\frac{100,000 - 600,000}{997} + 9.81(-50) + 25.81 = w$$

or

$$w = -966 \text{ J/kg}$$

The negative sign indicates that the water is doing work. The power obtained is thus

$$\text{power} = w\dot{M} = 966 \times 1000$$
$$= 9.66 \times 10^5 \text{ W}$$

Alternatively, locating plane 3 at the inlet to the turbine gives the pressure at the inlet as

$$P_3 = P_1 - \rho[g(z_2 - z_1) + E_f]$$
$$= 600,000 - 997[9.81(-50) + 25.81]$$
$$= 1.063 \times 10^6 \text{ Pa}$$

and

$$\frac{P_2 - P_3}{\rho} = w$$

gives

$$w = 997 \left(\frac{10^5 - 1.063 \times 10^6}{997} \right)$$
$$= -966 \text{ J/kg}$$

The friction loss in the turbine is expressed in terms of the efficiency with which the energy extracted from the water is converted to shaft energy. The efficiency of the turbine, η, is thus defined as

$$\eta = \frac{\text{shaft energy delivered by turbine}}{\text{energy extracted from water}}$$

Thus if the turbine has an efficiency of 90%, the shaft energy delivered would be $0.9 \times 9.66 \times 10^5 = 9.70 \times 10^5$ W.

EXAMPLE 5.3

Consider the gravity-induced flow of a fluid downward through the inclined circular pipe shown in Fig. 5.3(a). Both the entrance and exit are open to the atmosphere and $v_1 = v_2$. As no work is done by the fluid, Eq. (5.7) becomes

$$g(z_2 - z_1) + E_f = 0$$

or from Eq. (5.10),

$$g(z_1 - z_2) = 2f \frac{L}{D} \bar{v}^2 \tag{i}$$

that is, the steady state is that at which the rate of decrease of the potential energy of the fluid per unit mass equals the rate of dissipation of the energy per unit mass by friction loss, and the average velocity, $\bar{v}$, is determined by this energy balance. Alternatively, Eq. (i) can be obtained from the force balance:

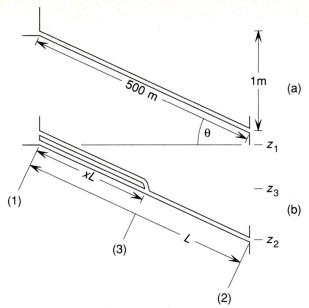

FIGURE 5.3 (a) Gravity-induced fluid flow in an inclined circular pipe considered in Example 5.3; (b) doubling of the pipe shown in (a).

gravitational force acting on fluid in pipe in flow direction
$$= \text{frictional force exerted on fluid by wall}$$

In Fig. 5.3(a) the component of gravity in the flow direction is

$$g \sin \theta = g \frac{z_2 - z_1}{L}$$

and hence the force acting on the fluid in the flow direction is

$$\rho \left(\frac{\pi D^2}{4} L \right) g \frac{z_2 - z_1}{L}$$

The frictional force exerted on the fluid is

$$\tau_0 \pi D L$$

and thus

$$\frac{\rho \pi D^2 g (z_2 - z_1)}{4L} = \tau_0 \pi D L$$

or from Eq. (4.10),

$$\frac{\rho \pi D^2 g (z_2 - z_1)}{4L} = \frac{1}{2} f \rho \bar{v}^2 \pi D L$$

which, on rearrangement, yields Eq. (i). Consider water of density 997 kg/m^3

and viscosity 8.57×10^{-4} Pa·s flowing through a smooth pipe of diameter 0.05 m and length 500 m, the entrance to which is 1 m higher than the exit. Equation (i) becomes

$$f\bar{v}^2 = \frac{g(z_1 - z_2)D}{2L} = \frac{9.81 \times 1 \times 0.05}{2 \times 500}$$
$$= 4.905 \times 10^{-4} \ \mathrm{m^2/s^2}$$

or

$$\bar{v} = \left(\frac{4.905 \times 10^{-4}}{f}\right)^{1/2} = \frac{0.0221}{f^{1/2}} \ \mathrm{m/s}$$

Thus

$$\mathrm{Re} = \frac{\bar{v}D\rho}{\eta} = \frac{0.0221 \times 0.05 \times 997}{8.57 \times 10^{-4} \times f^{1/2}} = \frac{1288}{f^{1/2}}$$

and from Eq. (4.18), assuming turbulent flow,

$$f^{-1/2} = -3.6 \log_{10} \frac{6.9}{\mathrm{Re}} = -3.6 \log_{10} \frac{6.9 f^{1/2}}{1288}$$

rearrangement of which gives

$$f^{-1/2} = 8.176 - 0.781 \ln f \qquad (ii)$$

Equation (ii) has the solution $f = 6.87 \times 10^{-3}$, and hence $\mathrm{Re} = 1.55 \times 10^4$ and $\bar{v} = 0.266$ m/s. The volume flow rate in the pipe is thus

$$\dot{V} = \bar{v}A = 0.266 \times \pi \times (0.025)^2$$
$$= 5.22 \times 10^{-4} \ \mathrm{m^3/s}$$

Equation (i) shows that for given values of $z_1 - z_2$ and L, the flow velocity is determined by the diameter of the pipe. Calculate the diameter of smooth-walled pipe required to give a volume flow rate of 10^{-3} m^3/s.

In this case the linear flow velocity is

$$\bar{v} = \frac{4\dot{V}}{\pi D^2}$$

substitution of which into Eq. (i) gives

$$g(z_1 - z_2) = \frac{32 f L \dot{V}^2}{\pi^2 D^5}$$

or

$$D^5 = \frac{32 L \dot{V}^2 f}{\pi^2 g(z_1 - z_2)}$$
$$= \frac{32 \times 500 \times (10^{-3})^2}{\pi^2 \times 9.81 \times 1} f$$
$$= 1.653 \times 10^{-4} f \qquad (iii)$$

The Reynolds number is

$$\text{Re} = \frac{\bar{v}D\rho}{\eta} = \frac{4\dot{V}\rho}{\pi D\eta} = \frac{4 \times 10^{-3} \times 997}{\pi \times 8.57 \times 10^{-4}D}$$

$$= \frac{1481}{D} \tag{iv}$$

substitution of which into Eq. (4.18) gives

$$f^{-1/2} = -3.6 \log \frac{6.9}{\text{Re}} = -3.6 \log (4.658 \times 10^{-3}D) \tag{v}$$

Combination of Eqs. (iii) and (v) to eliminate f gives

$$f^{-1/2} = \left(\frac{D^5}{1.653 \times 10^{-4}}\right)^{-1/2} = \frac{0.0129}{D^{2.5}}$$

$$= -3.6 \log (4.658 \times 10^{-3}D)$$

which has the solution $D = 0.0635$ m. From both Eqs. (iii) and (iv), $f = 6.25 \times 10^{-3}$, and from Eq. (iv), Re $= 2.33 \times 10^4$. Thus $\bar{v} = 0.316$ m/s.

If the pipe was not smooth-walled, the increased friction loss requires that the pipe diameter be increased to give the same volume flow rate. Calculate the diameter of a pipe of absolute roughness of 0.01 cm required to give a volume flow rate of 10^{-3} m^3/s. In this case Eq. (4.18) is written as

$$f^{-1/2} = -3.6 \log \left[\left(\frac{0.0001}{3.7D}\right)^{1.11} + 4.658 \times 10^{-3}D\right] \tag{vi}$$

and combination of Eqs. (iii) and (vi) to eliminate f gives

$$\frac{0.0129}{D^{2.5}} = -3.6 \log \left[\left(\frac{0.0001}{3.7D}\right)^{1.11} + 4.658 \times 10^{-3}D\right]$$

which has the solution $D = 0.0651$ m. In this case $f = 7.07 \times 10^{-3}$, Re $= 2.27 \times 10^4$, and $\bar{v} = 0.3$ m/s.

If, after installation of the pipe, it is required that the volume flow rate be increased, the pipe can be "doubled" for a portion of its length, as shown in Fig. 5.3(b). Obviously, if the entire length L is doubled with pipe of the same diameter as the original, the volume flow rate is increased by 100%.

Consider that it is required that the flow rate in the original smooth-walled 0.05-m-diameter pipe be increased by 25% from 5.22×10^{-4} m^3/s to $1.25 \times 5.22 \times 10^{-4} = 6.53 \times 10^{-4}$ m^3/s, and calculate the fraction, x, of the 500 m of pipe that must be doubled with 0.05-m-diameter pipe. Application of Eq. (5.7) to one of the two pipes between 1 and 3 in Fig. 5.3(b) gives

$$g(z_1 - z_3) = 2f \frac{xL}{D} \bar{v}^2 \tag{vi}$$

The volume flow rate in each of the pipes between 1 and 3 is half of the required volume flow rate of 6.53×10^{-4} m^3/s in the single pipe between 3 and 2, and hence in each of the pipes between 1 and 3,

$$\bar{v} = \frac{6.53 \times 10^{-4}}{2 \times \pi \times (0.025)^2} = 0.166 \text{ m/s}$$

Therefore,

$$Re = \frac{0.166 \times 0.05 \times 997}{8.57 \times 10^{-4}} = 9656$$

and

$$f^{-1/2} = -3.6 \log \frac{6.9}{9656} \quad \text{or} \quad f = 7.80 \times 10^{-3}$$

Therefore, from Eq. (vi),

$$z_1 - z_3 = \frac{2f\bar{v}^2 L}{gD} x = \frac{2 \times 7.80 \times 10^{-3} \times 0.166^2 \times 500}{9.81 \times 0.05} x$$

$$= 0.440x$$

Between 3 and 2,

$$g(z_3 - z_2) = 2f \frac{(1 - x)L}{D} \bar{v}^2$$

and in this section of pipe, the velocity and volume flow rates are twice the values in each of the sections between 1 and 3 (i.e., $\bar{v} = 0.332$ m/s). Therefore,

$$Re = 2 \times 9656 = 19,312$$

and

$$f^{-1/2} = -3.6 \log_{10} \frac{6.9}{19312} \quad \text{or} \quad f = 6.494 \times 10^{-3}$$

Then, from Eq. (i),

$$z_3 - z_2 = \frac{2f\bar{v}^2 L}{gD} (1 - x)$$

$$= \frac{2 \times 6.494 \times 10^{-3} \times 0.332^2 \times 500}{9.81 \times 0.05} (1 - x)$$

$$= 1.459(1 - x) \tag{vii}$$

Addition of Eqs. (vi) and (vii) gives

$$(z_1 - z_3) + (z_3 - z_2) = 0.440x + 1.459(1 - x)$$

$$= z_1 - z_2$$

The distance, $z_1 - z_2$, is 1 m, and hence

$$x = 0.450 \tag{viii}$$

Thus $0.450 \times 500 = 225$ m of the 500-m length of pipe must be doubled to increase the flow rate to 125% of its original value.

If the fluid flow in the pipes is fully rough, the friction factor is independent of the Reynolds number and hence is independent of the average flow velocity. Consider fully rough flow in a horizontal pipe of diameter D_1 and length L over which the pressure drop is ΔP, and consider the following three cases:

1. The latter half of the pipe is doubled by another pipe of diameter D_1.
2. The latter half of the pipe is replaced by a pipe of diameter $D_2 = 2D_1$.
3. The latter half of the pipe is replaced by a pipe with a cross-sectional flow area of twice the original (i.e., $A_2 = 2A_1$).

If the pressure drop, ΔP, over the length L remains constant, which of the three cases produces the greatest increase in the volume flow rate of the fluid?

Locating plane 1 at the entrance to the pipe and plane 2 at the exit, in the original pipe

$$\frac{P_1 - P_2}{\rho} = E_f = 2f \frac{L}{D_1} \bar{v}_b^2 \qquad (ix)$$

Case 1: After doubling the latter half of the pipe and locating plane 3 at $x = L/2$, the volume flow rate in the single pipe between planes 1 and 3 is twice the volume flow rate in each of the pipes between 3 and 2. Consequently, the average linear flow velocity between 1 and 3, $\bar{v}_{a(1\to3)}$, is twice the value in each of the pipes between 3 and 2, $\bar{v}_{a(3\to2)}$:

$$\bar{v}_{a(1\to3)} = 2\bar{v}_{a(3\to2)} \qquad (x)$$

Therefore, between 1 and 3,

$$\frac{P_1 - P_3}{\rho} = \frac{2f}{D_1} \frac{L}{2} \bar{v}_{a(1\to3)}^2 \qquad (xi)$$

and between 3 and 2,

$$\frac{P_3 - P_2}{\rho} = \frac{2f}{D_1} \frac{L}{2} \bar{v}_{a(3\to2)}^2 = \frac{2f}{D_1} \frac{L}{2} \frac{\bar{v}_{a(1\to3)}^2}{4} \qquad (xii)$$

the sum of which gives

$$\frac{P_1 - P_2}{\rho} = \frac{2fL}{D_1} \left(\frac{1}{2} + \frac{1}{8} \right) \bar{v}_{a(1\to3)}^2 \qquad (xiii)$$

$$= \frac{2fL}{D_1} \left(\frac{5}{8} \right) \bar{v}_{a(1\to3)}^2 \qquad (xiv)$$

Therefore, from Eqs. (ix) and (xiv),

$$\frac{\bar{v}_{a(1\to3)}}{\bar{v}_b} = \sqrt{\frac{8}{5}} = 1.26 = \frac{\dot{V}_a}{\dot{V}_b}$$

and thus the doubling increases the volume flow rate by the factor 1.26.

The ratio of Eq. (xi) to Eq. (xiv) gives

$$\frac{P_1 - P_3}{P_1 - P_2} = \frac{8}{10}$$

which indicates that 80% of the total pressure drop ΔP occurs between 1 and 3 and 20% occurs between 2 and 3. Thus from 1 to 3 the rate of pressure drop is $0.8\,\Delta P/(L/2) = 1.6\,\Delta P/L$ and from 2 to 3 it is $0.2\,\Delta P/(L/2) = 0.4\,\Delta P/L$.

Case 2: The latter half of the pipe is replaced by a pipe of diameter $D_2 = 2D_1$. Thus

$$\dot{V}_a = \bar{v}_{a(1\to3)}\frac{\pi D_1^2}{4} = \bar{v}_{a(3\to2)}\frac{\pi D_2^2}{4} = \bar{v}_{a(3\to2)}\frac{4\pi D_1^2}{4}$$

or

$$\bar{v}_{a(3\to2)} = \frac{\bar{v}_{a(1\to3)}}{4} \tag{xv}$$

Therefore, between 1 and 3,

$$\frac{P_1 - P_3}{\rho} = \frac{2f}{D_1}\frac{L}{2}\bar{v}^2_{a(1\to3)} \tag{xvi}$$

and between 3 and 2,

$$\frac{P_3 - P_2}{\rho} = \frac{2f}{D_2}\frac{L}{2}\bar{v}^2_{a(3\to2)} = \frac{2f}{2D_1}\frac{L}{2}\frac{\bar{v}^2_{a(1\to3)}}{16} \tag{xvii}$$

the sum of which gives

$$\frac{P_1 - P_2}{\rho} = 2f\frac{L}{D_1}\left(\frac{1}{2} + \frac{1}{64}\right)\bar{v}^2_{a(1\to3)}$$

$$= 2f\frac{L}{D_1}\left(\frac{33}{64}\right)\bar{v}^2_{a(1\to3)} \tag{xviii}$$

Then, from Eqs. (ix) and (xviii),

$$\frac{\bar{v}_{a(1\to3)}}{\bar{v}_b} = \sqrt{\frac{64}{33}} = 1.39 = \frac{\dot{V}_a}{\dot{V}_b}$$

and thus the alteration increases the volume flow rate by the factor 1.39. The ratio of Eq. (xvi) to Eq. (xviii) gives

$$\frac{P_1 - P_3}{P_1 - P_2} = \frac{64}{66} = 0.97$$

and thus 97% of the total pressure drop occurs between 1 and 3.

Case 3: The latter half of the pipe is replaced by a pipe with cross-sectional flow area $A_2 = 2A_1$:

$$\frac{\pi D_2^2}{4} = \frac{2\pi D_1^2}{4}$$

or

$$D_2 = \sqrt{2}\, D_1$$

Thus

$$\dot{V}_a = \bar{v}_{a(1\to3)} \frac{\pi D_1^2}{4} = \bar{v}_{a(3\to2)} \frac{\pi D_2^2}{4} = \bar{v}_{a(3\to2)} \frac{2\pi D_1^2}{4}$$

or

$$\bar{v}_{a(3\to2)} = \frac{\bar{v}_{a(1\to3)}}{2}$$

Therefore, between 1 and 3,

$$\frac{P_1 - P_3}{\rho} = \frac{2f}{D_1} \frac{L}{2} \bar{v}^2_{a(1\to3)} \qquad \text{(xix)}$$

and between 3 and 2,

$$\frac{P_3 - P_2}{\rho} = \frac{2f}{D_2} \frac{L}{2} \bar{v}^2_{a(3\to2)} = \frac{2f}{\sqrt{2}D_1} \frac{L}{2} \frac{\bar{v}^2_{a(1\to3)}}{4} \qquad \text{(xx)}$$

Summing gives

$$\frac{P_1 - P_2}{\rho} = 2f \frac{L}{D_1} \left(\frac{1}{2} + \frac{1}{8\sqrt{2}}\right) \bar{v}^2_{a(1\to3)}$$

$$= 2f \frac{L}{D_1} (0.588)\, \bar{v}^2_{a(1\to3)} \qquad \text{(xxi)}$$

and Eqs. (ix) and (xxi) give

$$\frac{\bar{v}_{a(1\to3)}}{\bar{v}_b} = \sqrt{\frac{1}{0.588}} = 1.30 = \frac{\dot{V}_a}{\dot{V}_b}$$

Thus the flow rate is increased by the factor 1.30. From Eqs. (xix) and (xxi),

$$\frac{P_1 - P_3}{P_1 - P_2} = \frac{1}{2 \times 0.588} = 0.85$$

and thus 85% of the total pressure drop occurs between 1 and 3. Case (2) causes the greatest increase in the volume flow rate.

5.5 Concept of Head

The total energy of a fluid at any point in a flow system is

$$\frac{Pm}{\rho} + \frac{1}{2} m\bar{v}^2 + mgz \qquad \text{(i)}$$

in which the first term is the flow energy, the second term is the kinetic energy, and the third term is the potential energy relative to some chosen datum level. Dividing Eq. (i) by the mass of the fluid, m, gives the energy of the fluid per unit mass as

$$\frac{P}{\rho} + \frac{1}{2}\bar{v}^2 + gz \tag{ii}$$

and dividing Eq. (ii) by g gives

$$\frac{P}{\rho g} + \frac{1}{2}\frac{\bar{v}^2}{g} + z \tag{iii}$$

Equation (iii) has the units of length and is called the *head, h*, of the flow at the point of interest. In considering flow between plane 1 and plane 2 in the flow system, the *loss of head*, $-h_L$, is obtained from Eq. (iii) as

$$-h_L = h_2 - h_1 = \frac{P_2 - P_1}{\rho g} + \frac{1}{2}\frac{\bar{v}_2^2 - \bar{v}_1^2}{g} + z_2 - z_1 \tag{iv}$$

Comparison of Eq. (iv) with Eq. (5.7) for a flow system in which Q and w are zero between planes and 1 and 2 shows that

$$h_L = \frac{E_f}{g} \tag{v}$$

and hence the loss of head between two points in a flow system is an alternative measure of the dissipation of energy by friction loss between the two points.

Consider the flow of water at 300 K at an average flow velocity of 5 m/s in a 100-m length of horizontal pipe of diameter 0.01 m. The Reynolds number for the flow is

$$\text{Re} = \frac{\bar{v}D\rho}{\eta} = \frac{5 \times 0.01 \times 997}{8.57} \times 10^{-4} = 5.82 \times 10^4$$

The flow is turbulent, and if the pipe has a smooth wall, the friction factor is obtained from

$$f^{-1/2} = -3.6 \log \frac{6.9}{8.57 \times 10^{-4}}$$

as

$$f = 5.0 \times 10^{-3}$$

The friction loss is thus

$$E_f = 2f\frac{L}{D}\bar{v}^2 = 2 \times 5.0 \times 10^{-3} \times \frac{100}{0.01} \times 5^2 = 2500 \text{ m}^2/\text{s}^2$$

and from Eq. (v) the loss of head over the 100 m of pipe is

$$h_L = \frac{2500}{9.81} = 254 \text{ m}$$

In Eq. (iii) the first term is called the *pressure head*, the second term is called the *velocity head*, and the third term is called the *elevation head*.

Consider, now, that the pipe is inclined at an angle of 45° to the horizontal and the flow is in the upward direction. In this case $z_2 - z_1 = 100 \sin 45° = 70.7$ m, and hence the increase in the elevation head of the flow is

$$z_2 - z_1 = 70.7 \text{ m}$$

The pressure drop over the pipe is obtained from Eq. (5.7):

$$\frac{P_2 - P_1}{\rho} + g(z_2 - z_1) + E_f = 0$$

as

$$P_2 - P_1 = -9.81 \times 70.7 - 2500 = -3.18 \times 10^6 \text{ Pa}$$

The change in the pressure head is thus

$$\frac{P_2 - P_1}{\rho g} = -\frac{3.18 \times 10^6}{997 \times 9.81} = -325 \text{ m}$$

The total change in the head is thus the change in the elevation head plus the change in the pressure head, that is,

$$-h = 70.7 - 325 = -254 \text{ m}$$

which, by virtue of the head loss being caused only by dissipation, is the same value as that for flow in the horizontal pipe.

Application of the concept of head loss is restricted to the flow of incompressible fluids in flow systems in which no work is done on the fluid by a pump, no work is done by the fluid in a turbine, and heat is neither transferred to nor removed from the fluid.

5.6 Fluid Flow in an Open Channel

Consider the flow of fluid in the inclined open channel shown in Fig. 5.4. The channel, of length L and width W, is inclined at an angle θ to the horizontal, and under conditions of uniform flow, the depth h of the fluid is constant along the channel. The force balance is as follows.

The gravitational force acting on the fluid in the channel in the direction of flow equals the frictional force exerted on the fluid by the walls of the channel:

$$(\rho h W L)(g \sin \theta) = \tau_0 L(2h + W) = \tfrac{1}{2} f \rho \bar{v}^2 L(2h + W) \tag{i}$$

From Eq. (4.19), the equivalent diameter of the flow, D_e, is

$$D_e = \frac{4 \times \text{cross-sectional flow area}}{\text{wetted perimeter of channel}} = \frac{4hW}{2h + W}$$

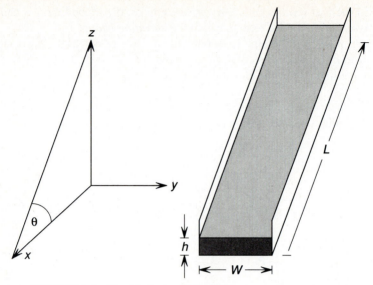

FIGURE 5.4 Gravity-induced fluid flow in an open channel.

substitution of which into Eq. (i) gives

$$\rho h W L g \sin\theta = \frac{1}{2} f \rho \bar{v}^2 L \times \frac{4hW}{D_e}$$

or

$$\bar{v}^2 = \frac{D_e g \sin\theta}{2f} \tag{ii}$$

For turbulent flow of Reynolds number greater than 8000, where

$$\text{Re} = \frac{\bar{v} D_e \rho}{\eta}$$

the friction factor is independent of the Reynolds number, being given by

$$f = \frac{1}{8}\left(\frac{\epsilon}{D_e}\right)^{1/3} \tag{iii}$$

where ϵ is the absolute roughness of the channel wall. Substitution of Eq. (iii) into Eq. (ii) gives

$$\bar{v} = 2\left(\frac{D_e^{4/3} g \sin\theta}{\epsilon^{1/3}}\right)^{1/2} \tag{5.15}$$

and the volume flow rate in the channel is

$$\dot{V} = \bar{v} h W = \frac{2hW D_e^{2/3}(g \sin\theta)^{1/2}}{\epsilon^{1/6}} \tag{5.16}$$

EXAMPLE 5.4

Liquid lead, tapped from a blast furnace, runs along an open channel of width 0.3 m and height 0.4 m into a kettle. The channel is inclined at 10° to the horizontal and the mass flow rate of lead in the channel is 500 kg/s. The absolute roughness of the channel wall is 3×10^{-4} m and the density and viscosity of liquid lead are 10,050 kg/m^3 and 1.82×10^{-3} Pa·s. Calculate the depth of the flow in the channel

$$D_e = \frac{4A}{P_w} = \frac{4hW}{2h + W} = \frac{1.2h}{2h + 0.3}$$

and

$$\dot{V} = \frac{\dot{M}}{\rho} = \frac{500}{10,050} = 0.0498 \text{ m}^3/\text{s}$$

Therefore, from Eq. (5.16),

$$\dot{V} = 2h \times 0.3 \times \left(\frac{1.2h}{2h + 0.3}\right)^{2/3} \times \frac{1}{(3 \times 10^{-4})^{1/6}} \times (9.81 \sin 10)^{1/2}$$

$$= \frac{3.418h^{5/3}}{(2h + 0.3)^{2/3}}$$

$$= 0.0498$$

which has the solution $h = 0.0554$ m. Thus

$$\bar{v} = \frac{\dot{V}}{hW} = \frac{0.0498}{0.0554 \times 0.3} = 3 \text{ m/s}$$

and

$$D_e = \frac{1.2 \times 0.0554}{(2 \times 0.0554) + 0.3} = 0.162 \text{ m}$$

The Reynolds number is

$$\text{Re} = \frac{\bar{v}D_e\rho}{\eta} = \frac{3 \times 0.162 \times 10,050}{1.82 \times 10^{-3}} = 2.68 \times 10^6$$

which, being greater than 8000, indicates that use of the correlation given by Eq. (iii) is valid.

5.7 Drainage from a Vessel

Figure 5.5 shows fluid draining through an opening in the base of a vessel. The fluid is being poured into the vessel at the same rate at which the vessel is draining, such that the depth of fluid in the vessel, h_0, remains constant. The diameter of the vessel and the orifice in the base are, respectively, D and d. If plane 1 is located at the

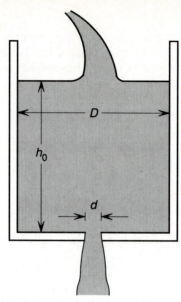

FIGURE 5.5   Fluid draining through an opening in the base of a vessel.

surface of the fluid in the vessel and plane 2 is located at the level of the orifice, Eq. (5.7) becomes

$$\frac{P_2}{\rho_2} - \frac{P_1}{\rho_1} = \frac{\bar{v}_2^2}{2\beta_2} - \frac{\bar{v}_1^2}{2\beta_1} + g(z_2 - z_1) + E_f = 0$$

As the fluid at both planes 1 and 2 is in contact with the atmosphere, $P_1 = P_2 = P_{atm}$. From mass balance considerations, the volume flow rate through the orifice equals that through the vessel:

$$\bar{v}_1 \frac{\pi D^2}{4} = \bar{v}_2 \frac{\pi d^2}{4}$$

or

$$\bar{v}_1 = \bar{v}_2 \left(\frac{d}{D}\right)^2 \tag{i}$$

such that if $d \ll D$, then $\bar{v}_1 \ll \bar{v}_2$. The friction loss is

$$E_f = 2f \frac{h_0}{D} \bar{v}_1^2 + \frac{1}{2} e_f \bar{v}_2^2$$

but with $\bar{v}_1 \ll \bar{v}_2$, all the terms containing $\bar{v}_1$ can be omitted, in which case Eq. (5.7) becomes

$$\frac{\bar{v}_2^2}{2\beta_2} + g(z_2 - z_1) + \frac{1}{2} e_f \bar{v}_2^2 = 0 \tag{ii}$$

or

$$\frac{\bar{v}_2^2}{2}\left(\frac{1}{\beta_2} + e_f\right) = g(z_1 - z_2) = gh_0$$

or

$$\bar{v}_2 = \frac{\sqrt{2gh_0}}{[(1/\beta_2) + e_f]^{1/2}} \tag{5.17}$$

The discharge coefficient, C_D, is defined as

$$C_D = \left(\frac{1}{\beta_2} + e_f\right)^{-1/2}$$

in which case Eq. (5.17) becomes

$$\bar{v}_2 = C_D\sqrt{2gh_0} \tag{5.18}$$

For turbulent flow through the orifice, $\beta_2 = 1$, and with $d \ll D$ and a sharp entrance to the orifice, $e_f = 0.5$, in which case C_D has the value 0.816. However, decreasing the sharpness of the entry to orifice decreases the value of e_f and hence increases the value of C_D toward a limiting value of unity. With $C_D = 1$,

$$\bar{v}_2 = \sqrt{2gh_0}$$

is the maximum possible value of $\bar{v}_2$.

5.8 Emptying a Vessel by Discharge Through an Orifice

In this case the depth of fluid in the vessel, h, decreases with time and Eq. (5.18) becomes

$$\bar{v}_2 = C_D\sqrt{2gh} \tag{iii}$$

The average flow velocity in the vessel, $\bar{v}_1$, is the rate of decrease of the level h in the vessel,

$$\bar{v}_1 = -\frac{dh}{dt} \tag{iv}$$

Combination of Eqs. (i), (iv), and (iii) thus gives

$$\bar{v}_2 = C_D\sqrt{2gh} = \bar{v}_1\left(\frac{D}{d}\right)^2 = -\left(\frac{D}{d}\right)^2\frac{dh}{dt}$$

or

$$\frac{dh}{\sqrt{h}} = -C_D\left(\frac{d}{D}\right)^2\sqrt{2g}\,dt$$

integration of which from $h = h_0$ at $t = 0$ to $h = h$ at $t = t$ gives

$$t = \frac{\sqrt{h_0} - \sqrt{h}}{C_D(d/D)^2\sqrt{g/2}} \tag{v}$$

as the time for the level of fluid to decrease from h_0 to h. Substitution of Eq. (v) into Eq. (iii) gives the variation of $\bar{v}_2$ with time as

$$\bar{v}_2 = C_D\sqrt{2gh_0} - C_D^2 \left(\frac{d}{D}\right)^2 gt$$

and Eq. (v) gives the time required to empty the vessel as

$$t = \frac{1}{C_D} \left(\frac{D}{d}\right)^2 \sqrt{\frac{2h_0}{g}} \tag{5.19}$$

In Eq. (5.19) the time required to empty the vessel is proportional to the square of the diameter of the vessel and to the square root of the initial depth of liquid in the vessel. For a given value of h_0 the volume of liquid to be drained increases in proportion to D^2, and hence for a constant d, the time required for the drainage increases in proportion to D^2. On the other hand, for a constant D, as the exit velocity is determined by the rate of decrease in the potential energy of the liquid, the exit velocity, $\bar{v}_2$, is proportional to the square root of the depth of liquid in the vessel and hence the time required to empty the vessel is proportional to the square root of the initial depth of the liquid.

The mass balance

$$\rho\bar{v}_1 \frac{\pi D^2}{4} = \rho\bar{v}_2 \frac{\pi d^2}{4}$$

comes from Eq. (3.1), namely, the rate of accumulation of mass in the vessel equals the difference between the rates at which mass enters and leaves the vessel. In the case of emptying a vessel by drainage there is no flow into the vessel, the rate at which mass leaves the vessel is

$$\rho\bar{v}_2 \frac{\pi d^2}{4}$$

and the rate of accumulation of mass in the vessel is

$$\frac{d}{dt} \int_0^h \rho \frac{\pi D^2}{4} dh$$

Thus

$$\frac{d}{dt} \int_0^h \rho \frac{\pi D^2}{4} dh = -\rho\bar{v}_2 \frac{\pi d^2}{4} \tag{vi}$$

or as D is a constant,

$$D^2 \frac{dh}{dt} = -D^2\bar{v}_1 = -d^2\bar{v}_2$$

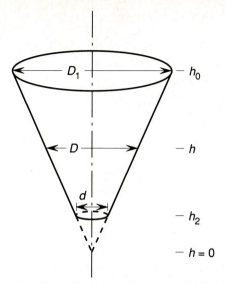

FIGURE 5.6   Fluid draining from a conical vessel.

Consider the draingage of fluid from the conical vessel shown in Fig. 5.6. From the geometry of a cone

$$\frac{D_1}{h_0} = \frac{D}{h} = \frac{d}{h_2} = \text{a constant} = k \qquad \text{(vii)}$$

and thus Eq. (vi) becomes

$$\frac{d}{dt} \int_{h_2}^{h} \frac{\rho \pi k^2 h^2 \, dh}{4} = -\rho \bar{v}_2 \frac{\pi d^2}{4}$$

or

$$\frac{k^2}{3} \frac{d}{dt} (h^3 - h_2^3) = -\bar{v}_2 d^2$$

which in combination with Eq. (iii) with $C_D = 1$ gives

$$\bar{v}_2 = -\frac{1}{3} \left(\frac{k}{d}\right)^2 \frac{d}{dt} (h^3 - h_2^3) = \sqrt{2g(h - h_2)} \qquad \text{(viii)}$$

For a small value of d, and hence h_2, $h \gg h_2$ during most of the draining and thus Eq. (viii) can be simplified as

$$-\frac{1}{3} \left(\frac{k}{d}\right)^2 \frac{d}{dt} h^3 = 9 \left(\frac{kh}{d}\right)^2 \frac{dh}{dt} = \sqrt{2gh}$$

or

$$\left(\frac{k}{d}\right)^2 \frac{1}{\sqrt{2g}} h^{3/2} \, dh = -dt \qquad \text{(ix)}$$

integration of which between the limits $h = h_0$ at $t = 0$ and $h = h$ at $t = t$ gives

$$\left(\frac{k}{d}\right)^2 \frac{1}{\sqrt{2g}} \left(\frac{2}{5}\right) (h_0^{5/2} - h^{5/2}) = t \qquad \text{(x)}$$

Assuming that $h_2 \approx 0$, the time required to drain the vessel is thus

$$t = \left(\frac{k}{d}\right)^2 \left(\frac{2}{g}\right)^{1/2} \frac{h_0^{5/2}}{5} \qquad \text{(xi)}$$

and

$$\bar{v}_1 = -\frac{dh}{dt} = \left(\frac{d}{k}\right)^2 \sqrt{2g}\, h^{-3/2}$$

$$= \frac{(d/R)^2 \sqrt{2g}}{[h_0^{5/2} - 5(d/k)^2 (g/2)^{1/2} t]^{3/5}} \qquad \text{(xii)}$$

EXAMPLE 5.5

Consider the draingage of a fluid from a conical vessel of $h_0 = 1$ m, $h_2 = 0.1$ m, and $d = 0.025$ m shown in Fig. 5.6. For this geometry

$$k = \frac{d}{h_2} = \frac{0.025}{0.1} = 0.25$$

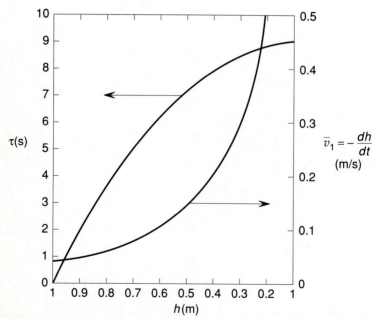

FIGURE 5.7 Relationships among depth, time, and flow velocity during drainage of fluid from the conical vessel considered in Example 5.5.

Therefore, from Eq. (xi), the time required to drain the vessel is

$$t = \left(\frac{0.25}{0.025}\right)^2 \left(\frac{2}{9.81}\right)^{1/2} \frac{1}{5} = 9.03 \text{ s}$$

From Eq. (x) the variation of h with t is

$$t = 9.03 \, (1 - h^{5/2})$$

which is shown in Fig. 5.7, and from Eqs. (x) and (xi), the variation of $\bar{v}_1$ with h is

$$\bar{v}_1 = \frac{(d/R)^2(2g)^{1/2}}{h^{3/2}} = \frac{0.0443}{h^{3/2}}$$

which is also shown in Fig. 5.7.

5.9 Drainage of a Vessel Using a Drainage Tube

Figure 5.8 shows fluid being poured into a vessel at a rate that maintains the level of fluid in the vessel at the constant value h_0, and fluid draining through a drainage tube of length L and diameter d attached to the bottom of the vessel. Plane 1 is located at the level of fluid in the vessel and plane 2 is located at the exit end of the drainage tube. Again, with $d \ll D$, all terms containing $\bar{v}_1$ can be omitted and in this case, the friction loss occurs at the contraction and in the drainage tube,

$$E_f = 2f \frac{L}{d} \bar{v}_2^2 + \frac{1}{2} e_f \bar{v}_2^2$$

Equation (ii) thus becomes

$$\frac{\bar{v}_2^2}{2\beta_2} + g(z_2 - z_1) + 2f \frac{L}{d} \bar{v}_2^2 + \frac{1}{2} e_f \bar{v}_2^2 = 0$$

or

$$\bar{v}_2 = \frac{\sqrt{2g(h_0 + L)}}{[(1/\beta_2) + e_f + 4f(L/d)]^{1/2}} \tag{xiii}$$

Assuming turbulent flow in the drainage tube and a sharp contraction of the flow into the drainage tube, $\beta_2 = 1$ and $e_f = 0.5$. For a smooth-walled drainage tube, Eq. (4.18) gives

$$f^{-1/2} = -3.6 \log \frac{6.9}{Re} = -3.6 \log \frac{6.9\eta}{\bar{v}_2 d\rho} \tag{xiv}$$

and $\bar{v}_2$ and f are obtained by simultaneous solution of Eqs. (xiii) and (xiv). For example, for water of $\rho = 997 \text{ kg/m}^3$ and $\eta = 8.57 \times 10^{-4}$ Pa·s draining from a vessel of $L = h_0 = D = 0.5$ m and $d = 0.005$ m, Eqs. (xiii) and (xiv) become

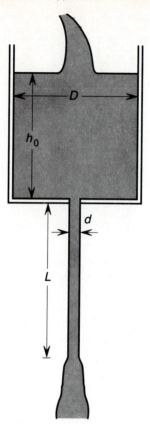

FIGURE 5.8   Drainage from a vessel using a drainage tube.

$$\bar{v}_2 = \frac{\sqrt{2 \times 9.81 \times 1}}{[1 + 0.5 + f(4 \times 0.5)/0.005)]^{1/2}} = \frac{4.429}{(1.5 + 400f)^{1/2}} \qquad \text{(xv)}$$

and

$$f^{-1/2} = -3.6 \log \frac{6.9 \times 8.57 \times 10^{-4}}{\bar{v}_2 \times 0.005 \times 997}$$

$$= 10.533 + 3.6 \log \bar{v}_2 \qquad \text{(xvi)}$$

solution of which gives

$$f = 0.00731 \qquad \text{and} \qquad \bar{v}_2 = 2.11 \text{ m/s}$$

The Reynolds number for flow in the drainage tube is thus 1.22×10^4, indicating that the flow is turbulent.

It is now of interest to examine how $\bar{v}_2$ is influenced by increasing L. Increasing L increases both the numerator in Eq. (xiii) (which tends to increase $\bar{v}_2$) and the denominator (which tends to decrease $\bar{v}_2$). Also, any variation of $\bar{v}_2$ changes the Reynolds number, which, in turn, changes the value of f in Eq. (xiii). Increasing L

from 0.5 m to 5 m in the example above gives

$$f = 0.00758 \quad \text{and} \quad \bar{v}_2 = 0.610 \text{ m/s}$$

with a Reynolds number of 3.55×10^3. Thus the increase in the friction loss, caused by increasing L, dominates and $\bar{v}_2$ decreases. The flow velocity, with the larger L, can be increased by increasing the diameter d. If it were required to increase $\bar{v}_2$ to the original value of 2.11 m/s, Eq. (xv) would be written as

$$\bar{v}_2 = 1.22 = \frac{\sqrt{2 \times 9.81 \times 5.5}}{[1 + 0.5 + 4 \times 5(f/d)]^{1/2}}$$

or

$$f = 1.142d \tag{xvii}$$

and Eq. (xvi) would be

$$f^{-1/2} = -3.6 \log \frac{6.9 \times 8.57 \times 10^{-4}}{2.11 \times d \times 997} \tag{xviii}$$

Simultaneous solution of Eqs. (xvii) and (xviii) gives

$$f = 0.00694 \quad d = 0.0061 \text{ m}$$

and a Reynolds number of 1.5×10^4.

5.10 Emptying a Vessel by Drainage Through a Drainage Tube

In this case h decreases with time and Eq. (xiii) becomes

$$\bar{v}_2 = \frac{\sqrt{2g(h + L)}}{[(1/\beta_2) + e_f + 4f(L/d)]^{1/2}} \tag{xix}$$

Proceeding as with the derivation of Eqs. (v) and (5.17) gives

$$t = \frac{2[(h_1 + L)^{1/2} - (h_2 + L)^{1/2}][(1/\beta_2) + e_f + 4f(L/d)]^{1/2}(D/d)^2}{\sqrt{2g}} \tag{xx}$$

as the time required for the level of the fluid in the vessel to decrease from h_1 to h_2. As h decreases, $\bar{v}_2$ and hence the Reynolds number decrease, and consequently, the value of f in Eq. (xx) increases. Consider the emptying of a vessel of $L = D = h_0 = 0.5$ m and $d = 0.005$ m. At any value of h, Eq. (xix) gives

$$\bar{v}_2 = \frac{\sqrt{2 \times 9.81 \times (h + 0.5)}}{[1 + 0.5 + ((4 \times 0.5)/0.005)f]^{1/2}} = \frac{4.439\sqrt{h + 0.5}}{(1.5 + 400f)^{1/2}} \tag{xxi}$$

and

$$f^{-1/2} = -3.6 \log \frac{6.9}{Re} = -3.6 \log \frac{6.9 \times 8.57 \times 10^{-4}}{\bar{v}_2 \times 0.005 \times 997}$$

TABLE 5.2

h (m)	f	$\bar{v}_2$ (m/s)	Re	$\dfrac{dh}{dt}$ (m/s)
0.5	0.00731	2.11	1.22×10^4	-2.11×10^{-4}
0.4	0.00742	1.99	1.16×10^4	-1.99×10^{-4}
0.3	0.00756	1.86	1.08×10^4	-1.86×10^{-4}
0.2	0.00771	1.73	1.01×10^4	-1.73×10^{-4}
0.1	0.00789	1.59	9.25×10^3	-1.59×10^{-4}
0	0.00812	1.43	8.35×10^3	-1.43×10^{-4}

or

$$\bar{v}_2 = \exp\left(\frac{f^{-1/2} - 10.533}{1.563}\right) \qquad \text{(xxii)}$$

At the start of draining (i.e., with $h = 0.5$), solution of Eqs. (xxi) and (xxii) gives

$$f = 0.00731 \quad \text{and} \quad \bar{v}_2 = 2.11 \text{ m/s}$$

When 20% of the vessel has been drained (i.e., $h = 0.4$),

$$f = 0.00742 \quad \text{and} \quad \bar{v}_2 = 1.99 \text{ m/s}$$

Following this procedure, Table 5.2 can be constructed.

Thus, during the course of drainage, f increases from 0.00731 to 0.00812 and the flow in the drainage tube is always turbulent. In this example, Eq. (xx) gives the time for the fluid level to decrease from h_1 to h_2 as

$$t = \frac{2[(h_1 + 0.5)^{1/2} - (h_2 + 0.5)^{1/2}][1.5 + 4((0.5/0.005)f)]^{1/2}(0.5/0.005)^2}{\sqrt{2 \times 9.81}}$$

$$= 4515(1.5 + 400f)^{1/2}[(h_1 + 0.5)^{1/2} - (h_2 + 0.5)^{1/2}] \qquad \text{(xxiii)}$$

During the time required for h decrease from 0.5 to 0.4, the arithmetically averaged value of f is 0.00737, insertion of which into Eq. (xxiii) gives $t = 489s = 8.14$ min. Following this procedure during draining gives the data shown in Table 5.3. Taking an average arithmetic average of f as $(0.00731 + 0.00812)/2 = 0.00722$, Eq. (xxiii) gives the time for h to decrease from 0.5 m to zero as 47.21 min.

TABLE 5.3

h_1	h_2	f (average)	t (min)
0.5	0.4	0.00737	8.14
0.4	0.3	0.00749	8.66
0.3	0.2	0.00764	9.28
0.2	0.1	0.00780	10.04
0.1	0.0	0.00801	11.01
		Total time =	47.13

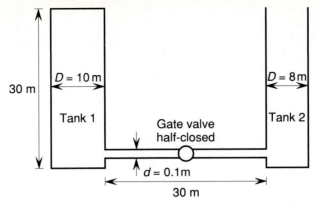

FIGURE 5.9 Two holding tanks considered in Example 5.6.

EXAMPLE 5.6 _____

Figure 5.9 shows two holding tanks connected by a 30-m length of 0.1-m-di-
ameter commercial steel pipe containing a gate valve. Initially, tank 1 is filled
with water and tank 2 is empty. Calculate the time required for water to flow
from tank 1 to tank 2 until the free surfaces in both tanks are at the same level
when the gate valve is half opened. Both free surfaces are in contact with the
atmosphere and the absolute roughness of the steel pipe is 0.046 mm. The water
is at a temperature of 300 K.

Locating plane 1 at the free surface in tank 1 and plane 2 at the free surface
in tank 2, with $P_2 = P_1$, Eq. (5.7) gives

$$\frac{v_2^2}{2} - \frac{v_1^2}{2} + g(z_2 - z_1) + E_f = 0 \tag{i}$$

in which v_1 and v_2 are, respectively, the velocities of the free surfaces of the water
in tanks 1 and 2. Designating the diameters of the tanks and the connecting pipe
as, respectively, D_1, D_2, and d, and the average linear velocity of flow through
the pipe as $\bar{v}$, the mass balance on the flow gives

$$v_1 D_1^2 = \bar{v} d^2 = v_2 D_2^2 \tag{ii}$$

The friction loss arises from sudden contraction, flow through the pipe, and the
gate valve (of $L_e/D = 190$) and sudden expansion, and hence

$$E_f = \left[2f \left(\frac{30}{0.1} + 190 \right) + \frac{1}{2} + \left(\frac{1}{2} \times \frac{1}{2} \right) \right] \bar{v}^2$$

When the gate valve is opened and $z_2 - z_1 = -30$ m, substitution of Eq. (ii)
into Eq. (i) gives

$$\frac{\bar{v}^2}{2} \left(\frac{d}{D_2} \right)^4 - \frac{\bar{v}^2}{2} \left(\frac{d}{D_1} \right)^4 - 30g + \left[2f \left(\frac{30}{0.1} + 190 \right) + 0.75 \right] \bar{v}^2 = 0$$

or

$$\frac{\bar{v}^2}{2}\left(\frac{0.1}{10}\right)^4 - \frac{\bar{v}^2}{2}\left(\frac{0.1}{8}\right)^4 - (9.81 \times 30) + (980f + 0.75)\bar{v}^2 = 0$$

or

$$\bar{v}^2(5 \times 10^{-9} - 1.22 \times 10^{-8} + 980f + 0.75) = 294.3 \tag{iii}$$

The friction factor is obtained from

$$f^{-1/2} = -3.6 \log\left[\left(\frac{\epsilon}{3.7d}\right)^{1.11} + \frac{6.9\eta}{\bar{v}d\rho}\right]$$

as

$$f = \frac{1}{12.96 \, [\log \, (4.623 \times 10^{-5} + (6.9 \times 8.57 \times 10^{-4})/(\bar{v} \times 0.1 \times 997))]^2} \tag{iv}$$

Substituting Eq. (iv) into Eq. (iii) and solving gives the initial value of $\bar{v}$ as 7.75 m/s. The initial value of the friction factor is thus

$$f = 4.23 \times 10^{-3}$$

and the Reynolds number for the initial flow is

$$\text{Re} = \frac{7.75 \times 0.1 \times 997}{8.57 \times 10^{-4}} = 9.02 \times 10^5$$

In Eq. (iii) the contribution of friction loss is

$$(980 \times 4.23 \times 10^{-3}) + 0.75 = 4.9$$

and the contribution of kinetic energy is

$$5 \times 10^{-9} - 1.22 \times 10^{-8} = -7.2 \times 10^{-9}$$

which shows that the kinetic energy contribution is small enough to be ignored. When flow has occurred to the extent that $z_2 - z_1 = -1$ m, the term on the right-hand side of Eq. (iii) is 9.81, which gives $\bar{v} = 1.35$ m/s and

$$f = 4.72 \times 10^{-3}$$

Similarly, when $z_2 - z_1 = -0.1$ m, $\bar{v}$ is 0.396 m/s and f is 5.61×10^{-3}. Thus, during most of the time that flow is occurring, the friction factor does not vary significantly and hence an average value of 4.5×10^{-3} can be assumed. When $z_2 - z_1 = h$, Eq. (i) can be written in terms of v_2 and h as

$$\frac{v_2^2}{2} - \frac{v_2^2}{2}\left(\frac{8}{10}\right)^4 + 9.81h + (980 \times 4.5 \times 10^{-3} + 0.75)v_2^2\left(\frac{8}{0.1}\right)^4$$

or

$$v_2^2(0.5 - 0.205 + 2.11 \times 10^8) = 9.81h$$

But

$$v_2 = -\frac{dh}{dt}$$

and hence

$$\frac{dh}{\sqrt{h}} = -2.16 \times 10^{-4}t$$

integration of which from

$$h = 0 \quad \text{at } t = t \quad \text{to} \quad h = -30 \quad \text{at } t = 0$$

gives

$$(2h^{1/2})_t - (2h^{1/2})_{t=0} = -2.16 \times 10^{-4}t$$

or

$$0 - 2 \times 30^{1/2} = -2.16 \times 10^{-4}t$$

or

$$t = 5.08 \times 10^4 \text{ s} = 14.1 \text{ h}$$

The level, z, to which the water rises in tank 2 is obtained from

$$30 \times \pi \times \frac{10^2}{4} = z \left(\pi \times \frac{10^2}{4} + \pi \times \frac{8^2}{4} \right)$$

as

$$z = 18.3 \text{ m}$$

and hence the volume of water transferred from tank 1 to tank 2 is

$$18.29 \times \pi \times \frac{8^2}{4} = 920 \text{ m}^3$$

The average linear flow rate through the pipe is thus obtained from

$$\bar{v}\pi \frac{0.1^2}{4} = \frac{920}{50,800}$$

as

$$\bar{v}_{av} = 2.30 \text{ m/s}$$

5.11 Bernoulli Equation for Flow of Compressible Fluids

Consideration of turbulent horizontal flow of a compressible fluid begins with Eq. (5.7) written in differential form,

$$\frac{dP}{\rho} + d\left(\frac{\bar{v}^2}{2}\right) + 2f\bar{v}^2 \frac{dx}{D} = 0 \qquad (5.20)$$

Although the local velocity and the density of the fluid vary with position along the flow, the mass flow rate $\dot{M}$ is constant:

$$\dot{M} = \rho\bar{v}A$$

or

$$\bar{v} = \frac{\dot{M}}{\rho A} \qquad (i)$$

For an ideal gas of molecular weight M,

$$\frac{1}{\rho} = V = \frac{RT}{PM} \qquad (ii)$$

substitution of which into Eq. (i) gives

$$\bar{v} = \frac{\dot{M}RT}{APM} \qquad (iii)$$

which is the variation of $\bar{v}$ with the local static pressure along the flow. Differentiation of Eq. (ii) gives the second term in Eq. (5.20) as

$$d\frac{\bar{v}^2}{2} = \frac{1}{2}\left(\frac{\dot{M}}{A}\right)^2 \left(\frac{RT}{M}\right)^2 d\frac{1}{P^2} \qquad (iv)$$

and the first term in Eq. (5.20) is obtained from Eq. (ii) as

$$\frac{dP}{\rho} = \frac{RT}{M}\frac{dP}{P} \qquad (v)$$

Equation (5.20) can thus be written as

$$\frac{RT}{M}\frac{dP}{P} + \frac{1}{2}\left(\frac{\dot{M}}{A}\right)^2 \left(\frac{RT}{M}\right)^2 d\frac{1}{P^2} + 2f\left(\frac{\dot{M}}{A}\right)^2 \left(\frac{RT}{PM}\right)^2 \frac{dx}{D} = 0 \qquad (vi)$$

and multiplying Eq. (vi) by $(PM/RT)^2$ gives

$$\frac{M}{RT}PdP + \frac{1}{2}\left(\frac{\dot{M}}{A}\right)^2 P^2 d\frac{1}{P^2} + 2f\left(\frac{\dot{M}}{A}\right)^2 \frac{dx}{D} = 0$$

or

$$\frac{M}{RT}PdP - \left(\frac{\dot{M}}{A}\right)^2 \frac{dP}{P} + 2f\left(\frac{\dot{M}}{A}\right)^2 \frac{dx}{D} = 0$$

integration of which between P_2 at $x = L$ and P_1 at $x = 0$ gives

$$\frac{M}{RT}\frac{P_2^2 - P_1^2}{2} + \left(\frac{\dot{M}}{A}\right)^2 \ln\frac{P_1}{P_2} + 2f\left(\frac{\dot{M}}{A}\right)^2 \frac{L}{D} = 0 \qquad (5.21)$$

EXAMPLE 5.7

Air, at 298 K, is pumped through a horizontal cylindrical conduit of length 100 m and diameter 1 m. The air enters the conduit at a pressure of 150 Pa and exits at a pressure of 50 Pa. Calculate the mass flow rate through the conduit. The molecular weight of air is $(0.21 \times 32) + (0.79 \times 28) = 28.84$ g/g mol or 0.02884 kg /g mol. Substitution into Eq. (5.21) gives

$$\frac{0.02884}{8.3144 \times 298} \left(\frac{50^2 - 150^2}{2} \right) + \frac{\dot{M}^2}{(\pi \times 0.5^2)} \ln \frac{150}{50}$$

$$+ 2f \frac{\dot{M}^2}{(\pi \times 0.5^2)^2} \frac{100}{1} = 0 \qquad \text{(vi)}$$

or

$$-0.1164 + 1.7796\dot{M}^2 + 323.9669\dot{M}^2 f = 0 \qquad \text{(vii)}$$

$\dot{M}$ and f are related through the Reynolds number:

$$f^{-1/2} = -3.6 \log \frac{6.9}{\text{Re}}$$

and

$$\text{Re} = \frac{4\dot{M}}{\pi \eta D} \qquad \text{(viii)}$$

gives

$$f^{-1/2} = -3.6 \log \frac{6.9 \, \pi \eta D}{4\dot{M}}$$

With $\eta = 1.85 \times 10^{-5}$ Pa·s and $D = 1$ m, this gives

$$f^{-1/2} = -3.6 \log \frac{1.003 \times 10^{-4}}{\dot{M}} \qquad \text{(ix)}$$

and simultaneous solution of Eqs. (vii) and (ix) gives

$$\dot{M} = 0.167 \text{ kg/s}$$

Equation (viii) then gives Re $= 1.15 \times 10^4$, which indicates that the flow is turbulent.

5.12 Pitot Tube

The pitot tube, shown in Fig. 5.10, is a device for measuring local velocities in flowing fluids. It is a double-walled tube with smooth-edged holes in the outer wall and is placed in the flow with its opening facing the flowing fluid. For the application of Eq. (5.7), plane 1 is placed a short distance in front of the entrance to the tube

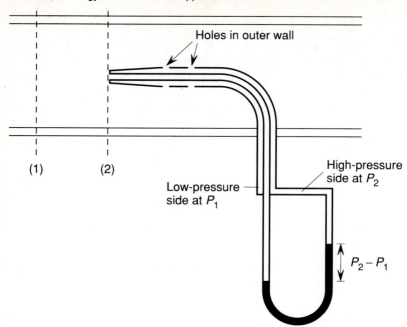

FIGURE 5.10 Schematic representation of a pitot tube.

and plane 2 is placed at the entrance. The local flow velocity at the entrance to the tube is zero and hence, for turbulent flow, Eq. (5.7) gives

$$\frac{P_2 - P_1}{\rho} = \frac{v_1^2}{2}$$

or

$$P_2 = P_1 + \frac{\rho v_1^2}{2} \tag{i}$$

Equation (i) shows that the increase in the static pressure at plane 2 is caused by the disappearance of the kinetic energy. As the pressure in the pitot tube is P_2 and the pressure between the double walls is P_1, the local flow velocity is obtained from measurement of their difference, as

$$v_1 = \sqrt{\frac{2(P_2 - P_1)}{\rho}}$$

or, in practice, where conditions might not be ideal, as

$$v_1 = C_p \sqrt{\frac{2(P_2 - P_1)}{\rho}} \tag{5.22}$$

in which C_p is the pitot tube coefficient. The pitot tube coefficient normally has a value in the range 0.98 to 1.00. If the flowing fluid is a compressible gas, Eq. (5.20) gives

$$\int_{P_1}^{P_2} \frac{dP}{\rho} = \frac{v_1^2}{2} \tag{5.23}$$

EXAMPLE 5.8 ————————————————————————————————————

A pitot tube is installed along the centerline of a 0.3-m-diameter tube in which air at a temperature of 350 K and a pressure of 175 kPa is flowing. What is the average linear velocity of the airflow when the pressure difference across the water manometer on the pitot tube is 0.5 in. of water?

At 300 K and 101.3 kPa the density of air is 1.009 kg/m³, and thus at 300 K and 175 kPa its density is

$$1.009 \times \frac{175}{101.3} = 1.74 \text{ kg/m}^3$$

From the formula

$$\Delta P = h\rho g$$

0.5 in. of water corresponds to a pressure difference of

$$\frac{0.5 \times 2.54}{100} \times 997 \times 9.81 = 124 \text{ Pa}$$

and thus, from Eq. (5.22), with $C_p = 1$,

$$v = \sqrt{\frac{2 \times 124}{1.74}} = 11.9 \text{ m/s}$$

which, being the local velocity at the centerline of the tube, is v_{max}. For turbulent flow in the range $10^4 < \text{Re} < 10^7$,

$$\frac{\bar{v}}{v_{max}} = 0.62 + 0.04 \log \frac{\text{Re } v_{max}}{\bar{v}}$$

$$= 0.62 + 0.04 \log \frac{D\rho v_{max}}{\eta} \tag{i}$$

The viscosity of air at 350 K is 2.08×10^{-5} Pa·s and hence

$$\frac{\bar{v}}{v_{max}} = 0.62 + 0.04 \log \frac{0.3 \times 1.74 \times 11.9}{2.08 \times 10^{-5}}$$

$$= 0.84$$

and thus

$$\bar{v} = 0.84 \times 11.9 = 10 \text{ m/s}$$

At this velocity the Reynolds number is

$$Re = \frac{0.3 \times 1.74 \times 10}{2.08} \times 10^{-5} = 2.51 \times 10^5$$

which shows that the use of Eq. (i) is valid.

For the adiabatic flow of an ideal gas

$$P \left(\frac{1}{\rho}\right)^\gamma = P_1 \left(\frac{1}{\rho_1}\right)^\gamma = \text{const}$$

where γ is the ratio of the molar heat capacity at constant pressure to the molar heat capacity at constant volume. Thus, for adiabatic flow, integration of Eq. (5.23) gives

$$\int_{P_1}^{P_2} P_1^{1/\gamma} \frac{1}{\rho_1} \frac{dP}{P^{1/\gamma}} = P_1^{1/\gamma} \frac{1}{\rho_1} \frac{\gamma}{\gamma - 1} (P_2^{(\gamma-1)/\gamma} - P_1^{(\gamma-1)/\gamma}) = \frac{v_1^2}{2}$$

or

$$v_1 = C_p \sqrt{\frac{2\gamma}{\gamma - 1} \frac{P_1^{1/\gamma}}{\rho_1} [P_2^{(\gamma-1)/\gamma} - P_1^{(\gamma-1)/\gamma}]} \qquad (5.24)$$

EXAMPLE 5.9

Consider the flow of air described in Example 5.8 to be adiabatic, in which case the velocity is obtained from Eq. (5.24). The constant-pressure *mass* heat capacity of air at 350 K is 1008 J/kg·K, and the difference between the constant-pressure *molar* heat capacity and the constant-volume *molar* heat capacity of an ideal gas is equal to the universal gas constant, $R = 8.3144$ J/g mol·K. The molecular weight of air is 28.84 g/g mol and thus the difference between the constant-pressure *mass* heat capacity and the constant-volume *mass* heat capacity of ideal gas air is

$$\frac{8.3144 \times 1000}{28.84} = 288 \text{ J/kg·K}$$

The constant-volume mass heat capacity of ideal gas air is thus

$$1008 - 288 = 720 \text{ J/kg·K}$$

and the ratio, γ, is

$$\gamma = \frac{1008}{720} = 1.4$$

Thus, in Eq. (5.24) with $C_p = 1$,

$$v_{max} = \sqrt{\left(\frac{2 \times 1.4}{1.4 - 1}\right) \frac{175,000^{1/1.4}}{1.74} (175,124^{(1.4-1)/1.4} - 175,000^{(1.4-1)/1.4})}$$

$$= 11.9 \text{ m/s}$$

5.13 Orifice Plate

The mass flow rate of a fluid in a pipe can be measured by means of an orifice plate placed in the pipe as shown in Fig. 5.11(a). An orifice plate is a thin circular disk containing a sharp-edged central orifice that restricts the cross-sectional area available to flow in the pipe. The sudden restriction causes turbulent eddies to form immediately in front of and behind the plate, with the consequence that the diameter of the central core of fluid moving in the flow direction decreases to a minimum value of D_2 at a location downstream from the plate called the *vena contracta*. Beyond the vena contracta the disturbing influence of the plate diminishes and the diameter of the central core of fluid increases to the original value of D_1. If friction loss through the orifice is ignored, application of Eq. (5.7) to location 1 (which is far enough upstream that the flow is not influenced by the plate) and the vena contracta (location 2) for fluid flow at a turbulent Reynolds number gives

$$\frac{P_2 - P_1}{\rho} + \frac{\bar{v}_2^2}{2} - \frac{\bar{v}_1^2}{2} = 0 \tag{i}$$

The mass balance

$$\bar{v}_1 D_1^2 = \bar{v}_2 D_2^2$$

shows that on passing through the orifice, the velocity of the fluid in the central core increases to a maximum value at the vena contracta given by

$$\bar{v}_2 = \bar{v}_1 \frac{D_1^2}{D_2^2}$$

substitution of which into Eq. (i) gives

$$\frac{P_1 - P_2}{\rho} = \frac{\bar{v}_2^2}{2}\left[1 - \left(\frac{D_2}{D_1}\right)^4\right]$$

or

$$\bar{v}_2 = \sqrt{\frac{2(P_1 - P_2)}{\rho[1 - (D_2/D_1)^4]}} \tag{ii}$$

As the friction loss through the orifice has been ignored, the velocity given by Eq. (ii) is the theoretical maximum value that the fluid can attain at the vena contracta. In addition to this approximation, neither the position nor the diameter of the vena contracta is known; the position, which is determined by the ratio D_1/D_0, can be found experimentaly, but all that is known about the diameter D_2 is that it is somewhat less than the diameter of the orifice D_0. These uncertainties are taken into consideration by defining a geometric factor, β, as

$$\beta = \frac{D_0}{D_1}$$

and introducing this and a *discharge coefficient* C_D into Eq. (ii) to give the velocity at the orifice, $\bar{v}_0$, as

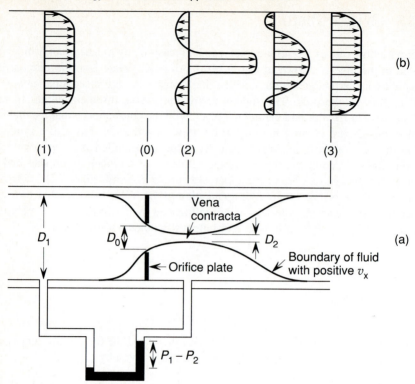

FIGURE 5.11 (a) Orifice plate for measurement of flow velocity; (b) variations in local flow velocity with position along a tube containing an orifice plate.

$$\bar{v}_0 = C_D \sqrt{\frac{2(P_1 - P_2)}{\rho(1 - \beta^4)}} \tag{5.25}$$

Alternatively, a *flow coefficient, K*, can be defined as

$$K = \frac{C_D}{\sqrt{1 - \beta^4}}$$

in which case Eq. (5.25) becomes

$$\bar{v}_0 = K \sqrt{\frac{2(P_1 - P_2)}{\rho}} \tag{5.26}$$

and the mass flow rate in the pipe is

$$\dot{M} = \bar{v}_0 \rho \pi \frac{D_0^2}{4}$$

$$= K \pi D_0^2 \sqrt{\frac{\rho(P_1 - P_2)}{8}} \tag{5.27}$$

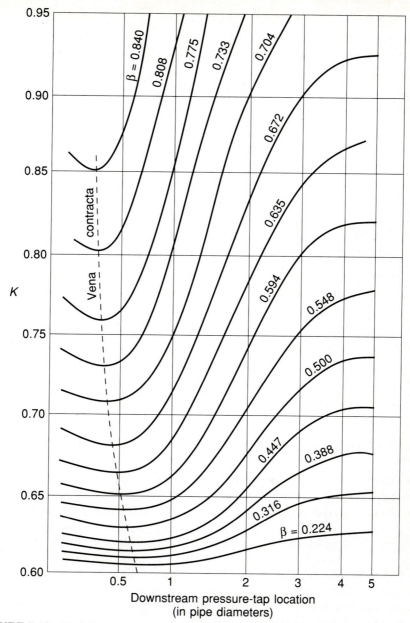

FIGURE 5.12 Variation of the flow coefficient with location of the downstream pressure tap and the geometric factor β for fluid flow through an orifice plate.

The experimentally measured variations of K with location of the downstream pressure tap and β for square-edged circular orifices with orifice Reynolds numbers of greater than 30,000 are shown in Fig. 5.12. The locus of the minimum in the curves in Fig. 5.12 defines the position of the vena contracta.

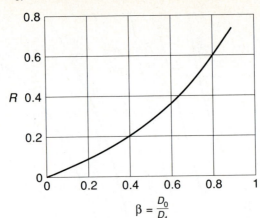

FIGURE 5.13 Variation of the fraction of the pressure drop recovered with geometric factor β for fluid flow through an orifice plate.

The increase in flow velocity through the orifice, shown in Fig. 5.11(b), causes a decrease in the static pressure in the fluid which is never fully recovered downstream from the plate. The fraction of the pressure drop recovered, R, defined as

$$R = \frac{P_3 - P_2}{P_1 - P_2}$$

varies with β as shown in Fig. 5.13.

5.14 Summary

The energy balance in flow systems is based on Bernoulli's equation, in which the flow work done on a flowing fluid is equal to the algebraic sum of the changes in kinetic and potential energy of the fluid plus the energy loss due to dissipation by friction. The energy loss in flow of a given fluid in a given flow geometry is determined by the drag force exerted at the walls of the conduit and is quantified in terms of the friction factor for the flow. In differing fluids and differing flow geometries the friction loss per unit mass of fluid per unit length of flow increases with increasing viscosity and velocity and decreases with increasing flow area and density. The energy loss in a flow system is increased by the presence of bends, fittings, and abrupt changes in flow area.

Fluid flow in pipes is generally facilitated by pumps that increase the static pressure of the fluid. The increase in pressure across a pump of given power is proportional to the density of the fluid and is inversely proportional to the mass flow rate. The potential energy of a fluid flowing under the influence of gravity can be converted to useful energy by means of a turbine, and in gravity-induced flow, steady state is reached when the rate of decrease in the potential energy of the fluid equals the rate of dissipation of energy by friction. This energy balance determines the rate at which vessels empty by drainage and the steady-state velocities attained in open channels and inclined pipes. The relationship between flow velocity and static pressure in Bernoulli's equation is the basis of devices that allow determination of the flow velocity from measurement of differences in pressure.

Problems

PROBLEM 5.1

Water is flowing in a cast iron pipe of diameter 0.5 m/s and absolute roughness 0.26 mm at an average flow velocity of 2 m/s. Calculate:

(a) The energy loss per 100 m
(b) The pressure drop in the pipe per 100 m

PROBLEM 5.2

Oil is transported through a pipeline at the rate of 5000 bbl/day. The static pressure in the oil at its point of emergence from one pumping station is 1850 kPa, and at the point of entry to the next pumping station in the line, the pressure is 900 kPa. The elevation of the second pumping station is 20 m higher than that of the first station. Calculate:

(a) The friction loss per kilogram of oil pumped
(b) The corresponding dissipation of power

The oil has a density of 770 kg/m^3. One barrel = 42 gallons and 1 gallon = 3.785 $\times$ 10^{-3} m^3.

PROBLEM 5.3

Power is generated by water at 300 K flowing at a volume flow rate of 0.2 m^3/s through a turbine. The water enters the turbine at a point that is 1 m above the point of exit and the pressure drop over the turbine is 190 kPa. Calculate the rate at which energy is transferred to the turbine.

PROBLEM 5.4

A pump rated at 8 kW pumps water from one holding tank to another at a rate of 0.015 m^3/s. Both tanks are open to the atmosphere and the free surface in the receiving tank is 10 m higher than that in the supply tank. The 0.05-m-diameter pipe connecting the tanks has an equivalent length of 40 m. What is the efficiency of the pump?

PROBLEM 5.5

Oil of density 870 kg/m^3 and viscosity 0.8 Pa·s is being pumped from the lower reservoir to the higher reservoir by a pump with a power rating of 150 kW and an efficiency of 80%. Calculate the mass flow rate of oil between the tanks if the energy loss by friction is 120 J/km.

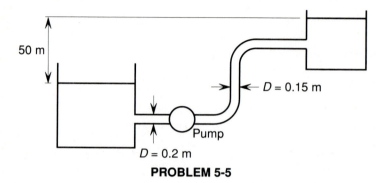

50 m

$D = 0.15$ m

Pump

$D = 0.2$ m

PROBLEM 5-5

PROBLEM 5.6

Calculate the flow rate of water at 300 K from tank 1 to tank 2.

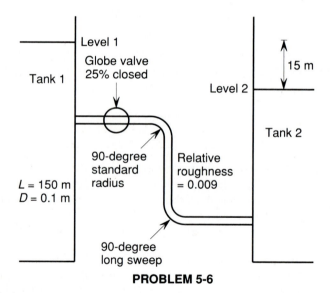

Level 1
Globe valve
25% closed

Tank 1

15 m

Level 2

Tank 2

90-degree
standard
radius

Relative
roughness
= 0.009

$L = 150$ m
$D = 0.1$ m

90-degree
long sweep

PROBLEM 5-6

PROBLEM 5.7

In the figure water at 300 K is being pumped from the lower to the higher reservoir through a pipe of 0.25 m diameter and equivalent length 75 m. The absolute roughness of the pipe is 0.0025 m. Calculate the power rating of the pump required to pump the water at a flow rate of 0.2 m³/s. The efficiency of the pump is 75%.

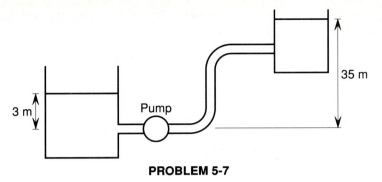

PROBLEM 5-7

PROBLEM 5.8 ──

A ladle with an inside diameter of 1.0 m and a depth of 1.5 m has a sharp-cornered tapping nozzle 0.1 m in diameter in its base. The ladle is initially filled with a liquid aluminum alloy of density 2410 kg/m^3 and viscosity 2.75×10^{-3} Pa·s. Calculate:

(a) The time required to empty the ladle by discharge through the nozzle
(b) The initial discharge rate in kg/s
(c) The discharge rate when the ladle is half-empty

PROBLEM 5.9 ──

The pipe in the figure is discharging gasoline to the atmosphere. The static pressure in the gasoline at a point 1000 m from the point of discharge is 2.5 kPa. Calculate the diameter of pipe of absolute roughness 0.0005 m required for a discharge rate of 0.1 m^3/s. For gasoline: $\rho = 720$ kg/m^3, and $\eta = 2.92 \times 10^{-4}$ Pa·s.

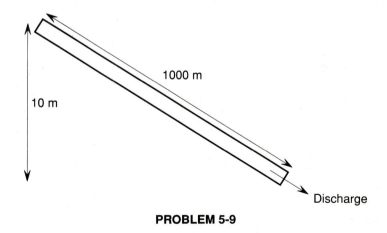

PROBLEM 5-9

PROBLEM 5.10 ──

Calculate the time required to fill the cylindrical mold with liquid metal from the tundish. The tundish is supplied with metal at a rate that maintains the free surface of metal in the tundish at a constant level. The density and viscosity of the metal are, respectively, 6420 kg/m³ and 0.00165 Pa·s.

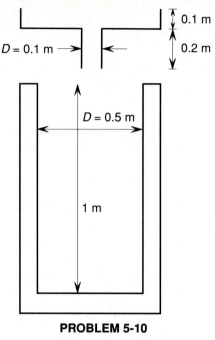

PROBLEM 5-10

PROBLEM 5.11

Calculate the time required to fill the cylindrical mold with liquid metal from the tundish. The level of the free surface of metal in the tundish remains constant. The viscosity and density of the liquid metal are, respectively, 6420 kg/m³ and 0.00165 Pa·s.

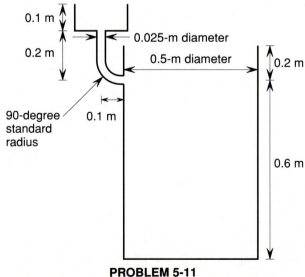

PROBLEM 5-11

PROBLEM 5.12 _____

The tank in the figure contains water that is open to the atmosphere

(a) Calculate the initial discharge rate from the tank when it is full of water.
(b) To what pressure would the tank have to be pressurized to obtain an initial discharge rate of 0.13 m³/s?
(c) When the tank is open to the atmosphere, how much time is required for the water level to fall from 30 m to 20 m?

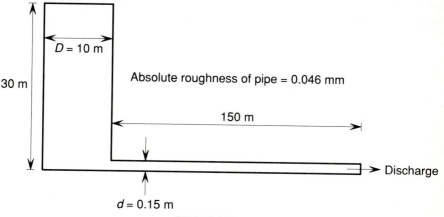

PROBLEM 5-12

PROBLEM 5.13 _____

Water at 300 K is being siphoned from a drum as shown in the figure. The siphon is a smooth-walled plastic tube of diameter 0.05 m and length 4.57 m. Calculate:

(a) The average linear velocity of water in the tube
(b) The static pressure in the water at point 2,

The additional friction loss caused by the bend in the tube can be ignored.

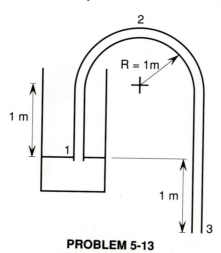

PROBLEM 5-13

PROBLEM 5.14

A pump rated at 1500 W with an efficiency of 60% pumps water at 300 K round a closed loop at a volume flow rate of 0.002 m^3/s. Calculate the pressure drop across the pump.

PROBLEM 5.15

Over a 30-m length of 0.15-m-diameter vacuum line carrying air at 300 K the pressure drops from 1500 Pa to 150 Pa. Calculate the mass flow rate if the relative roughness of the pipe is 0.002.

6

Transport of Heat by Conduction

6.1 Introduction

Consider the arrangement shown in Fig. 6.1, in which one end of a long metal rod is held over a gas flame. In this arrangement thermal energy (or heat) is transferred from the flame to the metal rod by *convection* and heat is transported along the metal rod by *conduction*. When the fuel is combusted, the heat of combustion appears as sensible heat in the gaseous products of the combustion, which, on the molecular level, appears as kinetic energy (and energy of vibration and rotation) in the individual molecules. In Chapter 2 it was seen that the average speed of the molecules in a gas is proportional to the square root of the temperature of the gas, and hence an increase in kinetic energy causes an increase in temperature. Furthermore, as the kinetic energy of the gas is proportional to the square of the average speed of the molecules, the temperature of the gas is proportional to its energy. The hot combustion gas is less dense than the surrounding air, and hence it rises, and during the upward flow of the hot gas, the energetic gas molecules collide with the surface of the metal rod. A metal can be considered as being a periodic array of positively charged ions embedded in a cloud of randomly moving free electrons, and the thermal energy and temperature of the metal are determined by the frequencies of vibration of the ions about their lattice points in the crystal and by the velocities of the free electrons. When the energetic gas molecules collide with the surface of the metal, they transfer energy that causes an increase in the frequency of vibration of the ions and an increase in the velocities of the electrons at the point of impact. The individual ions in the metal interact with ions on their neighboring lattice sites, and through this interaction, an increase in the frequency of vibration of one ion causes an increase in the frequency of vibration of its neighbors, which, in turn, causes an

235

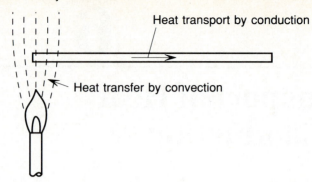

FIGURE 6.1 Transfer of heat from a flame to one end of a metal rod by convection and transfer of heat in the metal rod by conduction.

increase in the frequencies of vibration of their neighbors, and so on. Similarly, energy is transferred by collisions of electrons with themselves and with the ions in the lattice. Heat is also transported by vibratory motion of the crystal lattice as a whole and each vibration can be described as a traveling wave carrying energy. From the macroscopic point of view the transport of heat from the flame to the end of the metal rod sets up a temperature gradient in the rod and heat flows down this gradient. The questions of interest are thus: What determines the rate at which heat is transferred from the hot rising gas to the metal rod by convection, and what determines the rate at which heat is transported by conduction along the rod?

6.2 Fourier's Law and Newton's Law

Conduction is described by Fourier's law, which states that the rate at which heat is transported through a medium by conduction is proportional to the temperature gradient in the direction of the flow and to the cross-sectional area A through which the heat passes. With Q denoting a quantity of heat (J) and q denoting the rate of transport of heat (J/s), Fourier's law for unidirectional flow of heat in the x-direction is

$$q_x \propto -A_x \frac{dT}{dx}$$

or

$$q_x = -kA_x \frac{dT}{dx} \tag{6.1}$$

where the proportionality constant, k, is the *thermal conductivity* of the medium. The negative sign in Eq. (6.1) arises from the fact that for the flow of heat in the $+x$ direction, the temperature gradient in the x-direction must be negative. With reference to Fig. 6.2, which shows the temperature gradient through a slab of solid

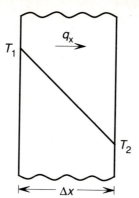

FIGURE 6.2 Transport of heat by conduction through a plane wall.

material of thickness Δx, the faces of which are maintained at the temperatures T_1 and T_2, Eq. (6.1) becomes

$$q_x = -kA_x \frac{T_2 - T_1}{\Delta x}$$

Equation (6.1) shows that the units of thermal conductivity are J/s·m·K or W/m·K. The thermal conductivities of materials vary over five orders of magnitude, from room-temperature values of 3400 W/m·K for isotopically pure diamond, to 400 W/m·K for copper, which is a good metallic thermal conductor, to 0.038 for fiberglass, which is a good thermal insulator.

Convection is described by Newton's law of cooling, which states that the rate at which heat is transferred from the surface of a solid to a fluid with which the solid is in contact is proportional to the difference between the temperature of the solid and the temperature of the fluid and to the area through which the flow occurs. With respect to Fig. 6.3, which shows the transfer of heat by convection from a solid surface at a temperature T_s to a fluid at a temperature T_∞, Newton's law of cooling is

$$q_x \propto A_s(T_s - T_\infty)$$

or

$$q_x = hA_s(T_s - T_\infty) \tag{6.2}$$

where h is the *heat transfer coefficient*. Unlike k in Eq. (6.1), which is a physical property, the value of h is dependent on several factors, such as the geometry of the surface, the nature of any flow in the fluid, and several physical properties of the fluid. Common experience indicates that a spoonful of hot soup is cooled more rapidly by being blown on than by being allowed to remain in still air. Blowing causes heat transfer by forced convection and heat transfer to still air occurs by natural convection; and the heat transfer coefficient for forced convection is larger

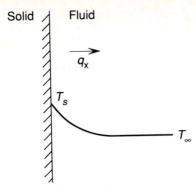

FIGURE 6.3 Transfer of heat by convection from the surface of a solid to a fluid.

than that for natural convection. Evaluation of h for any given system is one of the challenges in the subject of heat transport. The heat transfer coefficient has the units $W/m^2 \cdot K$.

6.3 Conduction

As seen in Fig. 6.4, the thermal conductivities of solids can vary significantly with temperature and integration of Eq. (6.1) must take this into account. If the solid slab shown in Fig. 6.2 is of a material that has a thermal conductivity which increases with increasing temperature, the temperature profile through the slab is shown by curve a in Fig. 6.5. As the values of q_x and A_x are constant across the slab, then, from Eq. (6.1), the product $k(dT/dx)$ is constant, and thus if k decreases with decreasing temperature, dT/dx must increase with decreasing temperature. Similarly, if k decreases with increasing temperature, the temperature profile is as shown by curve (b) in Fig. 6.5.

Any variation of the cross-sectional area, A_x, through which heat is conducted, with distance x, must also be taken into account in the integration of Eq. (6.1). Both effects are accounted for by writing Eq. (6.1) as

$$q_x = -A_x k(T) \frac{dT}{dx} \tag{6.3}$$

rearrangement of which gives

$$k(T)\, dT = -q_x \frac{dx}{A_x} \tag{6.4}$$

Integration of Eq. (6.4) from $T = T_2$ at $x = x_2$ to $T = T_1$ at $x = x_1$ gives

$$\int_{T_1}^{T_2} k(T)\, dT = -q_x \int_{x_1}^{x_2} \frac{dx}{A_x} \tag{6.5}$$

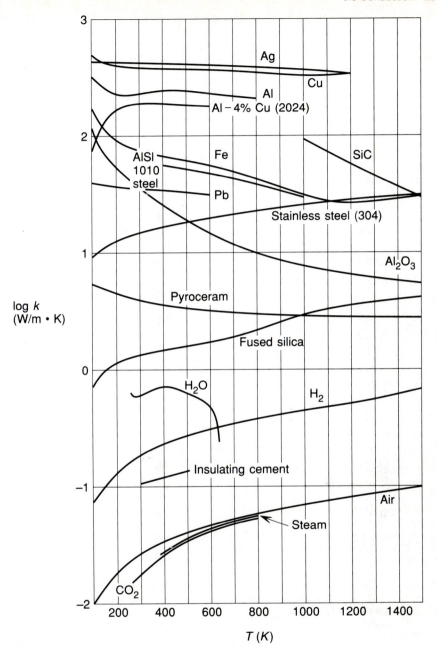

FIGURE 6.4 Variation, with temperature, of the thermal conductivities of several solids and gases.

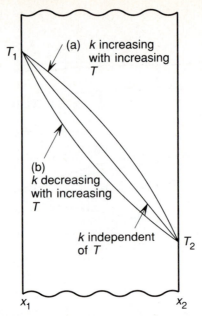

FIGURE 6.5 Influence of the temperature independence of thermal conductivity on the temperature profile in a plane wall during heat transport by conduction.

and the left side of this equation provides a definition of a mean value of the thermal conductivity, k_m, in the temperature range $T_1 - T_2$:

$$k_m(T_2 - T_1) = \int_{T_1}^{T_2} k(T) \, dT$$

Thus Eq. (6.5) can be written as

$$k_m(T_2 - T_1) = -q_x \int_{x_1}^{x_2} \frac{dx}{A_x} \tag{6.6}$$

EXAMPLE 6.1

Consider the unidirectional conduction of heat in a conically shaped conductor, the surfaces of which are thermally insulated (Fig. 6.6). The cross-sectional area of the cone normal to the direction of heat flow is $A_x = ax^2$, and thus Eq. (6.6) gives

$$k_m(T_2 - T_1) = -q_x \int_{x_1}^{x_2} \frac{dx}{ax^2} = \frac{q_x}{a} \left(\frac{1}{x_2} - \frac{1}{x_1} \right)$$

or

$$q_x = \frac{ak_m(T_2 - T_1)x_2 x_1}{(x_1 - x_2)} \tag{i}$$

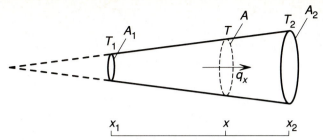

FIGURE 6.6 Unidirectional conduction of heat in the conically shaped conductor considered in Example 6.1.

Integration of Eq. (6.4) between the limits $T = T$ at $x = x$ and $T = T_1$ at $x = x_1$ gives

$$k_m(T - T_1) = \frac{q_x}{a}\left(\frac{1}{x} - \frac{1}{x_1}\right) = \frac{q_x(x_1 - x)}{ax_1x} \tag{ii}$$

and combination of Eqs. (i) and (ii) gives the variation of T with x as

$$T = T_1 - (T_1 - T_2)\frac{x_1 - x}{x_1 - x_2}\frac{x_2}{x} \tag{iii}$$

Consider a geometry in which $A_x = 0.1131x^2$, $T_1 = 1000$ K at $x_1 = 0.1$ m and $T_2 = 500$ K at $x_2 = 0.5$ m, and a material of $k_m = 5$ W/m·K. From Eq. (i),

$$q_x = \frac{0.1131 \times 5 \times (500 - 1000) \times 0.5 \times 0.1}{0.1 - 0.5}$$

$$= 35.3 \text{ W}$$

and from Eq. (iii),

$$T = 1000 - \frac{(1000 - 50) \times (0.1 - x) \times 0.5}{(0.1 - 0.5)x}$$

$$= 1000 + \frac{625(0.1 - x)}{x}$$

which is shown as curve (a) in Fig. 6.7. If $T_1 = 500$ K at $x_1 = 0.1$ m and $T_2 = 1000$ K at $x = 0.5$ m, Eq. (i) gives $q_x = -35.3$ W (heat flow in the $-x$-direction), and Eq. (iii) gives

$$T = 1000 - \frac{625(0.1 - x)}{x}$$

which is shown as curve (b) in Fig. 6.7. Figure 6.7 illustrates the influence of a change in cross-sectional area on the temperature gradient; for heat flow in the $+x$-direction, A_x increases with x and hence dT/dx decreases with x, and for heat flow in the $-x$-direction, A_x decreases in the direction of heat flow and hence

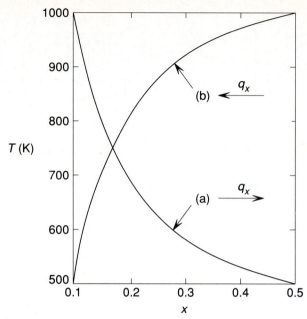

FIGURE 6.7 Temperature distributions in the conical conductor shown in Fig. 6.6 and considered in Example 6.1.

dT/dx increases in the direction of heat flow. If the variation of A_x with x is significant, the heat flow has a radial component and the problem can no longer be considered as one of simple undirectional flow.

Electric Analogy

The treatment of heat transport by conduction and convection is greatly facilitated by making an analogy with Ohm's law, which defines the resistance to the flow of electric current, R, as being the ratio of the driving force for the flow of current (the electrical voltage drop, ΔV) to the flow of electric current (amperes, I):

$$\frac{\Delta V}{I} = R$$

or

$$\frac{\text{driving force}}{\text{flux}} = \text{resistance to flow} \tag{6.7}$$

In heat transfer, with reference to Fig. 6.2, the driving force is the temperature difference $T_1 - T_2$ and the flux is the heat flow q_x. Thus a resistance to heat flow by conduction, R_k, is defined as

$$\frac{T_1 - T_2}{q_x} = R_k$$

and from Eq. (6.6), R_k is quantified as

$$\frac{T_1 - T_2}{q_x} = R_k = \frac{1}{k_m} \int_{x_1}^{x_2} \frac{dx}{A_x} \tag{6.8}$$

Similarly, a resistance to heat transport by convection, R_h is defined as

$$\frac{T_s - T_\infty}{q_x} = R_h$$

which from Eq. (6.2) can be written as

$$\frac{T_s - T_\infty}{q_x} = R_h = \frac{1}{Ah} \tag{6.9}$$

Heat Flow Through a Plane Slab

With reference to Fig. 6.2, A_x is independent of x and hence Eq. (6.8) gives

$$\frac{T_1 - T_2}{q_x} = \frac{x_2 - x_1}{Ak_m}$$

or

$$\frac{q_x}{A} = \frac{k_m(T_1 - T_2)}{x_2 - x_1} \tag{6.10}$$

Defining $q_x'(W/m^2)$ as the rate of heat transport per unit area or heat flux (q_x/A), Eq. (6.10) can be written as

$$q_x' = k_m \frac{T_1 - T_2}{x_2 - x_1} \tag{6.11}$$

Writing Eq. (6.8) as

$$\frac{T_1 - T}{q_x} = \frac{1}{k_m} \int_{x_1}^{x} \frac{dx}{A_x}$$

gives for heat flow through a plane slab,

$$\frac{T_1 - T}{q_x} = \frac{x - x_1}{Ak_m} \tag{6.12}$$

and combination of Eqs. (6.10) and (6.12) to eliminate q_x gives the temperature profile through the plane slab as

$$\frac{q_x}{Ak_m} = \frac{T_1 - T_2}{x_2 - x_1} = \frac{T_1 - T}{x - x_1}$$

or

$$T = (T_1 - T_2)\frac{x - x_1}{x_2 - x_1} \tag{6.13}$$

EXAMPLE 6.2 _____

The temperatures of the inner and outer surfaces of a glass window in a room are, respectively, 25°C and 0°C. The glass is 5 mm thick and has a mean thermal conductivity of $0.84 \ \text{W·m}^{-1}\text{·K}^{-1}$. Calculate the rate of loss of heat from the room by conduction through the glass window per unit area.

From Eq. (6.11), with the inner surface at $T_1 = 25°C$ and the outer surface at $T_2 = 0°C$,

$$q'_x = k_m \frac{T_1 - T_2}{x_1 - x_2}$$

$$= 0.84 \times \frac{25}{0.005} = 4200 \ \text{W/m}^2$$

Notice that doubling the thickness of the glass doubles the resistance to heat flow by conduction and, hence, decreases the rate of loss of heat by a factor of 2. What thickness of glass would be required to decrease that heat flux to 2000 W/m²?

$$x_1 - x_2 = k_m \frac{T_1 - T_2}{q'_x}$$

$$= 0.84 \times \frac{25}{2000} = 0.0105 \ \text{m or } 10.5 \ \text{mm}$$

Heat Flow Through the Wall of a Hollow Cylinder

Consider a hollow cylinder of inner radius R_1, outer radius R_2, and length L. If the inner surface is uniformly at the higher temperature T_1 and the outer surface is uniformly at a lower temperature T_2 heat flows in the radial direction down the temperature gradient. At a radius r, the area through which radial heat flow occurs is $2\pi rL$, and thus Eq. (6.8) becomes

$$\frac{T_1 - T_2}{q_r} = \frac{1}{k_m} \int_{R_1}^{R_2} \frac{dr}{2\pi rL} = \frac{1}{2\pi k_m L} \ln \frac{R_2}{R_1}$$

or

$$q_r = \frac{2\pi k_m L(T_1 - T_2)}{\ln (R_2/R_1)} \tag{6.14}$$

Between the limits $T = T$ at $r = r$ and $T = T_1$ at $r = R_1$, Eq. (6.8) gives

$$\frac{T_1 - T}{q_r} = \frac{1}{2\pi k_m L} \ln \frac{r}{R_1} \tag{6.15}$$

and combination of Eqs. (6.14) and (6.15) gives

$$\frac{q_r}{2\pi k_m L} = \frac{T_1 - T_2}{\ln(R_2/R_1)} = \frac{T_1 - T}{\ln(r/R_1)}$$

or

$$T = T_1 - (T_1 - T_2) \frac{\ln(r/R_1)}{\ln(R_2/R_1)} \tag{6.16}$$

as the temperature profile through the wall of the cylinder.

EXAMPLE 6.3

Hot water flows through a glass tube of inner radius 3 cm and outer radius 5 cm. The temperatures of the inner and outer surfaces of the tube are, respectively, 90°C and 85°C and the mean thermal conductivity of the glass is 0.84 $W \cdot m^{-1} \cdot K^{-1}$. Calculate the rate of heat loss from the tube per unit length, q_r''.

From Eq. (6.14), with

$$R_1 = 0.03 \text{ m}$$

$$R_2 = 0.05 \text{ m}$$

$$T_1 = 90°C$$

$$T_2 = 85°C$$

$$q_r'' = \frac{q_r}{L} = \frac{2\pi k_m(T_1 - T_2)}{\ln(R_2/R_1)}$$

$$= \frac{2\pi \times 0.84 \times (90 - 85)}{\ln(0.05/0.03)} = 51.7 \text{ W/m}$$

By how much is the rate of heat loss decreased if the wall thickness of the tube is doubled (i.e., $R_1 = 0.03$ m and $R_2 = 0.07$ m)?

$$q_r'' = \frac{2\pi \times 0.84 \times (90 - 85)}{\ln(0.07/0.03)} = 31.1 \text{ W/m}$$

Thus, in contrast with the case of a plane wall considered in Example 6.1, doubling the wall thickness of a cylindrical tube does not half the heat flux. From Eq. (6.15), increasing r from 0.05 m to 0.07 m increases the resistance to heat flow through the cylindrical wall, per unit length, from

$$R_k = \frac{1}{2\pi \times 0.84} \ln \frac{0.05}{0.03} = 0.0968$$

to

$$R_k = \frac{1}{2\pi \times 0.84} \ln \frac{0.07}{0.03} = 0.161$$

that is, the resistance increases by the factor 1.65 and hence the rate of heat loss decreases from 51.7 W/m with $R_2 = 0.05$ m to $51.7/1.65 = 31.1$ W/m with $R_2 = 0.07$ m. In the case of heat flow by conduction through a cylindrical wall, the area through which the heat flows increases with increasing radius, and hence the resistance is not the simple function of wall thickness that occurs in the case of heat flow through a plane wall.

Heat Flow Through the Wall of a Hollow Sphere

Consider a hollow sphere of inner radius R_1 and outer radius R_2. If the inner surface is at the higher temperature T_1 and the outer surface is at the lower temperature T_2, heat flows outward from the inner surface down the temperature gradient. As a radius r the area through which the heat flows is $4\pi r^2$, and thus Eq. (6.8) is

$$\frac{T_1 - T_2}{q_r} = \frac{1}{k_m} \int_{R_1}^{R_2} \frac{dr}{4\pi r^2} = \frac{1}{4\pi k_m}\left(-\frac{1}{R_2} + \frac{1}{R_1}\right)$$

or

$$q_r = \frac{4\pi k_m(T_1 - T_2)R_1 R_2}{R_2 - R_1} \tag{6.17}$$

Between the limits $T = T$ at $r = r$ and $T = T_1$ at $r = R_1$, Eq. (6.8) gives

$$\frac{T_1 - T}{q_r} = \frac{4\pi k_m(T_1 - T)R_1 r}{r - R_1} \tag{6.18}$$

and combination of Eqs. (6.17) and (6.18) gives

$$\frac{q_r}{4\pi k_m} = \frac{(T_1 - T_2)R_1 R_2}{R_2 - R_1} = \frac{(T_1 - T)R_1 r}{r - R_1}$$

or

$$T = T_1 - \frac{(T_1 - T_2)(r - R_1)R_2}{(R_2 - R_1)r} \tag{6.19}$$

as the temperature profile through the wall of the sphere.

The temperature profiles through a slab of thickness 0.4 m and the walls of a hollow cylinder and a hollow sphere each with $R_1 = 0.1$ m and $R_2 = 0.5$ m, with $T_1 = 500$ K and $T_2 = 300$ K, are shown in Fig. 6.8(a). For the plane slab, Eq. (6.13) gives the linear profile

$$T = 500 - \frac{200(x - 0.1)}{0.4}$$

For the hollow cylinder Eq. (6.16) gives

$$T = 500 - \frac{200 \ln(r/0.1)}{\ln 5}$$

which is shown as curve (a) in Fig. 6.8(a), and for the hollow sphere Eq. (6.19) gives

$$T = 500 - \frac{300(r - 0.1) \times 0.4}{0.4r}$$

which is shown as curve (b) in Fig. 6.8(a). The corresponding temperature profiles for $T_1 = 300$ K at $R_1 = 0.1$ m and $T_2 = 500$ K at $R_2 = 0.4$ m are shown in Fig. 6.8(b).

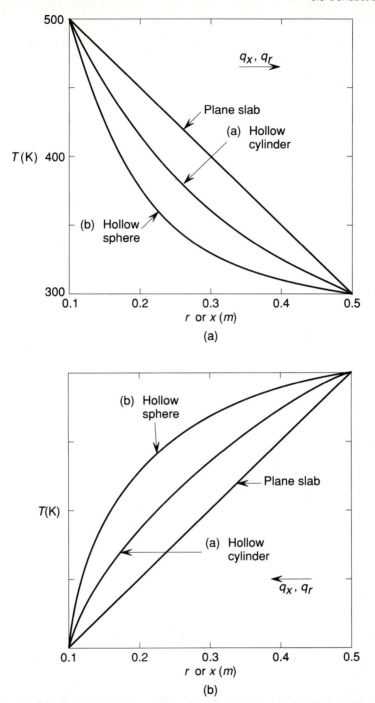

FIGURE 6.8 Temperature distributions in a plane slab, the wall of a hollow cylinder, and the wall of a hollow sphere during heat flow by conduction from (a) the inner radius to the outer radius and (b) the outer radius to the inner radius.

Heat Flow Through a Composite Wall

Figure 6.9 shows the temperature profile caused by the flow of heat from a hot fluid at T_i on one side of a composite plane wall to a cooler fluid at T_o on the other side of the wall. The composite plane wall consists of slabs of three different solid materials of thicknesses Δx_1, Δx_2, and Δx_3 and thermal conductivities k_1, k_2, and k_3. Heat is transferred by convection from the hot fluid to the inner surface of the wall, is transferred through the composite wall by conduction, and is transferred by convection from the outer surface of the wall to the cooler fluid. At the inner surface the heat transfer coefficient is h_i and at the outer surface the heat transfer coefficient is h_o. From the electric analogy

$$\frac{T_i - T_o}{q_x} = R$$

and in this example, there are five resistances to the flow of heat; $R_{h,i}$ associated with convection at the inner surface, $R_{k,1}$, $R_{k,2}$, and $R_{k,3}$ associated with conduction through the three solid layers and $R_{h,o}$ associated with convection at the outer surface. As these resistances occur in series the electric analogy gives

$$\frac{T_i - T_o}{q_x} = R_{h,i} + R_{k,1} + R_{k,2} + R_{k,3} + R_{h,o} \tag{6.20}$$

From Eq. (6.9)

$$R_{h,i} = \frac{1}{A h_i}$$

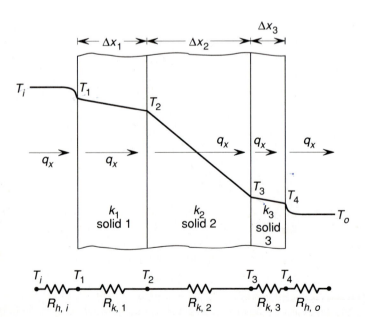

FIGURE 6.9 Heat transport by conduction through a composite plane wall.

and

$$R_{h,o} = \frac{1}{Ah_o}$$

and from Eq. (6.8), with A_x constant for a plane wall,

$$R_{k,1} = \frac{\Delta x_1}{k_1 A}$$

$$R_{k,2} = \frac{\Delta_2}{k_2 A}$$

and

$$R_{k,3} = \frac{\Delta x_3}{k_3 A}$$

and thus Eq. (6.20) becomes

$$\frac{T_i - T_o}{q_x} = \frac{1}{Ah_i} + \frac{\Delta x_1}{k_1 A} + \frac{\Delta x_2}{k_2 A} + \frac{\Delta x_3}{k_3 A} + \frac{1}{Ah_o} \qquad (6.21)$$

or in terms of heat flow per unit area,

$$\frac{T_i - T_o}{q'_x} = \frac{1}{h_i} + \frac{\Delta x_1}{k_1} + \frac{\Delta x_2}{k_2} + \frac{\Delta x_3}{k_3} + \frac{1}{h_o} \qquad (6.22)$$

EXAMPLE 6.4

Consider the composite wall to be a furnace wall consisting of a 15-cm thickness of silica brick (of $k = 1.1$ W/m·K) as material 1, a 5-cm thickness of glass fiber (of $k = 0.035$ W/m·K) as material 2, and a 1-cm thickness of steel (of $k = 41$ W/m·K) as material 3. $T_i = 500°C$, $h_i = 15$ W/m²·K and $T_o = 20°C$, $h_o = 20$ W/m²·K. Insertion into Eq. (6.22) gives

$$\begin{aligned}
\frac{500 - 20}{q'_x} &= \frac{1}{15} + \frac{0.15}{1.1} + \frac{0.05}{0.035} + \frac{0.01}{41} + \frac{1}{20} \\
&= (6.67 \times 10^{-2}) + (0.136) + (1.43) \\
&\quad + (2.44 \times 10^{-4}) + (0.05) \\
&= 1.682
\end{aligned}$$

Thus, of the total thermal resistance, 85% is provided by the glass fiber, 8% is provided by the silica brick, 4% is provided by convection at the inner wall, 3% is provided by convection at the outer wall, and virtually zero is provided by the steel. The heat flux through the furnace wall is

$$q'_x = \frac{500 - 20}{1.682} = 285.4 \text{ W/m}^2$$

The temperatures of the surfaces and interfaces are calculated by applying the electric analogy to each of the individual resistances. The temperature at the inner surface of the silica brick, T_1, is determined by $R_{h,i}$ and q'_x and is obtained as

$$\frac{T_i - T_1}{q'_x} = \frac{1}{h_i} = \frac{500 - T_1}{285.4} = \frac{1}{15}$$

or $T_1 = 481°C$. Similarly, the temperature at the interface between the silica brick and the glass fiber, T_2, is obtained as

$$\frac{T_1 - T_2}{q'_x} = \frac{\Delta x_1}{k_1} = \frac{481 - T_2}{285.4} = \frac{0.15}{1.1}$$

or $T_2 = 442°C$.

The temperature at the interface between the glass fiber and the steel, T_3, is obtained from

$$\frac{T_2 - T_3}{q'_x} = \frac{\Delta x_2}{k_2} = \frac{442 - T_3}{285.4} = \frac{0.05}{0.035}$$

as $34°C$. In similar fashion the temperature at the outer surface of the steel is calculated as approximately $34°C$. In this example the temperature at the interface between the silica brick and the glass fiber was calculated as being $442°C$. If it were required that the maximum temperature to be attained by the glass fiber be lower than this, say $400°C$, the furnace would have to be designed with a thicker layer of silica brick and the value of q'_x would be different. Application of the electric analogy to the resistances offered by convection to the silica brick and conduction through a thickness x of silica brick, for $T_i - T_2 = 500 - 400$, gives

$$\frac{500 - 400}{q'_x} = \frac{1}{15} + \frac{x}{1.1} \tag{i}$$

and for heat transfer from the inside to the outside of the furnace,

$$\frac{500 - 20}{q'_x} = \frac{1}{15} + \frac{x}{1.1} + \frac{0.05}{0.035} + \frac{0.01}{41} + \frac{1}{20} \tag{ii}$$

Simultaneous solution of Eqs. (i) and (ii) gives

$$x = 0.355 \text{ m}$$

and

$$q'_x = 257 \text{ W/m}^2$$

Thus, increasing the thickness of the silica brick from 0.15 m to 0.355 m decreases the temperature at the interface between the silica brick and the glass fiber from $442°C$ to $400°C$ and decreases the rate of heat transfer from the furnace from 285.4 W/m^2 to 257 W/m^2.

The electric analogy also holds for thermal resistances in parallel with one another. Consider the composite wall, shown in Fig. 6.10, constructed from four dif-

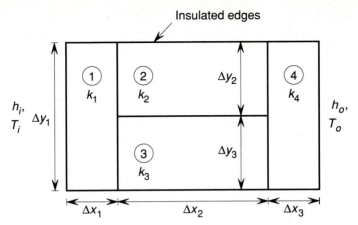

FIGURE 6.10 Transfer of heat by conduction through a plane composite wall in which the thermal resistances of materials 2 and 3 are in parallel.

ferent solid materials, in which the thermal resistances of materials 2 and 3 are in parallel. The thermal resistance to heat flow from the hotter fluid at T_i to the cooler fluid at T_o is

$$R = R_{h,i} + R_{k,1} + R_{k,2-3} + R_{k,4} + R_{h,o}$$

where the resistance $R_{k,2-3}$ is given by

$$\frac{1}{R_{k,2-3}} = \frac{1}{R_{k,2}} + \frac{1}{R_{k,3}}$$

or

$$R_{k,2-3} = \frac{R_{k,2} \times R_{k,3}}{R_{k,2} + R_{k,3}}$$

Thus

$$R = R_{h,1} + R_{k,1} + \frac{R_{k,2} \times R_{k,3}}{R_{k,2} + R_{k,3}} + R_{k,4} + R_{h,o}$$

or, for unit depth of wall ($\Delta z = 1$),

$$R = \frac{1}{\Delta y_1 h_i} + \frac{\Delta x_1}{\Delta y_1 k_1} + \frac{(\Delta x_2 / \Delta y_2 k_2)(\Delta x_2 / \Delta y_3 k_3)}{(\Delta x_2 / \Delta y_2 k_2) + (\Delta x_2 / \Delta y_3 k_3)} + \frac{\Delta x_3}{\Delta y_1 k_4} + \frac{1}{\Delta y_1 h_o}$$

For

$$\Delta x_1 = \Delta x_3 = 0.05 \text{ m}, \Delta x_2 = 0.1 \text{ m}$$

$$\Delta y_1 = 2 \text{ m}, \Delta y_2 = \Delta y_3 = 1 \text{ m}$$

$$k_1 = 20 \text{ W/m·K}, k_2 = 5 \text{ W/m·K}, k_3 = 4 \text{ W/m·K}, k_4 = 10 \text{ W/m·K}$$

$h_i = 20 \text{ W/m}^2 \cdot \text{K}, \ h_o = 10 \text{ W/m}^2 \cdot \text{K}$

$T_i = 50°C, \ T_o = 20°C$

$$R = \frac{1}{2 \times 20} + \frac{0.05}{2 \times 20}$$

$$+ \frac{(0.1/(1 \times 5))(0.1/(1 \times 4))}{0.1/(1 \times 5) + 0.1/(1 \times 4)} + \frac{0.05}{2 \times 10} + \frac{1}{2 \times 10}$$

$$= 2.5 \times 10^{-2} + 1.25 \times 10^{-3} + 1.11 \times 10^{-2} + 2.5 \times 10^{-3} + 0.05$$

$$= 8.99 \times 10^{-2}$$

Therefore,

$$q_x(\text{per unit depth}) = \frac{T_i - T_o}{R} = \frac{30}{8.99 \times 10^{-2}} = 334 \text{ W}$$

T_1 is obtained from

$$\frac{T_i - T_1}{q_x} = R_{h,i} + R_{k,1}$$

as

$$T_1 = 50 - 334(2.5 \times 10^{-2} + 1.25 \times 10^{-3}) = 41.2°C$$

and T_2 is obtained from

$$\frac{T_2 - T_o}{q_x} = R_{k,4} + R_{h,i}$$

as

$$T_2 = 20 + 334(2.5 \times 10^{-3} + 0.05) = 37.5°C$$

The heat flow (per unit depth) through material 2 is thus

$$q_x = \frac{k_2 A(T_1 - T_2)}{\Delta x_2} = \frac{5 \times 1 \times (4.12 - 37.5)}{0.1} = 185 \text{ W}$$

and the heat flow (per unit depth) through material 3 is

$$q_x = \frac{k_3 A(T_1 - T_2)}{\Delta x_2} = \frac{4 \times 1 \times (41.2 - 37.5)}{0.1} = 148 \text{ W}$$

The sum of these two flows is the total flow of 334 W, and their ratio, $185/148 = 5/4$, is the ratio k_2/k_3.

Heat Flow Through a Composite Cylindrical Wall

Figure 6.11 shows the temperature profile caused by the flow of heat from a hot fluid at T_i inside a hollow composite cylinder through the walls of the cylinder to a cooler fluid at T_o outside. This flow of heat encounters resistances to convection to

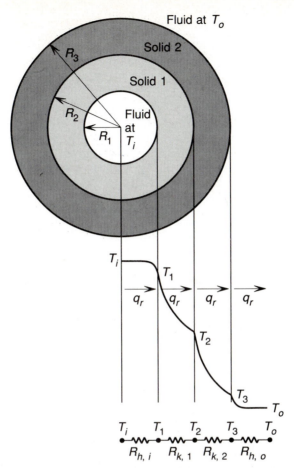

FIGURE 6.11 Heat transport by conduction through a composite cylindrical wall.

the inner surface, $R_{h,i}$, to convection from the outer surface, $R_{h,o}$ and resistance to conduction through material 1, $R_{k,1}$, and material 2, $R_{k,2}$. Again the electric analogy gives

$$\frac{T_i - T_o}{q_r} = R_{h,i} + R_{k,1} + R_{k,2} + R_{h,o} \qquad \text{(i)}$$

where for a cylinder of length L,

$$R_{h,1} = \frac{1}{Ah_i} = \frac{1}{2\pi R_1 L h_i}$$

$$R_{h,o} = \frac{1}{Ah_o} = \frac{1}{2\pi R_3 L h_o}$$

and from Eq. (6.14),

$$R_{k,1} = \frac{\ln(R_2/R_1)}{2\pi k_1 L}$$

and

$$R_{k,2} = \frac{\ln(R_3/R_2)}{2\pi k_2 L}$$

The heat flux per unit length of cylinder, q_r'', is thus given by

$$\frac{T_i - T_o}{q_r''} = \frac{1}{2\pi R_1 h_i} + \frac{\ln(R_2/R_1)}{2\pi k_1} + \frac{\ln(R_3/R_2)}{2\pi k_2} + \frac{1}{2\pi R_3 h_o} \tag{ii}$$

EXAMPLE 6.5

Consider the composite hollow cylinder to be a steel steam pipe of I.D. 3 cm and O.D. 3.5 cm covered with a 1.25-cm layer of insulating material of $k = 0.035$ W/m·K. The steam in the pipe is at 120°C and the temperature of the air, T_o, is 15°C. The thermal conductivity of the steel is 41 W/m·K and the values of h_i and h_o are, respectively, 150 W/m²·K and 30 W/m²·K.

From Eq. (ii),

$$\frac{120 - 15}{q_r''} = \frac{1}{2\pi \times 0.015 \times 150} + \frac{\ln(3.5/3)}{2\pi \times 41}$$

$$+ \frac{\ln(3/1.75)}{2\pi \times 0.035} + \frac{1}{2\pi \times 0.03 \times 30}$$

$$= (0.0707) + (5.98 \times 10^{-4}) + (2.450) + (0.1768)$$

$$= 2.698$$

and thus the rate of loss of heat per unit length of pipe is

$$q_r'' = \frac{120 - 15}{2.698} = 38.9 \text{ W/m}$$

Of the resistance, 2.450/2.698 or 91% is provided by the insulating material. The surface and interface temperatures are determined by the individual resistances

$$\frac{120 - T_1}{38.9} = \frac{1}{2\pi R_1 h_i} = 0.0707 \qquad \text{therefore,} \quad T_1 = 117°\text{C}$$

$$\frac{120 - T_2}{38.9} = \frac{1}{2\pi R_1 h_i} + \frac{\ln(R_2/R_1)}{2\pi k_1}$$

$$= 0.0707 + 5.09 \times 10^{-4} \qquad \text{therefore,} \quad T_2 \approx 117°\text{C}$$

$$\frac{120 - T_3}{38.9} = 0.0707 + 5.98 \times 10^{-4} + 2.450 \qquad \text{therefore,} \quad T_3 = 21.9°\text{C}$$

Inspection of Eq. (ii) shows that increasing the outer radius of the insulating material, R_3, increases the resistance to heat flow by conduction through the insulation [the third term on the right-hand side of Eq. (ii)] but decreases the resistance to heat transfer by convection from the outer surface of the insulation [the last term on the right side of Eq. (ii)]. The latter arises from the fact that increasing the thickness of the insulation increases the area of outer surface from which convection occurs. Thus there must exist some critical radius of insulation above or below which, with increasing radius, the decrease in the resistance to flow by convection is greater than the increase in resistance to flow by conduction through the insulation, in which case a situation can occur in which an increase in the thickness of the insulation actually increases the heat flux. Writing Eq. (ii) as

$$\frac{T_i - T_o}{q_r''} = R_{h,i} + R_{k,1} + \frac{\ln(R_3/R_2)}{2\pi k_2} + \frac{1}{2\pi R_3 h_o}$$

and differentiating with respect to R_3 gives

$$\frac{T_o - T_i}{q_r''^2}\frac{dq_r''}{dR_3} = \frac{1}{2\pi k_2 R_3} - \frac{1}{2\pi h_o R_3^2}$$

or

$$\frac{1}{q_r''^2}\frac{dq_r''}{dR_3} = \frac{1}{T_o - T_i}\frac{1}{2\pi R_3}\left(\frac{1}{k_2} - \frac{1}{h_o R_3}\right)$$

which is zero when

$$R_3 = \frac{k_2}{h_o} \tag{iii}$$

The second derivative of q_r'' with respect to R_3, evaluated at $R_3 = k_2/h_o$, is a negative quantity, which indicates that q_r'' has a maximum value when the radius of insulation, R_o, has the value

$$R_o = \frac{k_{\text{insulation}}}{h_o} \tag{6.23}$$

R_o is the *critical radius of insulation*, and for a typical $k_{\text{insulation}}$ of 0.04 W/m·K and the lowest attainable value of h_o of approximately 5 W/m²·K,

$$R_o = \frac{0.04}{5} = 0.008 \text{ m} = 8 \text{ mm}$$

Line (a) in Fig. 6.12 shows the influence of adding insulating material of thermal conductivity 0.04 W/m·K to a wire of radius 1 mm when the difference between the temperature of the surface of the wire and T_o is 20°C with $h_o = 5$ W/m²·K. The rate of heat loss per unit length of wire increases rapidly to a maximum value at $R_{\text{critical}} = 8$ mm and then decreases slowly as the thickness of insulation is increased further. Line (b) in the figure is for the case where the radius of the cylinder is 10 mm. Being of radius greater than the critical radius of insulation, the addition of insulation to this cylinder decreases the rate of heat loss per unit length.

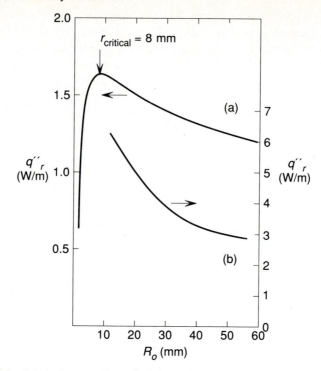

FIGURE 6.12 Dependence of heat flux through a cylinder on the radius of the cylinder and illustration of the critical radius of insulation.

6.4 Conduction in Heat Sources

The generation of heat within a medium has a significant influence on the temperature gradients that are established within the medium when steady state is attained. Heat can be generated in solid materials by such means as electrical resistance heating and nuclear fission. Consider the one-dimensional transport of heat by conduction in the x-direction through a solid in which heat is being generated at the rate of $\dot{q}(W/m^3)$ shown in Fig. 6.13. An energy balance on the differential volume $A_x\,dx$ indicates that the difference between the heat flux out of the volume and the heat flux into the volume is equal to the heat generated within the volume; that is, for heat transport by conduction in the x-direction,

$$dq_x = \dot{q}A_x\,dx$$

which, from Fourier's law, with a mean thermal conductivity, k_m, becomes

$$-d\left(k_m A_x \frac{dT}{dx}\right) = \dot{q}A_x\,dx$$

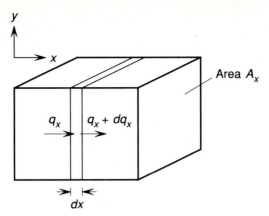

FIGURE 6.13 Heat flow by conduction through a heat-generating medium.

or

$$\frac{\dot{q}}{k_m} = -\frac{1}{A_x}\frac{d}{dx}\left(A_x\frac{dT}{dx}\right) \tag{6.24}$$

Heat Generation in a Plane Slab

As A_x is independent of x in a plane slab, Eq. (6.24) can be written as

$$\frac{d^2T}{dx^2} = -\frac{\dot{q}}{k_m}$$

the first integration of which gives

$$\frac{dT}{dx} = -\frac{\dot{q}x}{k_m} + C_1$$

and the second integration of which gives

$$T = -\frac{\dot{q}x^2}{2k_m} + C_1x + C_2 \tag{i}$$

The case of a plane slab is most conveniently dealt with by placing the origin of the x-axis at the centerline of the slab and placing the faces at $x = L$ and $x = -L$. If both faces of the slab are at the same temperature T_s, Eq. (i) gives

$$T_s \begin{cases} -\dfrac{\dot{q}L^2}{2k_m} + C_1L + C_2 & \text{at } x = L \\[2ex] -\dfrac{\dot{q}L^2}{2k_m} - C_1L + C_2 & \text{at } x = -L \end{cases}$$

which shows that

$$C_1 = 0$$

and

$$C_2 = T_s + \frac{\dot{q}L^2}{2k_m}$$

Equation (i) thus gives the temperature profile in the plane slab as

$$T = T_s + \frac{\dot{q}}{2k_m} (L^2 - x^2) \qquad (6.25)$$

(i.e., T is a parabolic function of x). The difference between the maximum temperature at the centerline ($x = 0$) and T_s is

$$T_{max} - T_s = \frac{\dot{q}L^2}{2k_m} \qquad (6.26)$$

which is proportional to the rate of generation of heat and is inversely proportional to the thermal conductivity of the solid. Equation (6.25) is analogous to the variation of the local flow velocity in a fluid flowing between two horizontal flate plates given by Eq. (2.10).

$$v_x = \frac{\Delta P}{2L\eta} (\delta^2 - y^2) \qquad (2.10)$$

In this analogy, $\dot{q}/k_m$, in the case of heat flow, is the equivalent of $\Delta P/L\eta$ in the case of fluid flow.

At steady state, the rate at which heat leaves the slab is equal to the rate at which it is generated in the slab. The latter is given by

$$\dot{q}A_x 2L \qquad (ii)$$

and the rate at which heat passes through area A_x at the surface at $x = L$ is

$$q_x = -k_m A_x \frac{dT}{dx}\bigg|_{x=L} \qquad (iii)$$

From Eq. (6.25)

$$\frac{dT}{dx}\bigg|_{x=L} = \frac{-\dot{q}L}{k_m}$$

and hence Eq. (iii) is

$$q_x = \dot{q}A_x L \qquad (iv)$$

Similarly, the rate at which heat passes through area A_x at the surface at $x = -L$ is

$$q_x = -k_m A_x \frac{dT}{dx}\bigg|_{x=-L}$$

$$= -k_m A_x \frac{\dot{q}L}{k_m}$$

$$= -\dot{q}A_x L \qquad \text{(v)}$$

where the negative sign indicates that heat flow is in the $-x$ direction. Thus, as required, the sum of the magnitudes of the heat fluxes given by Eqs. (iv) and (v) gives Eq. (ii).

EXAMPLE 6.6

Consider a plane slab of thickness $2L = 0.1$ m, which generates heat at the rate of 250,000 W/m³. Both faces are in contact with air at $T_\infty = 15°C$, the heat transfer coefficient at each face is 60 W/m²·K, and the thermal conductivity of the slab is 25 W/m·K. Calculate the temperature profile in the slab.

As the heat transfer conditions are identical at both faces of the slab, the rate at which heat is transferred through unit surface area of the slab, q'_x, is equal to the rate of heat generation in the volume L (length) × 1 (area):

$$q'_x = \dot{q}L$$

and from the electric analogy for heat transfer by convection from the surface to the air,

$$\frac{T_s - T_\infty}{q'_x} = \frac{1}{h}$$

which gives

$$T_s = T_\infty + \frac{q'_x}{h}$$

$$= 15 + \frac{250,000 \times 0.05}{60}$$

$$= 223.3°C$$

The temperature profile in the slab is then obtained from Eq. (6.25) as

$$T = 223 + \frac{250,000}{2 \times 25} (0.05^2 - x^2)$$

$$= 235.8 - 5000x^2$$

This variation is shown as line (a) in Fig. 6.14.

An asymmetric temperature profile occurs in the slab if the heat transfer conditions at the two faces of the slab are not identical. Consider the same value of T_∞ on both sides of the slab, but let h_1 at the left face at $x = -L$ be different from h_2 at the right face at $x = L$. The temperature of the left face, T_1 will now be different from the temperature of the right face T_2. Equation (i) gives

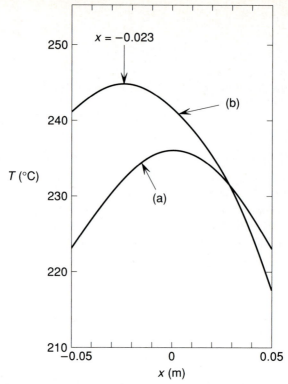

FIGURE 6.14 Temperature profiles through a heat-generating medium. Line *a* shows the temperature profile for identical heat transfer conditions at both faces of the slab and line *b* shows the temperature profile when the heat transfer conditions at the two faces differ from one another.

$$T_1 = \frac{-\dot{q}L^2}{2k_m} - C_1 L + C_2 \qquad \text{at } x = -L$$

and

$$T_2 = \frac{-\dot{q}L^2}{2k_m} + C_1 L + C_2 \qquad \text{at } x = L$$

The integration constants are evaluated as

$$C_1 = \frac{T_2 - T_1}{2L}$$

and

$$C_2 = T_1 + \frac{\dot{q}L^2}{2k_m} + \frac{T_2 - T_1}{2}$$

and thus from Eq. (i), the temperature profile is

$$T = T_1 + \frac{\dot{q}}{2k_m}(L^2 - x^2) + \frac{T_2 - T_1}{2}\left(1 + \frac{x}{L}\right) \tag{vi}$$

which is the combination of a parabola and a straight line. Differentiation of Eq. (vi) with respect to x gives

$$\frac{dT}{dx} = \frac{-\dot{q}x}{k_m} + \frac{T_2 - T_1}{2L} \tag{vii}$$

which locates the maximum temperature in the slab at

$$x = \frac{k_m(T_2 - T_1)}{2\dot{q}L} \tag{viii}$$

At each surface, the rate of conduction of heat at the surface equals the rate at which heat is transferred from the surface to the fluid by convection. Thus at the surface at $x = L$,

$$-k_m\frac{dT}{dx}\bigg|_{x=L} = h_2(T_2 - T_\infty) \tag{ix}$$

and at the surface at $x = -L$,

$$-k_m\frac{dT}{dx}\bigg|_{x=-L} = h_1(T_\infty - T_1)$$

Thus, from Eq. (vii), at $x = L$,

$$\dot{q}L - \frac{k_m(T_2 - T_1)}{2L} = h_2(T_2 - T_\infty) \tag{x}$$

and at $x = -L$,

$$-\dot{q}L - \frac{k_m(T_2 - T_1)}{2L} = h_1(T_\infty - T_1) \tag{xi}$$

Simultaneous solution of Eqs. (x) and (xi) gives

$$T_1 = \frac{2\dot{q}L(k_m + Lh_2) + T_\infty[(h_1 + h_2)k_m - 2Lh_1h_2]}{k_m(h_1 + h_2) + 2Lh_1h_2} \tag{xii}$$

and

$$T_2 = \frac{2\dot{q}L - h_1T_1 + T_\infty(h_1 + h_2)}{h_2} \tag{xiii}$$

In the previous example, with $T_\infty = 15°C$, $h_1 = 30$ W/m²·K and $h_2 = 90$ W/m²·K, Eqs. (xii) and (xiii) give

$$T_1 = 240.5°C$$

and

$$T_2 = 217.6°C$$

and the temeprature profile in the slab, as given by Eq. (vi), is shown as line b in Fig. 6.14. From Eq. (viii) the maximum temperature in the slab occurs at

$$x = \frac{25 \times (217.6 - 240.5)}{2 \times 250000 \times 0.05} = -0.023 \text{ m}$$

If an adiabatic barrier is placed on one side of the slab, all of the heat generated flows out of the other side. In this case side 1 at T_1 is located at $x = 0$ and side 2 at T_2 located at $x = L$. Equation (i) then gives

$$T_1 = C_2 \qquad\qquad \text{at } x = 0$$

and

$$T_2 = \frac{-\dot{q}L^2}{2k_m} + C_1L + C_2 \qquad \text{at } x = L$$

Therefore,

$$C_1 = \frac{T_2 - T_2}{L} + \frac{\dot{q}L}{2k_m}$$

and the temperature profile is

$$T = \frac{\dot{q}x}{2k_m}(L - x) + \frac{(T_2 - T_1)x}{L} + T_1 \qquad\qquad (xiv)$$

differentiation of which gives

$$\frac{dT}{dx} = \frac{-\dot{q}x}{k_m} + \frac{T_2 - T_1}{L} + \frac{\dot{q}L}{2k_m}$$

As there is no heat flow through side 1 at $x = 0$,

$$\left.\frac{dT}{dx}\right|_{x=0} = \frac{T_2 - T_1}{L} + \frac{\dot{q}L}{2k_m} = 0$$

and hence

$$T_1 = T_2 + \frac{\dot{q}L^2}{2k_m} \qquad\qquad (xv)$$

At side 2

$$\dot{q}L = h_2(T_2 - T_\infty)$$

or

$$T_2 = \frac{\dot{q}L}{h_2} + T_\infty$$

substitution of which into Eq. (xv) gives

$$T_1 = \frac{\dot{q}L}{h_2} + T_\infty + \frac{\dot{q}L^2}{2k_m} \tag{xvi}$$

Considering the slab of thickness 0.1 m, $\dot{q} = 250{,}000$ W/m³, and $k_m = 25$ W/m·K, with an adiabatic side 1 and $h_2 = 60$ W/m²·K and $T_\infty = 15°C$ at side 2, Eqs. (xv) and (xvi) give

$$T_1 = 481.6°C$$

$$T_2 = 431.7°C$$

and the temperature profile given by Eq. (xiv) is shown in Fig. 6.15.

Heat Generation in a Solid Cylinder

At steady state, the heat generated in a cylinder of radius R and length L flows down the temperature gradient in the radial direction toward the surface of the cylinder, which is at the temperature T_s. Again symmetry indicates that the temperature has a maximum value at the centerline or axis of the cylinder, $r = 0$, and at the radius r, the area through which heat flow occurs is $A_r = 2\pi r L$. Thus Eq. (6.24) is written as

$$\frac{\dot{q}}{k_m} = -\frac{1}{2\pi r L}\frac{d}{dr}\left(2\pi r L \frac{dT}{dr}\right)$$

or

$$\frac{d}{dr}\left(r \frac{dT}{dr}\right) = -\frac{\dot{q}r}{k_m}$$

integration of which gives

$$r \frac{dT}{dr} = -\frac{\dot{q}r^2}{2k_m} + C_1$$

From the boundary condition $dT/dr = 0$ at $r = 0$, $C_1 = 0$ and hence

$$\frac{dT}{dr} = -\frac{\dot{q}r}{2k_m}$$

A second integration gives

$$T = -\frac{\dot{q}r^2}{4k_m} + C_2$$

The boundary condition $T = T_s$ at $r = R$ gives

$$C_2 = T_s + \frac{\dot{q}R^2}{4k_m}$$

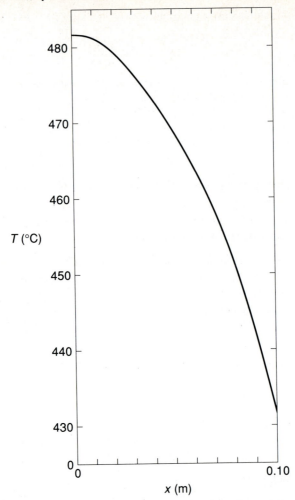

FIGURE 6.15 Temperature profile in a heat-generating plane slab when one surface is adiabatic.

and hence the temperature profile is

$$T = T_s + \frac{\dot{q}}{4k_m} (R^2 - r^2) \tag{6.27}$$

Equation (6.27) is analogous with Eq. (2.20) for the local flow velocity of a fluid in a circular pipe:

$$v_z = (\Delta P / L) + \frac{\rho g}{4\eta} (R^2 - r^2) \tag{2.20}$$

Again, the rate at which heat passes through the surface of the cylinder,

$$q = -k_m A_r \left. \frac{dT}{dr} \right|_R \bigg|_R$$

$$= -k_m 2\pi RL \frac{-\dot{q}R}{2k_m}$$

$$= \pi R^2 L\dot{q}$$

is equal to the rate at which the heat is generated in the cylinder.

Resistance Heating of Electrical Wires

Figure 6.16 shows an electrically conducting wire of radius R_1 covered with an electrical insulator of radius R_2. When an electric current is passed through the wire, the resistance to its passage generates heat at a rate given by the product of the voltage drop over the length of wire, V, and the current flow, I:

$$\text{rate of generation of heat (W)} = V \text{ (volts)} \times I \text{ (amperes)}$$

From Ohm's law, the electrical resistance of the wire, R, is defined by

$$\frac{V(\text{volts})}{I(\text{amperes})} = R(\text{ohms})$$

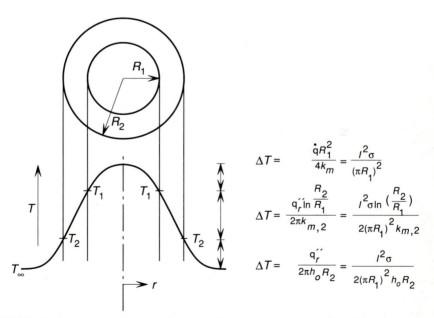

$$\Delta T = \frac{\dot{q}R_1^2}{4k_m} = \frac{I^2 \sigma}{(\pi R_1)^2}$$

$$\Delta T = \frac{q_r'' \ln \frac{R_2}{R_1}}{2\pi k_{m,2}} = \frac{I^2 \sigma \ln \left(\frac{R_2}{R_1} \right)}{2(\pi R_1)^2 k_{m,2}}$$

$$\Delta T = \frac{q_r''}{2\pi h_o R_2} = \frac{I^2 \sigma}{2(\pi R_1)^2 h_o R_2}$$

FIGURE 6.16 Temperature profile in a cylindrical composite consisting of a heat-generating inner cylinder and a non-heat-generating cylindrical covering.

and hence the rate of generation of heat in the wire is

$$I^2(\text{amperes}) \times R(\text{ohms})$$

The electrical resistance of the wire is proportional to the length of the wire, L, and is inversely proportional to the cross-sectional area, A, and the specific resistivity of the wire material, ρ, is defined by

$$R = \frac{\rho L}{A}$$

in which ρ has the units $\Omega \cdot m$. The rate of generation of heat in the wire is thus

$$\frac{I^2 \rho L}{A}$$

or the rate of generation of heat per unit volume of the wire is

$$\begin{aligned}
\dot{q} &= \frac{I^2 \rho L}{A} \times \frac{1}{AL} \\
&= \frac{I^2 \rho}{A^2}
\end{aligned} \tag{6.28}$$

With reference to Fig. 6.16,

$$\dot{q} = \frac{I^2 \rho}{(\pi R_1^2)^2} \tag{i}$$

The rate at which heat is generated per unit length of the wire is then

$$\dot{q}(\pi R_1^2) = \frac{I^2 \rho}{\pi R_1^2} \tag{ii}$$

and this heat is dissipated by conduction through the wire and the insulator in the radial direction and by convection from the outer surface of the insulator, and the radial flux per unit length, q_r'', is equal to the rate at which the heat is generated per unit length:

$$q_r'' = \frac{I^2 \rho}{\pi R_1^2} \tag{iii}$$

From the electric analogy, with reference to the temperature profile shown in Fig. 6.16,

$$\frac{T_1 - T_\infty}{q_r''} = \frac{\ln(R_2/R_1)}{2\pi k_{m,2}} + \frac{1}{2\pi h_o R_2}$$

or from Eq. (ii),

$$\frac{T_1 - T_\infty}{\ln(R_2/R_1)/2\pi k_{m,2} + 1/2\pi h_o R_2} = \frac{I^2 \rho}{\pi R_1^2} \tag{iv}$$

EXAMPLE 6.7

Consider, with respect to Fig. 6.16, that the wire is copper with $R_1 = 0.0005$ m, $k_{m,Cu} = 380$ W/m·K, $\rho_{Cu} = 1.96 \times 10^{-8}$ Ω·m, and that the insulation coating of $R_2 = 0.0015$ m is a plastic material of $k_m = 0.35$ W/m·K, $\sigma = 0$ Ω·m. The heat transfer coefficient, h_o, is 8 W/m²·K and $T_\infty = 30°C$. If the maximum temperature at which the plastic insulation can be operated is 100°C, calculate the maximum current that can be passed through the wire. Insertion into Eq. (iv) gives

$$\frac{100 - 30}{\ln 3/(2\pi \times 0.35) + 1/(2\pi \times 8 \times 0.0015)} = \frac{I^2 \times 1.96 \times 10^{-8}}{\pi(0.0005)^2}$$

or $I = 14.3$ A.

6.5 Thermal Conductivity and the Kinetic Theory of Gases

The thermal conductivity of a monatomic gas can be calculated from consideration of the kinetic theory of gases in a manner analogous to the calculation of viscosity. As was the case in its application to viscosity, some simplifying assumptions are made.

1. The atoms are rigid nonattracting spheres of diameter, d, which undergo perfectly elastic collisions with one another.
2. All of the atoms travel with the same speed, namely the average speed $\bar{c}$.
3. All of the atoms are moving in directions parallel to the x, y, or z axes, which means that one-sixth of them are moving in each of the $+x$, $-x$, $+y$, $-y$, $+z$, and $-z$ directions. ·

As shown in Chapter 2, the average speed of the atoms is $\bar{c} = (8\bar{k}T/\pi m)^{1/2}$, the number of atoms passing through unit area in unit time is $n\bar{c}/6$, and the mean free path (the average distance traveled between collisions) is $\sqrt{2}\,\pi d^2 n$, where n is the number of atoms per unit volume, m is the mass of an atom, and $\bar{k}$ is Boltzmann's constant (which has to be distinguished from k, the thermal conductivity). The internal energy of a population of gaseous atoms arises solely from the translational motions of the individual atoms and is equal to the average kinetic energy of the population. Consider a cubical box of side L containing Avogadro's number, N_0, of atoms at the temperature T. The velocity $\mathbf{v}$ of each atom can be broken into its components in the x, y, and z directions [as shown in Fig. 3.1(b)], where

$$\mathbf{v}^2 = v_x^2 + v_y^2 + v_z^2$$

Consider the x-component of motion of a single atom in the box. The time interval between collisions of the atom with the yz inside surfaces of the box is L/v_x, and as the direction of x-component motion is reversed by each collision, the x-component of the momentum of the atom is changed from mv_x to $-mv_x$ by each collision. The change in momentum is thus $(mv_x) - (-mv_x) = 2mv_x$ and the rate of change of momentum is $(2mv_x) \div (L/v_x) = 2mv_x^2/L$. In considering all N_0 atoms, the rate of change of the x-component of momentum is $2N_0m(\overline{v_x^2})/L$, where $(\overline{v_x^2})$ is the average of the squres of the x-components of velocities of the N_0 atoms. This rate of change of momentum is equal to the force exerted by the gas on the two yz faces of the box:

$$F = \frac{2N_0m(\overline{v_x^2})}{L} \tag{i}$$

and as the force, F, is the pressure, P, times the area on which it acts, $2L^2$, and the volume of the box, $V = L^3$, Eq. (i) can be written as

$$PV = N_0m(\overline{v_x^2}) \tag{ii}$$

As the directions of the motion of the atoms in the box are random,

$$\overline{v_x^2} = \overline{v_y^2} = \overline{v_z^2}$$

and hence

$$\overline{v^2} = 3\overline{v_x^2}$$

Substitution into Eq. (ii) gives

$$PV = \tfrac{1}{3}N_0m\overline{v^2} \tag{iii}$$

The quantity $\overline{v^2}$ is the mean-square speed of the atoms and the total translational kinetic energy of the mole of atoms, E_k, is

$$E_k = \tfrac{1}{2}N_0m\overline{v^2} \tag{iv}$$

From the ideal gas law for a mole of atoms, $PV = RT$, Eqs. (iii) and (iv) can be combined as

$$PV = \tfrac{2}{3}E_k = RT \tag{v}$$

or

$$E_k = \tfrac{3}{2}RT \tag{vi}$$

From Eqs. (iii) and (iv), with $mN_0 = M$ (the molecular weight of the gas),

$$\overline{v^2} = \frac{3RT}{M}$$

or the root-mean-square speed of the atoms, $(\overline{v^2})^{1/2}$, is given by

$$(\overline{v^2})^{1/2} = \left(\frac{3RT}{M}\right)^{1/2} = \left(\frac{3kT}{m}\right)^{1/2}$$

which is only slightly different from the average speed $\bar{c} = (8kT/\pi m)^{1/2}$.

Equation (vi) shows that the temperature of a gas, and hence its heat content, are determined by the velocities of the atoms. Per atom, Eq. (vi) can be written as

$$\tfrac{1}{2}m\bar{c}^2 = \tfrac{3}{2}kT \tag{vii}$$

Figure 6.17, which is similar to Fig. 2.22, shows three layers or sheets of atoms in a gas separated from one another by the mean free path λ. The atoms are moving in the x, y, and z directions with the average speed $\bar{c}$ and all of the collisions between atoms occur in the sheets. Consider that the temperature of the gas varies in the y-direction, with the temperatures of the planes at A, O, and B being, respectively, $T_{-\lambda}$, T_O, and $T_{+\lambda}$. When an atom jumps from one plane to another it transports a kinetic (and hence thermal) energy characteristic of that of the plane from which it jumped. The number of atoms jumping vertically from plane A to plane O per unit area is $\frac{1}{6}n\bar{c}$, and from Eq. (vii), each of these atoms transports a thermal energy $\frac{3}{2}kT_{-\lambda}$. Thus the rate of transport of thermal energy, per unit area, from plane A to plane O in the $+y$ direction is

$$(\tfrac{1}{6}n\bar{c})(\tfrac{3}{2}kT_{-\lambda})$$

Similarly, the rate of transport of thermal energy, per unit area, from plane B to plane O in the $-y$-direction is

$$-(\tfrac{1}{6}n\bar{c})(\tfrac{3}{2}kT_{+\lambda})$$

The net rate of transport of thermal energy, per unit area, is thus

$$
\begin{aligned}
q'_y &= \frac{1}{4}\,n\bar{c}k(T_{-\lambda} - T_{+\lambda}) \\
&= \frac{1}{4}\,n\bar{c}k\left[\left(T_O - \lambda\frac{dT}{dy}\right) - \left(T_O + \lambda\frac{dT}{dy}\right)\right] \\
&= -\frac{1}{2}\,n\bar{c}k\lambda\frac{dT}{dy}
\end{aligned}
$$

which, on comparison with Fourier's law, gives the thermal conductivity k of the gas as

$$k = \tfrac{1}{2}n\bar{c}k\lambda$$

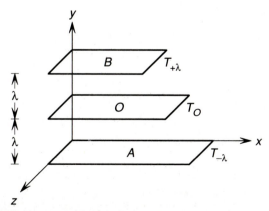

FIGURE 6.17 Construction for application of the kinetic theory of gases to theoretical calculation of the thermal conductivity of the gas.

or, by inserting the expressions for $\bar{c}$ and λ and simplifying, gives

$$k = \frac{1}{d^2}\left(\frac{k^3 T}{\pi^3 m}\right)^{1/2} \tag{6.29}$$

Thus, as is the case with viscosity [Eq. (2.30)], the thermal conductivity is independent of the pressure of the gas and is proportional to $T^{1/2}$. However, in contrast with viscosity, the thermal conductivity is inversely proportional to $m^{1/2}$.

Rigorous calculation of the thermal conductivity of a gas comprising noninteracting rigid spherical atoms gives

$$k = \frac{1.9891 \times 10^{-4}\sqrt{T/M}}{d^2} \tag{6.30}$$

where k is in $\text{cal·s}^{-1}\text{·cm}^{-1}\text{·K}^{-1}$

M is the molecular weight in g/g mol

T is in kelvin

d is in angstroms

Figure 6.18 shows a comparison between the thermal conductivities of helium and argon calculated from Eq. (6.30) and the experimental values, in which the hard-sphere diameters were obtained from the experimental values at 298 K. As is the case with viscosity (Fig. 2.23), the actual temperature dependence is more pronounced than $T^{1/2}$, due to the "softening" of the atom with increasing temperature. Chapman and Enskog considered this softening in terms of the collision integral Ω and derived

$$k = \frac{1.9891 \times 10^{-4}\sqrt{T/M}}{\sigma^2 \Omega} \tag{6.31}$$

in which σ is the collision diameter listed in Table 2.1, and Ω is a function of the energy parameter ϵ/k listed in Table 2.2. The variation of Ω with kT/ϵ is listed in Table 2.2.

EXAMPLE 6.8

Calculate the thermal conductivities of He and Ar at 300 K.
From Tables 2.1 and 2.2 the following data are obtained:

	M	$\sigma(\text{Å})$	$\dfrac{\epsilon}{k(T)}$	$\dfrac{kT}{\epsilon}$	Ω
He	4.003	2.576	10.2	29.41	0.7031
Ar	39.944	3.418	124	2.419	1.101

Thus for He, from Eq. (6.31),

$$k = \frac{1.9891 \times 10^{-4}\sqrt{300/4.003}}{(2.576)^2 \times 0.7031}$$

$$= 3.69 \times 10^{-4} \text{ cal·s}^{-1}\text{·cm}^{-1}\text{·K}^{-1}$$

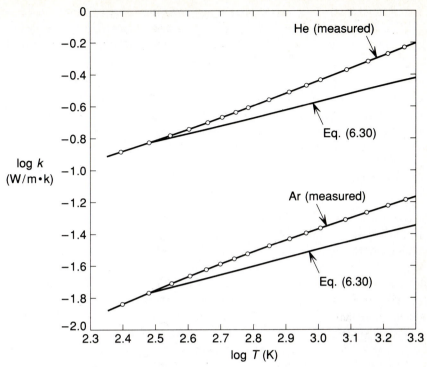

FIGURE 6.18   Comparison between the variations, with temperature, of the experimentally measured and calculated thermal conductivities of helium and argon.

and for Ar,

$$k = \frac{1.9891 \times 10^{-4}\sqrt{300/39.944}}{(3.418)^2 \times 1.101}$$

$$= 4.24 \times 10^{-5} \text{ cal·s}^{-1}\text{·cm}^{-1}\text{·K}^{-1}$$

As

$$1 \frac{\text{cal}}{\text{s·cm·K}} = 4.184 \frac{\text{J}}{\text{cal}} \times 100 \frac{\text{cm}}{\text{m}}$$

$$= 418.4 \frac{\text{J}}{\text{s·m·K}} \quad \text{or} \quad \frac{\text{W}}{\text{m·K}}$$

for He, $k = 3.69 \times 10^{-4} \times 418.4 = 0.154$ W/m·K, and for Ar, $k = 4.24 \times 10^{-5} \times 418.4 = 0.0177$ W/m·K. The respective measured values at 300K are 0.150 and 0.0177 W/m·K.

Combination of Eqs. (2.31) and (6.31) gives

$$\frac{k}{\eta} = \frac{7.4517}{M} \qquad (6.32)$$

In Eq. (6.32) the ratio k/η has the units $cal \cdot g^{-1} \cdot K^{-1}$ and M is g/g mol. The constant 7.4517 thus has the units $cal \cdot g\ mol^{-1} \cdot K^{-1}$, which are the units of molar heat capacity and the gas constant $R = 1.987\ cal \cdot g\ mol^{-1} \cdot K^{-1}$. Equation (6.32) can thus be written as

$$\frac{k}{\eta} = 3.750 \frac{R}{M}$$

or as the constant-volume molar heat capacity, C_v, of a monatomic gas, obtained by differentiation of Eq. (vi) with respect to T, is $1.5R$,

$$\frac{k}{\eta} = 2.5 \frac{C_v}{M} \tag{6.33}$$

The forgoing discussion pertains only to monatomic gases. In addition to having kinetic energies of translational motion, polyatomic gas molecules have energies of vibration and rotation, and any calculation of the thermal conductivities of such gases would have to take into account the exchange of these kinds of energies during collisions between molecules. Eucken [A. Eucken, *Phys. Z.* (1913), vol. 14, p. 324] developed a semiempirical method of treating energy exchanges between polyatomic molecules and derived

$$k = \left(C_{p(m)} + \frac{5}{4} \frac{R}{M} \right) \eta \tag{6.34}$$

as the relationship between thermal conductivity, viscosity, and the constant-pressure heat capacity (in $cal \cdot g^{-1} \cdot K^{-1}$).

Experimental data for diatomic O_2, triatomic CO_2, and tetratomic NH_3 at 300 K are listed in Table 6.1.

From Eq. (6.34) for O_2,

$$k = \left(0.220 + \frac{5}{4} \times \frac{1.987}{32} \right) \times 2.07 \times 10^{-4} = 6.16 \times 10^{-5}\ cal/s \cdot cm \cdot K$$

for CO_2,

$$k = \left(0.204 + \frac{5}{4} \times \frac{1.987}{44.01} \right) \times 1.50 \times 10^{-4} = 3.91 \times 10^{-5}\ cal/s \cdot cm \cdot K$$

TABLE 6.1

	M	$C_{p(m)}$ (cal/g·K)	η (g/cm·s)	k (cal/s·cm·K)
O_2	32.00	0.220	2.07×10^{-4}	6.38×10^{-5}
CO_2	44.01	0.204	1.50×10^{-4}	3.97×10^{-5}
NH_3	17.03	0.549	1.027×10^{-4}	5.88×10^{-5}

and for NH_3,

$$k = \left(0.549 + \frac{5}{4} \times \frac{1.987}{17.03}\right) \times 1.027 \times 10^{-4} = 7.14 \times 10^{-5} \text{ cal/s·cm·K}$$

The calculated thermal conductivities of O_2 and CO_2 are, respectively, within 3.4% and 1.5% of the measured values listed in Table 6.1. However, the calculated value for NH_3 differs from the measured value by 21%.

Thermal Conductivities of Gas Mixtures

The thermal conductivities of gas mixtures are calculated from a formula that is analogous to Wilke's formula for the viscosities of gas mixtures given by Eq. (2.36),

$$k_{mix} = \sum_{i=1}^{n} \frac{X_i k_i}{\sum_{j=1}^{n} X_j \Phi_{ij}}$$

in which Φ_{ij} is defined by Eq. (2.37).

EXAMPLE 6.9

Calculate the thermal conductivities of mixtures of H_2 and CO_2 at 300 K.
At 300 K,

	k(W/m·K)
H_2	0.182
CO_2	0.0166

In Chapter 2, $\Phi_{H_2-CO_2}$ was calculated as 2.455 and $\Phi_{CO_2-H_2}$ was calculated as 0.190. Thus, from Eq. (6.35),

$$k_{mix} = \frac{X_{H_2}k_{H_2}}{X_{H_2}\Phi_{H_2-H_2} + X_{CO_2}\Phi_{H_2-CO_2}} + \frac{X_{CO_2}k_{CO_2}}{X_{CO_2}\Phi_{CO_2-CO_2} + X_{H_2}\Phi_{CO_2-H_2}}$$

$$= \frac{0.182X_{H_2}}{X_{H_2} + 2.455X_{CO_2}} + \frac{0.0166X_{CO_2}}{X_{CO_2} + 0.190X_{H_2}}$$

which, with $X_{H_2} + X_{CO_2} = 1$, reduces to

$$k_{mix} = \frac{0.182X_{H_2}}{2.455 - 1.455X_{H_2}} + \frac{0.0166(1 - X_{H_2})}{1 - 0.81X_{H_2}}$$

This variation is shown in Fig. 6.19.

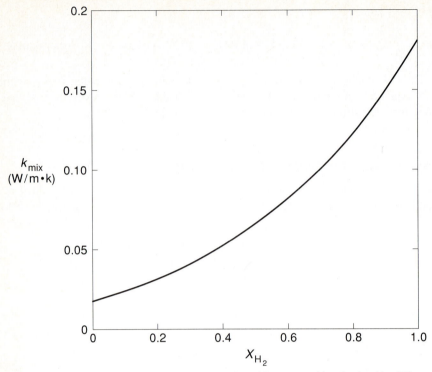

FIGURE 6.19 Variation of thermal conductivity with composition in the H_2–CO_2 system at 300 K.

6.6 General Heat Conduction Equation

The preceding discussion has been confined to unidirectional heat flow by conduction either in the x-direction through a plane wall, as shown in Fig. 6.20(a), or radially through the walls of a hollow cylinder or hollow sphere, as shown in Fig. 6.20(b). Both figures show that the heat flows in a direction that is normal to the lines of constant temperature (the isotherms) in the solid. In Fig. 6.20(a) the isotherms are straight lines parallel to the faces of the plane wall, and in Fig. 6.20(b) the isotherms are circles. Figure 6.20(c) shows the isotherms and the directions of heat flow in a square plate, the top edge of which is maintained at the constant temperature T_0 and the other three edges of which are maintained at $T = 0$. The directions of heat flow are such that they are always perpendicular to the isotherms, and hence the heat flow occurs in two dimensions. The positions of the isotherms, which are determined by the temperatures of the edges, define lanes along which heat flows and there is no transfer of heat between lanes. In two- or three-dimensional heat flow the temperature gradient in the direction i causes heat flow q_i' in that direction and q_i' must be considered in terms of its components in the x, y, and z directions. For an isotropic material in which the thermal conductivity, k, is independent of the direction of heat

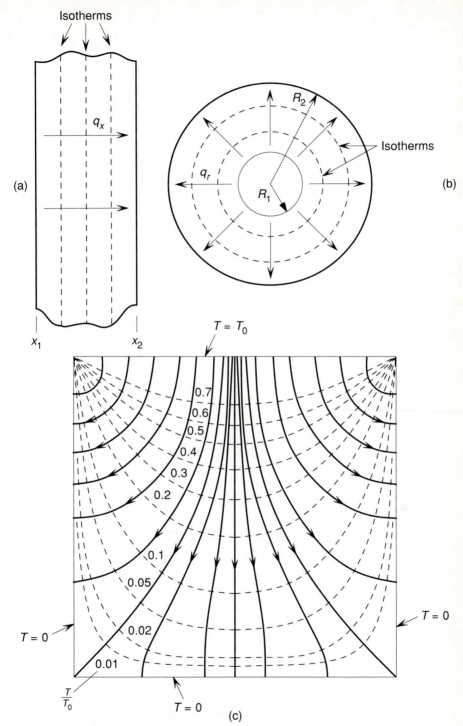

FIGURE 6.20 Isotherms in (a) plane wall, (b) a cylinder, and (c) a simple system undergoing heat flow in two dimensions.

flow, the components are

$$q_x' = -k\frac{\partial T}{\partial x} \tag{i}$$

$$q_y' = -k\frac{\partial T}{\partial y} \tag{ii}$$

and

$$q_z' = -k\frac{\partial T}{\partial z} \tag{iii}$$

Consider heat flow through the element of a solid shown in Fig. 6.21. An energy balance requires that

(heat conducted in) − (heat conducted out) + (heat generated within)
= (increase in thermal energy of element)

The x-component of the heat flow enters through the x-face of the element at the rate $q_x'|_x \Delta y\, \Delta z$ and leaves through the face at $x + \Delta x$ at the rate $q_x'|_{x+\Delta x}\, \Delta y\, \Delta z$. The y-component of the flow enters the y face at the rate $q_y'|_y\, \Delta x\, \Delta z$ and leaves through the face at $y + \Delta y$ at the rate $q_y'|_{y+\Delta y}\, \Delta x\, \Delta z$, and the z-component enters through the z face at the rate $q_z'|_z\, \Delta x\, \Delta y$ and leaves through the $z + \Delta z$ face at the rate

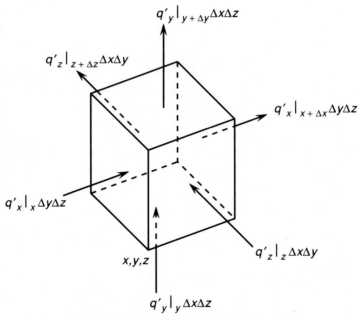

FIGURE 6.21 Volume element of a solid.

$q_z'|_{z+\Delta z} \, \Delta x \, \Delta y$. The net rate of conduction of heat into the element is thus

$$(q_x'|_x - q_x'|_{x+\Delta x}) \, \Delta y \, \Delta z + (q_y'|_y - q_y'|_{y+\Delta y}) \, \Delta x \, \Delta z + (q_z'|_z - q_z'|_{z+\Delta z}) \, \Delta x \, \Delta y$$

If heat is generated in the material at the rate of $\dot{q}$ per unit volume the rate of heat generation in the element is

$$\dot{q} \, \Delta x \, \Delta y \, \Delta z$$

For heat flow at constant pressure, the rate of increase of the thermal energy (enthalpy) of the element is

$$\frac{\partial H}{\partial t} \, \Delta x \, \Delta y \, \Delta z = \rho C_{p(m)} \frac{\partial T}{\partial t} \, \Delta x \, \Delta y \, \Delta z$$

where $C_{p(m)}$ is constant-pressure heat capacity of the material. Making the energy balance, dividing by the volume $\Delta x \, \Delta y \, \Delta z$ and allowing Δx, Δy, and Δz to approach zero, gives

$$-\left(\frac{\partial q_x'}{\partial x} + \frac{\partial q_y'}{\partial y} + \frac{\partial q_z'}{\partial z}\right) + \dot{q} = \rho C_{p(m)} \frac{\partial T}{\partial t} \tag{iv}$$

and substitution of Eqs. (i), (ii), and (iii) gives, for a material of constant k,

$$k\left(\frac{\partial^2 T}{\partial x^2} + \frac{\partial^2 T}{\partial y^2} + \frac{\partial^2 T}{\partial z^2}\right) + \dot{q} = \rho C_{p(m)} \frac{\partial T}{\partial t} \tag{6.35}$$

Equation (6.35) is known as the *general heat conduction equation*. The thermal diffusivity, α, of a material is defined as

$$\alpha = \frac{k}{\rho C_{p(m)}} \tag{6.36}$$

and hence Eq. (6.35) can be written as

$$\alpha\left(\frac{\partial^2 T}{\partial x^2} + \frac{\partial^2 T}{\partial y^2} + \frac{\partial^2 T}{\partial z^2}\right) + \frac{\dot{q}}{\rho C_{p(m)}} = \frac{\partial T}{\partial t}$$

Thermal diffusivity has the units

$$k\left(\frac{J}{s \cdot m \cdot K}\right)\frac{1}{\rho}\left(\frac{m^3}{kg}\right)\frac{1}{C_{p(m)}}\left(\frac{kg \cdot K}{J}\right) = \frac{m^2}{s}$$

which are the same as the units of kinematic viscosity, ν. Thermal diffusivity in heat transport is thus the analog of kinematic viscosity in momentum transport. For heat conduction at steady state, in a medium in which there is no generation of heat, Eq. (6.35) becomes

$$\frac{\partial^2 T}{\partial x^2} + \frac{\partial^2 T}{\partial y^2} + \frac{\partial^2 T}{\partial z^2} = 0 \tag{6.37}$$

which is known as *Laplace's equation*.

6.7 Conduction of Heat at Steady State in Two Dimensions

For steady-state conduction of heat in the xy plane, Eq. (6.37) becomes

$$\frac{\partial^2 T}{\partial x^2} + \frac{\partial^2 T}{\partial y^2} = 0 \tag{6.38}$$

and in principle, the variation of T with x and y is obtained by integration of Eq. (6.38) and knowledge of the appropriate boundary conditions. In practice, however, the mathematics required is extremely complicated and the analytical solutions are complicated. For example, for the relatively simple square, of sides of length L, illustrated in Fig. 6.20(c), the exact solution of Eq. (6.38) is

$$T(x,y) = \frac{4T_o}{\pi} \sum_{n=1,3,5,\dots}^{\infty} \frac{1}{n} \frac{\sinh(n\pi y/L)}{\sinh(n\pi)} \sin \frac{n\pi x}{L}$$

The difficulties caused by intractable mathematics can be circumvented by resorting to finite-difference techniques of analysis which allow the estimation of the temperatures at discrete points in the material.

In the finite-difference technique a grid of mesh size $\Delta x \times \Delta y$ is laid over the two-dimensional section of material to be analyzed, as shown in Fig. 6.22(a). The point of intersection of a vertical line in the grid and a horizontal line is a node and the technique permits estimation of the temperatures of the nodes. As illustrated in Fig. 6.22(b), the positions of the nodes on the x-axis are labeled m and the positions on the y-axis are labeled n. A central node at the temperature $T_{m,n}$ is surrounded by four nodes at the temperatures $T_{(m,n+1)}$, $T_{(m-1,n)}$, $T_{(m,n-1)}$, and $T_{(m+1,n)}$. In Fig. 6.22(c) the temperature gradient in the x-direction at the point $m + \frac{1}{2}, n$ is approximated as

$$\left. \frac{\partial T}{\partial x} \right|_{m+1/2,n} \approx \frac{T_{m+1,n} - T_{m,n}}{\Delta x} \tag{i}$$

and the corresponding gradient at the point $m - \frac{1}{2}, n$ is

$$\left. \frac{\partial T}{\partial x} \right|_{m-1/2,n} \approx \frac{T_{m,n} - T_{m-1,n}}{\Delta x} \tag{ii}$$

The second partial derivative of T with respect to x at the node m,n is then approximated as

$$\left. \frac{\partial^2 T}{\partial x^2} \right|_{m,n} \approx \frac{\left. \dfrac{\partial T}{\partial x} \right|_{m+1/2,n} - \left. \dfrac{\partial T}{\partial x} \right|_{m-1/2,n}}{\Delta x}$$

which from Eqs. (i) and (ii) can be written as

$$\left. \frac{\partial^2 T}{\partial x^2} \right|_{m,n} \approx \frac{T_{m+1,n} + T_{m-1,n} - 2T_{m,n}}{(\Delta x)^2} \tag{iii}$$

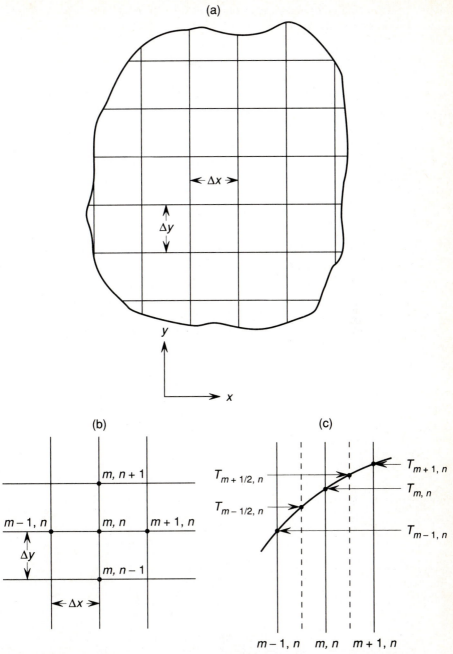

FIGURE 6.22 (a) Grid used in the finite-difference technique; (b) labeling of nodes in the grid; (c) temperatures at nodes identified in (b).

Similarly, the second partial derivative of T with respect to y at the node m,n can be approximated as

$$\left.\frac{\partial^2 T}{\partial y^2}\right|_{m,n} \approx \frac{T_{m,n+1} + T_{m,n-1} - 2T_{m,n}}{(\Delta y)^2} \tag{iv}$$

Thus, with a square mesh in which $\Delta x = \Delta y$, substitution of Eqs. (iii) and (iv) into Eq. (6.38) gives

$$T_{m,n+1} + T_{m-1,n} + T_{m,n-1} + T_{m+1,n} = 4T_{m,n} \tag{6.39}$$

that is, the temperature at the node m,n is four times the sum of the temperatures of the surrounding four nodes.

For steady-state conditions, the energy balance for the node at m,n in Fig. 6.23 requires that the net rate at which heat is conducted into the square, indicated by dashed lines, surrounding node m,n plus the rate at which heat is generated within the square is zero. Considering unit depth of plate (i.e., $\Delta z = 1$) the rates of conduction are

$$q_{(m,n+1)\to(m,n)} = -k \, \Delta x \, \frac{T_{m,n} - T_{m,n+1}}{\Delta y}$$

$$q_{(m-1,n)\to(m,n)} = -k \, \Delta y \, \frac{T_{m,n} - T_{m-1,n}}{\Delta x}$$

$$q_{(m,n-1)\to(m,n)} = -k \, \Delta x \, \frac{T_{m,n} - T_{m,n-1}}{\Delta y}$$

$$q_{(m+1,n)\to(m,n)} = -k \, \Delta y \, \frac{T_{m,n} - T_{m+1,n}}{\Delta x}$$

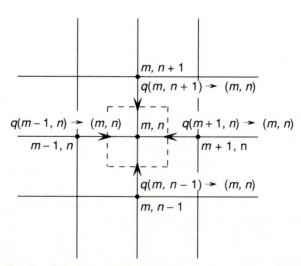

FIGURE 6.23 Energy balance for the node at m, n for steady-state conditions.

and heat is generated at the rate $\dot{q}\, \Delta x\, \Delta y$. With $\Delta x = \Delta y$, the energy balance is thus

$$T_{m,n+1} + T_{m-1,n} + T_{m,n-1} + T_{m+1,n} - 4T_{m,n} + \frac{\dot{q}\, \Delta x\, \Delta y}{k} = 0 \quad (6.40)$$

EXAMPLE 6.10

Calculate the temperature distribution at steady state in the square plate of side length L shown in Fig. 6.24. The upper edge of the plate is at 500°C and the other three sides are at 300°C. There is no generation of heat within the plate.

A square grid of mesh size $0.25L$ locates the positions of nine nodes, but as a result of the symmetry about the centerline containing nodes 4, 5, and 6, only one-half of the plate, which contains nodes 1 to 6, need be considered. The application of Eq. (6.39) to each node gives the following six equations containing the temperatures of the six nodes:

$$
\begin{aligned}
\text{Node 1:} & \quad 500 + 300 + T_2 + T_4 = 4T_1 \\
\text{Node 2:} & \quad T_1 + 300 + T_3 + T_5 = 4T_2 \\
\text{Node 3:} & \quad T_2 + 300 + 300 + T_6 = 4T_3 \\
\text{Node 4:} & \quad 500 + T_1 + T_5 + T_1 = 4T_4 \\
\text{Node 5:} & \quad T_4 + T_2 + T_6 + T_2 = 4T_5 \\
\text{Node 6:} & \quad T_5 + T_3 + 300 + T_3 = 4T_6
\end{aligned}
$$

Each of the equations is then written in the form

$$aT_1 + bT_2 + cT_3 + dT_4 + eT_5 + fT_6 = \text{constant}$$

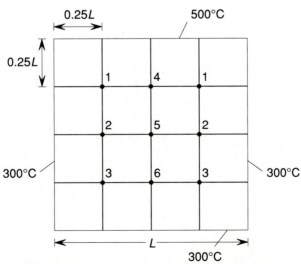

FIGURE 6.24   Grids considered in Example 6.10.

that is,

$$-4T_1 + T_2 + 0 + T_4 + 0 + 0 = -800$$

$$T_1 - 4T_2 + T_3 + 0 + T_5 + 0 = -300$$

$$0 + T_2 - 4T_3 + 0 + 0 + T_6 = -600$$

$$2T_1 + 0 + 0 - 4T_3 + T_5 + 0 = -500$$

$$0 + 2T_2 + 0 + T_4 - 4T_5 + T_6 = 0$$

$$0 + 0 + 2T_3 + 0 + T_5 - 4T_6 = -300$$

These equations are solved by matrix inversion $[\mathbf{A}][\mathbf{T}] = [\mathbf{C}]$, where

$$[\mathbf{A}] = \begin{bmatrix} -4 & 1 & 0 & 1 & 0 & 0 \\ 1 & -4 & 1 & 0 & 1 & 0 \\ 0 & 1 & -4 & 0 & 0 & 1 \\ 2 & 0 & 0 & -4 & 1 & 0 \\ 0 & 2 & 0 & 1 & -4 & 1 \\ 0 & 0 & 2 & 0 & 1 & -4 \end{bmatrix}$$

and

$$[\mathbf{C}] = \begin{bmatrix} -800 \\ -300 \\ -600 \\ -500 \\ 0 \\ -300 \end{bmatrix}$$

Matrix inversion, which can be done on a hand calculator, gives the temperatures as $[\mathbf{T}] = [\mathbf{A}]^{-1} [\mathbf{C}]$:

$$T_1 = 385°C$$

$$T_2 = 337°C$$

$$T_3 = 314°C$$

$$T_4 = 405°C$$

$$T_5 = 350°C$$

$$T_6 = 320°C$$

Each of these temperatures is the presumed uniform temperature in the square $\Delta x \, \Delta y$ at the center of which the node is located, and thus the plotting of these temperatures produces histograms or bar graphs. A perspective of the three-dimensional temperature histogram is shown in Fig. 6.25(a) and a color density map of the temperatures is shown in Fig. 6.25(b). The 500°C isotherm coincides with the upper edge of the plate and the 300°C isotherm coincides with the other three edges. Thus at the upper left and right corners of the plate, the temperature

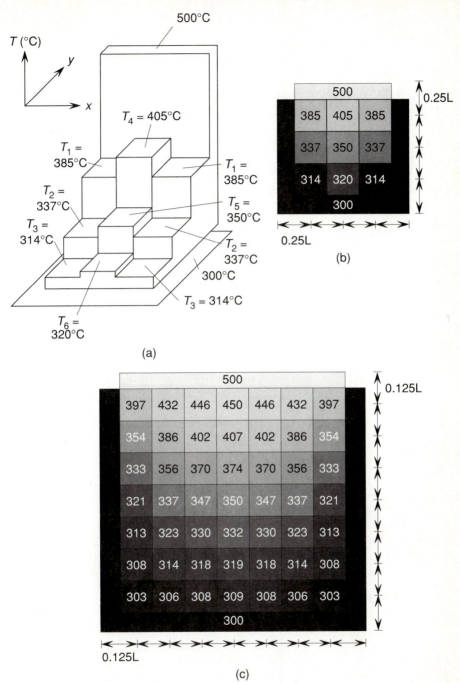

FIGURE 6.25 (a) Histogram of the temperatures obtained in Example 6.10; (b, c) color density map of histogram shown in (a).

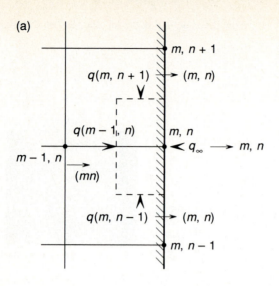

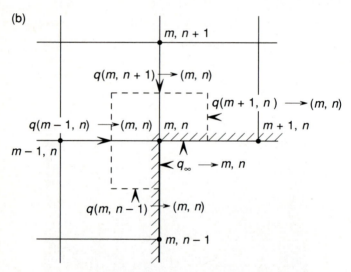

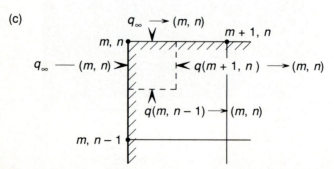

FIGURE 6.26 (a) Heat transfer to the node m, n located (a) at a surface, (b) at an inside corner, and (c) at an outer corner at which heat transfer is occurring by convection.

is in the range 300° to 500°C, so all of the isotherms emanate from these two corners. Note that the temperature, T_5, of the node in the center of the plate is 350°C. If the mesh size had been chosen as $\Delta x = \Delta y = 0.5L$, there would have been a single node at the center of the plate, the temperature of which would be obtained from

$$500 + 300 + 300 + 300 = 4T_{node}$$

as

$$T_{node} = 350°C$$

The color density map obtained with a mesh size of $\Delta x = \Delta y = 0.125L$ is shown in Fig. 6.25(c).

Heat Transfer by Convection at a Boundary

In Fig. 6.26(a) the node m,n is located at a surface at which heat is being transferred by convection. Heat transfer to the node m,n is thus the sum of

$$q_{(m,n+1)\rightarrow(m,n)} = -k\frac{\Delta x}{2}\frac{T_{m,n} - T_{m,n+1}}{\Delta y}$$

$$q_{(m-1,n)\rightarrow(m,n)} = -k\,\Delta y\,\frac{T_{m,n} - T_{m-1,n}}{\Delta x}$$

$$q_{(m,n-1)\rightarrow(m,n)} = -k\frac{\Delta x}{2}\frac{T_{m,n} - T_{m,n-1}}{\Delta y}$$

and

$$q_{(\infty)\rightarrow(m,n)} = h\,\Delta y(T_\infty - T_{m,n})$$

where T_∞ is the temperature of the fluid and h is the heat transfer coefficient. The sum of the four heat fluxes is zero, which with a square mesh of $\Delta x = \Delta y$ gives

$$2T_{m-1,n} + T_{m,n-1} + T_{m,n+1} + \frac{2h\,\Delta xT_\infty}{k} = \left(4 + \frac{2h\,\Delta x}{k}\right)T_{m,n} \quad (6.41)$$

Figure 6.26(b) shows the node, m,n at an inside corner at which heat transfer is occurring by convection. In this case the heat transferred to the node m,n is

$$q_{(m,n+1)\rightarrow(m,n)} = -k\,\Delta x\,\frac{T_{m,n} - T_{m,n+1}}{\Delta y}$$

$$q_{(m-1,n)\rightarrow(m,n)} = -k\,\Delta y\,\frac{T_{m,n} - T_{m-1,n}}{\Delta x}$$

$$q_{(m,n-1)\rightarrow(m,n)} = -k\frac{\Delta x}{2}\frac{T_{m,n} - T_{m,n-1}}{\Delta y}$$

$$q_{(m+1,n)\to(m,n)} = -k\frac{\Delta y}{2}\frac{T_{m,n} - T_{m+1,n}}{\Delta x}$$

$$q_{(\infty)\to(m,n)} = h\frac{\Delta x}{2}(T_\infty - T_{m,n}) + h\frac{\Delta y}{2}(T_\infty - T_{m,n})$$

which gives

$$2T_{m-1,n} + 2T_{m,n+1} + T_{m+1,n} + T_{m,n-1} + \frac{2h\,\Delta x}{k}T_\infty = \left(6 + \frac{2h\,\Delta x}{k}\right)T_{m,n} \tag{6.42}$$

For a node m,n at an outer corner, shown in Fig. 6.26(c), the heat balance gives

$$T_{m+1,n} + T_{m,n-1} + \frac{2\Delta x h}{k}T_\infty = \left(2 + \frac{2h\,\Delta x}{k}\right)T_{m,n} \tag{6.43}$$

EXAMPLE 6.11

Figure 6.27 shows a section of a square chimney through which a hot gas is being exhausted. Heat is transferred from the gas to inner surface by convection, is transferred through the wall by conduction, and is transferred from the outer surface to the atmosphere by convection. From the data given below, estimate the temperature distribution in the section.

$$T_{\infty,gas} = 300°C$$
$$h_i = 100 \text{ W/m}^2\cdot\text{K}$$
$$T_{\infty,atm} = 25°C$$
$$h_o = 10 \text{ W/m}^2\cdot\text{K}$$
$$k_{brick} = 1 \text{ W/m}\cdot\text{K}$$

In view of the symmetry of the square chimney, only one-eighth of the area, shown as the shaded area in Fig. 6.26, need be considered. A square mesh of size $\Delta x = \Delta y = 0.25$ m places 12 nodes in the area to be analyzed.

Node 1 is at an outside corner as shown in Fig. 6.26(c).
Nodes 2, 3, 4, 5, 11, and 12 are on surfaces as shown in Fig. 6.26(a).
Node 10 is at an inside corner as shown in Fig. 6.26(b).
Nodes 6, 7, 8, and 9 are inside nodes as shown in Fig. 6.23.

Thus for

node 1: $\quad T_2 + T_2 + \dfrac{2\Delta x h_o}{k}T_{\infty,atm} = \left(2 + \dfrac{2\Delta x h_i}{k}\right)T_1$

node 2: $\quad 2T_6 + T_1 + T_3 + \dfrac{2\Delta x h_o}{k}T_{\infty,atm} = \left(4 + \dfrac{2\Delta x h_o}{k}\right)T_2$

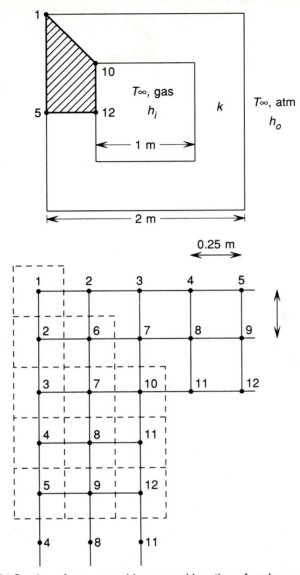

FIGURE 6.27 Section of a square chimney and location of nodes considered in Example 6.11.

node 3: $\quad 2T_7 + T_2 + T_4 + \dfrac{2\Delta x h_o}{k} T_{\infty,\text{atm}} \quad = \left(4 + \dfrac{2\Delta x h_o}{k}\right) T_3$

node 4: $\quad 2T_8 + T_3 + T_5 + \dfrac{2\Delta x h_o}{k} T_{\infty,\text{atm}} \quad = \left(4 + \dfrac{2\Delta x h_o}{k}\right) T_4$

node 5: $\quad 2T_9 + T_4 + T_4 + \dfrac{2\Delta x h_o}{k} T_{\infty,\text{atm}} \quad = \left(4 + \dfrac{2\Delta x h_o}{k}\right) T_5$

node 6: $T_2 + T_2 + T_7 + T_7$ $= 4T_6$

node 7: $T_6 + T_3 + T_8 + T_{10}$ $= 4T_7$

node 8: $T_7 + T_4 + T_9 + T_{11}$ $= 4T_8$

node 9: $T_8 + T_5 + T_8 + T_{12}$ $= 4T_9$

node 10: $2T_7 + 2T_7 + T_{11} + T_{11} + \dfrac{2\Delta x h_i}{k} T_{\infty,\text{gas}} = \left(6 + \dfrac{2\Delta x h_i}{k}\right) T_{10}$

node 11: $2T_8 + T_{10} + T_{12} + \dfrac{2\Delta x h_i}{k} T_{\infty,\text{gas}} = \left(4 + \dfrac{2\Delta x h_i}{k}\right) T_{11}$

node 12: $2T_9 + T_{11} + T_{11} + \dfrac{2\Delta x h_i}{k} T_{\infty,\text{gas}} = \left(4 + \dfrac{2\Delta x h_i}{k}\right) T_{12}$

$$\frac{2\Delta x h_o}{k} = \frac{2 \times 0.25 \times 10}{1} = 5$$

and

$$\frac{2\Delta x h_i}{k} = \frac{2 \times 0.25 \times 100}{1} = 50$$

This gives
and

$$[A] = \begin{bmatrix}
-7 & 2 & 0 & 0 & 0 & 0 & 0 & 0 & 0 & 0 & 0 & 0 \\
0 & -9 & 1 & 0 & 0 & 2 & 0 & 0 & 0 & 0 & 0 & 0 \\
0 & 1 & -9 & 1 & 0 & 0 & 2 & 0 & 0 & 0 & 0 & 0 \\
0 & 0 & 1 & -9 & 1 & 0 & 0 & 2 & 0 & 0 & 0 & 0 \\
0 & 0 & 0 & 2 & -9 & 0 & 0 & 0 & 2 & 0 & 0 & 0 \\
0 & 2 & 0 & 0 & 0 & -4 & 2 & 0 & 0 & 0 & 0 & 0 \\
0 & 0 & 1 & 0 & 0 & 1 & 4 & 1 & 0 & 1 & 0 & 0 \\
0 & 0 & 0 & 1 & 0 & 0 & 1 & -4 & 1 & 0 & 1 & 0 \\
0 & 0 & 0 & 0 & 1 & 0 & 0 & 2 & -4 & 0 & 0 & 1 \\
0 & 0 & 0 & 0 & 0 & 0 & 4 & 0 & 0 & -56 & 2 & 0 \\
0 & 0 & 0 & 0 & 0 & 0 & 0 & 2 & 0 & 1 & -54 & 1 \\
0 & 0 & 0 & 0 & 0 & 0 & 0 & 0 & 2 & 0 & 2 & -54
\end{bmatrix}$$

$$[C] = \begin{bmatrix}
-125 \\
-125 \\
-125 \\
-125 \\
-125 \\
0 \\
0 \\
0 \\
0 \\
-15{,}000 \\
-15{,}000 \\
-15{,}000
\end{bmatrix}$$

which from $[\mathbf{T}] = [\mathbf{A}]^{-1}[\mathbf{C}]$ gives

$$T_1 = 31°C$$
$$T_2 = 47°C$$
$$T_3 = 61°C$$
$$T_4 = 67°C$$
$$T_5 = 68°C$$
$$T_6 = 102°C$$
$$T_7 = 157°C$$
$$T_8 = 174°C$$
$$T_9 = 178°C$$
$$T_{10} = 290°C$$
$$T_{11} = 295°C$$
$$T_{12} = 295°C$$

6.8 Summary

The transport of heat by conduction is described by Fourier's law, which states that the heat transported in unit time is proportional to the cross-sectional area through which the heat flows and is proportional to the temperature gradient, with the proportionality constant being the thermal conductivity of the medium through which the conduction is occurring. Thermal conductivity, which has the units $W \cdot m^{-1} \cdot K^{-1}$, varies from room-temperature values of several hundreds for good thermal conductors such as copper, silver, and aluminum to several hundreths for good thermal insulators such as fiberglass. Although the thermal conductivity of a medium can vary significantly with temperature, an average value is normally used in heat conduction calculations.

Heat transfer to a fluid at a phase boundary occurs by convection and is described by Newton's law, which states that the heat transferred in unit time across the boundary is proportional to the area through which the flow occurs and is proportional to the difference between the temperature at the boundary and the temperature of the bulk of the fluid to which the heat is being transferred. In Newton's law the proportionality constant is the heat transfer coefficient, which, unlike thermal conductivity, is not a physical property; the heat transfer coefficient is a function of the physical properties of the fluid, the nature of any flow in the fluid and the geometry of the boundary across which the heat flow occurs.

Quantitative treatment of heat flow by conduction and/or convection is facilitated by the electric analogy, in which the resistence to flow is defined as the ratio of the temperature difference to the heat flux. Resistence to conduction is determined by thermal conductivity and resistence to heat transfer by convection is determined by the heat transfer coefficient. The electric analogy can be applied to flow in series and in parallel in multiphase linear and radial systems.

Application of the kinetic theory of gases to conduction in gases shows that the thermal conductivity of a gas is proportional to the square root of temperature and is inversely proportional to the molecular weight of the gas. This dependence on the molecular weight is in contrast to viscosity, which is proportional to the square root of the molecular weight.

Application of a heat balance to a control volume yields the general energy equation which describes the time dependence of heat flow in three-dimensional space, and treatment of this equation using the finite-difference technique avoids the complicated mathematics required for normal integration of the equation.

Problems

PROBLEM 6.1

The curved surface of a solid cylinder of length 0.2 m and diameter 0.02 m is covered with insulation. The ends are maintained at constant temperatures of 250°C and 150°C. Calculate the rate of heat flow through the cylinder if it is constructed of (a) pure copper, (b) pure aluminum, (c) carbon, (d) quartz, and (e) magnesia. (f) For the same end temperatures, how long would a copper rod have to be to have the same heat flux as occurs in the 0.2-m length of aluminum cylinder?

PROBLEM 6.2

The thermal conductivity of a material varies with temperature as

$$\ln k = 0.01T + 0.5$$

where k has the units $W \cdot m^{-1} \cdot K^{-1}$ and T has the units degrees Celsius. Heat flows by conduction through a plane slab of this material of thickness 0.1 m, the left face of which is at 100°C and the right face of which is at 0°C. Calculate:

(a) The mean thermal conductivity of the material in the range 0 to 100°C
(b) The heat flux through the slab
(c) The actual variation of T with distance through the slab
(d) The temperature at $x = 0.05$ m
(e) The temperature at $x = 0.05$ m calculated using the mean thermal conductivity
(f) The actual temperature gradient in the slab at $x = 0.05$ m
(g) The temperature gradient at $x = 0.05$ m calculated using the mean thermal conductivity

PROBLEM 6.3

A thermal window is constructed of two 5-mm thicknesses of window glass between which is a 5-mm gap containing stagnant air. The inside room temperature is 25°C, the outside temperature is $-15°C$, and the heat transfer coefficients on the inside and the outside are, respectively, $10 \text{ W·m}^{-2}\text{·K}^{-1}$ and $80 \text{ W·m}^{-2}\text{·s}^{-1}$.

(a) Calculate the heat flux through the window per unit area.
(b) Is the heat flux increased or decreased if the air gap is eliminated and a single pane of thickness 10 mm is used?
(c) What would the thickness of the air gap have to be to have a heat flux of 100 W·m^{-2}?

PROBLEM 6.4

A furnace is constructed of an inner layer of silica brick, a 5-cm thickness of an insulating material (of $k_m = 0.035 \text{ W·m}^{-1}\text{·K}^{-1}$) and an outer steel plate of thickness 1 cm. The furnace temperature is $1000°C$ and the outer wall of the steel plate is at $20°C$. The heat transfer coefficient inside the furnace is $25 \text{ W·m}^{-2}\text{·K}^{-1}$.

(a) If the maximum service temperature of the insulating material is $800°C$, what is the minimum thickness of silica brick that can be used
(b) What is the heat flux through the composite furnace wall with this thickness of silica brick?

PROBLEM 6.5

A cubical picnic chest of length 0.5 m, constructed of sheet Styrofoam of thickness 0.025 m, contains ice at $0°C$. The thermal conductivity of the Styrofoam is $0.035 \text{ W·m}^{-1}\text{·K}^{-1}$ and the ambient temperature is $25°C$. If the resistences to convection heat flow are negligible, calculate the rate at which the ice in the chest melts. The latent heat of melting of ice is $3.34 \times 10^5 \text{ J/kg}$.

PROBLEM 6.6

A cast-iron steam pipe of 0.11 m O.D. and 0.10 m I.D. carries steam at $200°C$. It is to be insulated with a 0.025 m thickness of an insulating material of $k_m = 0.035$ $\text{W·m}^{-1}\text{·K}^{-1}$. The ambient temperature is $0°C$ and the outer and inner heat transfer coefficients are, respectively, 75 and $225 \text{ W·m}^{-2}\text{·K}^{-1}$. Show that the heat flux through the wall of the pipe is less when the insulation is placed on the inner surface than when it is placed on the outer surface. What thickness of insulation, when placed on the outer surface, gives the same heat flux through the wall as a 0.025-m thickness placed on the inner wall?

PROBLEM 6.7

A stainless steel ball of 0.6 m O.D. and 0.4 m I.D. is cooled inside by a fluid at $15°C$. The heat transfer coefficient at the inner wall is $150 \text{ W·m}^{-2}\text{·K}^{-1}$ and the temperature of the outer wall is $50°C$. Calculate the temperature of the inner wall.

PROBLEM 6.8

A platinum wire of length 0.5 m and diameter 1 mm conducts a current of 10 A. The ambient temperature is $20°C$ and the heat transfer coefficient at the surface of the wire is 20 $W \cdot m^{-2} \cdot K^{-1}$. The thermal conductivity of platinum is 47 $W \cdot m^{-1} \cdot K^{-1}$ and the specific resistivity is 10.6 $\mu\Omega \cdot cm$. Calculate:

(a) The rate of generation of heat in the wire
(b) The temperature of the surface of the wire
(c) The difference between the temperatures at the centerline and the surface of the wire
(d) The wire is coated with a 0.1-mm layer of plastic insulating material of $k_m = 0.1 \ W \cdot m^{-1} \cdot K^{-1}$. Calculate the temperatures at the inner and outer surfaces of the insulation.
(e) If the maximum service temperature of the insulation is 150°C, what is the maximum current that can be passed through the wire?

PROBLEM 6.9

A 304 stainless steel slab of thickness 0.025 m conducts an electrical current that generates heat at the rate 6 × 10^6 W/m³. If the right-hand face of the slab is at 25°C, what must the temperature of the left face be in order that:

(a) All of the heat flux in the slab is from left to right?
(b) One-third of the heat generated in the slab leaves through the left face and two-thirds leave through the right face?

PROBLEM 6.10

A 304 stainless steel sphere of diameter 0.05 m generates heat at the rate 10^7 W/m³. Calculate the difference between the temperatures at the center and the surface of the sphere.

PROBLEM 6.11

A humidity box for laboratory use is constructed from panes of window glass of thickness 4 mm. The temperature inside the box is maintained at 90°C and the ambient room temperature is 25°C. The heat transfer coefficients at the inner and outer surfaces of the box are, respectively, 20 and 60 $W \cdot m^{-2} \cdot K^{-1}$.

(a) Calculate the maximum relative humidity that can be used in the box without condensation occurring on the inner surfaces.
(b) To increase the relative humidity, a thin, electrically heated film is attached to the inner surfaces of the box. Calculate the minimum power that must be supplied to the film in order that a relative humidity of 25% can be used without condensation occurring.

The saturated vapor pressure of water $p°$ (in Pa) varies with temperature (K) as

$$\ln p° = -\frac{6680}{T} - 4.65 \ln T + 56.97$$

PROBLEM 6.12

A cylindrical rod of a semiconducting material of length 5 cm and diameter 1 cm is well insulated on its curved surface. The ends of the rod as maintained at 25°C and 0°C and a current of 10 Amperes is passed. Calculate:

(a) The variation of temperature with distance along the rod
(b) The position at which the temperature has a maximum value
(c) The heat fluxes at both ends of the rod

The specific resistivity of the material is 2×10^{-5} Ω·m and the thermal conductvity is 2 W·M^{-1}·s^{-1}.

PROBLEM 6.13

An oil-fired boiler, operating at 80% efficiency, produces steam that is fed to a steam pipe of 0.08 m O.D. On combustion the fuel oil produces 2×10^7 J per gallon. The temperature of the inner wall of the pipe is 175°C and the ambient temperature is 20°C. The pipe is lagged with an insulating material of $k = 0.35$ W·m^{-1}·K-1 and the heat transfer coefficient at the outer surface of the lagging is 11 W·m^{-2}·K-1. The fuel oil costs $1.00 per gallon and 15% of the cost of the insulation is required for amortization and maintenance. Determine the most economic thickness of insulation if the insulation costs $350 per cubic meter and has to last for 1 year.

PROBLEM 6.14

Calculate the thermal conductivity of an equimolar mixture of He and CO at 300 K.

PROBLEM 6.15

Three sides of the square plate are maintained at 0°C, and the fourth side is in contact with a fluid at 200°C. The thermal conductivity of the plate material is 1 W/m·K. Calculate the temperatures of nodes 1 to 6.

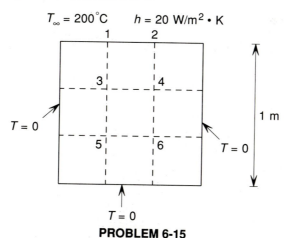

PROBLEM 6-15

PROBLEM 6.16

The plate in the figure is 1 m² with edges at the constant temperatures of 100, 200, and 400°C. Calculate the temperatures of nodes 1 to 9.

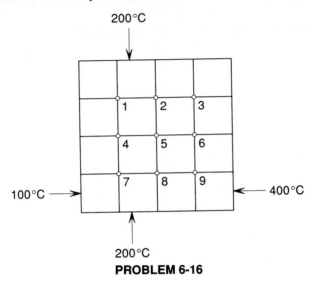

PROBLEM 6-16

PROBLEM 6.17

Two sides of the plate are adiabatic and the other two are at the constant temperatures 100 and 0°C. Calculate the temperatures of nodes 1 to 4.

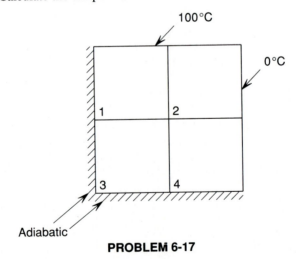

PROBLEM 6-17

7

Transport of Heat by Convection

7.1 Introduction

When a fluid at one temperature is in contact with the surface of a solid at another temperature, heat is transferred from the hotter to the cooler medium by convection. Although the magnitude of the heat transfer depends on many factors, such as the properties of the fluid, the geometry of the solid surface and the velocity of the fluid relative to that of the solid, two distinct regimes can be identified: forced convection and free convection. The difference between the two can be understood in terms of the familiar phenomenon of the "wind chill factor," wherein a windy cold day seems to be colder than a calm cold day at the same air temperature. The movement of a cold wind across the skin increases the rate of heat transfer from the skin and hence decreases the surface skin temperature to a lower value than would be obtained by contact with still air at the same temperature.

When a solid surface at the temperature T_s is in contact with a flowing fluid at a different temperature T_∞, the temperature gradient from the surface into the bulk of the fluid gives rise to a thermal boundary layer on the solid that is analogous to the momentum, or velocity, boundary layer discussed in Chapter 3. Consideration of heat transfer by convection begins with a consideration of thermal boundary layers.

7.2 Heat Transfer by Forced Convection from a Horizontal Flat Plate at a Uniform Constant Temperature

Consider a fluid at the temperature T_∞ flowing across a horizontal flat plate whose surface is uniformly at the higher temperature T_s, as shown in Fig. 7.1. The rate of

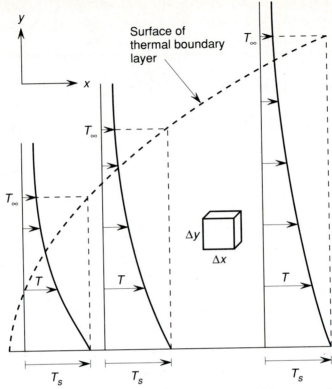

FIGURE 7.1. Temperature profiles in the thermal boundary layer on a flat plate of uniform surface temperature, T_s immersed in a flowing fluid of temperature T_∞.

transfer of heat, per unit area, from the plate to the flowing fluid is given by Newton's law:

$$q_y' = h(T_s - T_\infty) \tag{7.1}$$

and the problem now is to evaluate the heat transfer coefficient, h. As a packet of fluid, flowing at a fixed value of y, moves over the plate it is warmed by heat transfer from the plate, and hence the difference between T_s and the temperature of the packet of fluid decreases with increasing distance x along the plate. This decreases the heat flux per unit area from the plate to the fluid and hence, from Eq. (7.1), decreases the heat transfer coefficient. Thus it is expected that the local heat transfer coefficient at any location x on the surface, h_x, will decrease with increasing x. This phenomenon is analogous to the local shear stress, τ_0, exerted by the fluid on the surface, decreasing with increasing x. For laminar flow over a horizontal plate, Eq. (4.22) gave this variation as

$$\tau_0 = 0.323 \left(\frac{\rho \eta v_\infty^3}{x} \right)^{1/2} \tag{4.22}$$

As the temperature gradient from the plate into the fluid decreases with increasing x, the thickness of the thermal boundary layer increases. This thickness, δ_T, is defined as the value of y at which the difference between the temperature of the fluid and T_s is 99% of the difference $T_\infty - T_s$. The questions to be answered include: How does δ_T vary with x, what are the influences of the physical properties on δ_T, how is δ_T related to the momentum boundary layer thickness δ, and how does the thermal boundary layer influence the local heat transfer coefficient, h_x? Answers to these questions require an examination of the energy balance on a control volume in the thermal boundary layer.

Consider the control volume of dimensions Δx, Δy, and Δz in Fig. 7.1, shown as an enlargement in Fig. 7.2. At steady state the net rate at which energy enters the control volume is zero. Energy is transported into and out of the control volume by conduction and convection in both the x and y directions. The enthalpy per unit mass of fluid is H.

Energy is conducted in across the face at x at the rate $q_x|_x \, \Delta y \, \Delta z$.
Energy is conducted out across the face at $x + \Delta x$ at the rate $q_x|_{x+\Delta x} \, \Delta y \, \Delta z$.
Energy is conducted in across the face at y at the rate $q_y|_y \, \Delta x \, \Delta z$.
Energy is conducted out across the face at $y + \Delta y$ at the rate $q_y|_{y+\Delta y} \, \Delta x \, \Delta z$.
Energy enters the face at x by fluid flow at the rate $\rho v_x \, \Delta y \, \Delta z H|_x$.
Energy leaves through the face at $x + \Delta x$ by fluid flow at the rate $\rho v_x \, \Delta y \, \Delta z H|_{x+\Delta x}$.
Energy enters the face at y by fluid flow at the rate $\rho v_y \, \Delta x \, \Delta z H|_y$.

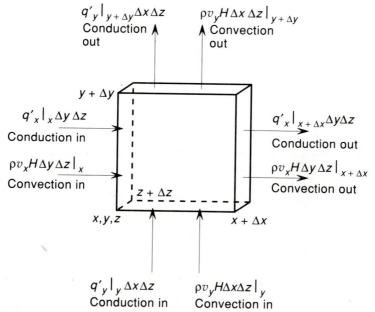

FIGURE 7.2. Control volume for consideration of an energy balance in the thermal boundary layer.

Energy leaves through the face at $y + \Delta y$ by fluid flow at the rate $\rho v_y \, \Delta x$ $\Delta z H|_{y+\Delta y}$.

Dividing the algebraic sum of these terms by $\Delta x \, \Delta y \, \Delta z$ and allowing Δx, Δy, and Δz to approach zero gives the differential energy balance as

$$\frac{\partial q_x}{\partial x} + \frac{\partial q_y}{\partial y} + \frac{\partial (\rho v_x H)}{\partial x} + \frac{\partial (\rho v_y H)}{\partial y} = 0$$

For a fluid of constant thermal conductivity, k, and density, this can be written as

$$-k \left(\frac{\partial^2 T}{\partial x^2} + \frac{\partial^2 T}{\partial y^2} \right) + \rho \left[v_x \frac{\partial H}{\partial x} + v_y \frac{\partial H}{\partial x} + H \left(\frac{\partial v_x}{\partial x} + \frac{\partial v_y}{\partial y} \right) \right] = 0 \quad (7.2)$$

For $v_z = 0$ and a fluid of constant density, the equation of continuity, Eq. (3.3), gives

$$\frac{\partial v_x}{\partial x} + \frac{\partial v_y}{\partial y} = 0$$

and from the definition of the constant-pressure heat capacity of the fluid as

$$C_p = \left(\frac{\partial H}{\partial T} \right)_P$$

Equation (7.2) can be written as

$$\rho C_p \left(v_x \frac{\partial T}{\partial x} + v_y \frac{\partial T}{\partial y} \right) = k \left(\frac{\partial^2 T}{\partial x^2} + \frac{\partial^2 T}{\partial y^2} \right)$$

or

$$v_x \frac{\partial T}{\partial x} + v_y \frac{\partial T}{\partial y} = \alpha \left(\frac{\partial^2 T}{\partial x^2} + \frac{\partial^2 T}{\partial y^2} \right) \quad (7.3)$$

where

$$\alpha = \frac{k}{\rho C_p}$$

is the thermal diffusivity of the fluid. The corresponding momentum boundary equation, presented in Chapter 3, for zero pressure gradient in the y-direction is

$$v_x \frac{\partial v_x}{\partial x} + v_y \frac{\partial v_x}{\partial y} = \nu \left(\frac{\partial^2 v_x}{\partial x^2} + \frac{\partial^2 v_x}{\partial y^2} \right)$$

which is analogous with Eq. (7.3). If the thermal and momentum boundary conditions are similar and $\alpha = \nu$, the temperature and velocity profiles are identical (i.e., $\delta = \delta_T$). The relative magnitudes of ν and α are quantified by their ratio, which is defined as the Prandtl number, Pr:

$$\text{Pr} = \frac{\nu}{\alpha} = \frac{\eta}{\rho} \frac{\rho C_p}{k} = \frac{\eta C_p}{k} \quad (7.4)$$

The Prandtl numbers of some common fluids are calculated and listed in Table 7.1.

TABLE 7.1

Fluid	T	η (Pa·s)	C_p (J/kg·K)	k (W/m·K)	Pr
Sodium	400 K	4.1×10^{-4}	1371	82	0.0097
Mercury	300 K	1.5×10^{-3}	139	8.9	0.0237
Air	300 K	1.85×10^{-5}	1005	0.0261	0.712
Water	300 K	8.57×10^{-4}	4177	0.608	5.88
Alcohol	20°C	1.2×10^{-3}	2395	0.168	17
Engine oil	20°C	8	1880	0.145	10,400
Glycerene	20°C	1.49	2386	0.287	12,400

The low heat capacities and high thermal conductivities of liquid metals cause the Prandtl numbers of liquid metals to be very much less than unity, and the high viscosities of oils cause the Prandtl numbers of oils to be very much greater than unity. The Prandtl numbers of gases are in the range 0.7 to 1.0 and the Prandtl numbers of common aqueous and organic liquids are in the range 2 to 50.

The exact solution of the thermal and momentum boundary equations, for laminar flow of fluids of $0.6 \leq \mathrm{Pr} \leq 50$, gives the local heat transfer coefficient, h_x, as

$$h_x = 0.332k \, \mathrm{Pr}^{0.343} \sqrt{\frac{v_\infty}{vx}} \qquad (7.5)$$

which as is the case with τ_0, is inversely proportional to $x^{1/2}$. The *average* heat transfer coefficient, $\overline{h}_L$, is obtained by averaging the values of h_x between $x = 0$ and $x = L$, as

$$\overline{h}_L = \frac{1}{L} \int_0^L h_x \, dx$$

which from Eq. (7.5) gives

$$\overline{h}_L = 0.664k \, \mathrm{Pr}^{0.343} \sqrt{\frac{v_\infty}{vL}} \qquad (7.6)$$

Thus the average heat transfer coefficient for heat transfer from the length of plate from $x = 0$ to $x = L$ is twice the value of the local heat transfer coefficient at $x = L$. Recall that the average friction factor for the length of plate from $x = 0$ to $x = L$, $\overline{f}_L$, is twice the local friction factor at $x = L$, and the average shear friction stress exerted by the fluid on the length of plate from $x = 0$ to $x = L$ is twice the local shear friction stress at $x = L$.

In heat transfer the Nusselt number, Nu, is a dimensionless parameter defined as

$$\mathrm{Nu} = \frac{h \times (\text{a characteristic distance})}{k}$$

If heat transfer from a solid surface at the temperature T_s to a fluid at the temperature T_∞ occurred only by convection, the heat flux per unit area of surface would be obtained as

$$q'_{convection} = h(T_s - T_\infty)$$

In contrast, if the heat transfer were only by conduction through a stagnant fluid, over a distance L from T_s to T_∞, the heat flux per unit area of the surface would be given, by Fourier's law, as

$$q'_{conduction} = -k \frac{T_\infty - T_s}{L} = \frac{k(T_s - T_\infty)}{L}$$

The ratio of the two,

$$\frac{q'_{convection}}{q'_{conduction}} = \frac{hL}{k} = Nu$$

where L is the characteristic distance. Thus the Nusselt number is a measure of the relative importance of the contributions of the convection and conduction mechanisms in the transport of heat between a solid surface and a fluid. A large Nusselt number indicates the predominance of convection. In the current application, the characteristic distance is the distance x from the leading edge of the plate. The local Nusselt number, Nu_x, is thus

$$Nu_x = \frac{h_x x}{k} \tag{7.7}$$

and the average Nusselt number for heat transfer from the length of plate from $x = 0$ to $x = L$, $\overline{Nu}_L$, is

$$\overline{Nu}_L = \frac{\overline{h}_L L}{k} \tag{7.8}$$

From Eqs. (7.5) and (7.7),

$$Nu_x = 0.332 Pr^{0.343} \sqrt{\frac{v_\infty x}{\nu}}$$

$$= 0.332 Pr^{0.343} \, Re_x^{0.5} \tag{7.9}$$

and from Eqs. (7.6) and (7.8),

$$\overline{Nu}_L = 0.664 Pr^{0.343} \sqrt{\frac{v_\infty L}{\nu}}$$

$$= 0.664 Pr^{0.343} Re_L^{0.5} \tag{7.10}$$

Thus for heat transfer by forced convection to or from a horizontal flat plate, the Nusselt number is a function only of the Prandtl number and the Reynolds number.

Approximate Integral Method

The application of the approximate integral method to heat transport in the thermal boundary layer is similar to that of its application to the transport of momentum in

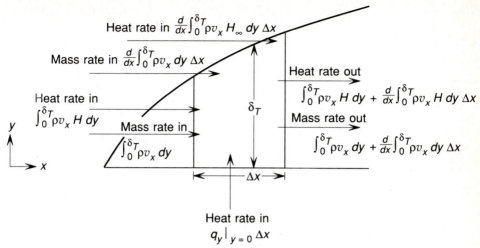

Heat rate in $\dfrac{d}{dx}\int_0^{\delta_T}\rho v_x H_\infty\, dy\, \Delta x$

Mass rate in $\dfrac{d}{dx}\int_0^{\delta_T}\rho v_x\, dy\, \Delta x$

Heat rate in
$\int_0^{\delta_T}\rho v_x H\, dy$

Mass rate in
$\int_0^{\delta_T}\rho v_x\, dy$

δ_T

Heat rate out
$\int_0^{\delta_T}\rho v_x H\, dy + \dfrac{d}{dx}\int_0^{\delta_T}\rho v_x H\, dy\, \Delta x$

Mass rate out
$\int_0^{\delta_T}\rho v_x\, dy + \dfrac{d}{dx}\int_0^{\delta_T}\rho v_x\, dy\, \Delta x$

Δx

Heat rate in
$q_y\big|_{y=0}\, \Delta x$

FIGURE 7.3. Control volume used in the approximate integral method for determining the properties of the thermal boundary.

the velocity boundary layer discussed in Chapter 3. The control volume of length Δx and unit width has the thermal boundary layer as its upper surface as shown in Fig. 7.3, and at steady state, thermal energy enters and leaves the control volume at equal rates. Mass enters the front face at the rate

$$\int_0^{\delta_T} \rho v_x\, dy$$

and leaves through the rear face at the rate

$$\int_0^{\delta_T} \rho v_x\, dy + \frac{d}{dx}\int_0^{\delta_T} \rho v_x\, dy\, \Delta x$$

Thus mass enters through the upper surface at the rate

$$\frac{d}{dx}\int_0^{\delta_T} \rho v_x\, dy\, \Delta x$$

Heat enters through the front face at the rate

$$\int_0^{\delta_T} \rho v_x H\, dy$$

and leaves through the rear face at the rate

$$\int_0^{\delta_T} \rho v_x H\, dy + \frac{d}{dx}\int_0^{\delta_T} \rho v_x H\, dy\, \Delta x$$

Heat enters through the upper surface at the rate

$$\frac{d}{dx}\int_0^{\delta_T} \rho v_x H_\infty\, dy\, \Delta x$$

where H_∞ is the enthalpy per unit mass of the fluid at the temperature T_∞. As the fluid in contact with the plate at $y = 0$ is motionless, heat is conducted into the control volume from the plate at the rate

$$q_y|_{y=0} \, \Delta x$$

The heat balance is thus

rate at which heat enters = rate at which heat leaves

$$\frac{d}{dx} \int_0^{\delta_T} \rho v_x H_\infty \, dy \, \Delta x + q_y|_{y=0} \, \Delta x = \frac{d}{dx} \int_0^{\delta_T} \rho v_x H \, dy \, \Delta x$$

Canceling the Δx terms and rearranging gives

$$\frac{d}{dx} \int_0^{\delta_T} (H - H_\infty)\rho v_x \, dy = q_y|_{y=0} = -k \left(\frac{\partial T}{\partial y}\right)_{y=0} \tag{7.11}$$

From the definition of the constant-pressure heat capacity,

$$H - H_\infty = C_p(T - T_\infty)$$

and for constant values of ρ, C_p, and k, Eq. (7.11) becomes

$$\frac{d}{dx} \int_0^{\delta_T} (T_\infty - T)v_x \, dy = \frac{k}{\rho C_p} \left(\frac{\partial T}{\partial y}\right)_{y=0} = \alpha \left(\frac{\partial T}{\partial y}\right)_{y=0} \tag{7.12}$$

In the consideration of the momentum boundary layer it was assumed that the velocity profile was a cubic parabola of the form

$$\frac{v_x}{v_\infty} = \frac{3}{2}\left(\frac{y}{\delta}\right) - \frac{1}{2}\left(\frac{y}{\delta}\right)^3 \tag{3.19}$$

It is now also assumed that the temperature profile is similar, being of the form

$$T_s - T = ay + by^3 \tag{7.13}$$

The boundary conditions are

$$T = T_\infty \qquad \text{at } y = \delta_T \tag{i}$$

and

$$\frac{\partial T}{\partial y} = 0 \qquad \text{at } y = \delta_T \tag{ii}$$

From boundary condition (i)

$$T_s - T_\infty = a\delta_T + b\delta_T^3 \tag{iii}$$

and from boundary condition (ii)

$$-\frac{\delta T}{\delta y} = 0 = a + 3b\delta_T^2$$

or

$$a = -3b\delta_T^2 \tag{iv}$$

substitution of which in Eq. (iii) gives

$$b = -\frac{1}{2}\frac{T_s - T_\infty}{\delta_T^3} \tag{v}$$

and hence

$$a = \frac{3}{2}\frac{T_s - T_\infty}{\delta_T} \tag{vi}$$

Substitution of the values of a and b into Eq. (7.13) gives the normalized temperature profile as

$$\frac{T_s - T}{T_s - T_\infty} = \frac{3}{2}\left(\frac{y}{\delta_T}\right) - \frac{1}{2}\left(\frac{y}{\delta_T}\right)^3 \tag{7.14}$$

From Eq. (7.14)

$$T = T_s - (T_s - T_\infty)\left[\frac{3}{2}\left(\frac{y}{\delta_T}\right) - \frac{1}{2}\left(\frac{y}{\delta_T}\right)^3\right]$$

or

$$T_\infty - T = T_\infty - T_s + (T_s - T_\infty)\left[\frac{3}{2}\left(\frac{y}{\delta_T}\right) - \frac{1}{2}\left(\frac{y}{\delta_T}\right)^3\right]$$

$$= (T_\infty - T_s)\left[1 - \frac{3}{2}\left(\frac{y}{\delta_T}\right) + \frac{1}{2}\left(\frac{y}{\delta_T}\right)^3\right] \tag{7.15}$$

and substitution of Eqs. (7.15) and (3.19) into Eq. (7.12) gives

$$\frac{d}{dx}\int_0^{\delta_T}(T_\infty - T_s)\left[1 - \frac{3}{2}\left(\frac{y}{\delta_T}\right) + \frac{1}{2}\left(\frac{y}{\delta_T}\right)^3\right]v_\infty$$

$$\times\left[\frac{3}{2}\left(\frac{y}{\delta}\right) - \frac{1}{2}\left(\frac{y}{\delta}\right)^3\right]dy = \alpha\left(\frac{\partial T}{\partial y}\right)_{y=0}$$

The left-hand side of this equation is

$$\frac{d}{dx}\left[(T_\infty - T_s)v_\infty\int_0^{\delta_T}\left[\frac{3}{2}\left(\frac{y}{\delta}\right) - \frac{1}{2}\left(\frac{y}{\delta}\right)^3 - \frac{9}{4}\left(\frac{y}{\delta_T}\right)\left(\frac{y}{\delta}\right) + \frac{3}{4}\left(\frac{y}{\delta_T}\right)\left(\frac{y}{\delta}\right)^3\right.\right.$$

$$\left.\left.+ \frac{3}{4}\left(\frac{y}{\delta_T}\right)^3\left(\frac{y}{\delta}\right) - \frac{1}{4}\left(\frac{y}{\delta_T}\right)^3\left(\frac{y}{\delta}\right)^3\right]dy\right]$$

which, when integrated with respect to y and evaluated between $y = \delta_T$ and $y = 0$, gives

$$\frac{d}{dx}\left\{(T_\infty - T_s)v_\infty\delta\left[0.15\left(\frac{\delta_T}{\delta}\right)^2 - 0.0107\left(\frac{\delta_T}{\delta}\right)^4\right]\right\} = \alpha\left(\frac{\partial T}{\partial y}\right)\bigg|_{y=0} \tag{7.16}$$

The occurrence of the term $(\delta_T/\delta)^4$ leads to mathematical difficulties in the solution of the equation and hence, for mathematical convenience, it is assumed that $\delta_T \leq \delta$, in which case the term $0.0107\,(\delta_T/\delta)^4$ is negligibly small and can be dropped from the equation. This limits the application of the final solution to fluids of certain Prandtl numbers. The right-hand side of Eq. (7.16) is obtained from Eq. (7.14) as

$$\alpha \left(\frac{\partial T}{\partial y}\right)\bigg|_{y=0} = \alpha(T_\infty - T_s)\frac{1.5}{\delta_T}$$

and hence

$$0.15(T_\infty - T_s)v_\infty \frac{d}{dx}\left[\delta\left(\frac{\delta_T}{\delta}\right)^2\right] = \frac{1.5\alpha(T_\infty - T_s)}{\delta_T}$$

$$= \frac{1.5\alpha(T_\infty - T_s)}{(\delta_T/\delta)\,\delta}$$

or

$$v_\infty \left(\frac{\delta_T}{\delta}\right)\delta\frac{d}{dx}\left[\delta\left(\frac{\delta_T}{\delta}\right)^2\right] = 10\alpha$$

or

$$v_\infty \left[\left(\frac{\delta_T}{\delta}\right)\delta^2\frac{d}{dx}\left(\frac{\delta_T}{\delta}\right)^2 + \left(\frac{\delta_T}{\delta}\right)^3\delta\frac{d\delta}{dx}\right] = 10\alpha \qquad (7.17)$$

Taking $(\delta_T/\delta) \equiv f$, the term in brackets becomes

$$f\delta^2 \frac{d}{dx}(f^2) + f^3\delta\frac{d\delta}{dx}$$

Now $fd(f^2) = 2f^2df$ and $d(f^3) = 3f^2df$; therefore,

$$fd(f^2) = \tfrac{2}{3}d(f^3)$$

and Eq. (19.17) becomes

$$v_\infty \left[\frac{2}{3}\delta^2\frac{d}{dx}\left(\frac{\delta_T}{\delta}\right)^3 + \left(\frac{\delta_T}{\delta}\right)^3\delta\frac{d\delta}{dx}\right] = 10\alpha \qquad (7.18)$$

From Eq. (3.21),

$$\delta = 4.64\left(\frac{\nu}{v_\infty}\right)^{1/2}x^{1/2}$$

and thus

$$\frac{d\delta}{dx} = \frac{4.64}{2}\left(\frac{\nu}{v_\infty}\right)^{1/2}x^{-1/2}$$

substitution of which into Eq. (7.18) gives

$$v_\infty \left[\frac{2}{3} (4.64)^2 \frac{v}{v_\infty} x \frac{d}{dx} \left(\frac{\delta_T}{\delta} \right)^3 \right.$$

$$\left. + 4.64 \left(\frac{v}{v_\infty} \right)^{1/2} x^{1/2} \frac{4.64}{2} \left(\frac{v}{v_\infty} \right)^{1/2} x^{-1/2} \left(\frac{\delta_T}{\delta} \right)^3 \right] = 10\alpha$$

or

$$14.35x \frac{d}{dx} \left(\frac{\delta_T}{\delta} \right)^3 + 10.76 \left(\frac{\delta_T}{\delta} \right)^3 = \frac{10\alpha}{v} = \frac{10}{Pr}$$

or

$$\frac{4}{3} x \frac{d}{dx} \left(\frac{\delta_T}{\delta} \right)^3 + \left(\frac{\delta_T}{\delta} \right)^3 = \frac{0.929}{Pr}$$

or

$$\frac{d(\delta_T/\delta)^3}{0.929/Pr - (\delta_T/\delta)^3} = \frac{3}{4} \frac{dx}{x} \tag{7.19}$$

For the purpose of integrating this expression it is considered that an initial length of the plate, from $x = 0$ to $x = x_0$, is unheated (i.e., is at the temperature of the fluid). In this case the thermal boundary layer begins at x_0, where $\delta_T = 0$. Integration of Eq. (7.19) and evaluation between the limits δ_T at x and $\delta_T = 0$ at x_0 gives

$$\ln \frac{0.929/Pr - (\delta_T/\delta)^3}{0.929/Pr} = -\frac{3}{4} \ln \frac{x}{x_0}$$

or

$$1 - \frac{Pr(\delta_T/\delta)^3}{0.929} = \left(\frac{x_0}{x} \right)^{3/4}$$

The length of the initial unheated length can now be decreased to zero (i.e., $\delta_T = 0$ at x_0), in which case

$$\left(\frac{\delta_T}{\delta} \right)^3 = \frac{0.929}{Pr}$$

or

$$\frac{\delta_T}{\delta} = \frac{1}{1.025 Pr^{1/3}} \tag{7.20}$$

For $\delta_T = \delta$, $Pr = 0.929$ and hence the elimination of the fourth-power term in Eq. (7.16) limits the application of Eq. (7.20) fluids of Prandtl number greater than 0.929. In the exact solution $\delta_T = \delta$ when $\alpha = v$ (i.e., when $Pr = 1$) and the exact solution gives

$$\frac{\delta_T}{\delta} = \frac{1}{Pr^{1/3}} \tag{7.21}$$

The exact solution is limited to fluids of $0.6 \leq \text{Pr} \leq 50$. The similarity between Eq. (7.20) and Eq. (7.21) is such that insignificant error is introduced when the equations, derived using the approximate integral method, are applied to fluids with Prandtl numbers as low as 0.6. This allows the approximate equations to be applied to gases (but not to liquid metals).

As the layer of fluid in contact with the surface of the plate is stagnant, heat is transported through this layer by conduction at the rate

$$q'_y = -k \left(\frac{\partial T}{\partial y} \right) \Big|_{y=0}$$

and in the boundary layer heat is transferred by convection at the rate

$$q'_y = h(T_s - T_\infty)$$

Thus

$$h = \frac{-k(\partial T/\partial y)|_{y=0}}{T_s - T_\infty} \tag{7.22}$$

Differentiation of Eq. (7.15) gives

$$\left(\frac{\partial T}{\partial y} \right) \Big|_{y=0} = \frac{1.5(T_\infty - T_s)}{\delta_T}$$

and thus the local heat transfer coefficient at x is

$$
\begin{aligned}
h_x &= \frac{1.5k}{\delta_T} \\
&= \frac{(1.5k) \times 1.025 \text{Pr}^{1/3}}{\delta} \\
&= (1.5k) \times 1.025 \text{Pr}^{1/3} \times \frac{1}{4.64} \left(\frac{v_\infty}{vx} \right)^{1/2} \\
&= 0.331k \, \text{Pr}^{1/3} \left(\frac{v_\infty}{vx} \right)^{1/2} \tag{7.23}
\end{aligned}
$$

which is in excellent agreement with exact solution given by Eq. (7.5). Averaging the heat transfer coefficient over the length of plate from $x = 0$ to $x = L$ gives the average heat transfer coefficient as

$$\bar{h}_L = 0.662k \, \text{Pr}^{1/3} \left(\frac{v_\infty}{vL} \right)^{1/2} \tag{7.24}$$

Thus the local Nusselt number is

$$\text{Nu}_x = \frac{h_x x}{k} = 0.331 \text{Pr}^{1/3} \text{Re}_x^{1/2} \tag{7.25}$$

and the average Nusselt number for heat transfer from the length of plate from $x = 0$ to $x = L$ is

$$\overline{\mathrm{Nu}}_L = \frac{\overline{h}_L L}{k} = 0.662 \mathrm{Pr}^{1/3} \mathrm{Re}_L^{1/2} \tag{7.26}$$

EXAMPLE 7.1

Consider air at 20°C flowing over a flat plate at $v_\infty = 5$ m/s. The plate is 1 m in length, 0.5 m in width, and is at the constant temperature $T_s = 100$°C. The values of the physical properties of the fluid to be used in convection heat transfer problems are those for the film temperature, T_f, which is the arithmetic average of T_s and T_∞,

$$T_f = \frac{T_s + T_\infty}{2}$$

In the present example, $T_f = (100 + 20)/2 = 60$°C, at which temperature the properties for air are

$$\mathrm{Pr} = 0.708$$

$$v = 1.89 \times 10^{-5} \ \mathrm{m^2/s}$$

$$k = 0.0284 \ \mathrm{W/m \cdot K}$$

The first requirement is the determination of whether the velocity boundary layer on the plate is laminar or turbulent. In Chapter 3 it was seen that at a Reynolds number of 3×10^5, a transition from a laminar boundary layer to a turbulent boundary layer begins, and that at $\mathrm{Re}_x = 3 \times 10^6$, the transition to turbulence is complete. For the purpose of convection heat transfer calculations it is assumed that a sharp transition from laminar to turbulent boundary flow occurs at $\mathrm{Re}_x = 5 \times 10^5$. In the present example the Reynolds number at the trailing edge of the plate at $x = 1$ m is

$$\mathrm{Re}_L = \frac{v_\infty L}{v} = \frac{5 \times 1}{1.89 \times 10^{-5}} = 2.65 \times 10^5$$

Therefore, the boundary layer is laminar over the entire length of the plate. The velocity, or momentum boundary layer, δ, is

$$\delta = 4.64 \left(\frac{v}{v_\infty} \right)^{1/2} x^{1/2} = 4.64 \left(\frac{1.89 \times 10^{-5}}{5} \right)^{1/2} x^{1/2}$$

$$= 9.02 \times 10^{-3} x^{1/2}$$

This variation is shown in Fig. 7.4(a). From Eq. (7.20),

$$\delta_T = \frac{\delta}{1.025 \mathrm{Pr}^{1/3}} = \frac{9.02 \times 10^{-3}}{1.025 \times 0.708^{1/3}} x^{1/2}$$

$$= 9.87 \times 10^{-3} x^{1/2}$$

This variation is shown in Fig. 7.4(a). The low value of the Prandtl number causes the thermal boundary layer to develop more rapidly than the momentum boundary layer. From Eq. (7.23),

$$h_x = 0.331k \, \text{Pr}^{1/3} \left(\frac{v_\infty}{\nu x}\right)^{1/2}$$

$$= 0.331 \times 0.0284 \times 0.708^{1/3} \times \left(\frac{5}{1.89 \times 10^{-5}}\right)^{1/2} x^{-1/2}$$

$$= 4.31x^{-1/2}$$

This is shown in Fig. 7.4(a). It can be noted that $h_x \to \infty$ as $x \to 0$. This artifact arises from Eq. (7.22), in which

$$\left(\frac{\partial T}{\partial y}\right)\bigg|_{y=0} = \frac{1.5(T_\infty - T_s)}{\delta_T}$$

approaches infinity as x, and hence δ_T, $\to 0$. The local heat flux from the plate is thus

$$q_y' = h_x(T_s - T_\infty)$$
$$= 80h_x$$
$$= 344.8x^{-1/2}$$

The heat flux from the entire plate is calculated as follows. At the trailing edge of the plate

$$\text{Re}_L = 2.65 \times 10^5$$

Therefore,

$$\overline{\text{Nu}}_L = 0.662\text{Pr}^{1/3}\text{Re}_L^{1/2} = 0.662 \times (0.708)^{1/3} \times (2.65 \times 10^5)^{1/2}$$
$$= 303$$

Thus

$$\overline{h}_L = \frac{\overline{\text{Nu}}_L k}{L} = \frac{303 \times 0.0284}{1} = 8.62 \text{ W/m}^2\cdot\text{K}$$

and

$$q_y = \overline{h}_L WL(T_s - T_\infty) = 8.62 \times 0.5 \times 1 \times (100 - 20)$$
$$= 345 \text{ W}$$

that is, the heat flux from the plate to the air is 345 W. The question can be raised; what fraction of this heat flux is from the first 0.5 m of plate length, and what fraction is from the last 0.5 m of plate length? The average heat transfer coefficient for heat transfer from the length of plate between $x = x_1$ and $x = x_2$ is obtained from

$$\overline{h}(x_2 - x_1) = \int_{x_1}^{x_2} h_x \, dx$$

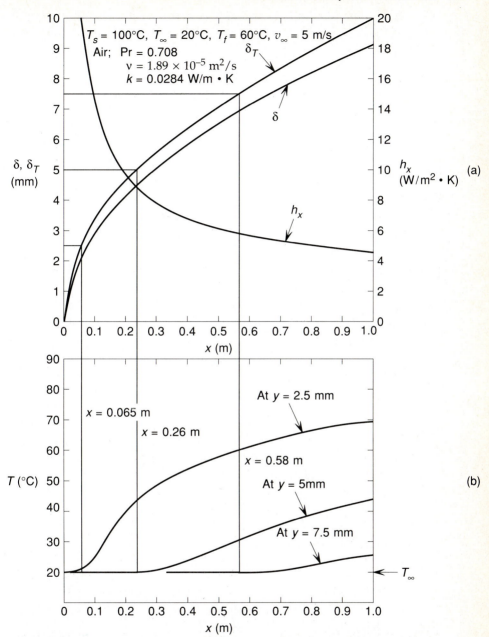

FIGURE 7.4. (a) The variation, with distance from the leading edge, of the thicknesses of the momentum and thermal boundary layers and the local heat transfer coefficient considered in Example 7.1; (b) variation, with x, of T at three values of y.

$$= 0.331k \, \text{Pr}^{1/3} \left(\frac{v_\infty}{\nu}\right)^{1/2} \int_{x_1}^{x_2} \frac{dx}{x^{1/2}}$$

$$= 0.662 \times 0.0284 \times (0.708)^{1/3} \left(\frac{5}{1.89 \times 10^{-5}}\right)^{1/2} (x_2^{1/2} - x_1^{1/2})$$

or

$$\bar{h} = \frac{8.62(x_2^{1/2} - x_1^{1/2})}{x_2 - x_1}$$

Thus, for the first 0.5 m of plate, with $x_2 = 0.5$ and $x_1 = 0$,

$$\bar{h} = \frac{8.62}{0.5^{1/2}} = 12.19 \text{ W/m}^2 \cdot \text{K}$$

and for the second 0.5 m of plate, with $x_2 = 1$ and $x_1 = 0.5$,

$$\bar{h} = 8.62 \frac{1 - 0.5^{1/2}}{1 - 0.5} = 5.05 \text{ W/m}^2 \cdot \text{K}$$

Note that the average of these two values, $(12.19 + 5.05)/2 = 8.62$, which is the average for the entire plate. Thus the heat flux from the first 0.5 m is

$$q_y = 12.19 \times 0.5 \times 0.5 \times (100 - 20) = 244 \text{ W}$$

and from the last 0.5 m is

$$q_y = 5.05 \times 0.5 \times 0.5 \times (100 - 20) = 101 \text{ W}$$

Thus $244/345 \equiv 70.7\%$ comes from the first 0.5 m and $101/345 = 29.3\%$ comes from the last 0.5 m.

The variations of the temperature of the air, at constant y, with variation of x through the thermal boundary layer are determined as follows. From Eq. (7.15),

$$T = T_s - (T_s - T_\infty) \left[1.5 \frac{y}{\delta_T} - 0.5 \left(\frac{y}{\delta_T}\right)^3 \right]$$

and as determined in the present example,

$$\delta_T = 9.78 \times 10^{-3} x^{1/2}$$

insertion of which into Eq. (7.15) gives the variations of the local temperature T with x and y within the thermal boundary layers. The variations of T with x for constant y-values of 0.0025, 0.005, and 0.0075 m are shown in Fig. 7.4(b). At $y = 0.0025$ m, the air enters the thermal boundary layer at $x = 0.065$ m and, thereafter, is progressively heated by convection from the plate. At $y = 0.005$ m, the air enters the thermal boundary layer at $x = 0.26$ m and at $y = 0.0075$ m entrance occurs at $x = 0.59$ m.

The Stanton number, St, is defined as

$$St = \frac{Nu}{Re \cdot Pr} \tag{7.27}$$

and hence the local Stanton number at x is

$$St_x = \frac{Nu_x}{Re_x Pr}$$

which from Eq. (7.25) is

$$St_x = \frac{0.331 Pr^{1/3} Re_x^{1/2}}{Pr \cdot Re_x} = 0.331 Pr^{-2/3} Re^{-1/2}$$

The exact solution gives the local friction factor for laminar boundary flow as

$$f_x = \frac{0.662}{Re_x^{1/2}}$$

[using the approximate integral method, the local friction factor was obtained in Eq. (4.23a) as $f_x = 0.646/Re_x^{1/2}$] and thus

$$St_x = 0.5 f_x \, Pr^{-2/3} \tag{7.28}$$

which is known as the Reynolds analogy between momentum transport and heat transport in laminar boundary layers.

Turbulent Boundary Flow

At the position along the plate at which $Re_x = 5 \times 10^5$, it is considered that an abrupt transition occurs from laminar to turbulent boundary flow, and both the friction shear stress exerted by the fluid on the plate and the heat transfer coefficient are greater in a turbulent boundary layer than in a laminar one. A traditional formula for Nu_x in a turbulent boundary layer, which is restricted to $0.7 \leq Pr \leq 3$, is

$$Nu_x = 0.0296 Re_x^{4/5} Pr^{1/3} \tag{7.29}$$

A formula that is valid for any turbulent Reynolds number and $Pr \geq 0.5$ is

$$St_x = \frac{f_x/2}{1 + 12.7(Pr^{2/3} - 1)(f_x/2)^{1/2}} \tag{7.30}$$

where St_x is the local Stanton number at x and f_x is the local friction factor at x given by

$$f_x = \frac{0.455}{\ln^2(0.06 Re_x)} \tag{4.36}$$

Correlation with experimental measurement has shown that the average Stanton number, $\overline{St}_L$, is

$$\overline{St}_L \approx 1.15 St_x \tag{7.31a}$$

and hence

$$\overline{\mathrm{Nu}}_L \approx 1.15 \mathrm{Nu}_x \qquad\qquad (7.31\mathrm{b})$$

Thus, in considering heat transfer by convection from a flat plate to a turbulent boundary layer:

1. f_x is determined from Re_x.
2. St_x is found from f_x and Pr.
3. $\overline{\mathrm{St}}_L$ is found from St_x.
4. $\overline{\mathrm{Nu}}_L$ is found from $\overline{\mathrm{St}}_L$.
5. $\overline{h}_L$ is found from $\overline{\mathrm{Nu}}_L$.
6. q_y is found from $\overline{h}_L$, area, and $T_s - T_\infty$.

EXAMPLE 7.2

Consider water at 20°C flowing over a flat plate at $v_\infty = 1$ m/s. The plate is 1 m in length, 0.5 m in width, and at constant temperature $T_s = 100$°C. At the film temperature of 60°C, the physical properties of water are

$$\mathrm{Pr} = 3.08$$

$$\nu = 4.9 \times 10^{-7} \ \mathrm{m^2/s}$$

$$k = 0.651 \ \mathrm{W/m \cdot K}$$

At the trailing edge of the plate,

$$\mathrm{Re}_L = \frac{v_\infty L}{\nu} = \frac{1}{4.9 \times 10^{-7}} = 2.04 \times 10^6$$

which, being greater than 5×10^5, indicates that a transition from laminar boundary flow to turbulent boundary flow occurs somewhere on the plate. At x_{trans},

$$\mathrm{Re}_x = 500{,}000 = \frac{v_\infty x_{\mathrm{trans}}}{\nu} = \frac{x_{\mathrm{trans}}}{4.9 \times 10^{-7}}$$

or

$$x_{\mathrm{trans}} = 0.245 \ \mathrm{m}$$

Thus the boundary layer is laminar over approximately the first 25% of the plate and is turbulent over the other 75%. Up to $x = 0.245$ m,

$$h_x = 0.331k \ \mathrm{Pr}^{1/3} \left(\frac{v_\infty}{\nu}\right)^{1/2} \frac{1}{x^{1/2}}$$

$$= 0.331 \times 0.651 \times 3.08^{1/3} \times \left(\frac{1}{4.9 \times 10^{-7}}\right)^{1/2} \frac{1}{x^{1/2}}$$

$$= \frac{448}{x^{1/2}}$$

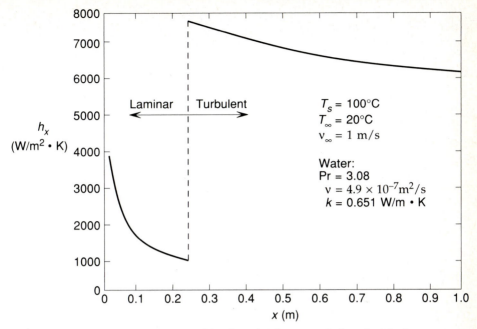

FIGURE 7.5. Influence of the transition from laminar to turbulent flow in the boundary layer on the local heat transfer coefficient.

This is shown in Fig. 7.5 along with the variation of h_x with x in the turbulent boundary layer. Heat transfer from the entire plate is calculated using an average Nusselt number obtained as

$$\overline{Nu} = \overline{Nu}_{trans,lam} + (\overline{Nu}_L - \overline{Nu}_{trans})_{turb}$$

in which $\overline{Nu}_{trans,lam}$ is the average Nusselt number for heat transport from the plate from $x = 0$ to x_{trans} (at $Re = 500{,}000$) assuming a laminar boundary layer, $\overline{Nu}_{trans,turb}$ is the average Nusselt number for heat transport from the plate from $x = 0$ to x_{trans} assuming a turbulent boundary layer, and $\overline{Nu}_{L,turb}$ is the average Nusselt number for heat transfer from the plate from $x = 0$ to $x = L$ assuming a turbulent boundary layer.

$$\overline{Nu}_{trans,lam} = 0.662 Pr^{1/3}(500{,}000)^{1/2}$$
$$= 0.662 \times (3.08)^{13} \times (5 \times 10^5)^{1/2}$$
$$= 681$$

Calculation of $\overline{Nu}_{L,turb}$: At the trailing edge of the plate, $x = 1$, $Re_x = 2.04 \times 10^6$, and therefore, from Eq. (4.36),

$$f_x = \frac{0.455}{\ln^2(0.06\ Re)} = \frac{0.455}{\ln^2(0.06 \times 2.04 \times 10^6)} = 3.315 \times 10^{-3}$$

$$St_x = \frac{f_x/2}{1 + 12.7 \, (Pr^{2/3} - 1)(f_x/2)^{1/2}}$$

$$= \frac{1.658 \times 10^{-3}}{1 + 12.7(3.08^{2/3} - 1)(1.658 \times 10^{-3})^{1/2}}$$

$$= 1.051 \times 10^{-3}$$

$$\overline{St}_L = 1.15 \times 1.051 \times 10^{-3}$$

$$= 1.208 \times 10^{-3}$$

and

$$\overline{Nu}_L = \overline{St}_L Re_L Pr$$

$$= 1.208 \times 10^{-3} \times 2.04 \times 10^6 \times 3.08$$

$$= 7596$$

$$= \overline{Nu}_{L,turb}$$

Calculation of $\overline{Nu}_{trans,turb}$: At $x = 0.245$ m, Re $= 500,000$; therefore,

$$f_x = \frac{0.455}{\ln^2(0.06 \times 500,000)} = 4.281 \times 10^{-3}$$

$$St_x = \frac{2.19 \times 10^{-3}}{1 + 12.7(3.08^{2/3} - 1)(4.626 \times 10^{-2})}$$

$$= 1.292 \times 10^{-3}$$

$$\overline{St}_L = 1.15 \times 1.292 \times 10^{-3}$$

$$= 1.486 \times 10^{-3}$$

$$\overline{Nu}_L = 1.486 \times 10^{-3} \times 500,000 \times 3.08$$

$$= 2289$$

$$\overline{Nu} = 681 + (7593 - 2289) = 5988$$

$$\overline{h} = \frac{\overline{Nu} \cdot k}{L} = 5988 \times 0.651 = 3898 \text{ W/m}^2 \cdot \text{K}$$

and

$$q_y = \overline{h}WL(T_s - T_\infty)$$

$$= 3898 \times 0.5 \times 1 \times (100 - 20)$$

$$= 156,000 \text{ W}$$

Alternatively, the average heat transfer coefficient from the plate can be obtained as

$$\overline{h}_L = \frac{1}{L} \left(\int_0^{x_{trans}} h_{x,lam} \, dx + \int_{x_{trans}}^L h_{x,turb} \, dx \right)$$

For fluids of $0.7 \leq Pr \leq 3$, $h_{x,turb}$ is obtained from Eq. (7.27) as

$$h_{x,\text{turb}} = 0.0296k \; \text{Pr}^{1/3} \left(\frac{v_\infty}{v}\right)^{4/5} x^{-1/5}$$

and thus

$$\bar{h}_L = \frac{1}{L}\left[0.662k \; \text{Pr}^{1/3}\text{Re}_{\text{trans}}^{1/2} + \frac{5}{4} \times 0.0296k \; \text{Pr}^{1/3}(\text{Re}_L^{4/5} - \text{Re}_{\text{trans}}^{4/5}) \right]$$

$$= \frac{k \; \text{Pr}^{1/3}}{L}(0.037\text{Re}_L^{4/5} - 873)$$

Examples 7.1 and 7.2 illustrate the influence of the physical properties of the fluid on heat transfer by convection from the surface of a horizontal plate. For heat transfer to a laminar boundary layer,

$$h_x = 0.331 \frac{k \; \text{Pr}^{1/3}}{v} v_\infty^{1/2} x^{-1/2}$$

which for given values of v_∞ and x, increases with increasing thermal conductivity and Prandtl number and decreasing kinematic viscosity. For air;

$$h_x = 0.331 \left[\frac{0.0284 \times 0.708^{1/3}}{(1.89 \times 10^{-5})^{1/2}}\right] v_\infty^{1/2} x^{-1/2}$$

$$= 1.93 v_\infty^{1/2} x^{-1/2}$$

and for water;

$$h_x = 0.331 \left[\frac{0.651 \times 3.08^{1/3}}{(4.9 \times 10^{-7})^{1/2}}\right] v_\infty^{1/2} x^{-1/2}$$

$$= 448 v_\infty^{1/2} x^{-1/2}$$

Thus the larger values of k and Pr and smaller value of v for water give, for the same values of v_∞ and x,

$$\frac{h_{x(\text{water})}}{h_{x(\text{air})}} = 232$$

7.3 Heat Transfer from a Horizontal Flat Plate with Uniform Heat Flux Along the Plate

For heat transfer by forced convection from a horizontal flat plate with uniform q'_y to a laminar boundary layer, the Nusselt number is given by

$$\text{Nu}_x = 0.453\text{Pr}^{1/3}\text{Re}_x^{1/2} \tag{7.32}$$

Thus, as with transfer from a horizontal plate at the uniform temperature T_s, the Nusselt number varies with distance along the plate as

$$\mathrm{Nu}_x \propto x^{1/2}$$

and hence the local heat transfer coefficient varies with distance along the plate as

$$h_x = \frac{\mathrm{Nu}_x k}{x} \propto \frac{1}{x^{1/2}}$$

Thus from Newton's equation

$$q'_y = h_x(T_{S(x)} - T_\infty) \tag{7.33}$$

With a constant value of q'_y,

$$(T_{S(x)} - T_\infty) \propto x^{1/2}$$

where $T_{S(x)}$ is the local surface temperature of the plate at x. From Eq. (7.32),

$$h_x = 0.453k \, \mathrm{Pr}^{1/3} \left(\frac{v_\infty}{v}\right)^{1/2} x^{-1/2}$$

and for constant q'_y, Eq. (7.33) can be written as

$$T_{S(x)} - T_\infty = \frac{q'_y}{0.453k \, \mathrm{Pr}^{1/3}} \left(\frac{v}{v_\infty}\right)^{1/2} x^{1/2} \tag{7.34}$$

Thus, for heat transfer by forced convection from a flat plate to a flowing fluid,

1. With T_s constant, q'_y decreases with distance along the plate as $q'_y \propto x^{-1/2}$.
2. With q'_y constant, $T_{S(x)}$ increases with distance along the plate as given by Eq. (7.34).

EXAMPLE 7.3 _____

In Example 7.1, air at 20°C flowed at a velocity $v_\infty = 5$ m/s over a horizontal flat plate 1 m in length and 0.5 m in width with a constant surface temperature $T_s = 100$°C and it was found that the total heat flux from the plate was 345 W. Consider, now, the flow of air at $T_\infty = 20$°C and $v_\infty = 5$ m/s over the same plate and let the heat flux, q'_y, have the uniform value

$$q'_y = \frac{q_y}{A} = \frac{345}{1 \times 0.5} = 690 \text{ W/m}^2$$

From Eq. (7.34),

$$T_{S(x)} = 20 + \frac{690}{0.453 \times 0.0284 \times (0.708)^{1/3}} \left(\frac{1.89 \times 10^{-5}}{5}\right)^{1/2} x^{1/2}$$

$$= 20 + 117x^{1/2}$$

Thus at the leading edge, $T_{S(x=0)} = 20$°C, and at the trailing edge, $T_{S(x=1)} =$

137°C. The local surface temperature of the plate is 100°C at $x = 0.468$ m and the average surface temperature is

$$T_{aver} = \frac{1}{L} \int_0^{x=L} T_{S(x)} \, dx$$

$$= \int_0^1 (20 + 117x^{1/2}) \, dx$$

$$= 20x + \tfrac{2}{3} \times 117x^{3/2}\big|_0^1$$

$$= 98°C$$

which is also the local surface temperature at $x = 0.444$ m.

7.4 Heat Transfer During Fluid Flow in Cylindrical Pipes

If the temperature of a fluid flowing in a cylindrical pipe is different from that of the inner surface of the pipe, heat transfer occurs between the fluid and the wall of the pipe and this gives rise to a variation in the temperature of the fluid with distance along the pipe and a variation in the temperature of the fluid in the radial direction. Consider the control volume in a fluid undergoing fully developed laminar flow in a circular pipe shown in Fig. 7.6. Fluid flow is in the x direction and hence heat enters the control volume through the surface at x and leaves through the surface at $x + \Delta x$ by convection due to the fluid flow and by conduction. However, because there is no component of fluid flow in the radial direction, heat transport in the radial direction can occur only by conduction down the temperature gradient in the radial direction. If the temperature of the fluid is higher than that of the surface of the pipe, heat is conducted into the control volume through the inner surface at r and out of

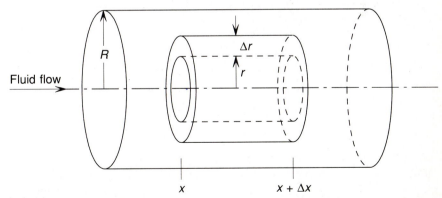

FIGURE 7.6. Control volume for consideration of heat transport to or from a fluid flowing in a cylindrical pipe.

the outer surface at $r + \Delta r$ Thus heat is conducted into the control volume through the surface at x at the rate

$$q'_x|_x \, 2\pi r \, \Delta r$$

and is conducted out of the control volume through the surface at $x + \Delta x$ at the rate

$$q'_x|_{x+\Delta x} \, 2\pi r \, \Delta r$$

Heat is transferred into the control volume by convection through the surface at x at the rate

$$\rho v_x H|_x \, 2\pi r \, \Delta r$$

and is transferred out of the control volume by convection through the surface at $x + \Delta x$ at the rate

$$\rho v_x H|_{x+\Delta x} \, 2\pi r \, \Delta r$$

Heat is conducted into the control volume through the inner surface at the rate

$$(q'_r \cdot 2\pi r \, \Delta x)|_r$$

and is conducted out of the control volume through the outer surface at the rate

$$(q'_r \cdot 2\pi r \, \Delta x)|_{r+\Delta r}$$

Thus, at steady state,

$$(q'_x|_x - q'_x|_{x+\Delta x}) \, 2\pi r \, \Delta r + (\rho v_x H|_x - \rho v_x H|_{x+\Delta x}) \, 2\pi r \, \Delta r$$
$$+ \, 2\pi[(rq'_r)_r - (rq'_r)_{r+\Delta r}] \, \Delta x = 0$$

Dividing through by $2\pi \, \Delta r \, \Delta x$ and allowing Δr and Δx to approach zero gives the differential heat balance as

$$r \frac{\partial q'_x}{\partial x} + r \frac{\partial(\rho v_x H)}{\partial x} + \frac{\partial(rq'_r)}{\partial r} = 0$$

Again, the constant-pressure heat capacity is defined as $dH = C_p \, dT$, and thus for a fluid of constant density and constant thermal conductivity,

$$-rk \frac{\partial^2 T}{\partial x^2} + r\rho C_p v_x \frac{\partial T}{\partial x} - k \frac{\partial}{\partial r}\left(r \frac{\partial T}{\partial r}\right) = 0$$

or

$$v_x \frac{\partial T}{\partial x} = \frac{\alpha}{r}\left[\frac{\partial}{\partial r}\left(r \frac{\partial T}{\partial r}\right) + r \frac{\partial^2 T}{\partial x^2}\right] \tag{7.35}$$

where $\alpha = k/\rho C_p$ is the thermal diffusivity of the fluid.

If q'_r is uniform along the pipe, $\partial^2 T/\partial x^2$ is zero, in which case Eq. (7.35) can be written as

$$v_x \frac{\partial T}{\partial x} = \frac{\alpha}{r}\left[\frac{\partial}{\partial r}\left(r \frac{\partial T}{\partial r}\right)\right] \tag{7.36}$$

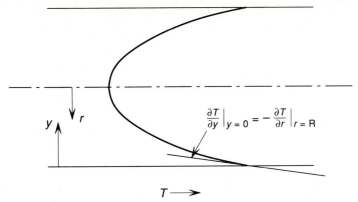

$$\frac{\partial T}{\partial y}\Big|_{y=0} = -\frac{\partial T}{\partial r}\Big|_{r=R}$$

FIGURE 7.7. Schematic representation of the radial distribution of temperature in a fluid flowing in a cylindrical pipe.

The local temperature of the fluid varies with both x and r and hence the concept of "the temperature of the fluid" has to be defined. If a thin slice of fluid at the location x is removed and is adiabatically mixed to eliminate the temperature gradients, the resulting uniform temperature that the fluid attains is called the *mean temperature of the fluid*, T_m. Consider a uniform flux, q'_r, from the surface of the pipe to the fluid (such as would be produced by winding electrically heated wire round the pipe). As the fluid moves through the pipe, both the surface temperature, T_s, and the mean temperature of the fluid, T_m, increase, and after some entry length into the pipe, a *fully developed temperature profile* is attained, in which the ratio $(T_s - T)/(T_s - T_m)$ is a unique function of r/R and is independent of x. In this relationship, T is the local temperature of the fluid at the radius r and R is the radius of the pipe. The heat transfer coefficient is then defined as

$$q'_r = h(T_s - T_m) \tag{7.37}$$

and, as the transport of heat across the stagnant layer of fluid in contact with the surface of the pipe occurs only by conduction, q'_r can be related to the limiting temperature gradient in the $-r$-direction via Fourier's law. For heat transfer from the pipe to the fluid, the fully developed temperature profile at any given x is as shown in Fig. 7.7. Considered as heat flow from the surface to the fluid in the $+y$-direction, Fourier's law gives

$$q'_y = -k \left(\frac{\partial T}{\partial y}\right)_{y=0}$$

and as $(\partial T/\partial y)_{y=0} < 0$, q'_y is a positive quantity. However, as

$$\left(\frac{\partial T}{\partial y}\right)_{y=0} = -\left(\frac{\partial T}{\partial r}\right)_{r=R}$$

a positive value of q' for heat flow from the surface to the fluid requires that

$$q'_r = -k \left(\frac{\partial T}{\partial y}\right)_{y=0} = k \left(\frac{\partial T}{\partial r}\right)_{r=R} \tag{7.38}$$

and thus combination of Eqs. (7.37) and (7.38) gives the heat transfer coefficient as

$$h = \frac{k(\partial T/\partial r)_{r=R}}{T_s - T_m} \tag{7.39}$$

As $(T_s - T)/(T_s - T_m)$ is a unique function of r/R, the partial derivative

$$\frac{\partial}{\partial(r/R)} \left(\frac{T_s - T}{T_s - T_m}\right)\bigg|_{r=R} = \frac{-R}{T_s - T_m} \left(\frac{\partial T}{\partial R}\right)\bigg|_{r=R}$$

has a unique value that is independent of x. Therefore, in Eq. (7.39), h is a constant, and as q'_r is also constant, Eq. (7.37) shows that $T_s - T_m$ is constant. Thus T_s and T_m increase with increasing x at the same rate, or

$$\frac{\partial T_s}{\partial x} = \frac{\partial T_m}{\partial x} \tag{7.40}$$

As $(T_s - T)/(T_s - T_m)$ is not a function of x,

$$\frac{\partial}{\partial x} \frac{T_s - T}{T_s - T_m} = 0$$

or

$$\frac{T_s - T_m}{(T_s - T_m)^2} \left(\frac{\partial T_s}{\partial x} - \frac{\partial T}{\partial x}\right) - \frac{T_s - T}{(T_s - T_m)^2} \left(\frac{\partial T_s}{\partial x} - \frac{\partial T_m}{\partial x}\right) = 0$$

which in comparison with Eq. (7.40) shows that

$$\frac{\partial T_s}{\partial x} = \frac{\partial T_m}{\partial x} = \frac{\partial T}{\partial x} \tag{7.41}$$

and hence Eq. (7.36) can be written as

$$v_x \frac{\partial T_m}{\partial x} = \frac{\alpha}{r} \left[\frac{\partial}{\partial r}\left(r \frac{\partial T}{\partial r}\right)\right] \tag{7.42}$$

In fully developed laminar flow in a cylindrical pipe,

$$v_x = \left(\frac{\Delta P}{L} \pm \rho g\right) \frac{R^2 - r^2}{4\eta} \tag{2.20}$$

and

$$\bar{v}_x = \left(\frac{\Delta P}{L} \pm \rho g\right) \frac{R^2}{8\eta} \tag{2.22}$$

combination of which gives

$$v_x = 2\bar{v}_x \left(1 - \frac{r^2}{R^2}\right) \tag{7.43}$$

and substitution of Eq. (7.43) into Eq. (7.42) gives

$$2\bar{v}_x \left(1 - \frac{r^2}{R^2}\right) \frac{\partial T_m}{\partial x} = \frac{\alpha}{r}\left[\frac{\partial}{\partial r}\left(r\frac{\partial T}{\partial r}\right)\right]$$

or

$$\partial\left(r\frac{\partial T}{\partial r}\right) = \frac{2\bar{v}_x}{\alpha}\frac{\partial T_m}{\partial x}\left(r - \frac{r^3}{R^2}\right) dr \qquad (7.44)$$

From the symmetry shown in Fig. 7.7, $(\partial T/\partial r) = 0$ at $r = 0$ and thus integrating Eq. (7.44) from $(\partial T/\partial r)$ at r to $(\partial T/\partial r)_{r=0}$ at $r = 0$ gives

$$r\frac{\partial T}{\partial r} = \frac{2\bar{v}_x}{\alpha}\frac{\partial T_m}{\partial x}\left(\frac{r^2}{2} - \frac{r^4}{4R^2}\right)$$

or

$$\frac{\partial T}{\partial r} = \frac{2\bar{v}_x}{\alpha}\left(\frac{\partial T_m}{\partial x}\right)\left(\frac{r}{2} - \frac{r^3}{4R^2}\right) \qquad (7.45)$$

and integrating Eq. (7.45) from $T = T_s$ at $r = R$ to $T = T$ at $r = r$ gives

$$T_s - T = \frac{2\bar{v}_x}{\alpha}\frac{\partial T_m}{\partial x}\left[\frac{r^2}{4} - \frac{r^4}{16R^2}\right]_r^R$$

$$= \frac{\bar{v}_x}{\alpha}\frac{\partial T_m}{\partial x}\frac{1}{8R^2}(3R^4 - 4R^2r^2 + r^4) \qquad (7.46)$$

as the variation of T with r. The difference $T_s - T_m$ is obtained from

$$(T_s - T_m)\int_0^{2\pi}\int_0^R v_x r\,dr\,d\theta = \int_0^{2\pi}\int_0^R (T_s - T)v_x r\,dr\,d\theta \qquad (7.47)$$

From Eq. (7.43) the left side of Eq. (7.47) is

$$(T_s - T_m)2\pi\cdot 2\bar{v}_x\int_0^R\left(r - \frac{r^3}{R^2}\right) = (T_s - T_m)\pi\bar{v}_x R^2$$

and from Eqs. (7.43) and (7.46), the right side is

$$2\pi\frac{\bar{v}_x}{\alpha}\left(\frac{\partial T_m}{\partial x}\right)\frac{1}{8R^2}2\bar{v}_x\int_0^R\left(1 - \frac{r^2}{R^2}\right)(3R^4 - 4R^2r^2 + r^4)r\,dr$$

The integral is $0.458R^6$ and hence the term is

$$0.229\pi\frac{\bar{v}_x^2}{\alpha}\frac{\partial T_m}{\partial x}R^4$$

Thus

$$T_s - T_m = 0.229\frac{\bar{v}_x}{\alpha}\frac{\partial T_m}{\partial x}R^2 \qquad (7.48)$$

and from Eq. (7.45),

$$\left.\frac{\partial T}{\partial r}\right|_{r=R} = \frac{\bar{v}_x}{\alpha}\frac{\partial T_m}{\partial x}\frac{R}{2} \tag{7.49}$$

Substitution of Eqs. (7.48) and (7.49) in Eq. (7.39) gives

$$h = \frac{k}{2 \times 0.229R}$$

and therefore

$$\mathrm{Nu}_D = \frac{hD}{k} = \frac{2R}{2 \times 0.229R} = 4.36$$

Thus for laminar flow in a cylindrical pipe with uniform q'_r,

$$\mathrm{Nu}_D = 4.36 \tag{7.50}$$

Analysis of laminar flow in a cylindrical pipe with uniform T_s gives

$$\mathrm{Nu}_D = 3.66 \tag{7.51}$$

For turbulent flow in a cylindrical pipe, for conditions of both uniform T_s and uniform q'_r,

$$\mathrm{Nu}_D = \mathrm{St}\cdot\mathrm{Re}_D\cdot\mathrm{Pr}$$

where when $\mathrm{Pr} > 0.7$, St is obtained from Eq. (7.30) as

$$\mathrm{St} = \frac{f/2}{1 + 12.7(\mathrm{Pr}^{2/3} - 1)(f/2)^{1/2}}$$

in which f is the friction for *pipe flow* obtained from Eq. (4.18) for flow in smooth pipes as

$$f^{-1/2} = -3.6 \log_{10}\frac{6.9}{\mathrm{Re}_D}$$

7.5 Energy Balance in Heat Transfer by Convection Between a Cylindrical Pipe and a Flowing Fluid

Consider Fig. 7.8, in which a fluid is flowing at the mass flow rate $\dot{M}$ through a cylindrical pipe of diameter D and length L. The heat flux from the pipe to the fluid, q', is uniform along the length of pipe and the fluid enters the pipe at the mean temperature $T_{m,\text{in}}$ and exists at the mean temperature $T_{m,\text{out}}$. The energy balance is based on

heat transferred to the fluid = (mass of fluid) × (heat capacity of fluid)
× (increase in the temperature of the fluid)

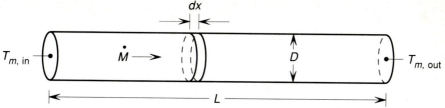

FIGURE 7.8. Volume element for consideration of the energy balance for fluid flow in a cylindrical pipe.

or, introducing time,

$$q'\pi DL = \dot{M}C_p(T_{m,\text{out}} - T_{m,\text{in}}) \qquad (7.52)$$

The differential energy balance on the element of thickness dx is

$$q'\pi D\, dx = \dot{M}C_p\, dT_m$$

or

$$dT_m = \frac{q'\pi D}{\dot{M}C_p}\, dx \qquad (7.53)$$

integration of which between $T_{m,x}$ at x and $T_{m,\text{in}}$ at $x = 0$ gives

$$T_{m,x} = T_{m,\text{in}} + \frac{q'\pi D}{\dot{M}C_p}\, x \qquad (7.54)$$

which for a uniform q' is a linear variation of T_m with x Thus for given values of q', D, $\dot{M}$, and C_p, the rate of increase of T_m with x, $(\partial T_m/\partial x)$, in Eq. (7.46) is constant and $(\partial^2 T/\partial x^2)$ in Eq. (7.35) is zero.

The entry length of pipe required for the establishment of a fully developed temperature profile is analogous with the entry length required for establishment of fully developed hydrodynamic flow shown in Fig. 3.7. In Fig. 3.7, L_E, is the entry length between two flat plates required for the velocity boundary layers developing on the plates to merge with one another; the entry length required for establishment of a fully developed temperature profile, $L_{E,\text{th}}$, is that at which the thermal boundary layers merge with one another. From the exact solution for laminar flow, Eq. (7.21),

$$\frac{\delta_T}{\delta} = \frac{1}{\text{Pr}^{1/3}}$$

and as the entry length decreases with increasing rate of development of the boundary layer,

$$\frac{L_E}{L_{E,\text{th}}} = \frac{\delta_T}{\delta} = \frac{1}{\text{Pr}^{1/3}}$$

and hence for laminar flow,

$$L_{E,\text{th}} = L_E\cdot\text{Pr}^{1/3}$$

For laminar flow in a cylindrical pipe the entry length required for establishment of a fully developed temperature profile is

$$L_{E,th} = 0.05 \text{Re} \cdot \text{Pr} \cdot D \tag{7.55}$$

Thus, although Eq. (7.54) shows that T_m is a linear function of x along the entire length of pipe, T_s does not become a linear function of x, as required by Eq. (7.40), until a fully developed temperature profile has been established, which occurs at $x \geq L_{E,th}$. The variations of T_s and T_m with x, for a uniform q', are shown in Fig. 7.9(a).

Equation (7.53) is also applicable to heat transfer with a uniform T_s, in which case

$$q' = h_x(T_s - T_m)$$

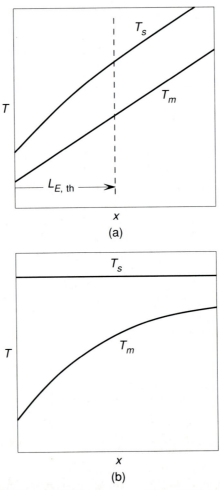

(a)

(b)

FIGURE 7.9. (a) Variations with distance, x, of the surface temperature of the pipe and the mean temperature of the fluid during fluid flow in a pipe under conditions of (a) a uniform heat flux from the wall, and (b) a uniform wall temperature.

Noting that for uniform T_s

$$d(T_s - T_m) = d \, \Delta T = -dT_m$$

Eq. (7.53) can be written as

$$-d \, \Delta T = \frac{\pi D h_x \, \Delta T}{\dot{M} C_p} \, dx$$

or

$$\frac{d \, \Delta T}{\Delta T} = -\frac{\pi D h_x}{\dot{M} C_p} \, dx$$

Integrating from $\Delta T_{in} = (T_s - T_{m,in})$ at $x = 0$ to $\Delta T_{out} = (T_s - T_{m,out})$ at $x = L$, and noting that

$$\bar{h} L = \int_0^L h_x \, dx$$

gives

$$\ln \frac{\Delta T_{out}}{\Delta T_{in}} = -\frac{\pi D \bar{h} L}{\dot{M} C_p} \tag{7.56}$$

or

$$\frac{\Delta T_{out}}{\Delta T_{in}} = \exp \left(-\frac{\pi D \bar{h} L}{\dot{M} C_p} \right)$$

in which $\bar{h}$ is the average heat transfer coefficient for heat transfer from the surface of the pipe over the length from 0 to L. The logarithmic variation of T_m with x is shown in Fig. 7.9(b). For a constant T_s, Eq. (7.52) can be written as

$$q' \pi D L = q_{conv} = \dot{M} C_p \, [(T_s - T_{m,in}) - (T_s - T_{m,out})]$$
$$= \dot{M} C_p (\Delta T_{in} - \Delta T_{out}) \tag{7.57}$$

From Eq. (7.56)

$$\dot{M} C_p = \frac{\pi D \bar{h} L}{\ln (\Delta T_{in}/\Delta T_{out})}$$

substitution of which into Eq. (7.57) gives

$$q_{conv} = \pi D L \bar{h} \frac{\Delta T_{in} - \Delta T_{out}}{\ln (\Delta T_{in}/\Delta T_{out})}$$
$$= \bar{h} A \, \Delta T_{lm} \tag{7.58}$$

in which the log mean temperature difference, ΔT_{lm}, is defined as

$$\Delta T_{lm} = \frac{\Delta T_{in} - \Delta T_{out}}{\ln (\Delta T_{in}/\Delta T_{out})}$$

The fact that $(\partial^2 T/\partial x^2) \neq 0$ when T_s is constant makes the derivation of Eq. (7.51) from Eq. (7.35) significantly more difficult than the derivation of Eq. (7.50).

EXAMPLE 7.4

Water flows through a 10-m length of pipe of 0.05 m I.D. at a flow rate of 0.01 kg/s. If the water enters the pipe at a mean temperature of 22°C, calculate the uniform heat flux from the pipe to produce a mean temperature of 72°C at the exit from the pipe.

The average temperature of the water in the pipe is $(22 + 72)/2 = 47°C$, at which temperature the properties are

$$\eta = 5.79 \times 10^{-4} \text{ Pa·s}$$

$$\rho = 989 \text{ kg/m}^3$$

$$C_p = 4176 \text{ J/kg·K}$$

$$k = 0.637 \text{ W/m·K}$$

$$\nu = 5.9 \times 10^{-7} \text{ m}^2/\text{s}$$

$$\text{Pr} = 3.79$$

From Eq. (7.54)

$$
\begin{aligned}
q' &= (T_{m,\text{out}} - T_{m,\text{in}}) \frac{\dot{M}C_p}{\pi DL} \\
&= \frac{(72 - 22) \times 0.01 \times 4176}{\pi \times 0.05 \times 10} \\
&= 1329 \text{ W/m}^2
\end{aligned}
$$

The Reynolds number for the flow is

$$
\text{Re} = \frac{\bar{v}_x D\rho}{\eta} = \frac{4\dot{M}}{\pi D\eta} = \frac{4 \times 0.01}{\pi \times 0.05 \times 5.79 \times 10^{-4}}
$$
$$
= 400
$$

which indicates that the flow is laminar. From Eq. (7.55) the entry length required for the establishment of a fully developed temperature profile is

$$
L_{E,\text{th}} = 0.05\text{Re·Pr·}D = 0.05 \times 440 \times 3.79 \times 0.05
$$
$$
= 4.17 \text{ m}
$$

and beyond this distance, $T_s - T_m$ is constant. From Eq. (7.50),

$$
\text{Nu}_D = 4.36 = \frac{hD}{k}
$$

or

$$
h = \frac{4.36 \times 0.637}{0.05} = 55.5 \text{ W/m}^2\text{·K}
$$

Thus over the length of fully developed temperature profile

$$
T_s - T_m = \frac{q'}{h} = \frac{1329}{55.5} = 23.9°C
$$

and hence, at the exit from the tube, $T_s = 72 + 23.9 = 95.9°C$. From Eq. (7.46) the normalized temperature profile is

$$\frac{T_s - T}{T_s - T_m} = \frac{\bar{v}_x}{\alpha} \frac{\partial T_m}{\partial x} \frac{1}{8R^2} \frac{3R^4 - 4R^2r^2 + r^4}{T_s - T_m}$$

in which

$$\bar{v}_x = \frac{4\dot{M}}{\pi D^2 \rho} = \frac{4 \times 0.01}{\pi \times 0.05^2 \times 989} = 5.15 \times 10^{-3} \text{ m/s}$$

$$\alpha = \frac{k}{\rho C_p} = \frac{0.637}{989 \times 4176} = 1.54 \times 10^{-7} \text{ m}^2/\text{s}$$

$$\frac{\partial T_m}{\partial x} = \frac{72 - 22}{10} = 5 \text{ K/m}$$

and

$$T_s - T_m = 23.9$$

Therefore,

$$\frac{T_s - T}{T_s - T_m} = \frac{5.15 \times 10^{-3} \times 5}{1.54 \times 10^{-8} \times 8 \times (0.025)^2}$$

$$\cdot \left[\frac{3 \times (0.025)^4 - 4 \times (0.025)^2 r^2 + r^4}{23.9} \right]$$

$$= 1.64 - 3498r^2 + 1.4 \times 10^6 r^4$$

This variation is shown in Fig. 7.10, which shows that the difference between T_s and $T|_{r=0}$ is $1.64 \times 23.9 = 39.2°C$.

EXAMPLE 7.5

Consider the same flow situation as in Example 7.4 except that T_s is maintained constant at 100°C. In this case, from Eq. (7.56),

$$\bar{h} = \frac{\dot{M}C_p}{\pi DL} \ln \frac{\Delta T_{in}}{\Delta T_{out}}$$

$$= \frac{0.01 \times 4176}{\pi \times 0.05 \times 10} \ln \frac{100 - 22}{100 - 72}$$

$$= 27.2 \text{ W/m}^2\text{·K}$$

and from Eq. (7.58),

$$q_{conv} = \bar{h}A \Delta T_{lm}$$

$$= \frac{27.2 \times \pi \times 0.01 \times 10 \times (78 - 28)}{\ln (78/28)}$$

$$= 417 \text{ W}$$

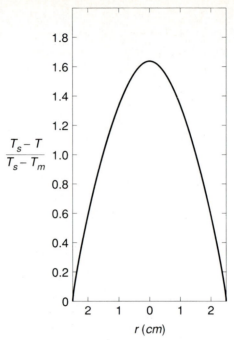

$$\frac{T_s - T}{T_s - T_m}$$

r (cm)

FIGURE 7.10. Radial distribution of the normalized temperature calculated in Example 7.4.

In turbulent flow in cylindrical pipes, to a first approximation, the flow becomes hydrodynamically fully developed after an entry length of between 10 and 60 diameters and becomes thermally fully developed after an entry length of between 20 and 40 diameters.

EXAMPLE 7.6 _____

Water at 10°C is passed through an electrically heated cylindrical pipe of diameter 0.04 m and length 5 m at a mass flow rate of 1 kg/s. The heat flux q' from the surface of the pipe to the water is uniform along the length of the pipe. In order that boiling of the water be prevented, the maximum allowable surface temperature, T_s, is 95°C. Calculate (a) the maximum allowable value of q', (b) the mean outlet temperature of the water, $T_{m,o}$, and (c) the electrical power supplied to the pipe.

As the mean outlet temperature is not known, a guess must be made as to the average temperature of the water in the pipe. As the value of T_s at the tube exit is set at 95°C, the average temperature will be guessed as $(10 + 90)/2 = 50$°C, at which temperature water has the properties

$$\rho = 989 \text{ kg/m}^3$$

$$C_p = 4176 \text{ J/kg·K}$$

$$k = 0.637 \text{ W/m·K}$$

$$\nu = 0.59 \times 10^{-6} \text{ m}^2/\text{s}$$

$$\text{Pr} = 3.79$$

With $\dot{M} = 1$ kg/s,

$$\bar{\upsilon} = \frac{4\dot{M}}{\pi D^2 \rho} = \frac{4 \times 1}{\pi \times 0.04^2 \times 989} = 0.804 \text{ m/s}$$

Thus

$$\text{Re}_D = \frac{\bar{\upsilon}D}{\nu} = \frac{0.804 \times 0.04}{0.59 \times 10^{-6}} = 5.453 \times 10^4$$

The flow is thus turbulent and the friction factor, f, is obtained from

$$f^{-1/2} = -3.6 \log_{10} \frac{6.9}{\text{Re}} = -3.6 \log_{10} \frac{6.9}{5.453 \times 10^4}$$

as

$$f = 5.08 \times 10^{-3}$$

As the flow is turbulent and $L/D = 125$ the entrance development region can be neglected, in which case

$$\begin{aligned}
\text{St} &= \frac{f/2}{1 + 12.7(\text{Pr}^{2/3} - 1)(f/2)^{1/2}} \\
&= \frac{2.54 \times 10^{-3}}{1 + 12.7(3.79^{2/3} - 1)(2.54 \times 10^{-3})^{1/2}} \\
&= 1.326 \times 10^{-3}
\end{aligned}$$

and

$$\begin{aligned}
\text{Nu}_D &= \text{St·Re}_D\text{·Pr} \\
&= 1.326 \times 10^{-3} \times 5.453 \times 10^4 \times 3.79 = 274
\end{aligned}$$

Thus

$$h_D = \frac{274 \times 0.637}{0.04} = 4362 \text{ W/m}^2\text{·K}$$

Therefore,

$$q' = h_D(T_{s,o} - T_{m,o}) = 4362 \times (95 - T_{m,o}) = 414,390 - 4,362T_{m,o} \quad \text{(i)}$$

and from Eq. (7.54),

$$T_{m,o} = T_{m,i} + \frac{q'\pi DL}{\dot{M}C_p} = 10 + \frac{q'\pi \times 0.04 \times 5}{1 \times 4176}$$

or

$$T_{m,o} = 10 + 1.505 \times 10^{-4}q' \quad \text{(ii)}$$

Simultaneous solution of Eqs. (i) and (ii) gives

$$q' = 223.8 \text{ kW/m}^2 \quad \text{and} \quad T_{m,o} = 44°C$$

Using this value of $T_{m,o}$ to recalculate the average temperature of the water gives $(10 + 44)/2 = 27°C$ (300 K), at which temperature the properties of water are

$$\rho = 997 \text{ kg/m}^3$$
$$C_p = 4177 \text{ J/kg·K}$$
$$k = 0.608 \text{ W/m·K}$$
$$v = 0.86 \times 10^{-6} \text{ m}^2/\text{s}$$
$$\text{Pr} = 5.88$$

With these new data,

$$\text{Re} = 3.74 \times 10^4$$
$$f = 5.534 \times 10^{-3}$$
$$\text{St} = 1.103 \times 10^{-3}$$
$$\text{Nu}_D = 243$$
$$h_D = 3687 \text{ W·m}^{-2}\text{·K}^{-1}$$

and

$$q' = 201.6 \text{ kW/m}^2 \quad \text{and} \quad T_{m,o} = 40°C$$

The electrical power supplied to the pipe is calculated from

$$q = q'\pi DL = 201,600 \times \pi \times 0.04 \times 5 = 126.7 \text{ kW}$$

or from

$$q = \bar{v}\,\frac{\pi D^2}{4}\,\rho C_p(T_{m,o} - T_{m,i})$$

$$= 0.804 \times \frac{\pi \times 0.04^2}{4} \times 997 \times (40 - 10) = 126.7 \text{ kW}$$

If, now, T_s is maintained constant at 95°C, calculate q and $T_{m,o}$.

As the entry region can be ignored, h_D is obtained as above as 3687 W·m^{-2}·K^{-1}, and thus, from Eq. (7.56),

$$\ln\frac{95 - T_{m,o}}{95 - T_{m,i}} = -\frac{\pi D h_D L}{\dot{M}C_p}$$

$$= \frac{\pi \times 0.04 \times 3687 \times 5}{1 \times 4177} = -0.555$$

Therefore,

$$\frac{95 - T_{m,o}}{95 - 10} = 0.574 \quad \text{and} \quad T_{m,o} = 46°C$$

The power requirements are then calculated from Eq. (7.58) as

$$q = \pi D L h_D \frac{\Delta T_{in} - \Delta T_{out}}{\ln (\Delta T_{in}/\Delta T_{out})}$$

$$= \pi \times 0.04 \times 5 \times 3687 \times \frac{85 - 49}{\ln (85/49)} = 151.5 \text{ kW}$$

or as

$$q = 0.804 \times \frac{\pi \times 0.04^2}{4} \times 997 \times 4177 \times (46 - 10) = 151.5 \text{ kW}$$

7.6 Heat Transfer by Forced Convection from Horizontal Cylinders

The Nusselt number for forced convection from a horizontal cylinder to a fluid flowing in a direction that is normal to the axis of the cylinder is correlated with the Reynolds number and the Prandtl number as

$$\text{Nu}_D = C \cdot \text{Re}_D^m \cdot \text{Pr}^{1/3} \tag{7.59}$$

The values of C and m to be used in Eq. (7.59) depend on the Reynolds number as listed in Table 7.2.

EXAMPLE 7.7

Consider a horizontal platinum wire of length 0.1 m and diameter 0.0005 m which is heated by the passage of an electric current and is in an airstream at $T_\infty = 25°C$ flowing with a velocity, v, of 100 m/s in a direction normal to the length of the wire. Let the resistance heating of the wire be such that its surface temperature, T_s, is 300°C. The properties of air at the film temperature of $(25 + 300)/2 = 163°C$ are

$$\rho = 0.812 \text{ kg/m}^3$$

$$k = 0.0354 \text{ W/m·K}$$

TABLE 7.2

Range of Re_D	C	m
0.4–4	0.989	0.330
4–40	0.911	0.385
40–4000	0.683	0.466
$4 \times 10^3 - 4 \times 10^4$	0.193	0.618
$4 \times 10^4 - 4 \times 10^5$	0.027	0.805

$$\nu = 3.02 \times 10^{-5} \text{ m}^2/\text{s}$$

$$\text{Pr} = 0.701$$

The specific resistivity of platinum varies with temperature as

$$\rho(\mu\Omega\cdot\text{cm}) = 10.52 + 3.93 \times 10^{-3}T(°C)$$

and thus the electrical resistance of the wire varies with temperature as

$$R = \frac{[10.52 + 3.93 \times 10^{-3}T(°C)] \times 10^{-8} \times 0.1}{\pi \times 0.00025^2}$$

$$= .0536 + 2.00 \times 10^{-5}T \quad \Omega \tag{i}$$

The Reynolds number for the airflow is

$$\text{Re}_D = \frac{\upsilon D}{\nu} = \frac{100 \times 0.0005}{3.02 \times 10^{-5}} = 1655$$

and hence from Eq. (7.59) and Table 4.1, the Nusselt number for heat transfer by forced convection from the wire is

$$\text{Nu}_D = 0.683(1655)^{0.466} \times (0.701)^{1/3}$$

$$= 19.19$$

$$= \frac{\bar{h}D}{k}$$

Therefore,

$$\bar{h} = \frac{19.19 \times 0.0354}{0.0005} = 1359 \text{ W/m}^2\cdot\text{K}$$

and the rate of heat transfer from the wire is

$$q = \bar{h}A_s(T_s - T_\infty) = 1359 \times \pi \times 0.0005 \times 0.1 \times (300 - 25)$$

$$= 58.7 \text{ W}$$

At steady state, this heat is generated by the resistance heating of the wire,

$$I^2R = 58.7 \text{ W}$$

From Eq. (i), at 300°C, $R = 0.0596 \ \Omega$ and hence the current flowing in the wire must be

$$I = \left(\frac{5.87}{0.0596}\right)^{1/2} = 31.4 \text{ A}$$

and the voltage drop across the wire must be $58.7/31.4 = 1.87$ V.

The relationship between the velocity of the flowing fluid and the resistance of the wire is the basis for construction of an anemometer. In this example, from Eq. (i),

$$T_s = 50{,}000R - 2680$$

substitution of which into

$$q = \bar{h} A_s (T_s - 25)$$

gives

$$q = \bar{h} \times \pi \times 0.0005 \times 0.1(50{,}000R - 2680 - 25)$$
$$= \bar{h}(7.857R - 0.4251)$$
$$= I^2 R$$

Therefore,

$$\bar{h} = \frac{I^2 R}{7.857R - 0.4251} = \frac{\mathrm{Nu}_D k}{D} = 70.8 \mathrm{Nu}_D$$

or from Eq. (7.59),

$$\mathrm{Nu}_D = \frac{0.01412 \times I^2 R}{7.857R - 0.4251} = C \left(\frac{\upsilon D}{\nu} \right)^m \mathrm{Pr}^{1/3}$$

$$= C \upsilon^m \left(\frac{0.0005}{3.02 \times 10^{-5}} \right)^m \times 0.701^{1/3}$$

$$= C \upsilon^m \times 16.56^m \times 0.888$$

Thus

$$C \upsilon^m 16.56^m = \frac{0.0159 I^2 R}{7.857R - 0.4251} \tag{ii}$$

which is the relationship between the velocity of the airstream and the measured current in the wire passing through a constant voltage drop of 1.87 V.

Calculate the velocity of the airstream if the measured current is 28 A.

$$VI = 1.87 \times 28 = 52.36 = 28^2 \times R$$

Thus

$$R = 0.0668 \ \Omega$$

substitution of which into Eq. (ii) gives

$$C \upsilon^m 16.56^m = 8.346 \tag{iii}$$

Assuming that the Reynolds number is in the range 40 to 4000, Table 7.2 gives $C = 0.683$ and $m = 0.466$. Therefore, from Eq. (iii),

$$\upsilon = \left(\frac{8.346}{0.683} \right)^{1/0.466} \times \frac{1}{16.56} = 13 \ \mathrm{m/s}$$

and

$$\mathrm{Re} = \frac{\upsilon D}{\nu} = \frac{13 \times 0.0005}{3.02 \times 10^{-5}} = 215$$

which is in the range 40 to 4000, and hence the calculation is valid.

7.7 Heat Transfer by Forced Convection from a Sphere

For heat transfer by conduction in spherical coordinates the general heat conduction equation corresponding to Eq. (6.35) is

$$\frac{1}{r^2}\frac{d}{dr}\left(\alpha r^2 \frac{dT}{dr}\right) + \frac{\dot{q}}{\rho C_p} = \frac{dT}{dt} \tag{7.60}$$

which for steady-state conduction with no heat generation gives

$$\frac{d}{dr}\left(r^2 \frac{dT}{dr}\right) = 0 \tag{7.61}$$

The first integration of Eq. (7.61) gives

$$r^2 \frac{dT}{dr} = C_1 \tag{i}$$

and the second integration gives

$$T = -\frac{C_1}{r} + C_2 \tag{ii}$$

With the boundary conditions

$$T = T_1 \qquad \text{at } r = R_1$$

and

$$T = T_2 \qquad \text{at } r = R_2$$

Equation (ii) gives

$$T_1 = -\frac{C_1}{R_1} + C_2$$

and

$$T_2 = -\frac{C_1}{R_2} + C_2$$

Thus

$$C_1 = \frac{T_1 - T_2}{1/R_2 - 1/R_1}$$

and

$$C_2 = T_2 + \frac{C_1}{R_2} = T_2 + \frac{T_1 - T_2}{1/R_2 - 1/R_1}\frac{1}{R_2}$$

substitution of which into Eq. (ii) gives

$$T = T_2 + \frac{T_1 - T_2}{1/R_2 - 1/R_1}\left(\frac{1}{R_2} - \frac{1}{r}\right) \tag{7.62}$$

If, now, $T_1 = T_s$ at the surface of a solid sphere of radius R ($R = R_1$) and $T_2 = T_\infty$ at $R_2 = \infty$, where T_∞ is the ambient temperature of the fluid in which the sphere is immersed, Eq. (7.62) becomes

$$T = T_\infty + (T_s - T_\infty)\frac{R}{r}$$

Thus with

$$\left.\frac{dT}{dr}\right|_{r=R} = \frac{T_\infty - T_s}{R}$$

the heat transfer coefficient is obtained as

$$h_D = \frac{-k\left.\dfrac{dT}{dr}\right|_{r=R}}{T_s - T_\infty} = \frac{k}{R}$$

and hence the Nusselt number for heat transfer by conduction from a sphere is

$$\mathrm{Nu}_D = \frac{h_D D}{k} = \frac{D}{R} = 2 \tag{7.63}$$

For heat transfer by forced convection from a sphere, Whitaker [S. Whitaker, *AIChE J.* (1972), vol. 18, pp. 361–371] has proposed the correlation

$$\mathrm{Nu}_D = 2 + (0.4\mathrm{Re}_D^{0.5} + 0.06\mathrm{Re}_D^{0.667})\mathrm{Pr}^{0.4}\left(\frac{\eta_\infty}{\eta_s}\right)^{0.25} \tag{7.64}$$

in which all of the fluid properties except η_s are evaluated at T_∞. This correlation is valid for Prandtl numbers in the range 0.7 to 380 and Reynolds numbers in the range 3.5 to 8×10^4. For heat transfer by convection from freely falling liquid drops, Ranz and Marshall [W. Ranz and W. Marshall, *Chem. Eng. Prog.* (1952), vol. 48, p. 141] have proposed

$$\mathrm{Nu}_D = 2 + 0.6\mathrm{Re}_D^{0.5}\mathrm{Pr}^{0.333} \tag{7.65}$$

7.8 General Energy Equation

Consider the flow of fluid, and hence energy, through the control volume shown in Fig. 3.1, in which the energy is the sum of the temperature-dependent internal energy per unit mass of the fluid, U, and the kinetic energy per unit mass of the fluid, $\frac{1}{2}v^2$.

The velocity of the fluid, $\mathbf{v}$, can be expressed in terms of its components v_x, v_y, and v_z as

$$\mathbf{v}^2 = v_x^2 + v_y^2 + v_z^2$$

An energy balance on the control volume gives

$$\begin{pmatrix} \text{rate of accumulation of} \\ \text{internal and kinetic energy} \\ \text{in control volume} \end{pmatrix} = \begin{pmatrix} \text{net rate at which thermal} \\ \text{energy enters control volume} \\ \text{by conduction} \end{pmatrix}$$

$$+ \begin{pmatrix} \text{net rate at which thermal} \\ \text{energy is transported into} \\ \text{control volume by convection} \end{pmatrix}$$

$$- (\text{net rate at which fluid does work})$$

The rate of accumulation of internal and kinetic energy is

$$\frac{\partial}{\partial t}\left[\rho\left(U + \frac{1}{2}v^2\right)\right] \Delta x \, \Delta y \, \Delta z$$

From consideration of Fig. 6.21, the net rate at which thermal energy, or heat, is transferred into the control volume by conduction is

$$(q_x'|_x - q_x'|_{x+\Delta x}) \Delta y \, \Delta z + (q_y'|_y - q_y'|_{y+\Delta y}) \Delta x \, \Delta z + (q_z'|_z - q_z'|_{z+\Delta z}) \Delta x \, \Delta y$$

From consideration of Fig. 3.1, the net rate at which mass is transported into the control volume in the x-direction is

$$(\rho v_x|_x - \rho v_x|_{x+\Delta x}) \Delta y \, \Delta z$$

The net rate at which internal energy and kinetic energy are transported into the control volume by convection in the x-direction is thus

$$[\rho(U + \tfrac{1}{2}v^2)v_x|_x - \rho(U + \tfrac{1}{2}v^2)v_x|_{x+\Delta x}] \Delta y \, \Delta z$$

Similarly, the net rates at which internal energy and kinetic energy are transported into the control volume by convection in the y and z directions are, respectively,

$$[\rho(U + \tfrac{1}{2}v^2)v_y|_y - \rho(U + \tfrac{1}{2}v^2)v_y|_{y+\Delta y}] \Delta x \, \Delta z$$

and

$$[\rho(U + \tfrac{1}{2}v^2)v_z|_z - \rho(U + \tfrac{1}{2}v^2)v_z|_{z+\Delta z}] \Delta x \, \Delta y$$

The fluid in the control volume does work against the gravitational force, the "pressure" forces, and the viscous forces.

The gravitational force is $-\rho(g_x + g_y + g_z) \Delta x \, \Delta y \, \Delta z$, and hence the rate at which work is done against the gravitational force is

$$-\rho(v_x g_x + v_y g_y + v_z g_z) \Delta x \, \Delta y \, \Delta z$$

The rate at which work is done against the pressure force is

$$[(Pv_x)|_{x+\Delta x} - (Pv_x)|_x] \Delta y \, \Delta z + [(Pv_y)|_{y+\Delta y} - (Pv_y)|_y] \Delta x \, \Delta z$$
$$+ [(Pv_z)|_{z+\Delta z} - (Pv_z)|_z] \Delta x \, \Delta y$$

The viscous forces arising from the x-component of the velocity are, as illustrated in Fig. 3.2, the normal stress τ_{xx} on the $\Delta y\, \Delta z$ faces, the shear stress τ_{yx} on the $\Delta x\, \Delta z$ faces arising from the transport of momentum in the y-direction, and the shear stress τ_{zx} on the $\Delta x\, \Delta y$ faces arising from the transport of momentum in the z-direction. Thus the rate at which work is done against the viscous forces caused by the x-component of the velocity is

$$(v_x\tau_{xx}|_{x+\Delta x} - v_x\tau_{xx}|_x)\, \Delta y\, \Delta z + (v_x\tau_{yx}|_{y+\Delta y} - v_x\tau_{yx}|_y)\, \Delta x\, \Delta z$$
$$+ (v_x\tau_{zx}|_{z+\Delta z} - v_x\tau_{zx}|_z)\, \Delta x\, \Delta y$$

Similarly, the rate at which work is done against the viscous forces caused by the y-component of the velocity is

$$(v_y\tau_{yy}|_{y+\Delta y} - v_y\tau_{yy}|_y)\, \Delta x\, \Delta z + (v_y\tau_{xy}|_{x+\Delta x} - v_y\tau_{xy}|_x)\, \Delta y\, \Delta z$$
$$+ (v_y\tau_{zy}|_{z+\Delta z} - v_y\tau_{zy}|_z)\, \Delta x\, \Delta y$$

and that for the z-component is

$$(v_z\tau_{zz}|_{z+\Delta z} - v_z\tau_{zz}|_z)\, \Delta x\, \Delta y + (v_z\tau_{xz}|_{x+\Delta x} - v_z\tau_{xz}|_x)\, \Delta y\, \Delta z$$
$$+ (v_z\tau_{yz}|_{y+\Delta y} - v_z\tau_{yz}|_y)\, \Delta x\, \Delta z$$

Dividing throughout by $\Delta x\, \Delta y\, \Delta z$ and allowing Δx, Δy, and Δz to approach zero gives the energy balance as

$$\frac{\partial}{\partial t}\left[\rho\left(U + \frac{1}{2}v^2\right)\right] = -\frac{\partial q'_x}{\partial x} - \frac{\partial q'_y}{\partial y} - \frac{\partial q'_z}{\partial z} - \frac{\partial}{\partial x}\left[\rho v_x\left(U + \frac{1}{2}v^2\right)\right]$$

$$-\frac{\partial}{\partial y}\left[\rho v_y\left(U + \frac{1}{2}v^2\right)\right] - \frac{\partial}{\partial z}\left[\rho v_z\left(U + \frac{1}{2}v^2\right)\right]$$

$$+ \rho\left(v_x g_x + v_y g_y + v_z g_z\right) - \frac{\partial}{\partial x}(v_x P) - \frac{\partial}{\partial y}(v_y P)$$

$$-\frac{\partial}{\partial z}(v_z P) - \frac{\partial}{\partial x}(v_x\tau_{xx} + v_y\tau_{xy} + v_z\tau_{xz})$$

$$-\frac{\partial}{\partial y}(v_x\tau_{yx} + v_y\tau_{yy} + v_z\tau_{yz})$$

$$-\frac{\partial}{\partial z}(v_x\tau_{zx} + v_y\tau_{zy} + v_z\tau_{zz}) \qquad (7.66)$$

The terms containing $U + \frac{1}{2}v^2$ can be moved to the left-hand side of Eq. (7.66):

$$\frac{\partial}{\partial t}\left[\rho\left(U + \frac{1}{2}v^2\right)\right] + \frac{\partial}{\partial x}\left[\rho v_x\left(U + \frac{1}{2}v^2\right)\right]$$

$$+ \frac{\partial}{\partial y}\left[\rho v_y\left(U + \frac{1}{2}v^2\right)\right] + \frac{\partial}{\partial z}\left[\rho v_z\left(U + \frac{1}{2}v^2\right)\right]$$

and can be rearranged as

$$\rho\left[\frac{\partial}{\partial t}(U + \frac{1}{2}v^2) + v_x\frac{\partial}{\partial x}\left(U + \frac{1}{2}v^2\right) + v_y\frac{\partial}{\partial y}\left(U + \frac{1}{2}v^2\right)\right.$$

$$+ v_z \frac{\partial}{\partial z} \left(U + \frac{1}{2} v^2 \right) \Bigg]$$

$$+ \left(U + \frac{1}{2} v^2 \right) \left[\frac{\partial \rho}{\partial t} + \frac{\partial}{\partial x} (\rho v_x) + \frac{\partial}{\partial y} (\rho v_y) + \frac{\partial}{\partial z} (\rho v_z) \right]$$

The term in the first set of brackets is the substantial derivative of $U + \frac{1}{2} v^2$ (i.e., $D(U + \frac{1}{2}v^2)/Dt$), and by virtue of the equation of continuity given by Eq. (3.2), the term in the second set of brackets is zero. Thus in shorthand notation the energy balance is

$$\rho \frac{D(U + \frac{1}{2}v^2)}{Dt} = -(\nabla q') + \rho(vg) - (\nabla Pv) - (\nabla[\tau v]) \tag{7.67}$$

The substantial derivative of the kinetic energy times the density has been derived as

$$\rho \frac{D(\frac{1}{2}v^2)}{Dt} = P(\nabla v) - (\nabla Pv) + \rho(vg) - (\nabla[\tau v]) + \eta \Phi \tag{7.68}$$

In this expression Φ is the dissipation function, given by

$$\Phi = 2 \left[\left(\frac{\partial v_x}{\partial x} \right)^2 + \left(\frac{\partial v_y}{\partial y} \right)^2 + \left(\frac{\partial v_z}{\partial z} \right)^2 \right]$$

$$+ \left[\left(\frac{\partial v_x}{\partial y} + \frac{\partial v_y}{\partial x} \right)^2 + \left(\frac{\partial v_y}{\partial z} + \frac{\partial v_z}{\partial y} \right)^2 + \left(\frac{\partial v_z}{\partial x} + \frac{\partial v_x}{\partial z} \right)^2 \right]$$

$$- \frac{2}{3} \left(\frac{\partial v_x}{\partial x} + \frac{\partial v_y}{\partial y} + \frac{\partial v_z}{\partial z} \right)^2 \tag{7.69}$$

and the quantity $\eta \Phi$ is the rate at which mechanical energy is degraded to thermal energy by the irreversible nature of the mass flow.

Subtraction of Eq. (7.68) from Eq. (7.67) gives the substantial derivative of the internal energy as

$$\rho \frac{DU}{Dt} = -(\nabla q') - P(\nabla v) + \eta \Phi \tag{7.70}$$

Taking the internal energy of the fluid, U, as a function of the independent thermodynamic variables, T and V, gives

$$dU = \left(\frac{\partial U}{\partial V} \right)_T dV + \left(\frac{\partial U}{\partial T} \right)_V dT \tag{7.71}$$

Combination of the first and second laws of thermodynamic gives

$$dU = T\, dS - P\, dV$$

Therefore,

$$\left(\frac{\partial U}{\partial V} \right)_T = T \left(\frac{\partial S}{\partial V} \right)_T - P$$

which from Maxwell's equation, $(\partial S/\partial V)_T = (\partial P/\partial T)_V$, gives

$$\left(\frac{\partial U}{\partial V}\right)_T = T\left(\frac{\partial P}{\partial T}\right)_V - P \tag{7.72}$$

and by definition, the constant-volume heat capacity of the fluid, C_v, is

$$C_v = \left(\frac{\partial U}{\partial T}\right)_V \tag{7.73}$$

The substitution of Eqs. (7.72) and (7.73) into Eq. (7.71) gives

$$dU = \left[T\left(\frac{\partial P}{\partial T}\right)_V - P\right] dV + C_v \, dT$$

and thus the substantial derivative of U is

$$\frac{DU}{Dt} = \left[T\left(\frac{\partial P}{\partial T}\right)_V - P\right] \frac{DV}{Dt} + C_v \frac{DT}{Dt} \tag{7.74}$$

Multiplication of Eq. (7.74) by density ρ and comparison with Eq. (7.70) gives

$$\rho C_v \frac{DT}{Dt} = -(\nabla q') - P\,(\nabla v) - \rho\left[T\left(\frac{\partial P}{\partial T}\right)_V - P\right] \frac{DV}{Dt} + \eta\Phi$$

$$= -(\nabla q') - P\,(\nabla v) + \frac{1}{\rho}\left[T\left(\frac{\partial P}{\partial T}\right)_V - P\right] \frac{D\rho}{Dt} + \eta\Phi \tag{7.75}$$

In the third term on the right-hand side of Eq. (7.75),

$$\frac{D\rho}{Dt} = \frac{\partial\rho}{\partial t} + v_x \frac{\partial\rho}{\partial x} + v_y \frac{\partial\rho}{\partial y} + v_z \frac{\partial\rho}{\partial z}$$

and from the continuity equation, Eq. (3.2),

$$\frac{\partial\rho}{\partial t} = -\frac{\partial}{\partial x}(\rho v_x) - \frac{\partial}{\partial y}(\rho v_y) - \frac{\partial}{\partial_z}(\rho v_z)$$

$$= -\rho\left(\frac{\partial v_x}{\partial x} + \frac{\partial v_y}{\partial y} + \frac{\partial v_z}{\partial z}\right) - v_x\frac{\partial\rho}{\partial x} - v_y\frac{\partial\rho}{\partial y} - v_z\frac{\partial\rho}{\partial z}$$

Therefore,

$$\frac{D\rho}{Dt} = -\rho\left(\frac{\partial v_x}{\partial x} + \frac{\partial v_y}{\partial y} + \frac{\partial v_z}{\partial z}\right) = -\rho(\nabla v) \tag{7.76}$$

and substitution of Eq. (7.76) into Eq. (7.75) gives

$$\rho C_v \frac{DT}{Dt} = -(\nabla q') - P(\nabla v) - \left[T\left(\frac{\partial P}{\partial T}\right)_V - P\right](\nabla v) + \eta\Phi$$

$$= -(\nabla q') - T\left(\frac{\partial P}{\partial T}\right)_V (\nabla v) + \eta\Phi \tag{7.77}$$

and substitution from Eq. (7.76) into Eq. (7.77) gives

$$\rho C_v \frac{DT}{Dt} = -(\nabla q') + T \left(\frac{\partial P}{\partial T}\right)_V \frac{1}{\rho} \frac{D\rho}{Dt} + \eta\Phi \tag{7.78}$$

The isobaric thermal expansivity of the fluid is defined as

$$\beta = \frac{1}{V} \left(\frac{\partial V}{\partial T}\right)_P = -\frac{1}{\rho} \left(\frac{\partial \rho}{\partial T}\right)_P$$

and the isothermal compressibility of the fluid is defined as

$$\alpha = -\frac{1}{V} \left(\frac{\partial V}{\partial P}\right)_T = \frac{1}{\rho} \left(\frac{\partial \rho}{\partial P}\right)_T$$

As P, V, and T are thermodynamic state functions,

$$\left(\frac{\partial P}{\partial T}\right)_V \left(\frac{\partial T}{\partial V}\right)_P \left(\frac{\partial V}{\partial P}\right)_T = -1$$

or

$$\left(\frac{\partial P}{\partial T}\right)_V = -\frac{(\partial V/\partial T)_P}{(\partial V/\partial P)_T} = \frac{\beta}{\alpha}$$

and thus Eq. (7.78) becomes

$$\rho C_v \frac{DT}{Dt} = -(\nabla q') + \frac{T\beta}{\alpha\rho} \frac{D\rho}{Dt} + \eta\Phi \tag{7.79}$$

The difference between C_p, the constant-pressure heat capacity, and C_v, the constant-volume heat capacity, is given by

$$C_p - C_v = \frac{VT\beta^2}{\alpha} = \frac{T\beta^2}{\alpha\rho}$$

and hence

$$\rho C_p \frac{DT}{Dt} = \rho C_v \frac{DT}{Dt} + \frac{T\beta^2}{\alpha} \frac{DT}{Dt}$$

$$= -(\nabla q') + \frac{T\beta}{\alpha\rho} \frac{D\rho}{Dt} + \frac{T\beta^2}{\alpha} \frac{DT}{Dt} + \eta\Phi$$

$$= -(\nabla q') + T\beta \left(\frac{1}{\alpha\rho} \frac{D\rho}{Dt} + \frac{\beta}{\alpha} \frac{DT}{Dt}\right) + \eta\Phi \tag{7.80}$$

Taking P as a function of T and V,

$$dP = \left(\frac{\partial P}{\partial V}\right)_T dV + \left(\frac{\partial P}{\partial T}\right)_V dT$$

$$= -\frac{1}{V\alpha} dV + \frac{\beta}{\alpha} dT$$

$$= \frac{1}{\alpha \rho} \, d\rho + \frac{\beta}{\alpha} \, dT$$

or

$$\frac{DP}{Dt} = \frac{1}{\alpha \rho} \frac{D\rho}{Dt} + \frac{\beta}{\alpha} \frac{DT}{Dt} \tag{7.81}$$

which is the second term in parentheses in Eq. (7.80). Substitution of Eq. (7.81) into Eq. (7.80) thus gives

$$\rho C_p \frac{DT}{Dt} = -(\nabla q') + T\beta \frac{DP}{Dt} + \eta \Phi \tag{7.82}$$

which is the *general energy equation*. Equation (7.82) is modified for application to specific situations. If the thermal conductivity, k, is constant,

$$\nabla q' = \frac{\partial q'_x}{\partial x} + \frac{\partial q'_y}{\partial y} + \frac{\partial q'_z}{\partial z}$$

$$= -k \left(\frac{\partial^2 T}{\partial x^2} + \frac{\partial^2 T}{\partial y^2} + \frac{\partial^2 T}{\partial z^2} \right)$$

$$= -k\nabla^2 T$$

and the viscous dissipation term is negligible except in certain situations (see Example 7.8).

1. For an ideal gas $\beta = 1/T$, in which case Eq. (7.82) becomes

$$\rho C_p \frac{DT}{Dt} = k\nabla^2 T + \frac{DP}{Dt} \tag{7.83}$$

2. For a fluid at constant pressure

$$\rho C_p \frac{DT}{Dt} = k\nabla^2 T \tag{7.84}$$

3. For heat flow in a solid, $v = 0$ and hence

$$\rho C_p \frac{\partial T}{\partial t} = k\nabla^2 T \tag{7.85}$$

which is the general heat conduction equation derived as Eq. (6.35).

Dissipation Factor

In Chapter 2, Couette flow was described as occurring in a fluid contained by two flat plates moving at differing velocities. With reference to Fig. 2.5, in which the upper plate is moving at the velocity V, the lower plate is stationary and the distance between the plates is Y, Eq. (2.7) gave

$$v_x = \frac{V}{Y} \, y \tag{2.7}$$

and Eq. (2.8) gave

$$\tau_{yx} = -\frac{V}{Y}\eta \tag{2.8}$$

The general energy equation including the dissipation factor is

$$\rho C_p \left(\frac{\partial T}{\partial t} + v_x \frac{\partial T}{\partial x} + v_y \frac{\partial T}{\partial y} + v_z \frac{\partial T}{\partial z} \right)$$

$$= k \left(\frac{\partial^2 T}{\partial x^2} + \frac{\partial^2 T}{\partial y^2} + \frac{\partial^2 T}{\partial z^2} \right) + T\beta \frac{DP}{Dt}$$

$$+ 2\eta \left[\left(\frac{\partial v_x}{\partial x} \right)^2 + \left(\frac{\partial v_y}{\partial y} \right)^2 + \left(\frac{\partial v_z}{\partial z} \right)^2 \right]$$

$$+ \eta \left[\left(\frac{\partial v_x}{\partial y} + \frac{\partial v_y}{\partial x} \right)^2 + \left(\frac{\partial v_y}{\partial z} + \frac{\partial v_z}{\partial y} \right)^2 + \left(\frac{\partial v_z}{\partial x} + \frac{\partial v_x}{\partial z} \right)^2 \right]$$

$$- \frac{2}{3} \eta \left(\frac{\partial v_x}{\partial x} + \frac{\partial v_y}{\partial y} + \frac{\partial v_z}{\partial z} \right)^2 \tag{7.86}$$

In Couette flow, at steady state with a fully developed velocity profile, v_y, v_z, and $\partial v_x/\partial x$ are zero. Also, as the fluid flow arises from the relative motion of the plates, there are no pressure gradients within the fluid and hence $DP/Dt = 0$. If the surface temperatures of the upper and lower plates are, respectively, at the uniform temperatures T_Y and T_0, a fully developed temperature profile occurs and hence $\partial T/\partial x$ and $\partial T/\partial z$ are zero, and under these circumstances, Eq. (7.86) reduces to

$$0 = k \frac{\partial^2 T}{\partial y^2} + \eta \left(\frac{\partial v_x}{\partial y} \right)^2$$

or, from differentiation of Eq. (2.7),

$$\frac{\partial^2 T}{\partial y^2} = -\frac{\eta}{k} \left(\frac{V}{Y} \right)^2$$

Thus

$$\frac{\partial T}{\partial y} = -\frac{\eta}{k} \left(\frac{V}{Y} \right)^2 y + C_1 \tag{7.87}$$

and

$$T = -\frac{\eta}{k} \left(\frac{V}{Y} \right)^2 \frac{y^2}{2} + C_1 y + C_2$$

From the boundary conditions,

$$T = T_0 \quad \text{at } y = 0 \quad \text{and} \quad T = T_Y \quad \text{at } y = Y$$

$$C_2 = T_0$$

and

$$C_1 = \frac{\eta}{k}\frac{V^2}{2Y} + \frac{T_Y - T_0}{Y} \tag{7.88}$$

Therefore,

$$T = \frac{\eta}{k}\left(\frac{V}{Y}\right)^2\frac{yY - y^2}{2} + (T_Y - T_0)\frac{y}{Y} + T_0 \tag{7.89}$$

which if $T_Y = T_0$ is a parabolic variation of T with y. Substitution of Eq. (7.88) into Eq. (7.87) gives

$$\frac{\partial T}{\partial y} = -\frac{\eta}{k}\left(\frac{V}{Y}\right)^2 y + \frac{\eta V^2}{2kY} + \frac{T_Y - T_0}{Y}$$

$$= \frac{\eta}{k}\left(\frac{V}{Y}\right)^2\frac{Y - 2y}{2} + \frac{T_Y - T_0}{Y} \tag{7.90}$$

From the condition $(\partial T/\partial y) = 0$, the value of y at which T has a maximum value is

$$y = \frac{T_Y - T_0}{Y}\frac{k}{\eta}\left(\frac{Y}{V}\right)^2 + \frac{Y}{2}$$

For a given difference $T_Y - T_0$, the value of y at which T has a maximum value depends on V. With $T_Y > T_0$, the maximum temperature occurs within the fluid when

$$y = \frac{T_Y - T_0}{Y}\frac{k}{\eta}\left(\frac{Y}{V}\right)^2 + \frac{Y}{2} < Y$$

or

$$V > \left[\frac{2(T_Y - T_0)k}{\eta}\right]^{1/2} \tag{7.91}$$

The rate of heat transfer per unit area from the fluid to the upper plate is

$$q' = -k\left(\frac{\partial T}{\partial y}\right)_{y=Y}$$

$$= -k\left[-\frac{\eta}{k}\left(\frac{V}{Y}\right)^2\frac{Y}{2} + \frac{T_Y - T_0}{Y}\right]$$

$$= \frac{\eta V^2}{2Y} - \frac{k(T_Y - T_0)}{Y} \tag{7.92}$$

and the rate of heat transfer per unit area from the fluid to the lower plate is

$$q' = -k\left(\frac{\partial T}{\partial y}\right)_{y=0}$$

$$= -k\left[\frac{\eta}{R}\left(\frac{V}{Y}\right)^2\frac{Y}{2} + \frac{T_Y - T_0}{Y}\right]$$

$$= -\frac{\eta V^2}{2Y} - \frac{k(T_Y - T_0)}{Y} \tag{7.93}$$

Thus if $T_0 = T_Y$, and the plates are of area A_S, heat is transferred from the fluid to both plates at the rate

$$q = \frac{\eta V^2}{Y} A_S \qquad (7.94)$$

In the analysis in Example 2.1, the shear stress required to keep the upper plate in motion was

$$\tau_{yx} = \eta \frac{V}{Y}$$

and hence the shear force on the plate of area A_S was

$$F_S = \tau_{yx}(\text{area}) = \eta \frac{V}{Y} A_S$$

and the power required to maintain steady-state motion was

$$\text{power} = \text{force} \times \text{velocity}$$

$$= \frac{\eta V^2}{Y} A_S \qquad (7.95)$$

Comparison of Eqs. (7.94) and (7.95) shows that the mechanical energy, expended to match the retarding viscous force and keep the upper plate moving at the constant velocity V, is degraded to heat, which is then transported from the fluid to the plates.

EXAMPLE 7.8 _____

Consider Couette flow of an oil of thermal conductivity 0.14 W/m·K and viscosity 0.2 Pa·s between horizontal plates separated by a distance 1mm. The upper plate is moving at a velocity of 8 m/s and the lower plate is stationary: (a) both plates are at 30°C, and (b) the upper plate is at 40°C and the lower plate is at 20°C.

(a) From Eq. (7.89), with $T_Y = T_0 = 30°C$,

$$T = \frac{0.2}{0.14}\left(\frac{8}{0.001}\right)^2 \left(\frac{0.001y - y^2}{2}\right) + 30$$

$$= 4.571 \times 10^7(0.001y - y^2) + 30$$

This symmetrical variation is shown as curve (a) in Fig. 7.11 and the maximum temperature of 41.4°C occurs at $y = 0.005$ m.

(b) With $T_Y = 40°C$ and $T_0 = 20°C$,

$$T = 4.571 \times 10^7(0.001y - y^2) + \frac{40 - 20}{0.001}y + 20$$

which is shown as curve (b) in Fig. 7.11. In this situation the maximum temperature of 43.6°C occurs at $y = 0.00072$ m. Inspection of curve (b) shows that the limiting temperature gradient in the fluid at the warmer upper plate is less than

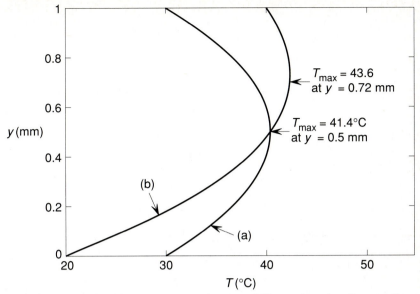

FIGURE 7.11. Temperature distributions in the oil film undergoing Couette flow considered in Example 7.8.

that at the cooler lower plate, and hence the rate of heat transfer from the fluid to the upper plate is less than that to the lower plate.

From Eq. (7.92) the rate of heat transfer to the upper plate is

$$q' = \frac{0.2 \times 8^2}{2 \times 0.001} - \frac{0.14(40 - 20)}{0.001}$$

$$= 6400 - 2800$$

$$= 3600 \text{ W/m}^2$$

(The positive sign indicates that heat flow is in the positive y-direction from the fluid to the plate.) From Eq. (7.93) the rate of heat transfer to the lower plate is

$$q' = -\frac{0.2 \times 8^2}{2 \times 0.001} - \frac{0.14(40 - 20)}{0.001}$$

$$= -6400 - 2800$$

$$= -9200 \text{ W/m}^2$$

(The negative sign indicates that heat flow is in the negative y-direction from the fluid to the plate.) Thus the total rate of heat transfer from the fluid to the plates is $3600 + 9200 = 12,800 \text{ W/m}^2$. In case (a) the rate of heat transfer to each plate is 6400 W/m^2, for a total of $12,800 \text{ W/m}^2$.

When the velocity of the upper plate at 8 m/s, the shear stress is

$$\tau_{yx} = \eta \frac{V}{Y} = 0.2 \times \frac{8}{0.001} = 1600 \text{ N/m}^2$$

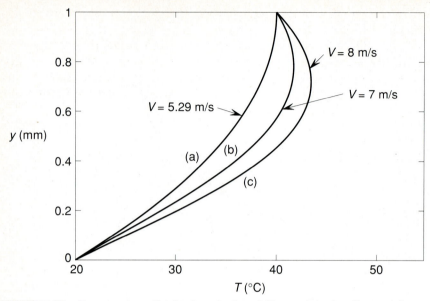

FIGURE 7.12. Temperature distributions in the oil film undergoing Couette flow considered in Example 7.8.

Thus, per unit area, the power requirement is

$$\tau_{yx}V = 1600 \times 8$$
$$= 12{,}800 \text{ W/m}^2$$

Consider the influence of a variation in V on the temperature profile in the oil. From Eq. (7.91)

$$V > \left[\frac{2(T_Y - T_0)}{\eta}\right]^{1/2} = \left[\frac{2(40 - 20) \times 0.14}{0.2}\right]^{1/2} = 5.29 \text{ m/s}$$

is the value of V at which $\partial T/\partial y = 0$ at $y = Y$. Equation (7.89) with $V = 5.29$ m/s is shown as line (a) in Fig. 7.12. The corresponding temperature profiles for $V = 7$ m/s and $V = 8$ m/s are shown, respectively, as lines (b) and (c). With $V < 5.29$ m/s heat flows from the upper plate to the oil and from the oil to the lower plate.

7.9 Heat Transfer from a Vertical Plate by Natural Convection

When a vertical plate at a uniform temperature, T_s, is immersed in a stagnant or quiescent fluid at a lower temperature, T_∞, the initial mechanism of heat transfer is conduction in the fluid. This heat transfer causes the development of a temperature gradient in the fluid adjacent to the plate, which, in turn, establishes a variation in the density of the fluid in the direction normal to the plate. In a gravitational field, differences in density give rise to buoyancy forces that cause fluid flow and hence establish, at steady state, a velocity boundary layer on the surface of the plate. In this situation heat is transferred from the plate by free convection or natural convection, and in contrast with forced convection, the velocity gradients in the boundary layer, which are caused by the buoyancy forces, are intimately linked with the temperature gradients, which give rise to the buoyancy forces.

Figure 7.13 shows the nature of the temperature and velocity gradients established at steady state. With $T_s > T_\infty$ the density of the fluid in the boundary layer is less than ρ_∞ and hence the buoyance forces cause the less dense fluid to rise. The local velocity, v_x, is zero at both the surface of the plate and at the surface of the velocity boundary layer and hence has a maximum value somewhere between these two locations. In the control volume shown in Fig. 7.13, at steady state and with $v_z = 0$, the equation of conservation of x-directed momentum is obtained from Eq. (3.15) as

$$\rho \left(v_x \frac{\partial v_x}{\partial x} + v_y \frac{\partial v_x}{\partial y} \right) = \eta \frac{\partial^2 v_x}{\partial y^2} - \frac{\partial P}{\partial x} + \rho g_x \qquad (7.96)$$

The significance of the terms $-\partial P/\partial x$ and ρg_x can be understood by examining the equation of conservation of momentum at a location outside the momentum boundary layer where $v_x = v_y = 0$ and $\rho = \rho_\infty$,

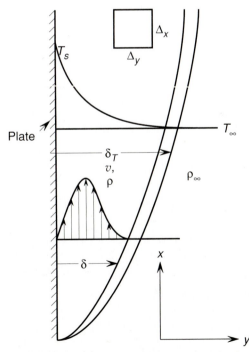

FIGURE 7.13. Temperature and velocity gradients at steady state in the boundary layers on a vertical plate undergoing heat transfer by natural convection.

$$0 = -\frac{dP}{dx} + \rho_\infty g$$

or

$$dP = \rho_\infty g\ dx$$

which is the barometric formula. Thus, in Eq. (7.96),

$$-\frac{\partial P}{\partial x} + \rho g_x = (\rho - \rho_\infty)g_x \tag{7.97}$$

and the difference between ρ and ρ_∞ is caused by the difference between T inside the boundary layer and T_∞. The isobaric thermal expansivity of the fluid is

$$\beta = \frac{1}{V}\left(\frac{\partial V}{\partial T}\right)_P = -\frac{1}{\rho}\left(\frac{\partial \rho}{\partial T}\right)_P$$

which for small differences in ρ and T can be written as

$$\beta\rho = -\frac{\rho - \rho_\infty}{T - T_\infty}$$

substitution of which into Eq. (7.97) gives

$$-\frac{\partial P}{\partial x} + \rho g_x = \beta\rho(T_\infty - T)g_x \tag{7.98}$$

and substitution of Eq. (7.98) into Eq. (7.96) gives the momentum balance as

$$\rho\left(v_x\frac{\partial v_x}{\partial x} + v_y\frac{\partial v_x}{\partial y}\right) = \eta\frac{\partial^2 v_x}{\partial y^2} + g_x\beta\rho(T_\infty - T)$$

or

$$v_x\frac{\partial v_x}{\partial x} + v_y\frac{\partial v_x}{\partial y} = v\frac{\partial^2 v_x}{\partial y^2} + g_x\beta(T_\infty - T) \tag{7.99}$$

From Eq. (7.82) the corresponding equation for conservation of energy is

$$v_x\frac{\partial T}{\partial x} + v_y\frac{\partial T}{\partial y} = \alpha\frac{\partial^2 T}{\partial y^2} \tag{7.100}$$

Equations (7.99) and (7.100) must be solved simultaneously. The occurrence of the term $(T_\infty - T)$ in Eq. (7.99) shows that unlike the case of forced convection, the velocity and temperature are coupled in free convection.

Approximate Integral Method

A momentum balance on a control volume in the boundary layer gives

$$\frac{d}{dx}\int_0^\delta v_x^2\ dy - \int_0^\delta g\beta(T - T_\infty)\ dy = -v\frac{\partial v_x}{\partial y}\bigg|_{y=0} \tag{7.101}$$

which is Eq. (3.18) with $v_\infty = 0$ and the buoyancy contribution included, and the energy balance is the same as given by Eq. (7.12):

$$\frac{d}{dx} \int_0^{\delta_T} (T_\infty - T)\, v_x \, dy = \alpha \left(\frac{\partial T}{\partial y}\right)_{y=0} \tag{7.102}$$

As in the previous applications of the approximate integral method, forms must be assumed for the velocity and temperature profiles. For mathematical convenience it is assumed that the Prandtl number for the fluid is approximately unity, in which case $\delta_T \approx \delta$. The velocity profile is assumed to be

$$\frac{v_x}{v_0} = \frac{y}{\delta} - 3\left(\frac{y}{\delta}\right)^2 + 3\left(\frac{y}{\delta}\right)^3 - \left(\frac{y}{\delta}\right)^4 \tag{7.103}$$

in which v_0 is a ''characteristic'' velocity that is a function only of x, and the temperature profile is assumed to be

$$\frac{T - T_\infty}{T_s - T_\infty} = 1 - 2\frac{y}{\delta} + \left(\frac{y}{\delta}\right)^2 \tag{7.104}$$

Equations (7.103) and (7.104) are shown in Fig. 7.14, which shows that they satisfy the boundary conditions. Substitution of Eqs. (7.103) and (7.104) into Eq. (7.101), gives

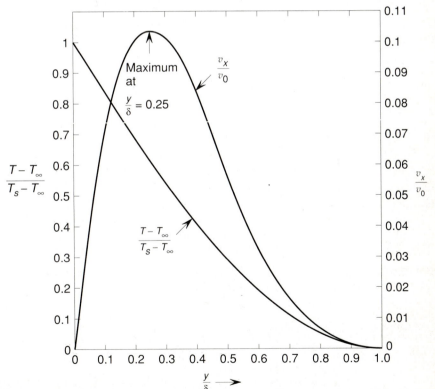

FIGURE 7.14. Assumed temperature and velocity profiles in the boundary layers on a vertical plate undergoing heat transfer by natural convection.

$$\frac{d}{dx}\left(\frac{v_0^2\delta}{252}\right) - g\beta\frac{(T_s - T_\infty)\delta}{3} = -\frac{vv_0}{\delta} \tag{7.105}$$

and substitution of Eqs. (7.103) and (7.104) into Eq. (7.102) gives

$$\frac{d}{dx}\left[\frac{v_0(T_s - T_\infty)\delta}{42}\right] = \frac{2\alpha(T_s - T_\infty)}{\delta} \tag{7.106}$$

It is now assumed that the characteristic velocity, v_0, and δ vary with x as

$$\delta = C_1 x^m \tag{7.107}$$

and

$$v_0 = C_2 x^n \tag{7.108}$$

Substituting Eqs. (7.107) and (7.108) into Eq. (7.105) and differentiating gives

$$\frac{C_1 C_2^2(2n + m)x^{2n+m-1}}{252} - \frac{g\beta(T_s - T_\infty)C_1 x^m}{3} = -\frac{vC_2 x^{n-m}}{C_1} \tag{7.109}$$

and substituting in Eq. (7.106) and differentiating gives

$$\frac{C_1 C_2(n + m)x^{n+m-1}}{42} = \frac{2\alpha x^{-m}}{C_1} \tag{7.110}$$

Identity of the exponents in Eqs. (7.109) and (7.110), respectively, requires that

$$m = n - m$$

and

$$n + m - 1 = -m$$

which gives $m = \frac{1}{4}$ and $n = \frac{1}{2}$, in which case Eqs. (7.109) and (7.110) become

$$\frac{5C_1 C_2^2}{4 \times 252} - \frac{g\beta(T_s - T_\infty)C_1}{3} = -v\frac{C_2}{C_1} \tag{7.111}$$

and

$$\frac{3C_1 C_2}{4 \times 42} = \frac{2\alpha}{C_1} \tag{7.112}$$

From Eq. (7.112)

$$C_2 = \frac{112\alpha}{C_1^2} \tag{7.113}$$

substitution of which into Eq. (7.111) gives

$$C_1 = \left[\frac{\dfrac{5 \times (112\alpha)^2}{4 \times 252} + 112\alpha v}{g\beta(T_s - T_\infty)/3}\right]^{1/4}$$

$$= \left[\frac{336(\alpha v + 0.556\alpha^2)}{g\beta(T_s - T_\infty)}\right]^{1/4}$$

$$= \left[\frac{336[(\alpha v/v^2) + 0.556(\alpha^2/v^2)]}{g\beta(T_s - T_\infty)/v^2}\right]^{1/4}$$

$$= \left[\frac{336(1/\text{Pr} + 0.556/\text{Pr}^2)}{g\beta(T_s - T_\infty)/v^2}\right]^{1/4}$$

$$= \left[\frac{336(\text{Pr} + 0.556)}{\text{Pr}^2}\right]^{1/4} \left[\frac{g\beta(T_s - T_\infty)}{v^2}\right]^{-1/4} \qquad (7.114)$$

and

$$C_2 = 112\alpha \left[\frac{336(\text{Pr} + 0.556)}{\text{Pr}^2}\right]^{-1/2} \left[\frac{g\beta(T_s - T_\infty)}{v^2}\right]^{1/2}$$

$$= 112[336(\text{Pr} + 0.556)]^{-1/2} [g\beta(T_s - T_\infty)]^{1/2} \qquad (7.115)$$

The term $g\beta(T_s - T_\infty)/v^2$, which appears in Eq. (7.114) and which arose from the buoyancy term in Eq. (7.99), leads to an important dimensionless number in free convection. In Eq. (7.99) the term $g\beta(T_s - T_\infty)$ has the units m/s². Multiplying by a characteristic length L and dividing by the square of a characteristic velocity, v_0^2, gives the dimensionless number

$$\frac{g\beta(T_s - T_\infty)L}{v_0^2}$$

and multiplying this dimensionless number by the square of the Reynolds number (Re $= v_0 L/v$) gives

$$\frac{g\beta(T_s - T_\infty)L^3}{v^2}$$

which defines the Grashof number, Gr:

$$\text{Gr}_L = \frac{g\beta(T_s - T_\infty)L^3}{v^2}$$

or

$$\text{Gr}_x = \frac{g\beta(T_s - T_\infty)x^3}{v^2} \qquad (7.116)$$

Substitution of Eq. (7.114) in Eq. (7.107) gives

$$\delta = \frac{[336(\text{Pr} + 0.556)/\text{Pr}^2]^{1/4} x^{1/4}}{[g\beta(T_s - T_\infty)/v^2]^{1/4}}$$

$$= \frac{[336(\text{Pr} + 0.556)/\text{Pr}^2]^{1/4} x}{g\beta(T_s - T_\infty)x^3/v^2]^{1/4}}$$

which in view of the definition of the Grashof number, Gr_x, becomes

$$\frac{\delta}{x} = \left[\frac{336(\text{Pr} + 0.556)}{\text{Pr}^2\text{Gr}_x}\right]^{1/4} \tag{7.117}$$

The local heat transfer coefficient is obtained from

$$q'|_{y=0} = -k\left(\frac{\partial T}{\partial y}\right)_{y=0} = h_x(T_s - T_\infty)$$

From Eq. (7.104), $(\partial T/\partial y)_{y=0}$ is obtained as

$$\left(\frac{\partial T}{\partial y}\right)_{y=0} = -\frac{2(T_s - T_\infty)}{\delta}$$

and hence

$$h_x = \frac{2k}{\delta} \tag{7.118}$$

The local Nusselt number is thus

$$\text{Nu}_x = \frac{h_x x}{k} = \frac{2x}{\delta}$$

which from Eq. (7.117) becomes

$$\text{Nu}_x = \frac{2\text{Pr}^{1/2}\text{Gr}_x^{1/4}}{[336(\text{Pr} + 0.556)]^{1/4}} \tag{7.119}$$

The average heat transfer coefficient, $\overline{h}_L$, is obtained as

$$\overline{h}_L = \frac{1}{L}\int_0^L h_x \, dx \tag{7.120}$$

In Eq. (7.118), $h_x \propto 1/\delta$, and in Eq. (7.117),

$$\frac{1}{\delta} \propto \frac{\text{Gr}^{1/4}}{x} \propto x^{-1/4}$$

Therefore, in Eq. (7.120)

$$\overline{h}_L \propto \frac{1}{L}\int_0^L x^{-1/4} \, dx$$

or

$$\overline{h}_L = \tfrac{4}{3}h_x \qquad (\text{at } x = L)$$

and consequently, the average Nusselt number, $\overline{\text{Nu}}_L$, is

$$\overline{\text{Nu}}_L = \tfrac{4}{3}\,\text{Nu}_x \qquad (\text{at } x = L)$$

$$= \frac{8\text{Pr}^{1/2}\text{Gr}_L^{1/4}}{3[336(\text{Pr} + 0.0556)]^{1/4}} \tag{7.121}$$

Equations (7.119) and (7.121) show that, in free convection, the Nusselt number is a function of the Prandtl number and the Grashof number, which is in contrast with Eqs. (7.25) and (7.26), which show that in forced convection, the Nusselt number is a function of the Prandtl number and the Reynolds number. The Grashof number is a measure of the ratio of the buoyance forces to the viscous forces in the boundary layer and the Reynolds number is a measure of the ratio of the inertial forces to the viscous forces in the boundary layer.

The exact solutions to Eqs. (7.99) and (7.100) obtained by Ostrach (S. Ostrach, "An Analysis of Laminar Free Convection and Heat Transfer About a Flat Plate Parallel to the Direction of the Generating Body Force," *NACA Report 1111*, 1953) using a numerical technique are shown in Fig. 7.15(a) and (b).

As is the case with the velocity boundary layer in forced convection, the velocity boundary layer in free convection can become turbulent, and a common illustration is given by a lighted cigarette in a horizontal position in an ashtray in a draft-free room. Initially, the smoke rises as a well-defined laminar column, but when it reaches some critical height the laminar column breaks down and becomes turbulent. From correlation with experimental observations, the criterion for transition from laminar to turbulent flow in a free convection boundary layer on a vertical plate is given in terms of the Rayleigh number, Ra, where

$$Ra = Pr \cdot Gr \qquad (7.122)$$

The variation with $\overline{Nu}_L$ with Ra_L for heat transfer by free convection from a vertical plate is shown in Fig. 7.16. The transition from laminar to turbulent flow occurs at $Ra_L = 10^9$ and correlation of $\overline{Nu}_L$ with Ra_L in the range $10^4 \leq Ra_L \leq 10^9$ gives

$$\overline{Nu}_L = 0.59 Ra_L^{1/4} \qquad \text{(laminar flow)} \qquad (7.123)$$

and with Ra_L in the range $10^9 \leq Ra_L \leq 10^{14}$ gives

$$\overline{Nu}_L = 0.10\ Ra_L^{1/3} \qquad \text{(turbulent flow)} \qquad (7.124)$$

With $Ra_L = 10^9$, Eq. (7.123) gives $\overline{Nu}_L = 105$ and Eq. (7.124) gives $\overline{Nu}_L = 100$.

EXAMPLE 7.9

Consider heat transfer by free convection from a vertical surface of length 0.25 m and width 0.25 m to a fluid of Pr = 0.72. Air has a Prandtl number of 0.72 at $-23°C$, so to make the film temperature $-23°C$, T_s will be set at 27°C and T_∞ will be set at $-73°C$. At $T_f = -23°C$, the properties of air are

$$Pr = 0.72$$

$$\nu = 1.114 \times 10^{-5}\ m^2/s$$

$$\frac{g\beta}{\nu^2} = 3.02 \times 10^8\ K^{-1} \cdot m^{-3}$$

$$k = 0.0223\ W/m \cdot K$$

The Grashof number at the upper edge of the plate is thus

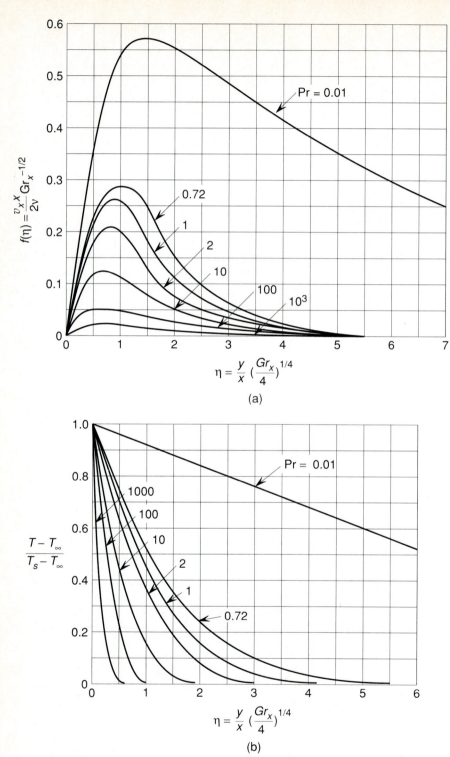

FIGURE 7.15. Exact solutions to (a) Eq. (7.99) and (b) Eq. (7.100).

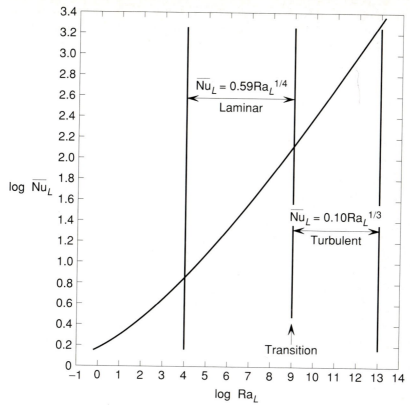

FIGURE 7.16. Variation of the average Nusseldt number with the Rayleigh number for heat transfer by natural convection from a vertical plate.

$$Gr_L = \frac{g\beta(T_s - T_\infty)L^3}{\nu^2} = 3.02 \times 10^8 \times 100 \times 0.25^3 = 4.719 \times 10^8$$

and the Rayleigh number is

$$Ra_L = Gr_L Pr = 4.719 \times 10^8 \times 0.72 = 3.398 \times 10^8$$

which, being less than 10^9, indicates that the boundary layer is laminar over the entire plate. Thus from Eq. (7.117) the thickness of the boundary layer at the upper edge is

$$\delta = \left[\frac{336(Pr + 0.556)}{Pr^2 Gr_L} \right]^{1/4}$$

$$L = \left[\frac{336(0.72 + 0.556)}{0.72^2 \times 4.719 \times 10^8} \right]^{1/4} \times 0.25 = 9.1 \times 10^{-3} \text{ m}$$

In Fig. 7.15(a) the ordinate, $f(\eta)$, is proprotional to v_x and the abscissa, η, is proportional to y. Each curve is thus a measure of the variation of v_x with y for a given Prandtl number, increasing from $f(\eta) = 0$, $\eta = 0$, passing through a maximum value of $f(\eta)$ and decreasing to zero at what, by definition, is the value of η corresponding to the boundary layer thickness δ, at the given value of x. For Pr $= 0.72$, $f(\eta)$ falls to zero at $\eta = 5.5$; therefore,

$$\eta = 5.5 = \frac{y}{x}\left(\frac{Gr_x}{4}\right)^{1/4} = \frac{y}{0.25}\left(\frac{4.719 \times 10^8}{4}\right)^{1/4}$$

which gives $y = 1.32 \times 10^{-2}$ m as the value of δ at $x = 0.25$ m. The value obtained from Eq. (7.117) is thus approximately two-thirds of the value obtained from the exact solution.

The maximum value of $f(\eta)$ on the curve for Pr $= 0.72$ is 0.28, and thus the maximum value of v_x in the boundary layer is obtained as

$$f(\eta) = 0.28 = \frac{v_x x}{2v}(Gr_x)^{-1/2} = \frac{0.25v_x}{2 \times 1.14 \times 10^{-5}}(4.719 \times 10^8)^{-1/2}$$

which gives $v_{x,max} = 0.55$ m/s at $x = 0.25$ m. This maximum occurs at $\eta = 1$ (i.e., at $y/\delta = 1/5.5 = 0.18$), in comparison with $y/\delta = 0.25$ shown in Fig. 7.14.

Equation (7.121) gives the average Nusselt number as

$$\overline{Nu}_L = \frac{8Pr^{1/2}Gr_L^{1/4}}{3[336(Pr + 0.556)]^{1/4}} = \frac{8 \times 0.72^{1/2} \times (4.719 \times 10^8)^{1/4}}{3[336(0.72 + 0.556)]^{1/4}} = 73.3$$

and the empirical correlation Eq. (7.123) gives

$$\overline{Nu}_L = 0.59Ra_L^{1/4} = 0.59 \times (3.398 \times 10^8)^{1/4} = 80.1$$

With $\overline{Nu}_L = 73.3$,

$$\overline{h}_L = \frac{\overline{Nu}_L k}{L} = \frac{73.3 \times 0.0223}{0.25} = 6.83 \text{ W/m}^2\cdot\text{K}$$

and

$$q = \overline{h}_L LW(T_s - T_\infty) = 6.83 \times 0.25 \times 0.25 \times 100 = 42.7 \text{ W}$$

If the air were flowing over the plate at $v_\infty = 5$ m/s, heat transport would occur by forced convection, and from Eq. (7.26)

$$\overline{Nu}_L = 0.662 \, Pr^{1/3}Re_L^{1/2} = 0.662 \times 0.72^{1/3} \times \left(\frac{5 \times 0.25}{1.14 \times 10^{-5}}\right)^{1/2}$$
$$= 196$$

Therefore,

$$\overline{h}_L = \frac{196 \times 0.0223}{0.25} = 18.3 \text{ W/m}^2\cdot\text{K}$$

and

$$q = 18.3 \times 0.25 \times 0.25 \times 100 = 114 \text{ W}$$

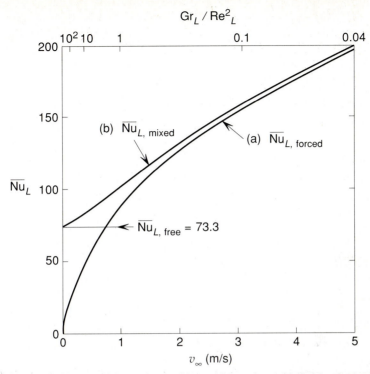

FIGURE 7.17. Variation of the average Nusseldt number with fluid velocity for heat transfer from a vertical plate. Line (a) shows the Nusseldt number for force convection and line (b) shows the Nusseldt number for mixed convection.

The relative magnitudes of the Reynolds number and the Grashof number determine whether heat transfer is occurring by forced convection or by free convection. Free convection is negligible if $Re_L^2 >> Gr_L$, and forced convection is negligible if $Gr_L >> Re_L^2$. In the example above, with $v_\infty = 5$ m/s, the Reynolds number is 1.096×10^5 and the Grashof number is 4.719×10^8. Thus as

$$(1.096 \times 10^5)^2 = 1.2 \times 10^{10} >> 4.719 \times 10^8$$

the contribution of free convection to the heat transfer process is negligible. For $Gr_L = Re_L^2$,

$$Re_L = \frac{v_\infty \times 0.25}{1.14 \times 10^{-5}} = (4.719 \times 10^8)^{1/2}$$

or $v_\infty = 1$ m/s, and in this situation heat transfer would occur by mixed convection. If v_∞ is in the same direction as the velocity induced by free convection, the Nusselt number is given by the approximation

$$\overline{Nu}_{mixed} \doteq (\overline{Nu}_{L,free}^3 + \overline{Nu}_{L,forced}^3)^{1/3} \qquad (7.125)$$

With $v_\infty = 1$ m/s, $Re_L = 2.17 \times 10^4$, and

$$\overline{\text{Nu}}_{L,\text{forced}} = 0.662 \times 0.72^{1/3} \times (2.17 \times 10^4)^{1/2} = 87.5$$

and hence, from Eq. (7.125),

$$\overline{\text{Nu}}_{\text{mixed}} \doteq (73.3^3 + 87.5^3)^{1/3} = 102$$

The influence of an externally imposed airstream on heat transfer by convection from the plate is shown in Fig. 7.17. Line (a) is calculated assuming that only forced convection occurs:

$$\overline{\text{Nu}}_{L,\text{forced}} = 0.662\text{Pr}^{1/3}\text{Re}_L^{1/2} = 0.662 \times 0.72^{1/3} \times \left(\frac{0.25}{1.14 \times 10^{-5}}\right)^{1/2} v_\infty^{1/2}$$

$$= 87.87v_\infty^{1/2}$$

and line (b) is calculated from Eq. (7.125) with $\overline{\text{Nu}}_{L,\text{free}} = 73.3$. With $\text{Gr}_L/\text{Re}_L^2 = 10$ (at $v_\infty = 0.3$ m/s) $\overline{\text{Nu}}_{L,\text{mixed}}$ is 28% larger than $\overline{\text{Nu}}_{L,\text{free}}$, and with $\text{Gr}_L/\text{Re}_L^2 = 0.1$ (at $v_\infty = 3.1$ m/s) $\overline{\text{Nu}}_{L,\text{mixed}}$ is 3% larger than $\overline{\text{Nu}}_{L,\text{forced}}$. As $\text{Gr}_L/\text{Re}_L^2$ increases above 10, $\overline{\text{Nu}}_{L,\text{mixed}}$ approaches $\overline{\text{Nu}}_{L,\text{free}}$, and as $\text{Gr}_L/\text{Re}_L^2$ decreases below 0.1, $\overline{\text{Nu}}_{L,\text{mixed}}$ approaches $\overline{\text{Nu}}_{L,\text{forced}}$.

7.10 Heat Transfer from Cylinders by Natural Convection

The empirical correlations given by Eqs. (7.123) and (7.124) are applicable to vertical plates with a uniform heat flux and also to vertical cylinders of diameter D and length L, provided that the velocity boundary layer on the cylinder is significantly less than the diameter of the cylinder. From Eq. (7.117) the maximum boundary layer thickness on a vertical cylinder of length L and sufficiently large diameter is

$$\frac{\delta}{L} = \left[\frac{336(\text{Pr} + 0.556)}{\text{Pr}^2}\right]^{1/4} \frac{1}{\text{Gr}_L^{1/4}}$$

which for a fluid of $\text{Pr} = 1$ is

$$\frac{\delta}{L} = \frac{4.78}{\text{Gr}_L^{1/4}}$$

The diameter of the vertical cylinder is "sufficiently large" for the application of Eqs. (7.123) and (7.124) if

$$\frac{D}{L} \geq \frac{35}{\text{Gr}_L^{1/4}} \tag{7.126}$$

From correlation with experimental observations, the correlation of Nu_D with Ra_D for free convection from a horizontal cylinder is given by

$$\text{Nu}_D = C \cdot \text{Ra}_D^n \tag{7.127}$$

and the values of C and n to be used depend on the Rayleigh number as listed in Table 7.3.

TABLE 7.3

Ra_D	C	n
$10^{-10} - 10^{-2}$	0.675	0.058
$10^{-2} - 10^{2}$	1.02	0.148
$10^{2} - 10^{4}$	0.850	0.188
$10^{4} - 10^{7}$	0.480	0.250
$10^{7} - 10^{12}$	0.125	0.333

EXAMPLE 7.10

Consider the initial cooling, by free convection, of cylindrical cans of $D = 0.08$ m and $L = 0.12$ m, at a temperature of 26°C placed in a refrigerator at 0°C. Heat transfer through the ends of the cans will be ignored. For air at $T_f = 13$°C,

$$Pr = 0.715$$

$$\frac{g\beta}{v^2} = 1.803 \times 10^8 \ K^{-1}\cdot m^{-3}$$

$$k = 0.025 \ W/m\cdot K$$

If placed vertically, from Eq. (7.116),

$$Gr_L = \frac{g\beta(T_s - T_\infty)L^3}{v^2} = 1.803 \times 10^8 \times 26 \times 0.12^3 = 8.1 \times 10^6$$

and from Eq. (7.122),

$$Ra_L = Pr\cdot Gr_L = 0.715 \times 8.1 \times 10^6 = 5.79 \times 10^6$$

With respect to the criterion given by Eq. (7.126),

$$\frac{D}{L} = \frac{0.08}{0.12} = 0.667 \quad \text{and} \quad \frac{35}{Gr_L^{1/4}} = 0.65$$

Therefore,

$$\frac{D}{L} > \frac{35}{Gr_L^{1/4}}$$

and Eq. (7.123) can be used. Thus

$$\overline{Nu}_L = 0.59(5.79 \times 10^6)^{1/4} = 28.9$$

and

$$\overline{h}_L = \frac{\overline{Nu}_L k}{L} = \frac{28.9 \times 0.025}{0.12} = 6.03 \ W/m^2\cdot K$$

If placed horizontally,

$$Gr_D = \frac{g\beta(T_s - T_\infty)D^3}{v^2} = 1.803 \times 10^8 \times 26 \times 0.08^3 = 2.4 \times 10^6$$

and

$$\text{Ra}_D = 0.715 \times 2.4 \times 10^6 = 1.72 \times 10^6$$

Therefore, from Eq. (7.127) and Table 7.3,

$$\text{Nu}_D = 0.48(1.72 \times 10^6)^{1/4} = 17.4$$

and

$$h_D = \frac{\text{Nu}_D k}{D} = \frac{17.4 \times 0.025}{0.08} = 5.42 \text{ W/m}^2\text{·K}$$

As $\overline{h}_L > h_D$, the cans cool more rapidly if placed vertically. To achieve the maximum rate of cooling, the cans should be placed sufficiently far apart that their velocity boundary layers do not interfere with one another. From Eq. (7.117) at the upper ends of the vertical cans

$$\delta = \left[\frac{336(\text{Pr} + 0.556)}{\text{Pr}^2 \text{Gr}_L} \right]^{1/4} L$$

$$= \left[\frac{336(0.715 + 0.556)}{0.715^2 \times 8.1 \times 10^6} \right]^{1/4} \times 0.12 = 0.012 \text{ m}$$

Therefore, the cans should be placed at least 2×0.012 m, or 2.4 cm, apart from one another.

7.11 Summary

When a fluid at one temperature flows past a flat plate at a different temperature, a thermal boundary layer forms on the surface of the plate and heat transfer between the fluid and the plate occurs by forced convection. In the thermal boundary layer the temperature varies from the temperature of the plate to the temperature of the bulk fluid, and if the thermal and momentum boundary conditions are similar and the thermal diffusivity of the fluid, α, is equal to the kinematic viscosity, ν, the temperature and velocity profiles in the boundary layers are identical. The local heat transfer coefficient is determined by the velocity, thermal conductivity, and kinematic viscosity of the fluid, and it is inversely proportional to the square root of the distance from the leading edge of the plate. The average heat transfer coefficient over the distance from $x = 0$ to $x = L$ is twice the local value at L. The Nusselt number, Nu, is a measure of the relative contributions of convection and conduction to the heat transport between the fluid and the plate, with large Nusselt numbers indicating the predominance of convection. In forced convection the Nusselt number is a function of the Reynolds number and the Prandtl number, with the latter being the ratio of the kinematic viscosity of the fluid to its thermal diffusivity. The Reynolds analogy between momentum transport and heat transport in laminar boundary layers gives the dependence of the Stanton number on the friction factor and the Prandtl number and knowledge of the Stanton number and the Reynolds number allows determination of the Nusselt number, and hence the magnitude of the heat flux between the fluid and the plate. Transition from laminar flow to turbulent flow in the boundary layer causes a significant increase in the heat flux.

In heat transfer by forced convection between a fluid and a cylindrical pipe, a fully developed temperature profile is developed when the thermal boundary layers, which form at the entrance to the pipe, merge and the thermal entry length is determined by the diameter of the pipe and by the Reynolds number of the flow and the Prandtl number of the fluid. In a fully developed temperature profile in laminar pipe flow in which the heat flux from the pipe to the fluid is uniform along the pipe, the ratio $(T_s - T)/(T_s - T_m)$ is a function only of radial position in the flow and is independent of x. Thus the surface temperature of the pipe, T_s and the mean temperature of the fluid, T_m, increase linearly along the pipe and the Nusselt number has the constant value of 4.36. In laminar pipe flow in which the surface temperature of the pipe is uniform, the mean temperature of the fluid is a logarithmic function of distance along the pipe and the Nusselt number has the constant value of 3.66.

In free or natural convection, the transfer of heat from a vertical plate to a fluid causes the development of a temperature gradient, which, in turn, establishes a variation in the density of the fluid in the direction normal to the plate. In a gravitational field differences in density give rise to buoyancy forces that cause fluid flow and hence establish, at steady state, a velocity boundary layer on the surface of the plate in which the velocity increases from zero at the plate to a maximum value in the boundary layer and then decreases to zero at the surface of the boundary layer. In contrast with forced convection the velocity gradients in the boundary layer, which are caused by the buoyancy forces, are coupled with the temperature gradients which give rise to the buoyancy forces, and determination of temperature and velocity profiles requires simultaneous solution of the momentum and energy balance equations. The influence of buoyancy gives rise to the definition of the Grashof number, Gr, and the Nusselt number for free convection is a function of the Prandtl number and the Grashof number. A transition from laminar flow to turbulent flow in the free convection boundary layer occurs when the Rayleigh number reaches a value of 10^9, where the Rayleigh number, Ra, is the product of the Reynolds number and the Prandtl number.

The relative magnitudes of the Reynolds number and the Grashof number determine whether heat transfer is occurring by forced convection or by free convection. Free convection is negligible if the square of the Reynolds number is significantly larger than the Grashof number, and forced convection is negligible if the Grashof number is significantly larger than the square of the Reynolds number. Mixed convection occurs when the two quantities are of similar magnitude.

Problems

PROBLEM 7.1

A steel billet of length 5 m and width 0.5 m with a surface temperature of 329°C is rolled along a conveyer at a velocity of 5 m/s through still air. Calculate the rate of heat loss from the upper surface of the billet if the air temperature is 25°C.

PROBLEM 7.2

A hot steel plate of length 1 m and width 0.5 m at a temperature of 629°C is cooled in an airstream at 25°C. Calculate the velocity of the airstream which removes 10 kW from both sides of the plate combined.

PROBLEM 7.3

A steel plate of length 1 m and width 0.5 m at a temperature of 329°C is cooled in a horizontal 25°C airstream of $v_\infty = 50$ m/s. Calculated the rate of heat loss from one side of the plate.

PROBLEM 7.4

When immersed in an airstream flowing at 5 m/s at a temperature of 25°C, an electrically heated flat plate of 0.3 m length and 0.6 m width dissipates 90 W from each side. Calculate the temperature of the plate at its trailing edge.

PROBLEM 7.5

A flat surface is made up of electrically heated strips of length 1 m and width 0.05 m. The strips, which are electrically insulated from one another, are laid with their long sides in contact and an airstream at 25°C flows over the surface in a direction normal to long sides of the strips at a velocity of 50 m/s. Electrical power is supplied individually to each of the strips such that they all have a surface temperature of 229°C. Determine which strip requires the most power.

PROBLEM 7.6

Water flowing at 5 kg/s through a tube of diameter 0.03 m and constant $T_s = 80°C$ has to be heated from 10°C to 50°C

(a) Calculate the required length of the tube.
(b) Derive an expression for the variation, with average linear velocity, of the length required to heat the water from 10°C to 30°C.

PROBLEM 7.7

Water to 50°C is passed through a cylindrical tube of diameter 0.03 m, length 4 m, and constant T_s of 5°C. Calculate the mass flow rate that gives a mean outlet temperature of 30°C.

PROBLEM 7.8

Water at 30°C is passed through an electrically heated cylindrical tube of diameter 0.03 m and length 4 m. The uniform heat flux from the tube is 2000 W/m². Calculate:

(a) The mass flow rate that gives a mean outlet temperature of 50°C
(b) The value of T_s at the outlet

PROBLEM 7.9

Helium at 400°C is passed through a cylindrical tube of diameter 0.05 m, length 1 m, and uniform surface temperature of 100°C. Calculate the mean outlet temperature when the mass flow rate is 10^{-3} kg/s.

PROBLEM 7.10

Air at 400 K flows over a horizontal cylinder at a velocity of 15 m/s. The diameter of the cylinder is 0.05 m and its surface temperature is 300 K. Calculate the rate of heat transfer to the cylinder per unit length.

PROBLEM 7.11

Air at 20°C flows over a 0.003-m-diameter platinum wire in an anemometer. The readings indicate that the surface temperature of the hot wire is 80°C and the heat flux from the wire is 5000 W/m². Calculate the velocity of the air.

PROBLEM 7.12

In a Couette flow the upper plate is moving at a velocity of 10 m/s relative to the lower plate and the plates are separated by a film of oil of thickness 0.003 m. The upper plate is at 30°C and the lower plate is at 10°C. Calculate:

(a) The maximum temperature of the oil
(b) The position of the maximum
(c) The heat flux to the upper and the lower plates

PROBLEM 7.13

A long steel plate with a surface temperature of 130°C is suspended vertically in still air at 25°C. Calculate:

(a) The distance from the lower edge of the plate at which the boundary layer becomes turbulent
(b) The thickness of the momentum boundary layer at a distance of 0.25 m from the lower edge of the plate
(c) The maximum local velocity in the boundary layer at this point
(d) The position in the boundary layer at which the maximum local velocity occurs

PROBLEM 7.14

Aluminum plates of length 0.2 m are cooled by being placed in a rack and immersed vertically in water at 25°C. Calculate:

(a) The heat flux from the plates when their surfaces are at 69°C
(b) The minimum distance between the plates at which the boundary layers do not interfere with one another

PROBLEM 7.15

(a) Calculate the rate of heat loss per meter by free convection from a horizontal steam pipe of diameter 0.05 m and constant surface temperature 140°C to still air at 15°C.

(b) If the pipe is lagged with an insulating material of $k = 0.035$ W/(m·K), calculate the thickness of insulation required to half the rate of heat loss from the bare pipe.

PROBLEM 7.16

Water at 10°C flows upward at 0.35 m/s past a vertical plate of dimensions 0.3 m × 0.3 m. Calculate the rate of heat loss from the plate if its surface temperature is 84°C.

PROBLEM 7.17

A cylindrical 150-W electric immersion heater is 0.01 m in diameter and 0.1 m in length.

(a) Calculate the temperature which the surface of the heater reaches when it is immersed horizontally in still water at 47°C and plugged into an electric outlet.

(b) Show why the heater should not be left unplugged in still air at room temperature. For arithmetic convenience ignore the influence of temperature on the thermophysical properties of the fluids.

8

Transient Heat Flow

8.1 Introduction

If a solid object at some uniform temperature is suddenly immersed in a fluid that is at a uniform lower temperature, the consequent transfer of heat from the solid to the liquid causes the solid to cool, and during the period of cooling (the transient period) the local temperature in the solid varies with both position and time. In principle, these variations are obtained from integration of the general heat conduction equation, Eq. (6.35), for the appropriate boundary conditions. Although these integrations generally involve complicated mathematics, one situation does exist in which the mathematical analysis is simple. In this situation the temperature gradients developed in the solid during cooling are small enough to be neglected, in which case the transient behavior can be examined by the lumped capacitance method of analysis.

8.2 Lumped Capacitance Method; Newtonian Cooling

During cooling, heat is transported from the interior of the solid to the surface by conduction and is then transferred from the surface to the fluid by convection. There are thus two thermal resistances to the cooling process—a resistance to conduction and a resistance to convection—and the relative magnitudes of these resistances are of great importance. If the resistance to convection is significantly greater than the resistance to conduction, a "bottleneck" in the heat transport process occurs at the surface of the solid. Heat is easily transported to the surface by conduction but experiences difficulty in being transferred away from the surface to the fluid. In such a situation the temperature gradients developed in the solid can be small enough to be ignored. In the lumped capacitance method it is assumed that there are no temperature gradients in the solid (i.e., that, during cooling, the temperature throughout the solid is the same as that of the surface). Although this situation is clearly impossible—heat transport by conduction requires a temperature gradient—it is an acceptable assumption for certain ratios of the convection resistance to the conduction resistance. If, during cooling, the temperature of the solid is uniformly equal to the surface temperature, T, the rate at which heat is transferred from the surface to the fluid, at temperature T_∞, is equal to the rate of decrease of the thermal energy (enthalpy) of the solid,

$$hA_s(T - T_\infty) = -m\frac{dH}{dt} = -\rho V C_p \frac{dT}{dt} \tag{8.1}$$

where A_s, m, and V are the surface area, mass, and volume of the solid and C_p is its constant pressure heat capacity per unit mass. Rearrangement of Eq. (8.1) gives

$$\frac{dT}{T - T_\infty} = -\frac{hA_s}{\rho V C_p}\, dt \tag{8.2}$$

which, on integration between T at t and $T = T_i$ and $t = 0$ (where T_i is the initial uniform temperature of the solid), gives

$$\ln\frac{T - T_\infty}{T_i - T_\infty} = -\frac{hA_s}{\rho V C_p}\, t$$

or

$$\frac{T - T_\infty}{T_i - T_\infty} = \exp\left(-\frac{hA_s}{\rho V C_p}\, t\right) \tag{8.3}$$

Thus, under the conditions that significant temperature gradients are not developed within the solid during cooling, the temperature of the solid decreases exponentially with time. This is known as Newtonian cooling, and the experimental observation of this behavior lead to the introduction of the constant h as defined in Eq. (6.2).

The nature of the exponential decrease is determined by the magnitude of the term $hA_s/\rho V C_p$, and, mathematically, this term serves as a time constant, that is, if

$$\frac{hA_s}{\rho V C_p} = \frac{1}{\tau} \tag{8.4}$$

then the normalized temperature in Eq. (8.3) varies with time as

$$\frac{T - T_\infty}{T_i - T_\infty} = \exp\left(-\frac{t}{\tau}\right) \tag{8.5}$$

Equation (8.5) is shown in Fig. 8.1 for time constants of 1, 2, 5, and 10 s. As is seen, the rate of cooling decreases with increasing τ and the time required to cool to a given value of the normalized temperature is a linear function of τ:

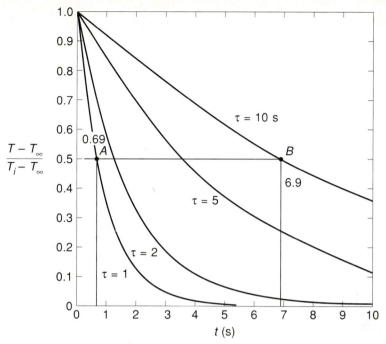

FIGURE 8.1. Variation of the normalized temperature of an isothermal body with time and time constant during Newtonian cooling.

$$t = -\tau \ln \frac{T - T_\infty}{T_i - T_\infty}$$

This is shown by points A and B in Fig. 8.1. With $\tau = 1$, 0.69 s are required for the normalized temperature to decrease to 0.5 (point A) and with $\tau = 10$, 6.9 s are required for the same decrease (point B). The validity of the assumption that the temperature gradients developed in the solid are small enough to be ignored increases with decreasing cooling rate, and hence with increasing value of τ, and the value of τ is determined by properties of the fluid and by properties of the solid. From Eq. (6.9), $R_h = 1/hA_s$, and hence Eq. (8.4) can be written as

$$\tau = (\rho V C_p) R_h \tag{8.6}$$

The quantity in parentheses is determined only by the properties of the solid and is called the lumped thermal capacitance of the solid, C_t, and therefore

$$\tau = C_t R_h$$

and an increase in either C_t or R_h decreases the rate of cooling of the solid. For a given mass of solid, C_t is determined by the heat capacity of the solid, which is defined as the ratio of the heat added to or withdrawn from unit mass of the solid to the consequent change in the temperature of the solid. Thus more heat must be withdrawn from a high C_p solid than from a low C_p solid to effect a given decrease in temperature.

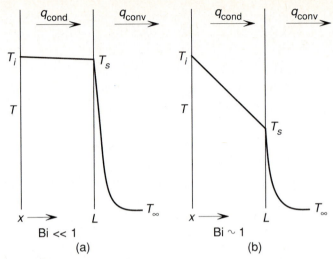

FIGURE 8.2. (a) Temperature gradient that develops in a solid during cooling when the resistance to heat transfer by convection at the surface is sufficiently larger than the resistance to heat transport by conduction in the solid that the Biot number is less than 0.1; (b) temperature gradient that develops in a solid during cooling when the resistances to heat transfer by conduction and heat transfer by convection are of similar magnitude.

Figure 8.2 shows heat transfer from a slab of thickness $2L$ to a fluid at T_∞. In Fig. 8.2(a) the resistance to convection is very much larger than the resistance to conduction and hence the temperature gradient in the slab is negligibly small. In Fig. 8.2(b) the resistances are comparable with one another and hence a temperature gradient has developed in the solid. Assuming, for convenience, a linear temperature gradient over the distance L, the heat balance is

$$q_x' = \frac{k(T_i - T_s)}{L} = h(T_s - T_\infty)$$

or

$$\frac{T_i - T_s}{T_s - T_\infty} = \frac{hL}{k} = \frac{L/kA_s}{1/hA_s} = \frac{R_k}{R_h}$$

The ratio of the conduction resistance to the convection resistance defines the Biot number,

$$\text{Bi} = \frac{R_k}{R_h} = \frac{hL}{k} \tag{8.7}$$

and it has been found that Newtonian cooling requires that

$$\text{Bi} \leq 0.1$$

The Biot number should not be confused with the Nusselt number defined in Eq. (7.8). In the former, k is the thermal conductivity of the solid, and in the latter it is the thermal conductivity of the fluid. The ratio V/A_s occurring in Eq. (8.2) defines a characteristic length, L_c, which is used in calculating the Biot number. For a slab with two faces of area A_s and a thickness $2L$

$$L_c = \frac{V}{A_s} = \frac{2LA_s}{2A_s} = L \tag{8.8}$$

For a cylinder of radius R and length L,

$$L_c = \frac{V}{A_s} = \frac{\pi R^2 L}{2\pi R} = \frac{R}{2} \tag{8.9}$$

and for a sphere of radius R

$$L_c = \frac{V}{A_s} = \frac{\frac{4}{3}\pi R^3}{4\pi R^2} = \frac{R}{3} \tag{8.10}$$

Therefore, in Eq. (8.5),

$$\frac{t}{\tau} = \frac{hA_s t}{\rho V C_p} = \frac{ht}{\rho C_p L_c} = \frac{hL_c}{k}\frac{k}{\rho C_p}\frac{t}{L_c^2}$$

$$= \text{Bi} \times \frac{\alpha t}{L_c^2}$$

The normalized time, $\alpha t/L_c^2$, defines the Fourier number, Fo:

$$\text{Fo} = \frac{\alpha t}{L_c^2} \tag{8.11}$$

and hence Eq. (8.5) becomes

$$\frac{T - T_\infty}{T_i - T_\infty} = \exp(-\text{Bi} \cdot \text{Fo}) \tag{8.12}$$

EXAMPLE 8.1

An aluminum plate of dimensions 0.1 m × 0.5 m × 0.5 m at an initial temperature of 400°C is cooled in an airstream at 25°C. If, during cooling, $h = 400$ W/m²·K, calculate (a) the time required for the plate to cool to 50°C, (b) the initial rate of cooling, and (c) the rate of cooling after 3 min. For aluminum, in the range 25 to 400°C,

$$\rho = 2700 \text{ kg/m}^3$$
$$C_p = 900 \text{ J/kg·K}$$
$$k = 238 \text{ W/m}^2\text{·K}$$

With $L_c = 0.05$ m,

$$\text{Bi} = \frac{hL_c}{k} = \frac{400 \times 0.05}{238} = 0.084$$

which, being less than 0.1, indicates that Newtonian cooling can be assumed.

$$\text{Fo} = \frac{\alpha t}{L_c^2} = \frac{kt}{\rho C_p L_c^2} = \frac{238}{2700 \times 900 \times 0.05^2} t = 3.92 \times 10^{-2} t$$

(a) From Eq. (8.12),

$$\frac{T - T_\infty}{T_i - T_\infty} = \exp\left(-\text{Bi}\cdot\text{Fo}\right)$$

$$\frac{50 - 25}{400 - 25} = \exp\left(-0.084 \times 3.92 \times 10^{-2} t\right) \tag{i}$$

which gives $t = 823$ s (13.7 min).

(b) With $T = 400°C$, Eq. (8.2) gives the initial cooling rate as

$$\frac{dT}{dt} = -\frac{hA_s}{\rho V C_p}(T - T_\infty) = -\frac{h}{L_c \rho C_p}(T - T_\infty)$$

$$= \frac{-400(400 - 25)}{0.05 \times 2700 \times 900}$$

$$= 1.23°C/s \tag{ii}$$

(c) At $t = 3$ min $= 180$ s, the temperature of the plate is obtained from Eq. (i) as

$$\frac{T - 25}{400 - 25} = \exp\left(-0.084 \times 3.92 \times 10^{-2} \times 180\right)$$

or

$$T = 232°C$$

and from Eq. (ii), when $T = 232$,

$$\frac{dT}{dt} = \frac{-400(232 - 25)}{0.05 \times 2700 \times 900}$$

$$= 0.68°C/s$$

The maximum thickness of aluminium sheet that can be considered to undergo Newtonian cooling is calculated from

$$\text{Bi} = \frac{hL_c}{k} = \frac{400}{238} L_c \leq 0.1$$

as $L_c = 0.0595$ m or thickness $= 2L_c = 0.119$ m.

EXAMPLE 8.2 _____

After being annealed at 700°C, spherical steel balls of diameter 0.01 m are allowed to cool slowly in still air at 50°C under conditions in which the heat transfer coefficient is 20 W·m^{-2}·K^1. Calculate the time required for the temperature of the balls to fall to 100°C.

The data for steel are

$$C_p = 830 \text{ J(kg·K)}$$
$$\rho = 7800 \text{ kg/m}^3$$
$$k = 45 \text{ W/(m·K)}$$

From Eq. (8.10), the characteristic length, L_c, is

$$L_c = \frac{R}{3} = \frac{0.01}{2 \times 3} = 0.00167 \text{ m}$$

and thus the Biot number is

$$\text{Bi} = \frac{hL_c}{k} = \frac{20 \times 0.00167}{45} = 7.41 \times 10^{-4}$$

$$\alpha = \frac{k}{\rho C_p} = \frac{45}{7800 \times 830} = 6.95 \times 10^{-6}$$

and thus

$$\text{Fo} = \frac{\alpha t}{L_c^2} = \frac{6.95 \times 10^{-6}t}{0.00167^2} = 2.49t$$

Then, from Eq. (8.12),

$$\frac{T - T_\infty}{T_i - T_\infty} = \exp\left(-\text{Bi·Fo}\right)$$

or

$$\ln \frac{100 - 50}{700 - 50} = -2.565 = -7.41 \times 10^{-4} \times 2.49t$$

The required time is thus

$$t = 1390 \text{ s} = 23.2 \text{ min}$$

EXAMPLE 8.3 _____

Copper spheres of diameter 0.01 m at an initial uniform temperature of 360 K are dropped vertically into a tank of water at 300 K in which the depth of the water is 1 m. On the assumption that a sphere attains its constant terminal velocity as soon as it enters the water, calculate the temperature of a sphere when it reaches the bottom of the tank.

For water at 300 K,

$$\rho = 997 \text{ kg/m}^3$$
$$\nu = 8.6 \times 10^{-7} \text{ m}^2/\text{s}$$
$$k = 0.608 \text{ W·m}^{-1}\text{·K}^{-1}$$
$$\text{Pr} = 5.88$$

$$\eta_\infty \text{ (at 300 K)} = 8.57 \times 10^{-4} \text{ Pa·s}$$

$$\eta_s \text{ (at 360 K)} = 3.2 \times 10^{-4} \text{ Pa·s}$$

For copper,

$$\rho = 8933 \text{ kg/m}^3$$

$$k = 410 \text{ W·m}^{-1}\cdot\text{K}^{-1}$$

$$\alpha = 11.6 \times 10^{-5} \text{ m}^2/\text{s}$$

From Eq. (4.41),

$$\tfrac{4}{3}\pi R^3(\rho_s - \rho)g = f(\pi R^2)(\tfrac{1}{2}\rho v_t^2) \tag{4.41}$$

or

$$
\begin{aligned}
fv_t^2 &= \frac{8R(\rho_s - \rho)g}{3\rho} \\
&= \frac{8 \times 0.005 \times (8933 - 997) \times 9.81}{3 \times 997} \\
&= 1.04
\end{aligned}
$$

Assuming that Newton's law is valid (i.e., $f = 0.44$),

$$v_t = \sqrt{\frac{1.04}{0.44}} = 1.54 \text{ m/s}$$

and

$$\text{Re}_D = \frac{v_t D}{\nu} = \frac{1.54 \times 0.01}{8.6 \times 10^{-7}} = 1.79 \times 10^4$$

which being in the range 500 to 2×10^5 indicates that Newton's law is valid. The Reynolds number is also within the range of applicability of Eq. (7.64) for determination of the Nusselt number

$$
\begin{aligned}
\text{Nu}_D &= 2 + (0.4\text{Re}_D^{0.5} + 0.06\text{Re}_D^{0.667})\text{Pr}^{0.4}\left(\frac{\eta_\infty}{\eta_s}\right)^{0.25} \\
&= 2 + [0.4 \times (1.79 \times 10^4)^{0.5} + 0.06 \times (1.79 \times 10^4)^{0.667}] \\
&\quad \times 5.88^{0.4} \times \left(\frac{8.57}{3.2}\right)^{0.25} \\
&= 2 + 246 = 248 \\
&= \frac{h_D D}{k} \tag{7.64}
\end{aligned}
$$

Thus the heat transfer coefficient is

$$h_D = \frac{248 \times 0.608}{0.01} = 15,080 \text{ W·m}^{-2}\cdot\text{K}^{-1}$$

and the Biot number is

$$\mathrm{Bi} = \frac{h_D R}{3k} = \frac{15080 \times 0.005}{3 \times 410} = 0.061$$

which, being less than 0.1, indicates that Newtonian cooling occurs. The Fourier number is

$$\mathrm{Fo} = \frac{\alpha t}{L_c^2} = \frac{11.6 \times 10^{-5} t}{0.005^2/3} = 41.8t$$

and hence, from Eq. (8.12),

$$\frac{T - T_\infty}{T_i - T_\infty} = \exp(-\mathrm{Bi \cdot Fo}) = \exp(-0.061 \times 41.8t)$$

With a terminal velocity of 1.54 m/s, a sphere reaches the bottom of the tank in $1/1.54 = 0.649$ s, and hence the uniform temperature of the sphere when it reaches the bottom is

$$T = 300 + 60 \exp(-0.061 \times 41.8 \times 0.649)$$
$$= 311 \text{ K}$$

8.3 Non-Newtonian Cooling in Semi-infinite Systems

Analysis of non-Newtonian cooling by heat flow in one dimension requires the integration of the general heat conduction equation, Eq. (6.35), which for one-dimensional heat flow and no internal generation of heat is written as

$$\frac{\partial T}{\partial t} = \alpha \frac{\partial^2 T}{\partial x^2} \tag{8.13}$$

The variations of temperature with position and time in two semi-infinite, one-dimensional systems will be discussed. In a one-dimensional system, "semi-infinite" means that the system is sufficiently long that within any time interval of interest, the temperature at one end of the system is not influenced by variations in temperature caused by heat transfer at the other end. The following systems are examined:

1. One end of a semi-infinite rod, along which the temperature is uniformly T_i, is instantaneously raised to the higher constant temperature T_0 and heat flows into the rod.
2. One end of a semi-infinite rod, along which the temperature is uniformly T_i, is contacted by a fluid at the higher temperature T_∞ and heat is transferred to the rod by convection in which the heat transfer coefficient, h_0, is constant.

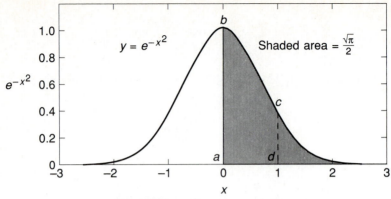

FIGURE 8.3. The "bell" curve.

The Error Function

The "bell curve," shown in Fig. 8.3 is generated by the equation

$$y = e^{-x^2}$$

The curve, which has $y = 1$ at $x = 0$ and $y \to 0$ as $x \to \pm\infty$, has the interesting property that it encloses an area equal to the square root of π:

$$\int_{-\infty}^{\infty} e^{-x^2}\, dx = \sqrt{\pi}$$

or for positive values of x,

$$\int_{0}^{\infty} e^{-x^2}\, dx = \frac{\sqrt{\pi}}{2}$$

The error function of x, erf(x), is defined as the ratio of the area under the curve between $x = 0$ and $x = x$ to the area under the curve between $x = 0$ and $x = \infty$. Thus in Fig. 8.3, the error function of 1 is the area $abcd$ divided by the crosshatched area. The area $abcd$ given by

$$\int_{0}^{1} e^{-x^2}\, dx$$

has the value 0.7468, and hence the error function of 1 is

$$\mathrm{erf}(1) = \frac{\displaystyle\int_{0}^{1} e^{-x^2}\, dx}{\displaystyle\int_{0}^{\infty} e^{-x^2}\, dx} = \frac{0.7648}{\sqrt{\pi}/2} = 0.8427$$

and generally,

$$\mathrm{erf}(x) = \frac{2}{\sqrt{\pi}} \int_{0}^{x} e^{-x^2}\, dx \tag{8.14}$$

Thus

$$\text{erf}(0) = 0 \quad \text{and} \quad \text{erf}(\infty) = 1$$

Furthermore,

$$\text{erf}(-x) = \frac{\displaystyle\int_0^{-x} e^{-x^2}\, dx}{\displaystyle\int_0^{\infty} e^{-x^2}\, dx} = \frac{-\displaystyle\int_0^{x} e^{-x^2}\, dx}{\displaystyle\int_0^{\infty} e^{-x^2}\, dx} = -\text{erf}(x)$$

The complement of the error function of x, $\text{erfc}(x)$, is defined by

$$\text{erf}(x) + \text{erfc}(x) = 1$$

or

$$\text{erfc}(x) = \frac{\displaystyle\int_x^{\infty} e^{-x^2}\, dx}{\displaystyle\int_0^{\infty} e^{-x^2}\, dx} = \frac{2}{\sqrt{\pi}} \int_x^{\infty} e^{-x^2}\, dx \tag{8.15}$$

and thus $\text{erfc}(0) = 1$ and $\text{erfc}(\infty) = 0$.

The error function and its complement, which cannot be expressed in exact closed analytical forms, are shown in Fig. 8.4. Traditionally, the values of $\text{erf}(x)$ and $\text{erfc}(x)$ have been obtained either from graphs such as Fig. 8.4 or from tabulations such as

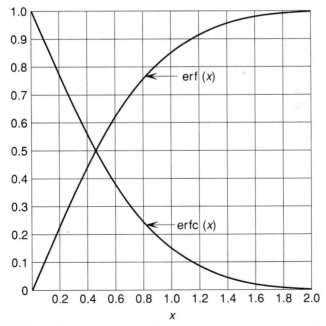

FIGURE 8.4.   Error function and the complementary error function of x.

TABLE 8.1

x	erfc(x)	Eq. (ii)	x	erfc(x)	Eq. (ii)
0.0	1.0	1.0	1.1	0.1198	0.1155
0.05	0.9436	0.9436	1.2	0.0897	0.0851
0.1	0.8875	0.8875	1.3	0.0660	0.0617
0.15	0.8320	0.8320	1.4	0.0477	0.0445
0.2	0.7773	0.7773	1.5	0.0339	0.0328
0.25	0.7237	0.7237	1.6	0.0237	0.0259
0.3	0.6714	0.6713	1.7	0.0162	
0.35	0.6206	0.6206	1.8	0.0109	
0.4	0.5716	0.5715	1.9	0.0072	
0.5	0.4795	0.4792	2.0	0.0047	
0.6	0.3962	0.3955	2.5	0.0004	
0.7	0.3222	0.3210	∞	0	
0.8	0.2579	0.2560			
0.9	0.2031	0.2004			
1.0	0.1573	0.1537			

Table 8.1. However, it is now possible to evaluate them by performing the integration in Eq. (8.14) on a hand-held calculator to obtain erf(x) and evaluating erfc(x) as $1 - \text{erf}(x)$. Alternatively, erf(x) is remarkably well approximated up to $x = 1.6$ by the simple expression

$$\text{erf}(x) \doteq \frac{2}{\sqrt{\pi}} \frac{3x}{x^2 + 3} \tag{i}$$

Equation (i) exhibits a maximum value at $x = 1.73$. The error function complement for values of $x \leq 1.6$ is thus approximated by

$$\text{erfc}(x) \doteq 1 - \frac{2}{\sqrt{\pi}} \frac{3x}{x^2 + 3} \tag{ii}$$

The values of erfc(x) and the approximations given by Eq. (ii) are listed in Table 8.1.

Heat Flow in Semi-infinite Systems

One end of a semi-infinite rod, along which the temperature is uniformly T_i, is instantaneously increased to the higher constant temperature T_0. The initial condition is thus

$$T = T_i \qquad \text{at } t = 0 \quad \text{and} \quad x \geq 0$$

and the boundary condition is

$$T = T_0 \qquad \text{at } x = 0 \quad \text{and} \quad t > 0$$

The general heat conduction equation for one-dimensional heat flow,

$$\frac{\partial T}{\partial t} = \alpha \frac{\partial^2 T}{\partial x^2} \tag{8.13}$$

shows that T is a function of both x and t (i.e., the local temperature at any position varies with time) and at any time the temperature is a function of position. It is thus necessary to find a single variable that is an appropriate combination of x and t which will convert Eq. (8.13) from a partial differential equation to an ordinary differential equation. This function is

$$\eta = \frac{x}{2\sqrt{\alpha t}} \tag{8.16}$$

which is seen to be dimensionless.
  Equation (8.16) gives

$$\frac{\partial \eta}{\partial x} = \frac{1}{2\sqrt{\alpha t}} \quad \text{and} \quad \frac{\partial \eta}{\partial t} = -\frac{x}{4\sqrt{\alpha t^3}} \tag{i}$$

and thus

$$\frac{\partial T}{\partial t} = \frac{dT}{d\eta}\frac{\partial \eta}{\partial t} = -\frac{x}{4\sqrt{\alpha t^3}}\left(\frac{dT}{d\eta}\right)$$

and

$$\frac{\partial^2 T}{\partial x^2} = \frac{\partial}{\partial x}\left[\frac{dT}{d\eta}\left(\frac{\partial \eta}{\partial x}\right)\right] = \frac{1}{4\alpha t}\left(\frac{d^2 T}{d\eta^2}\right)$$

substitution of which into Eq. (8.13) gives

$$\frac{-x}{4\sqrt{\alpha t^3}}\left(\frac{dT}{d\eta}\right) = \frac{1}{4t}\left(\frac{d^2 T}{d\eta^2}\right)$$

or

$$\frac{dT}{d\eta} = -\frac{\sqrt{\alpha t}}{x}\left(\frac{d^2 T}{d\eta^2}\right)$$

or

$$\frac{dT}{d\eta} = -\frac{1}{2\eta}\left(\frac{d^2 T}{d\eta^2}\right) \tag{ii}$$

As both x and t have been eliminated, Eq. (ii) is an ordinary differential equation. Setting

$$y = \frac{dT}{d\eta}$$

Eq. (ii) can be rewritten as

$$y = -\frac{1}{2\eta}\left(\frac{dy}{d\eta}\right)$$

or

$$-2\eta \, d\eta = \frac{dy}{y}$$

integration of which gives

$$-\eta^2 = \ln y + \text{constant}$$

or if the constant is set equal to $-\ln A$,

$$y = \frac{dT}{d\eta} = A \exp(-\eta^2)$$

A second integration gives

$$\int dT = A \int e^{-\eta^2} d\eta \tag{iii}$$

and setting the limits as T_i at $\eta = \infty$ ($t = 0$, any x) and T_0 at $\eta = 0$ ($x = 0, t > 0$) gives

$$T_i - T_0 = A \int_0^\infty e^{-\eta^2} d\eta \tag{iv}$$

From Fig. 8.3, the evaluated integral is $\sqrt{\pi}/2$ and hence

$$A = \frac{2}{\sqrt{\pi}} (T_i - T_0)$$

Setting the limits for integration of Eq. (iii) as T at η and $T = T_i$ at $\eta = \infty$ gives

$$T - T_i = \frac{2}{\sqrt{\pi}} (T_i - T_0) \int_\infty^\eta e^{-\eta^2} d\eta$$

$$= -\frac{2}{\sqrt{\pi}} (T_i - T_0) \int_\eta^\infty e^{-\eta^2} d\eta \tag{v}$$

From the definition of the error function compliment given by Eq. (8.15),

$$\text{erfc}(\eta) = \frac{2}{\sqrt{\pi}} \int_\eta^\infty e^{-\eta^2} d\eta$$

and hence Eq. (v) becomes

$$\frac{T - T_i}{T_0 - T_i} = \text{erfc}(\eta) = \text{erfc}\left(\frac{x}{2\sqrt{\alpha t}}\right) \tag{8.17}$$

which is the complementary error function shown in Fig. 8.4. If $T_0 < T_i$, the normalized temperature is

$$\frac{T - T_0}{T_i - T_0} = 1 - \frac{T - T_i}{T_0 - T_i}$$

$$= 1 - \text{erfc}(\eta)$$

$$= \text{erf}(\eta) \tag{8.18}$$

which is the error function shown in Fig. 8.4.

EXAMPLE 8.4

Consider the application of Eq. (8.17) to the one-dimensional conduction of heat in a semi-infinite bar of steel of $\alpha = 1.2 \times 10^{-5}$ m^2/s under the conditions $T_i = 25°C$, $T_0 = 100°C$. (a) What is the temperature at point 0.1 m from the face of the bar ($x = 0.1$ m) after 5 min? (b) After what time interval has the temperature at $x = 0.01$ m increased to 75°C?

(a) With $x = 0.1$ m and $t = 5 \times 60 = 300$ s,

$$\eta = \frac{x}{2\sqrt{\alpha t}} = \frac{0.1}{2(1.2 \times 10^{-5} \times 300)^{1/2}} = 0.833$$

From Fig. 8.4, erfc(0.833) is approximately 0.23, or from integration, by hand-held calculator

$$\text{erfc}(0.833) = 1 - \frac{2}{\sqrt{\pi}} \int_0^{0.833} e^{-\eta^2} \, d\eta = 0.239$$

and hence

$$\frac{T - 25}{100 - 25} = 0.239$$

which gives $T = 42.9°C$.

(b) With $T = 75°C$,

$$\frac{T - T_i}{T_0 - T_i} = \frac{75 - 25}{100 - 25} = 0.667$$

and from Fig. 8.4, erfc(η) = 0.667 gives $\eta = 0.3$. Therefore,

$$\eta = 0.3 = \frac{x}{2\sqrt{\alpha t}} = \frac{0.01}{2(1.2 \times 10^{-5}t)^{1/2}}$$

which gives $t = 23$ s.

Figure 8.5 shows the temperature profiles in the steel rod occurring at 1, 5, and 10 min, and Fig. 8.6 shows the temperatures at $x = 0.01, 0.025,$ and 0.05 m as functions of time. From Fig. 8.4, considered as the variation of erfc(η) with η, $\eta = 0.5$ gives erfc(0.5) = 0.48, which is close to 0.50. A value of 0.5 for η requires that

$$x = \sqrt{\alpha t} \tag{vi}$$

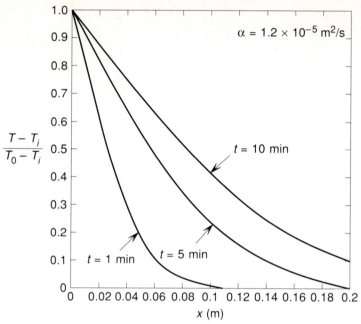

FIGURE 8.5.   Normalized temperature profiles in the semi-infinite steel rod after the three times considered in Example 8.4.

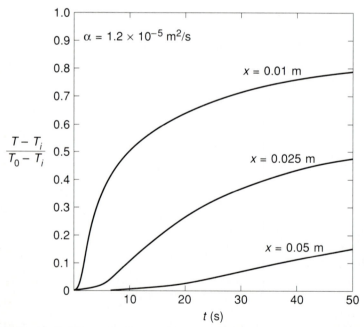

FIGURE 8.6.   Variations, with time, of the normalized temperatures at three locations in the semi-infinite steel rod considered in Example 8.4.

and hence Eq. (vi) is a good approximation of the time required for the value of $(T - T_i)/(T_0 - T_i)$ to reach 0.5 at any value of x, or the value of x at which $(T - T_i)/(T_0 - T_i)$ equals 0.5 in time t. In the present example, $(T - T_i)/(T_0 - T_i) = 0.5$ corresponds to $T = 62.5°C$, and hence from

$$x = \sqrt{1.2 \times 10^{-5}} \, t^{1/2} = 3.46 \times 10^{-3} t^{1/2}$$

the time required for the temperature at $x = 0.1$ m to reach $62.5°C$ is

$$t = \left(\frac{0.1}{3.46 \times 10^{-3}} \right)^2 = 835 \text{ s} = 13.9 \text{ min}$$

and at $t = 5$ min $= 300$ s, the local temperature is $62.5°C$ at

$$x = 3.46 \times 10^{-3}(300)^{1/2} = 0.06 \text{ m}$$

The second semi-infinite system to be examined is that in which one end of a semi-infinite rod, along which the temperature is uniformly T_i, is contacted by a fluid at the higher temperature T_∞ and heat is transferred to the rod by convection. In this case Eq. (8.13) has to be integrated using the initial condition

$$T = T_i \text{ at } t = 0 \quad \text{and} \quad x \geq 0$$

and the boundary condition

$$q'_x = -k\frac{\partial T}{\partial x} = h_0(T_0 - T) \quad \text{at } x = 0 \quad \text{and} \quad t \geq 0$$

The solution is

$$\frac{T - T_i}{T_0 - T_i} = \text{erfc}(\eta) - \exp[\lambda(2\eta + \lambda)]\text{erfc}(\eta + \lambda) \tag{8.19}$$

where

$$\lambda = \frac{h_0}{k}\sqrt{\alpha t}$$

EXAMPLE 8.5

One end of a semi-infinite steel bar of $k = 42$ W/m·K and $\alpha = 1.2 \times 10^{-5} \text{m}^2/\text{s}$, which is initially $25°C$ is placed in contact with a fluid at $400°C$ and $h_0 = 100$ W/m²·K. Figure 8.7 shows the temperature profiles in the bar after 1, 5, 10, and 30 min obtained from Eq. (8.19) and Fig. 8.8 shows the corresponding variations with time at the values of $x = 0$ and 0.1 m.

With $x = 0$, Eq. (8.18) is simplified to

$$\frac{T - T_i}{T_0 - T_i} = 1 - e^{\lambda^2} \text{erfc}(\lambda)$$

and points a, b, c, and d in Figs. 8.7 and 8.8 are the surface temperature at, respectively, 1, 5, 10, and 30 min.

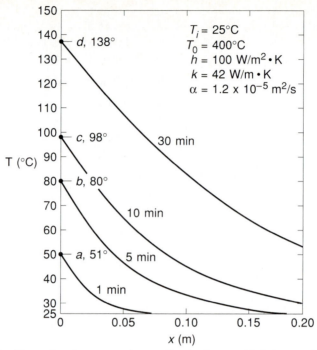

FIGURE 8.7. Normalized temperature profiles in the semi-infinite steel rod after the four times considered in Example 8.5.

8.4 Non-Newtonian Cooling in a One-Dimensional Finite System

The intractable mathematics required for the integration of the general heat conduction equation can be avoided by using the finite-difference method to determine the variation of temperature with time and position during the heating or cooling of a finite system. The cooling of a plate of finite thickness involves one-dimensional heat flow for which the general heat conduction equation, Eq. (6.38), reduces to

$$\frac{\partial T}{\partial t} = \alpha \frac{\partial^2 T}{\partial x^2} \tag{8.20}$$

In Fig. 8.9 nodes are placed at a distance Δx apart across the thickness, L, of the plate. If, initially, the temperature is uniform throughout the plate, and if once heating or cooling is started, the heat transfer conditions are the same on both sides of the plate, the temperature profile in the plate is symmetrical about the centerline and hence only one half of the thickness need be considered. The thermal energy balance for each node requires that the net rate at which heat is transported into the node be equal to the rate at which thermal energy (enthalpy) is accumulated in the node.

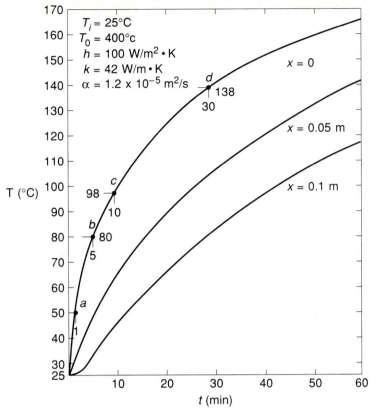

FIGURE 8.8. Variations, with time, of the normalized temperatures at three locations in the semi-infinite steel rod considered in Example 8.5. Points *a*, *b*, *c*, and *d* correspond to the points shown in Fig. 8.7.

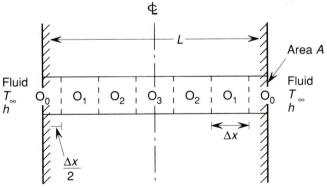

FIGURE 8.9. Location of nodes for the calculation, by the finite-difference technique, of the rate of heat transport in one dimension.

With reference to node 0 located at the surface, heat is transported from the fluid to the node 0 by convection at the rate

$$hA(T_\infty - T_0)$$

and heat is transported from node 1 to node 0 by conduction at the rate

$$-kA \frac{T_0 - T_1}{\Delta x}$$

In these expressions, if $T_0 > T_\infty$, heat is transported from node 0 to the fluid, and if $T_0 > T_1$, heat is transported by conduction from node 0 to node 1. The net rate at which thermal energy is accumulated in node 0 is thus

$$hA(T_\infty - T_0) - kA \frac{T_0 - T_1}{\Delta x}$$

In the finite-difference method, just as distance is broken down into discrete intervals of Δx, the passage of time is broken into discrete increments of Δt. Thus the passage of time t is considered as the passage of p intervals of Δt,

$$t = p\, \Delta t$$

so that the temperature of the node i is T_i^p after p increments of Δt and is T_i^{p+1} after $(p + 1)$ increments of time. Thus the rate of accumulation of thermal energy in the node i, which is given by the partial differential

$$\text{mass} \times \frac{\partial H_i}{\partial t} = \rho V C_p \frac{\partial T_i}{\partial t}$$

is approximated by the increase or decrease in the energy stored in the node during the increment of time from p to $p + 1$,

$$\rho V C_p \frac{\partial T_i}{\partial t} \doteq \rho V C_p \frac{T_i^{p+1} - T_i^p}{\Delta t}$$

The energy balance for node 0 is thus

$$hA(T_\infty - T_0^p) - kA \frac{T_0^p - T_1^p}{\Delta x} = \rho \frac{\Delta x}{2} A C_p \frac{T_0^{p+1} - T_0^p}{\Delta t} \tag{i}$$

Noting that $\alpha = k/\rho C_p$, Eq. (i) can be rearranged as

$$T_0^{p+1} = 2 \frac{\alpha\, \Delta t}{(\Delta x)^2} \frac{h\, \Delta x}{k} T_\infty + 2 \frac{\alpha\, \Delta t}{(\Delta x)^2} T_1^p$$

$$+ T_0^p \left[1 - 2 \frac{h\, \Delta x}{k} \frac{\alpha\, \Delta t}{(\Delta x)^2} - 2 \frac{\alpha\, \Delta t}{(\Delta x)^2} \right] \tag{8.21}$$

From comparison with Eqs. (8.7) and (8.11), the term

$$\frac{h\, \Delta x}{k} \equiv \frac{hL_c}{k}$$

in Eq. (8.21) is equivalent to a Biot number, and the term

$$\frac{\alpha \, \Delta t}{(\Delta x)^2} \equiv \frac{\alpha t}{L_c^2}$$

is equivalent to a Fourier number. Consequently, the quantities $h \, \Delta x/k$ and $\alpha \, \Delta t/(\Delta x)^2$ are referred to, respectively, as the mesh Biot number and the mesh Fourier number.

Mathematical stability of Eq. (8.21) requires that the third term on the right-hand side be greater than, or equal to, zero:

$$1 - 2 \left[\frac{h \, \Delta x}{k} \right] \left[\frac{\alpha \, \Delta t}{(\Delta x)^2} \right] - 2 \left[\frac{\alpha \, \Delta t}{(\Delta x)^2} \right] \geq 0$$

or

$$\frac{(\Delta x)^2}{\alpha \, \Delta t} \geq 2 \left(\frac{h \, \Delta x}{k} + 1 \right) \tag{8.22}$$

which places a limit on the allowed magnitude of Δt when a mesh size Δx has been selected.

The energy balance for the internal nodes involve only heat transport by conduction. For node 1, the rate at which heat is conducted in from node 0 is

$$-kA \, \frac{T_1^p - T_0^p}{\Delta x}$$

the rate at which heat is conducted in from node 2 is

$$-kA \, \frac{T_1^p - T_2^p}{\Delta x}$$

and the energy accumulated in the time increment from p to $p + 1$ is

$$\rho \, \Delta x \, AC_p \, \frac{T_1^{p+1} - T_1^p}{\Delta t}$$

Therefore,

$$-kA \, \frac{T_1^p - T_0^p}{\Delta x} - kA \, \frac{T_1^p - T_2^p}{\Delta x} = \rho \, \Delta x A C_p \, \frac{T_1^{p+1} - T_1^p}{\Delta t}$$

or

$$T_1^{p+1} = T_1^p + \frac{\alpha \, \Delta t}{(\Delta x)^2} \, (T_0^p + T_2^p - 2T_1^p) \tag{8.23}$$

Similarly, for node 2,

$$T_2^{p+1} = T_2^p + \frac{\alpha \, \Delta t}{(\Delta x)^2} \, (T_1^p + T_3^p - 2T_2^p) \tag{8.24}$$

and for node 3,

$$T_3^{p+1} = T_3^p + \frac{\alpha \, \Delta t}{(\Delta x)^2} (2T_2^p - 2T_3^p) \tag{8.25}$$

Equations (8.21) to (8.35) allow calculation of the variation of the temperatures of the nodes by marching forward in time increments of Δt, The values of T_i^{p+1} calculated are substituted back into the right-hand sides of the equations and the values of T_i^{p+2} are calculated, and so on.

EXAMPLE 8.6

In Example 8.1, an aluminum plate of thickness 0.1 m at an initial uniform temperature of 400°C was cooled in an airstream at 25°C. In this example the Biot number was less than 0.1, and hence the assumption of Newtonian cooling was valid. The cooling of the plate will be analyzed again using the finite-difference method. The required data are

$$\rho = 2700 \text{ kg/m}^3$$

$$C_p = 900 \text{ J/kg·K}$$

$$k = 238 \text{ W/m·K}$$

$$\alpha = 9.79 \times 10^{-5} \text{ m}^2/\text{s}$$

$$h = 400 \text{ W/m}^2\cdot\text{K}$$

$$T_\infty = 25°C$$

With $L = 0.1$ m and seven nodes placed as shown in Fig. 8.10, $\Delta x = 0.01667$ m. Therefore, the mesh Biot number is

$$\frac{h \, \Delta x}{k} = \frac{400 \times 0.01667}{238} = 0.028$$

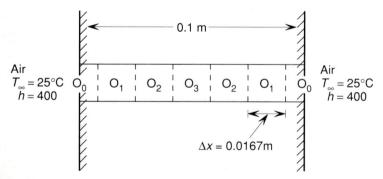

FIGURE 8.10. Nodes used for calculation of one-dimensional heat flow in the aluminum plate considered in Example 8.6.

and hence, from Eq. (8.22), the mesh Fourier number is restricted as

$$\frac{(\Delta x)^2}{\alpha\ \Delta t} \geq 2(1\ +\ 0.028)\ =\ 2.056$$

or

$$\Delta t \leq \frac{(\Delta x)^2}{2.056\alpha} = \frac{(0.01667)^2}{2.056\ \times\ 9.79\ \times\ 10^{-5}} = 1.38\ \text{s}$$

The time increment, Δt, can thus be conveniently selected as 1 s, in which case the mesh Fourier number is

$$\frac{\alpha\ \Delta t}{(\Delta x)^2} = \frac{9.79\ \times\ 10^{-5}\ \times\ 1}{(0.01667)^2} = 0.352$$

TABLE 8.2

For $p = 0$ to $p = 1$; $t = \Delta t = 1$ s:

$T_0^{(1)} = 400 + 2 \times 0.352(400 + 0.028 \times 25 - 1.028 \times 400) = 392.6°C$

$T_1^{(1)} = 400 + 0.352(400 + 400 - 800)$ $\qquad\qquad\qquad = 400°C$

$T_2^{(1)} = 400 + 0.352(400 + 400 - 800)$ $\qquad\qquad\qquad = 400°C$

$T_3^{(1)} = 400 + 0.352(800 - 800)$ $\qquad\qquad\qquad\qquad\quad = 400°C$

For $p = 1$ to $p = 2$; $t = 2\Delta t = 2$ s:

$T_0^{(2)} = 392.6 + 2 \times 0.352(400 + 0.028 \times 25 - 1.028 \times 392.6) = 390.6°C$

$T_1^{(2)} = 400 + 0.352(392.6 + 400 - 800)$ $\qquad\qquad\qquad = 397.4°C$

$T_2^{(2)} = 400 + 0.352(400 + 400 - 800)$ $\qquad\qquad\qquad = 400°C$

$T_3^{(2)} = 400 + 0.352(800 - 800)$ $\qquad\qquad\qquad\qquad\quad = 400°C$

For $p = 2$ to $p = 3$; $t = 3\ \Delta t = 3$ s:

$T_0^{(3)} = 390.6 + 2 \times 0.352(397.4 + 0.028 \times 25 - 1.028 \times 390.6) = 388.2°C$

$T_1^{(3)} = 397.4 + 0.352(390.6 + 400 - 2 \times 397.4)$ $\qquad\qquad = 395.9°C$

$T_2^{(3)} = 400 + 0.352(397.4 + 400 - 800)$ $\qquad\qquad\qquad = 399.1°C$

$T_3^{(3)} = 400 + 0.352(800 - 800)$ $\qquad\qquad\qquad\qquad\quad = 400°C$

For $p = 3$ to $p = 4$; $t = 4\ \Delta t = 4$ s:

$T_0^{(4)} = 388.2 + 2 \times 0.352(395.9 + 0.028 \times 25 - 1.028 \times 388.2) = 386.5°C$

$T_1^{(4)} = 395.9 + 0.352(388.2 + 399.1 - 2 \times 395.9)$ $\qquad\qquad = 394.3°C$

$T_2^{(4)} = 399.1 + 0.352(395.9 + 400 - 2 \times 399.1)$ $\qquad\qquad = 398.3°C$

$T_3^{(4)} = 400 + 0.352(2 \times 399.1 - 2 \times 400)$ $\qquad\qquad\quad = 399.4°C$

TABLE 8.3

t (s)	T_0	T_1	T_2	T_3	$T_{\text{Newtonian}}$
0	400	400	400	400	400
11	377.3	385.6	390.5	392.2	386.7
50	335.9	343.2	347.6	349.1	343.1
100	289.9	296.1	299.9	301.1	294.8
150	250.7	256.3	258.5	259.5	253.8
200	217.3	221.8	224.6	225.5	219.1
300	164.6	167.9	169.9	170.5	164.6
400	126.4	128.7	130.2	130.6	125.5
500	98.6	100.3	101.3	101.7	97.3
600	78.4	79.7	80.4	80.7	77.0
800	53.2	53.8	54.2	54.3	44.4
1000	39.9	40.2	40.4	40.5	38.9

and insertion of the values of the mesh Biot and Fourier numbers and T_∞ into Eqs. (8.21) to (8.25) gives

$$T_0^{p+1} = T_0^p + 2 \times 0.352(T_1^p + 0.028 \times 25 - 1.028T_0^p)$$
$$T_1^{p+1} = T_1^p + 0.352(T_0^p + T_2^p - 2T_1^p)$$
$$T_2^{p+1} = T_2^p + 0.352(T_1^p + T_3^p - 2T_2^p)$$
$$T_3^{p+1} = T_3^p + 0.352(2T_2^p - 2T_3^p)$$

For the first four increments of time the temperatures are calculated as shown in Table 8.2. The results obtained using a computer program are listed in Table 8.3.

The difference between T_3 and T_0 increases with time to a maximum of 14.9°C after 11 seconds and thereafter decreases. Thus during most of the cooling time the temperature variation in the cross section is small and the average temperature is very close to $T_{\text{Newtonian}}$ calculated, assuming Newtonian cooling, from Eq. (i) in Example 8.1.

EXAMPLE 8.7

Figure 8.11 shows 11 nodes, with $\Delta x = 0.03$ m, placed in an aluminum plate of thickness 0.3 m. From the data given in Example 8.6, the Biot number is

$$\text{Bi} = \frac{hL_c}{k} = \frac{400 \times 0.15}{238} = 0.252$$

which, being greater than 0.1, indicates that Newtonian cooling does not occur. In a finite-difference analysis the mesh Biot number is

$$\frac{h\,\Delta x}{k} = \frac{400 \times 0.03}{238} = 0.0504$$

and hence

$$\frac{(\Delta x)^2}{\alpha \, \Delta t} \geq 2 + 2 \times 0.0504 = 2.1008$$

and

$$\Delta t \leq \frac{(\Delta x)^2}{2.1008\alpha} = \frac{(0.03)^2}{2.1008 \times 9.79 \times 10^{-5}} = 4.38 \text{ s}$$

Selecting $\Delta t = 1$ s gives the mesh Fourier number as

$$\frac{\alpha \, \Delta t}{(\Delta x)^2} = \frac{9.79 \times 10^{-5} \times 1}{(0.03)^2} = 0.1088$$

The temperature profiles obtained by the finite-difference method at $t = 50, 100, 200, 500,$ and 1000 are shown in Fig. 8.12.
   If the plate is stainless steel with the properties

$$\alpha = 4.0 \times 10^{-6} \text{ m}^2/\text{s}$$

$$k = 18 \text{ W/m·K}$$

the Biot number is

$$\text{Bi} = \frac{hL_c}{k} = \frac{400 \times 0.15}{18} = 3.33$$

which shows that because of the relatively low thermal conductivity, the resistance to heat transport by conduction is significantly greater than that by convection and, consequently, significant temperature gradients can be expected to develop in the plate during cooling. The mesh Biot number is

$$\frac{h \, \Delta x}{k} = \frac{400 \times 0.03}{18} = 0.667$$

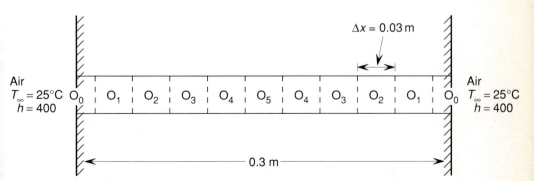

FIGURE 8.11. Locations of the nodes used in Example 8.7.

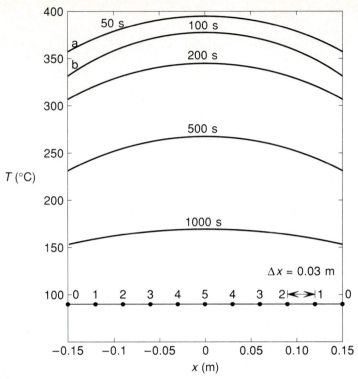

FIGURE 8.12.   Temperature profiles in the aluminum plate calculated in Example 8.7.

and hence

$$\Delta t \leq \frac{(0.03)^2}{0.667 \times 4.0 \times 10^{-6}} = 337 \text{ s}$$

Selecting $\Delta t = 1$ s gives the mesh Fourier number as

$$\frac{\alpha \Delta t}{(\Delta x)^2} = \frac{4.0 \times 10^{-6} \times 1}{(0.03)^2} = 0.00444$$

The temperature profiles at $t = 500$ and 1000 s are shown in Fig. 8.13, and comparison of Figs. 8.12 and 8.13 illustrates the influence of the magnitude of the Biot number on the temperature gradients which are developed in a slab during cooling.

In Example 8.6, the criterion for mathematical stability of Eq. (8.21), given by Eq. (8.22), required that $\Delta t \leq 4.38$ s. Figure 8.14 shows the variation with time of T_0 calculated by finite difference with $\Delta t = 1$ s (as in Example 8.6) and with $\Delta t = 5$ s. With Δt greater than the allowed maximum value the calculated temperatures oscillate with an increasing amplitude as time increases. Points a and b in Fig. 8.12 are the same as points a and b in Fig. 8.14.

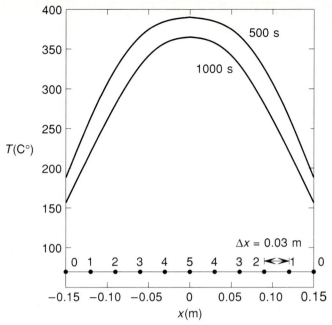

FIGURE 8.13. Temperature profiles in the stainless steel plate calculated in Example 8.7.

In the preceding discussion of the forward difference method the energy balance on a node produced an equation in which the temperature of a node after $p + 1$ increments of Δt is determined only by the temperature of the node and temperatures of its neighboring nodes after p increments of Δt. The equation thus contains only one unknown. In an alternative approach, called the backward difference approach, the energy balance leading to Eq. (8.23), namely,

$$-kA\frac{T_1^p - T_0^p}{\Delta x} - kA\frac{T_1^p - T_2^p}{\Delta x} = \rho\,\Delta x\,AC_p\frac{T_1^{p+1} - T_1^p}{\Delta t}$$

can be written as

$$-kA\frac{T_1^{p+1} - T_0^{p+1}}{\Delta x} - kA\frac{T_1^{p+1} - T_2^{p+1}}{\Delta x} = \rho\,\Delta x AC_p\frac{T_1^{p+1} - T_1^p}{\Delta t}$$

or

$$-T_0^{p+1} + \left[2 + \frac{(\Delta x)^2}{\alpha\,\Delta t}\right]T_1^{p+1} - T_2^{p+1} = \frac{(\Delta x)^2}{\alpha\,\Delta t}T_1^p$$

for node 1 in Fig. 8.9. Similarly for node 2,

$$-T_1^{p+1} + \left[2 + \frac{(\Delta x)^2}{\alpha\,\Delta t}\right]T_2^{p+1} - T_3^{p+1} = \frac{(\Delta x)^2}{\alpha\,\Delta t}T_2^p$$

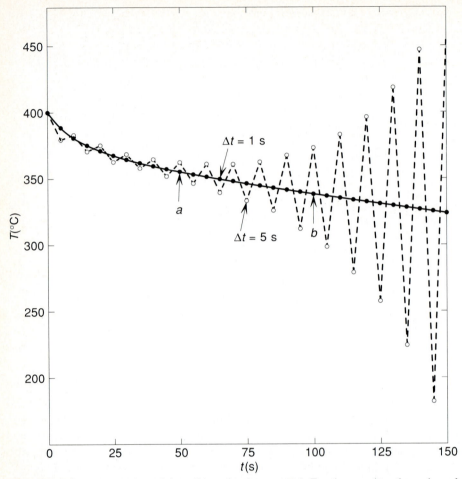

FIGURE 8.14.   Illustration of the effect of using a mesh Fourier number the value of which is larger than the maximum allowed by the criterion for mathematical stability.

for the central node 3,

$$-2T_2^{p+1} + \left[2 + \frac{(\Delta x)^2}{\alpha\,\Delta t}\right] T_3^{p+1} = \frac{(\Delta x)^2}{\alpha\,\Delta t}\, T_3^p$$

and for the surface node 0,

$$hA(T_\infty - T_0^{p+1}) - kA\,\frac{T_0^{p+1} - T_1^{p+1}}{\Delta x} = \rho A\,\frac{\Delta x}{2}\, C_p\,\frac{T_0^{p+1} - T_0^p}{\Delta t}$$

or

$$\left[1 + \frac{h\,\Delta x}{k} + \frac{(\Delta x)^2}{2\alpha\,\Delta t}\right] T_0^{p+1} - T_1^{p+1} = \frac{(\Delta x)^2}{2\alpha\,\Delta t}\, T_0^p + \frac{h\,\Delta x}{k}\, T_\infty$$

Thus, in the backward difference approach, the energy balances for n nodes produce a set of n equations with n unknown temperatures which can be solved simultaneously by the matrix inversion method. Application to the system dealt with by the forward difference approach in Example 8.6, in which $h \, \Delta x / k = 0.0504$ and $\alpha \, \Delta t / (\Delta x)^2 = 0.1088$, gives

$$\left(1 + 0.0504 + \frac{1}{2 \times 0.1088} \right) T_0^{p+1} - T_1^{p+1} = \frac{400}{2 \times 0.1088} + 0.0504 \times 25$$

or

$$5.646 T_0^{p+1} - T_1^{p+1} = 1839 \tag{i}$$

for node, 0,

$$- T_0^{p+1} + \left(2 + \frac{1}{0.1088} \right) T_1^{p+1} - T_2^{p+1} = \frac{400}{0.1088}$$

or

$$- T_0^{p+1} + 11.19 T_1^{p+1} - T_2^{p+1} = 3676 \tag{ii}$$

for node 1,

$$- T_1^{p+1} + 11.19 T_2^{p+1} - T_3^{p+1} = 3676 \tag{iii}$$

for node 2, and

$$- 2 T_2^{p+1} + 11.19 T_3^{p+1} = 3676 \tag{iv}$$

for node 3. Equations (i) to (iv) can be arranged as

$$
\begin{aligned}
5.646 T_0^{p+1} - \quad & T_1^{p+1} & & & = 1839 \\
- \quad T_0^{p+1} + 11.19 T_1^{p+1} - \quad & T_2^{p+1} & & & = 3676 \\
- \quad & T_1^{p+1} + 11.19 T_2^{p+1} - \quad & T_3^{p+1} & = 3676 \\
- \quad & 2 T_2^{p+1} + 11.19 T_3^{p+1} & & = 3676
\end{aligned}
$$

Solution by matrix inversion (i.e., $[A][T] = [C]$), where

$$[A] = \begin{bmatrix} 5.646 & -1 & 0 & 0 \\ -1 & 11.19 & -1 & 0 \\ 0 & -1 & 11.19 & -1 \\ 0 & 0 & -2 & 11.19 \end{bmatrix}$$

and

$$[C] = \begin{bmatrix} 1839 \\ 3676 \\ 3676 \\ 3676 \end{bmatrix}$$

TABLE 8.4

	By Forward Difference				By Backward Difference			
	T_0	T_1	T_2	T_3	T_0	T_1	T_2	T_3
0	400	400	400	400	400	400	400	400
1	392.6	400	400	400	394.6	398.8	399.7	399.9
2	390.6	397.4	400	400	391.5	397.4	399.2	399.6
3	388.2	395.9	399.1	400	389.1	395.9	398.5	399.2
4	386.5	394.3	398.3	399.4	387.1	394.5	397.7	398.6

gives

$$T_0^{p+1} = 394.6°C$$

$$T_1^{p+1} = 398.8°C$$

$$T_2^{p+1} = 399.7°C$$

$$T_3^{p+1} = 399.9°C$$

as the temperatures of the nodes after $t = \Delta t = 1$ s. These values are then substituted into [C] and the temperatures at $t = 2\Delta t = 2$ s are determined by a second matrix inversion, and so on. The temperatures calculated after the first 1, 2, 3 and 4 s are compared in Table 8.4 with the corresponding temperatures calculated by the forward difference listed in Table 8.2. There is no limitation on the magnitude of the mesh Fourier number in the backward difference technique, and hence Δx and Δt can be selected independently.

8.5 Non-Newtonian Cooling in a Two-Dimensional Finite System

The finite-difference equations for transient behavior in two-dimensional systems are obtained by including a term for the rate of accumulation of thermal energy in the equations derived in Section 6.7. Thus for a mesh size of $\Delta x \, \Delta y$ and a depth of $\Delta z = 1$, the energy balance for an internal node shown in Fig. 6.23, in which there is no heat generation, is obtained from the derivation leading to Eq. (6.40) as

$$-k \, \Delta x \, \frac{T^p_{m,n} - T^p_{m,n+1}}{\Delta y} - k \, \Delta y \, \frac{T^p_{m,n} - T^p_{m-1,n}}{\Delta x} - k \, \Delta x \, \frac{T^p_{m,n} - T^p_{m,n-1}}{\Delta y}$$

$$- k \, \Delta y \, \frac{T^p_{m,n} - T^p_{m+1,n}}{\Delta x} = \rho \, \Delta x \, \Delta y \, C_p \, \frac{T^{p+1}_{m,n} - T^p_{m,n}}{\Delta t} \quad (8.26)$$

which for a square mesh of $\Delta x = \Delta y$ gives

$$T^{p+1}_{m,n} = T^p_{m,n} + \frac{\alpha \, \Delta t}{(\Delta x)^2} \, (T^p_{m,n+1} + T^p_{m-1,n}$$

$$+ T^p_{m,n-1} + T^p_{m+1,n} - 4T^p_{m,n}) \quad (8.27)$$

Similarly, for the surface node in Fig. 6.26(a), the derivation leading to Eq. (6.41) is adapted to give

$$-k \frac{\Delta x}{2} \frac{T^p_{m,n} - T^p_{m,n+1}}{\Delta y} - k \Delta y \frac{T^p_{m,n} - T^p_{m-1,n}}{\Delta x} - k \frac{\Delta y}{2} \frac{T^p_{m,n} - T^p_{m,n-1}}{\Delta y}$$

$$+ h \Delta y(T_\infty - T^p_{m,n}) = \rho \Delta y \frac{\Delta x}{2} C_p \frac{T^{p+1}_{m,n} - T^p_{m,n}}{\Delta t} \qquad (8.28)$$

or with $\Delta x = \Delta y$,

$$T^{p+1}_{m,n} = \frac{\alpha \Delta t}{(\Delta x)^2} \left(T^p_{m,n+1} + T^p_{m,n-1} + 2T^p_{m-1,n} + \frac{2h \Delta x}{k} T_\infty \right)$$

$$+ T^p_{m,n} \left[1 - 2 \frac{h \Delta x}{k} \frac{\alpha \Delta t}{(\Delta x)^2} - 4 \frac{\alpha \Delta t}{(\Delta x)^2} \right] \qquad (8.29)$$

For the node at an inner corner shown in Fig. 6.26(b),

$$T^{p+1}_{m,n} = \frac{2\alpha \Delta t}{3(\Delta,x)^2} \left(2T^p_{m,n+1} + 2T^p_{m-1,n} + T^p_{m,n-1} + T^p_{m+1,n} + \frac{2h \Delta x}{k} T_\infty \right)$$

$$+ T^p_{m,n} \left[1 - \frac{4}{3} \frac{h \Delta x}{k} \frac{\alpha \Delta t}{(\Delta x)^2} - 4 \frac{\alpha \Delta t}{(\Delta x)^2} \right] \qquad (8.30)$$

and for the node at an outer corner shown in Fig. 6.26(c),

$$T^{p+1}_{m,n} = \frac{2\alpha \Delta t}{(\Delta x)^2} \left(T^p_{m+1,n} + T^p_{m,n-1} + \frac{2h \Delta x}{k} T_\infty \right)$$

$$+ T^p_{m,n} \left[1 - 4 \frac{h \Delta x}{k} \frac{\alpha \Delta t}{(\Delta x)^2} - 4 \frac{\alpha \Delta t}{(\Delta x)^2} \right] \qquad (8.31)$$

As was the case with Eq. (8.21), mathematical stability of Eqs. (8.29), (8.30), and (8.31) requires that the second term on the right-hand side of each of the equations be greater than, or equal to, zero. Thus in Eq. (8.29),

$$1 - 2 \frac{h \Delta x}{k} \frac{\alpha \Delta t}{(\Delta x)^2} - 4 \frac{\alpha \Delta t}{(\Delta x)^2} \geq 0$$

or

$$\frac{(\Delta x)^2}{\alpha \Delta t} \geq 4 + 2 \frac{h \Delta x}{k} \qquad (8.32)$$

in Eq. (8.30),

$$1 - \frac{4}{3} \frac{h \Delta x}{k} \frac{\alpha \Delta t}{(\Delta x)^2} - 4 \frac{\alpha \Delta t}{(\Delta x)^2} \geq 0$$

or

$$\frac{(\Delta x)^2}{\alpha \Delta t} \geq 4 + \frac{4}{3} \frac{h \Delta x}{k} \qquad (8.33)$$

and in Eq. (8.31),

$$\frac{(\Delta x)^2}{\alpha \, \Delta t} \geq 4 + 4 \frac{h \, \Delta x}{k} \tag{8.34}$$

EXAMPLE 8.8

A mild steel bar of square cross-sectional area 0.4 m × 0.4 m at an initial uniform temperature of 400°C is cooled in air at $T_\infty = 25°C$ under conditions for which $h = 300 \text{ W/m}^2\cdot\text{K}$. Determine the locations in the cross section at which the slowest and fastest rates of cooling occur.

The cross section, shown in Fig. 8.15, is overlaid with a 0.1 m × 0.1 m mesh to locate 25 nodes. However, from symmetry, only one-eighth of the section, which contains six nodes, need be considered. The energy balances for the nodes are calculated as follows.

Node 1:

$$h \, \Delta y (T_\infty - T_1^p) - k \frac{\Delta x}{2} \frac{T_1^p - T_2^p}{\Delta y} - k \, \Delta y \frac{T_1^p - T_4^p}{\Delta x} - k \frac{\Delta x}{2} \frac{T_1^p - T_2^p}{\Delta y}$$
$$= \rho \frac{\Delta x}{2} C_p \frac{T_1^{p+1} - T_1^p}{\Delta t}$$

or

$$T_1^{p+1} = T_1^p + \frac{2\alpha \, \Delta t}{(\Delta x)^2} \left[T_2^p + T_4^p - 2T_1^p + \frac{h \, \Delta x}{k} (T_\infty - T_1^p) \right] \tag{i}$$

Node 2:

$$h \frac{\Delta y}{2} (T_\infty - T_2^p) - k \frac{\Delta x}{2} \frac{T_2^p - T_1^p}{\Delta y} - k \, \Delta y \frac{T_2^p - T_5^p}{\Delta x} - k \frac{\Delta x}{2} \frac{T_2^p - T_3^p}{\Delta y}$$
$$= \rho \frac{\Delta x}{2} C_p \frac{T_2^{p+1} - T_2^p}{\Delta t}$$

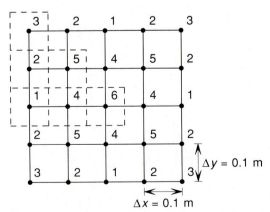

$\Delta y = 0.1$ m

$\Delta x = 0.1$ m

FIGURE 8.15. Location of the nodes used in Example 8.8.

or

$$T_2^{p+1} = T_2^p + \frac{2\alpha\,\Delta t}{(\Delta x)^2}\left[\frac{T_1^p}{2} + \frac{T_3^p}{2} - T_2^p + \frac{h\,\Delta x}{k}(T_\infty - T_2^p)\right] \qquad \text{(ii)}$$

Node 3:

$$h\,\frac{\Delta y}{2}(T_\infty - T_3^p) - \frac{k\,\Delta x}{2}\frac{T_3^p - T_2^p}{\Delta y} - k\,\frac{\Delta y}{2}\frac{T_3^p - T_2^p}{\Delta x} + h\,\frac{\Delta x}{2}(T_\infty - T_3^p)$$

$$= \rho\,\frac{\Delta x}{4}\,C_p\,\frac{T_3^{p+1} - T_3^p}{\Delta t}$$

or

$$T_3^{p+1} = T_3^p + \frac{4\alpha\,\Delta t}{(\Delta x)^2}\left[T_2^p - T_3^p + \frac{h\,\Delta x}{k}(T_\infty - T_3^p)\right] \qquad \text{(iii)}$$

Node 4:

$$-k\,\Delta y\,\frac{T_4^p - T_1^p}{\Delta x} - k\,\Delta x\,\frac{T_4^p - T_5^p}{\Delta y} - k\,\Delta y\,\frac{T_4^p - T_6^p}{\Delta x} - k\,\Delta x\,\frac{T_4^p - T_5^p}{\Delta y}$$

$$= \rho\,\Delta x\,C_p\,\frac{T_4^{p+1} - T_4^p}{\Delta t}$$

or

$$T_4^{p+1} = T_4^p + \frac{\alpha\,\Delta t}{(\Delta x)^2}(T_1^p + 2T_5^p + T_6^p - 4T_4^p) \qquad \text{(iv)}$$

Node 5:

$$T_5^{p+1} = T_5^p + \frac{\alpha\,\Delta t}{(\Delta x)^2}(2T_2^p + 2T_4^p - 4T_5^p) \qquad \text{(v)}$$

Node 6:

$$T_6^{p+1} = T_6^p + \frac{\alpha\,\Delta t}{(\Delta x)^2}(4T_4^p - 4T_6^p) \qquad \text{(vi)}$$

The data for mild steel are

$$\alpha = 1.2 \times 10^{-5}\ \text{m}^2/\text{s}$$

$$k = 42\ \text{W/m·K}$$

Thus with $\Delta x = \Delta y = 0.1$ m, the mesh Biot number is

$$\frac{h\,\Delta x}{k} = \frac{300 \times 0.1}{42} = 0.7143$$

Inspection of Eqs. (8.32), (8.33), and (8.34) shows that the mesh Fourier number is restricted by Eq. (8.34):

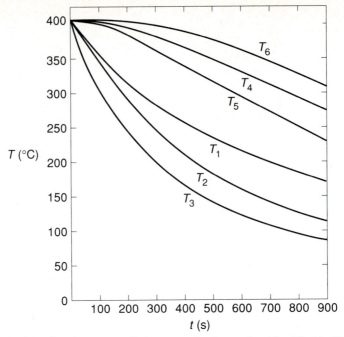

FIGURE 8.16. Continuous cooling curves for the nodes identified in Fig. 8.15.

$$\frac{(\Delta x)^2}{\alpha \, \Delta t} \geq 4 + 4 \frac{h \, \Delta x}{k} = 6.8571$$

and therefore $\Delta t \leq 121$ s. Selecting $\Delta t = 10$ s gives the mesh Fourier number as

$$\frac{\alpha \, \Delta t}{(\Delta x)^2} = \frac{1.2 \times 10^{-5} \times 10}{(0.1)^2} = 0.012$$

Substitution of the values of the mesh Biot number and the mesh Fourier number into Eqs. (i) to (iv) gives the results shown in Fig. 8.16. The corner node 3 cools most rapidly, followed by nodes 2 and 1. The difference between the temperatures of the corner node and the central node 6 increases to a maximum of 235°C after 630 s of cooling.

Heat Generation Within the Solid

If heat is generated within the solid during the temperature transient, the energy balance for a node is:

(net rate at which energy is transported into node)
+ (rate at which heat is generated in node)
= (rate of accumulation of thermal energy in node)

Thus the term $\dot{q}\,\Delta x\,\Delta y$ is included, where $\dot{q}$ is the rate of generation of heat per unit volume, and with $\Delta z = 1$, $\Delta x\,\Delta y$ is the volume of the node.

EXAMPLE 8.9

A nichrome wire of square cross section 2 mm × 2 mm is electrically heated by the passage of a current of 10 A. The initial temperature of the wire is 25°C and the current flows for 60 s. Calculate the temperature of the wire after 60 s and after 120 s.

The required data for nichrome are

$$k = 12 \text{ W/m·K}$$

$$\alpha = 3.4 \times 10^{-6} \text{ m}^2/\text{s}$$

$$\sigma = 1.5 \times 10^{-6} \text{ }\Omega\text{·m}$$

and the heat transfer coefficient, h, is 10 W/m²·K. From Eq. (6.28),

$$\dot{q} = \frac{I^2\sigma}{A^2} = \frac{10^2 \times 1.5 \times 10^{-6}}{(0.002)^4} = 9.375 \times 10^6 \text{ W/m}^3$$

A square mesh of $\Delta x = \Delta y = 0.001$ m is laid over the cross section as shown in Fig. 8.17, and again, from symmetry, only three nodes need be considered. The energy balance for node 1 is

$$h\frac{\Delta y}{2}(T_\infty - T_1^p) + h\frac{\Delta x}{2}(T_\infty - T_1^p) - k\frac{\Delta y}{2}\frac{T_1^p - T_2^p}{\Delta x} - k\frac{\Delta x}{2}\frac{T_1^p - T_2^p}{\Delta y}$$

$$+ \dot{q}\frac{\Delta x}{2}\frac{\Delta y}{2} = \rho\frac{\Delta x}{2}\frac{\Delta y}{2}C_p\frac{T_1^{p+1} - T_1^p}{\Delta T}$$

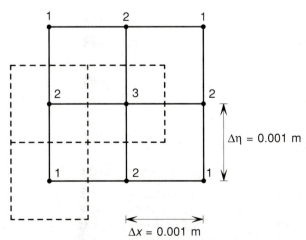

FIGURE 8.17. Mesh used in Example 8.9.

or

$$T_1^{p+1} = T_1^p + \frac{4\alpha \, \Delta t}{(\Delta x)^2} \left[T_2^p - T_1^p + \frac{\dot{q}(\Delta x)^2}{4k} + \frac{h \, \Delta x}{k} (T_\infty - T_1^p) \right] \quad (i)$$

The energy balance for node 2 is

$$h \, \Delta y \, (T_\infty - T_2^p) - k \frac{\Delta x}{2} \frac{T_2^p - T_1^p}{\Delta y} - k \, \Delta y \frac{T_2^p - T_3^p}{\Delta x} - k \frac{\Delta x}{2} \frac{T_2^p - T_1^p}{\Delta y}$$

$$+ \dot{q} \frac{\Delta x}{2} \Delta y = \rho \frac{\Delta x}{2} \Delta y \, C_p \frac{T_2^{p+1} - T_2^p}{\Delta t}$$

or

$$T_2^{p+1} = T_2^p + \frac{2\alpha \, \Delta t}{(\Delta x)^2} \left[T_1^p + T_3^p - 2T_2^p + \frac{\dot{q}(\Delta x)^2}{2k} + \frac{h \, \Delta x}{k} (T_\infty - T_2^p) \right]$$

$$(ii)$$

and the energy balance for node 3 is

$$- k \, \Delta y \frac{T_3^p - T_2^p}{\Delta x} - k \, \Delta x \frac{T_3^p - T_2^p}{\Delta y} - k \, \Delta y \frac{T_3^p - T_2^p}{\Delta x} - k \, \Delta x \frac{T_3^p - T_2^p}{\Delta y}$$

$$+ \dot{q} \, \Delta x \, \Delta y = \rho \, \Delta x \, \Delta y \, C_p \frac{T_3^{p+1} - T_3^p}{\Delta t}$$

or

$$T_3^{p+1} = T_3^p + \frac{\alpha \, \Delta t}{(\Delta x)^2} \left[4T_2^p - 4T_3^p + \dot{q} \frac{(\Delta x)^2}{k} \right] \quad (iii)$$

The mesh Biot number is

$$\frac{h \, \Delta x}{k} = \frac{10 \times 0.001}{12} = 8.33 \times 10^{-4}$$

and as the cross section contains an outside corner, Eq. (8.34) restricts the mesh Fourier number as

$$\frac{(\Delta x)^2}{\alpha \, \Delta t} \geq 4 + 4 \frac{h \, \Delta x}{k} = 4.0033$$

or

$$\Delta t \leq \frac{(\Delta x)^2}{4.0033\alpha} = 0.073 \text{ s}$$

Selecting $\Delta t = 0.05$ s gives the mesh Fourier number as

$$\frac{\alpha \, \Delta t}{(\Delta x)^2} = \frac{3.4 \times 10^{-6} \times 0.05}{(0.001)^2} = 0.17$$

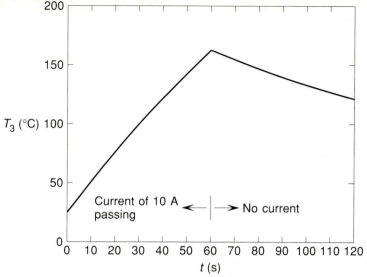

FIGURE 8.18.   Change in temperature, with time, of the electrically heated nichrome wire considered in Example 8.9.

A computer program in which $\dot{q} = 9.375 \times 10^6$ W/m^3 during $0 \le t \le 60$ s and $\dot{q} = 0$ when $t > 60$ s gives the variation of T_3 with time shown in Fig. 8.18. At $t = 60$ s, $T_3 = 161°C$ and the difference between T_3 and T_1 is 0.1°C. When the current is turned off, T_3 decreases to 121°C after 60 s of cooling.

8.6 Solidification of Metal Castings

The production of metal castings involves pouring the liquid metal into a mold and allowing solidification to occur. During solidification, heat is transported by conduction and convection from the bulk of the liquid metal to the solid–liquid interface, the latent heat of solidification is evolved as sensible heat at this interface, then heat is conducted through the solid metal to the metal–mold interface and through the wall of the mold. The rate of solidification, which is measured as the rate of advance of the solid–liquid interface, is thus determined by the relative magnitudes of the resistance to heat transport in the metal and the resistance to heat transport in the mold.

Sand Casting

The thermal conductivity of sand is approximately 0.3 W/m·K, which is orders of magnitude lower than the thermal conductivity of a typical metal, and hence it is to be expected that the rate of solidification of sand castings will be determined by conduction of heat through the mold wall. Consider that a pure liquid metal at its melting temperature T_M is poured into a sand mold at the uniform temperature T_0, and that when the mold is full of liquid metal, the temperature of the inner surface of the mold instantaneously increases to T_M at $t = 0$. A temperature gradient then develops in the mold wall and the transfer of heat from the metal to the mold causes freezing of the metal and hence evolution of the latent heat of solidification. The energy balance is thus

> rate at which heat is transferred from solid metal to mold
> = rate of evolution of latent heat of solidification (i)

and the rate of evolution of the latent heat determines the rate of advance of the solid–liquid interface into the liquid metal. If, during solidification, the temperature of the outer wall of the mold does not deviate from T_0, the mold can be considered to be semi-infinite in thickness, in which case, for unidirectional conduction of heat, the normalized temperature profile in the mold wall is given by Eq. (8.17) as

$$\frac{T - T_0}{T_M - T_0} = \text{erfc}(-\eta) \tag{8.35}$$

The negative sign in Eq. (8.35) arises from the fact that the inner surface of the mold is located at $x = 0$ and hence heat is conducted through the mold wall in the $-x$-direction. The conditions at $t = 0$ and at some time $t > 0$ are shown in Fig. 8.19(a) and (b).

In Fig. 8.19(b) the energy balance given in Eq. (i) is

$$-k_m A \left.\frac{\partial T}{\partial x}\right|_{x=0} = -\rho_s \frac{dV_s}{dt} \Delta H_s \tag{ii}$$

where ΔH_s is the latent heat of solidification of the metal per unit mass and dV_s/dt is the volume rate of freezing. The subscripts m and s denote, respectively, the mold material and the solid metal, and the negative sign on the right-hand side of the equation arises because the heat evolved is being conducted in the $-x$-direction.

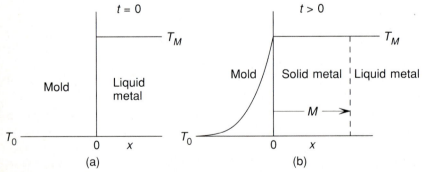

FIGURE 8.19. Temperature profiles during solidification of a liquid metal contained in a sand mold; (a) at $t = 0$ and (b) at $t > 0$.

For unit surface area of mold–metal interface, Eq. (ii) becomes

$$-k_m \frac{\partial T}{\partial x}\bigg|_{x=0} = -\rho_s \frac{dM}{dt} \Delta H_s \tag{iii}$$

where M is the thickness of the layer of solid metal at time t. Equation (8.35) is

$$\frac{T - T_0}{T_m - T_0} = \text{erfc}(-\eta) = -\text{erfc}(\eta) = \text{erf}(\eta) - 1$$

and from the definition of the error function given by Eq. (8.14),

$$dT = (T_M - T_0)\, d\,[\text{erfc}(\eta)] = (T_M - T_0)\frac{2}{\sqrt{\pi}} e^{-\eta^2}\, d\eta$$

which, with $\eta = x/2\sqrt{\alpha_m t}$ and $x = 0$, gives

$$\frac{\partial T}{\partial x}\bigg|_{x=0} = (T_M - T_0)\frac{2}{\sqrt{\pi}}\frac{1}{2\sqrt{\alpha_m t}} = \frac{T_M - T_0}{\sqrt{\pi \alpha_m t}} \tag{iv}$$

Substitution of Eq. (iv) into Eq. (iii) gives

$$\frac{k_m(T_M - T_0)}{\sqrt{\pi \alpha_m t}} = \rho_s\, \Delta H_s \frac{dM}{dt}$$

integration of which from $M = 0$ at $t = 0$ gives

$$M = \frac{2k_m(T_M - T_0)}{\sqrt{\pi \alpha_m}\, \rho_s\, \Delta H_s} t^{1/2}$$

or as $\alpha_m = k_m/(\rho_m C_{p,m})$,

$$M = \frac{2}{\sqrt{\pi}}\frac{T_M - T_0}{\rho_s\, \Delta H_s}(k_m C_{p,m}\rho_m)^{1/2}t^{1/2} \tag{8.36}$$

The rate of solidification is thus determined by:

1. The melting temperature of the metal, the latent heat of solidification, and the density of the solid metal.
2. The thermal conductivity, the heat capacity, and the density of the mold material.

The term $kC_p\rho$ is called the *heat diffusivity* of the mold material (which should not be confused with the thermal diffusivity $\alpha = k/\rho C_p$) and the heat diffusivity is a measure of the ability of the mold to absorb heat. Thus, in Eq. (8.36), the rate of solidification increases with increasing T_M [which, from Eq. (iv) increases the temperature gradient in the mold at $x = 0$], decreasing ΔH_s and increasing heat diffusivity.

In the sand casting of a flat plate of thickness $2L$, solidification is complete when $M = L$ (i.e., when the solid layers growing on each of the mold walls meet at the centerline of the casting), and hence from Eq. (8.36), the time for complete solidification, t_f, is

$$t_f = \frac{\pi L^2}{k_m C_{p,m} \rho_m} \left[\frac{\rho_s \Delta H_s}{2(T_M - T_0)} \right]^2 \tag{8.37}$$

Solidification is complete when all of the latent heat of solidification has been transported to the mold.

In Eq. (ii), if A is the surface area of the mold and V_s is volume of the solid casting, a characteristic length L_c can be defined as

$$L_c = \frac{V_s}{A}$$

For a slab of thickness $2L$ and surface area A_0,

$$L_c = \frac{V_s}{A} = \frac{2LA_0}{2A_0} = L$$

and hence, in Eq. (8.36), solidification is complete when M reaches the value L_c and Eq. (8.37) can be written as

$$t_f = CL_c^2 \tag{8.38}$$

which is *Chvorinov's rule*, in which C given by

$$C = \frac{\pi}{k_m C_{p,m} \rho_m} \left(\frac{\rho_s \Delta H_s}{T_M - T_0} \right)^2 \tag{8.39}$$

is known as Chvorinov's constant. Equation (8.38) thus permits comparison of the freezing rates in molds of different geometries, in which the rates of absorption of heat per unit area of mold surface are the same. In a cylindrical casting of radius R and length L, the characteristic length is

$$L_c = \frac{V_s}{A} = \frac{\pi R^2 L}{2\pi R L} = \frac{R}{2}$$

and in a spherical casting of radius R,

$$L_c = \frac{V_s}{A} = \frac{\frac{4}{3}\pi R^3}{4\pi R^2} = \frac{R}{3}$$

Thus, from Chvorinov's rule a cylindrical casting of radius L should require the same time for completion of freezing as does a slab casting of thickness L, and a spherical casting of radius L should freeze completely in four-ninths the time required for complete freezing of a cylindrical casting of radius R. However, the rate of absorption of heat by a mold varies with the curvature of the mold surface, and thus Chvorinov's rule should be restricted to different sizes of castings of similar geometries. The influence of curvature of the mold surface on its rate of absorption of heat is illustrated in Fig. 8.20.

The general heat conduction equation for conduction in the radial direction in a cylinder is

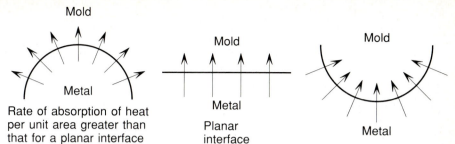

FIGURE 8.20.   Influence of the curvature of the mold surface on its rate of absorption of heat.

$$\frac{\partial T}{\partial t} = \alpha \left[\frac{1}{r} \frac{\partial}{\partial r} \left(r \frac{\partial T}{\partial r} \right) \right] = \alpha \left[\frac{\partial^2 T}{\partial r^2} + \frac{1}{r} \left(\frac{\partial T}{\partial r} \right)^2 \right] \tag{8.40}$$

and for radial conduction in a sphere is

$$\frac{\partial T}{\partial t} = \alpha \left[\frac{1}{r^2} \frac{\partial}{\partial r} \left(r^2 \frac{\partial T}{\partial r} \right) \right] = \alpha \left[\frac{\partial^2 T}{\partial r^2} + \frac{2}{r} \left(\frac{\partial T}{\partial r} \right)^2 \right] \tag{8.41}$$

Equations (8.40) and (8.41) can be combined as

$$\frac{\partial T}{\partial t} = \alpha \left[\frac{\partial^2 T}{\partial r^2} + \frac{n}{r} \left(\frac{\partial T}{\partial r} \right)^2 \right] \tag{v}$$

in which $n = 1$ for a cylinder and $n = 2$ for a sphere.

Integration of Eq. (v) gives the time required for complete solidification in a semi-infinite cylindrical mold and a semi-infinite spherical mold as

$$L_c = \frac{V_s}{A} = \frac{T_M - T_0}{\rho_s \, \Delta H_s} \left[\frac{2}{\sqrt{\pi}} (k_m C_{p,m} \rho_m)^{1/2} t_f^{1/2} + \frac{n k_m}{2R} t_f \right] \tag{8.42}$$

in which R is the radius of the cylindrical or spherical casting. With decreasing thermal diffusivity of the mold material (k_m decreasing or $\rho_m C_{p,m}$ increasing) and with increasing R, Eq. (8.42) approaches Chvorinov's rule.

EXAMPLE 8.10

Consider the casting of slabs of thickness 0.1 m, and cylinders and spheres of diameter 0.1 m of aluminum and copper, at their melting temperatures in sand molds at 25°C. For sand,

$$C_p = 800 \text{ J/kg·K}$$
$$\rho = 1500 \text{ kg/m}^3$$
$$k = 0.3 \text{ W/m·K}$$
$$\alpha = 2.5 \times 10^{-7} \text{ m}^2/\text{s}$$

For aluminum,

$$T_M = 660°C$$

$$\Delta H_s = 400,100 \text{ J/kg at } 660°C$$

$$\rho = 2390 \text{ kg/m}^3 \text{ at } 660°C$$

For copper,

$$T_M = 1083°C$$

$$\Delta H_s = 205,400 \text{ J/kg at } 1083°C$$

$$\rho = 6080 \text{ kg/m}^3 \text{ at } 1083°C$$

From the data it is seen that $\Delta H_{s,\text{Al}}$ is almost twice the value of $\Delta H_{s,\text{Cu}}$, which might seem strange in view of Richard's rule, which states that the ratio of the molar heat of solidification to the absolute melting temperature is virtually the same for all metals. Thus from Richard's rule the molar heat of solidification of copper should be 1.45 times the molar heat of solidification of aluminum. However, as the atomic weight of copper (63.54) is more than twice the atomic weight of aluminum (26.98), the heat of solidification of aluminum per unit mass is larger than the heat of solidification of copper per unit mass.

(1) *Casting of a 0.1-m-thick slab of aluminum*: From Eq. (8.36),

$$
\begin{aligned}
M &= \frac{2}{\sqrt{\pi}} \frac{T_M - T_0}{\rho_s \, \Delta H_s} (k_m C_{p,m} \rho_m)^{1/2} t^{1/2} \\
&= \frac{2}{\sqrt{\pi}} \left(\frac{600 - 25}{2390 \times 400,100} \right) (0.3 \times 800 \times 1500)^{1/2} t^{1/2} \\
&= 1.128 \times 6.64 \times 10^{-7} \times (360,000)^{1/2} t^{1/2} \\
&= 4.496 \times 10^{-4} t^{1/2}
\end{aligned}
\tag{i}
$$

$t_f = 12,368 \text{ s} = 3.44 \text{ h}$ and hence solidification is complete in 3.44 h.

(2) *Casting of a 0.1-m-thick slab of copper*: For copper,

$$\frac{T_M - T_0}{\rho_s \, \Delta H_s} = \frac{1083 - 25}{6080 \times 205,400} = 8.47 \times 10^{-7}$$

Therefore, from Eq. (i),

$$
\begin{aligned}
M &= (4.496 \times 10^{-4}) \times \frac{8.47 \times 10^{-7}}{6.64 \times 10^{-7}} t^{1/2} \\
&= 5.736 \times 10^{-4} t^{1/2}
\end{aligned}
$$

which gives $t_f = 7597 \text{ s} = 2.11 \text{ h}$ as the time for complete solidification of the copper casting. Although $\rho_s \, \Delta H_s$ is smaller for aluminum than for copper, the melting temperature of the metal has the major influence on the time required for complete solidification in a mold of given heat diffusivity.

(3) *Casting of a cylinder of aluminum of radius* 0.05 m: From Eq. (8.42), with $n = 1$,

$$L_c = \frac{R}{2} = \frac{T_M - T_0}{\rho_s \, \Delta H_s} \left[\frac{2}{\sqrt{\pi}} \, (k_m C_{p,m} \rho_m)^{1/2} t_f^{1/2} + \frac{k_m}{2R} \, t_f \right]$$

Therefore,

$$0.025 = 6.64 \times 10^{-7} \left[1.128 \times (360{,}000)^{1/2} t_f^{1/2} + \frac{0.3}{2 \times 0.05} \, t_f \right]$$

$$= 4.494 \times 10^{-4} t_f^{1/2} + 1.992 \times 10^{-6} t_f \tag{ii}$$

which has the solution $t_f = 2132 \text{ s} = 0.59$ h.

(4) *Casting of a cylinder of copper of radius* 0.05 m;

$$0.025 = 8.47 \times 10^{-7} \left[1.128 \times (360{,}000)^{1/2} t_f^{1/2} + \frac{0.3}{2 \times 0.05} \, t_f \right]$$

$$= 5.732 \times 10^{-4} t_f^{1/2} + 2.541 \times 10^{-6} t_f \tag{iii}$$

which has the solution $t_f = 1399 \text{ s} = 0.39$ h. Again, the higher melting temperature of copper is the major influence on t_f.

(5) *Casting of a sphere of aluminum of radius* 0.1 m: From Eq. (8.42), with $n = 2$, and Eq. (ii),

$$0.05/3 = 4.494 \times 10^{-4} t_f^{1/2} + 2 \times 1.992 \times 10^{-6} t_f$$

which has the solution $t_f = 865 \text{ s} = 0.24$ h.

(6) *Casting of a sphere of copper of radius* 0.1 m: From Eq. (8.41), with $n = 2$, and Eq. (iii),

$$0.05/3 = 5.732 \times 10^{-4} t_f^{1/2} + 2 \times 2.541 \times 10^{-6} t_f$$

which has the solution $t_f = 579 \text{ s} = 0.16$ h. Thus

$$\frac{t_f \text{ (Al slab)}}{t_f \text{ (Al cylinder)}} = 5.80 \qquad \frac{t_f \text{ (Cu slab)}}{t_f \text{ (Cu cylinder)}} = 5.43$$

$$\frac{t_f \text{ (Al cylinder)}}{t_f \text{ (Al sphere)}} = 3.62 \qquad \frac{t_f \text{ (Cu cylinder)}}{t_f \text{ (Cu sphere)}} = 2.42$$

$$\frac{t_f \text{ (Al slab)}}{t_f \text{ (Cu slab)}} = 1.62, \qquad \frac{t_f \text{ (Al cylinder)}}{t_f \text{ (Cu cylinder)}} = 1.52, \qquad \frac{t_f \text{ (Al sphere)}}{t_f \text{ (Cu sphere)}} = 1.49$$

The validity of the assumption of semi-infinitely thick mold walls can be examined as follows. Complete soldification of the aluminum slab required 12,368 s. Therefore,

$$\eta = \frac{x}{2\sqrt{\alpha_m t}} = \frac{x}{2(2.5 \times 10^{-7} \times 12{,}368)^{1/2}} = 8.99x$$

and thus the value of x at which the normalized temperature is 1% is obtained from

$$\frac{T - T_0}{T_M - T_0} = 0.01 = \text{erfc}(\eta)$$

and Table 8.1 as

$$\eta \doteq 1.8 = 8.99x$$

or $x = 0.2$m, which shows that the assumption is valid if the mold wall is more than 20 cm in thickness. At $x = 20$ cm from the inner surface of the mold wall, $T = 31.3°C$ after 12,368 s of freezing.

Casting into Cooled Metal Molds

In sand casting it was considered that the resistance to heat flow in the mold was so much larger than that in the metal that the temperature gradients which developed in the metal were small enough to be ignored in the analysis, and this is equivalent to the assumption made in the analysis of Newtonian cooling. The other extreme from that in which the inner surface of the mold remains at T_M during solidification is that in which the temperature of the inner surface of the mold remains at T_0 (i.e., the temperature gradient through the mold wall is negligibly small). This condition is approached with water-cooled metal molds of high thermal conductivity, and the energy balance is

> rate at which heat is conducted away from solid–liquid
> interface into solid metal
> = rate at which latent heat of solidification is
> evolved at interface   (i)

With liquid metal at its melting temperature T_M, the conditions at $t = 0$ and $t > 0$ are shown in Fig. 8.21(a) and (b). The conduction of heat from the solid–liquid

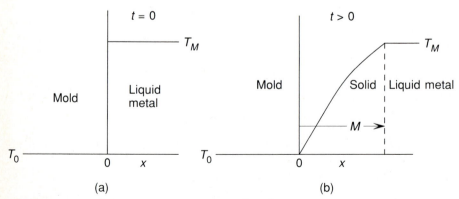

(a)   (b)

FIGURE 8.21.   Temperature profiles during solidification of a liquid metal contained in a cooled metal mold; (a) at $t = 0$ and (b) at $t > 0$.

interface to the solid–mold interface is governed by the general heat conduction equation for heat flow in one dimension,

$$\frac{\partial T}{\partial t} = \alpha_s \frac{\partial^2 T}{\partial x^2}$$

which in the derivation leading to Eq. (8.17) gave

$$\int dT = A \int e^{-\eta^2} \, d\eta$$

or

$$T = B + A' \, \text{erf}(\eta) \tag{ii}$$

The integration constants A' and B are obtained from the boundary conditions

$$T = \begin{cases} T_0 & \text{at } x = 0 \text{ and } t \geq 0 \\ T_M & \text{at } x = M \text{ and } t > 0 \end{cases}$$

At $x = 0$ (and $\eta = 0$), erf $(\eta) = 0$ and hence in Eq. (ii),

$$T_0 = B$$

Thus

$$T = T_0 + A' \, \text{erf}(\eta)$$

At $x = M$,

$$T_M = T_0 + A' \, \text{erf} \left(\frac{M}{2\sqrt{\alpha_s t}} \right)$$

or

$$A' = \frac{T_M - T_0}{\text{erf}(M/2\sqrt{\alpha_s t})} \tag{iii}$$

As T_m, T_0, and A' are constants, Eq. (iii) shows that

$$\frac{M}{2\sqrt{\alpha_s t}} = \text{constant} = \phi$$

or

$$M = 2\phi\sqrt{\alpha_s t} \tag{iv}$$

that is, as with the sand casting of a slab, M is a linear function of $t^{1/2}$. Insertion of the integration constants into Eq. (ii) gives

$$T = T_0 + \frac{T_M - T_0}{\text{erf}(\phi)} \, \text{erf}(\eta) \tag{8.43}$$

The energy balance given in Eq. (i) is

$$-k_s \frac{\partial T}{\partial x} \bigg|_M = -\rho_s \, \Delta H_s \frac{dM}{dt} \tag{v}$$

and from Eq. (8.43),

$$dT = \frac{T_M - T_0}{\text{erf}(\phi)} \frac{2}{\sqrt{\pi}} e^{-\eta^2} d\eta$$

or with $\eta = x/2\sqrt{\alpha_s t}$,

$$\left.\frac{\partial T}{\partial x}\right|_M = \frac{T_M - T_0}{\text{erf}(\phi)} e^{-M^2/4\alpha_s t} \frac{1}{(\pi\alpha_s t)^{1/2}} = \frac{T_M - T_0}{\text{erf}(\phi)} \frac{e^{-\phi^2}}{(\pi\alpha_s t)^{1/2}} \tag{vi}$$

From Eq. (iv),

$$\frac{dM}{dt} = \phi\sqrt{\frac{\alpha_s}{t}} \tag{vii}$$

and insertion of Eqs. (vi) and (vii) with Eq. (v) gives

$$\frac{k_s(T_M - T_0)}{\text{erf}(\phi)} \frac{e^{-\phi^2}}{(\pi\alpha_s t)^{1/2}} = \frac{\rho_s \, \Delta H_s \phi \sqrt{\alpha_s}}{\sqrt{t}}$$

or

$$\phi e^{\phi^2} \text{erf}(\phi) = \frac{T_M - T_0}{\sqrt{\pi}} \frac{C_{p,s}}{\Delta H_s} \tag{8.44}$$

which determines the value of ϕ in Eq. (iv). The variation of $\phi e^{\phi^2} \text{erf}(\phi)$ with ϕ is shown in Fig. 8.22.

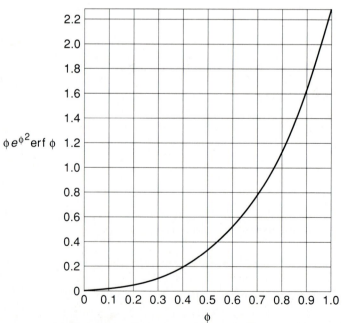

FIGURE 8.22. Variation of $\phi e^{\phi^2} \text{erf}(\phi)$ with ϕ.

EXAMPLE 8.11

Consider the casting of (a) aluminum, and (b) copper slabs of thickness 0.1 m in chilled metal molds maintained at 25°C. For aluminum,

$$T_M = 660°C$$

$$\Delta H_s = 400,100 \text{ J/kg at } 660°C$$

$$C_p = 1048 \text{ J/kg·K}$$

$$\alpha = 9.18 \times 10^{-5} \text{ m}^2/\text{s}$$

For copper,

$$T_M = 1083°C$$

$$\Delta H_s = 205,400 \text{ J/kg at } 1083°C$$

$$C_p = 439 \text{ J/kg·K}$$

$$\alpha = 1.38 \times 10^{-4} \text{ m}^2/\text{s}$$

(a) *Aluminum*: From Eq. (8.44),

$$\frac{T_M - T_0}{\sqrt{\pi}} \frac{C_{p,s}}{\Delta H_s} = \frac{660 - 25}{\sqrt{\pi}} \times \frac{1048}{400,100} = 0.938$$

$$= \phi e^{\phi^2} \text{ erf}(\phi)$$

and from Fig. 8.22, ϕ is obtained as 0.75. Therefore, from Eq. (iv),

$$M = 2\phi\sqrt{\alpha_s t} = 2 \times 0.75 \times (9.18 \times 10^{-5})^{1/2} t^{1/2} = 1.437 \times 10^{-2} t^{1/2}$$

and

$$M = L_c = 0.05 \qquad \text{at } t_f = 12.1 \text{ s}$$

Copper: From Eq. (8.44),

$$\frac{T_M - T_0}{\sqrt{\pi}} \frac{C_{p,s}}{\Delta H_s} = \frac{1083 - 25}{\sqrt{\pi}} \times \frac{439}{205,400} = 1.276$$

$$= \phi e^{\phi^2} \text{ erf}(\phi)$$

and from Fig. 8.22, ϕ is obtained as 0.835. Therefore, from Eq. (iv),

$$M = 2\phi\sqrt{\alpha_s t} = 2 \times 0.835 \times (1.38 \times 10^{-4})^{1/2} t^{1/2}$$
$$= 1.961 \times 10^{-2} t^{1/2}$$

and $M = L_c = 0.05$ at $t = 6.5$ s.

The higher melting temperature of copper, which increases ϕ, and the higher thermal diffusivity of copper, cause the copper casting to solidify more rapidly than the aluminum casting. The temperature profiles in the solidified aluminum after 1, 5, 12.1, and 20 s obtained from Eq. (8.43) with $\phi = 0.75$ [erf(ϕ) = 0.7112] are shown in Fig. 8.23. After 12.1 s the solidified layer is 0.05 m thick.

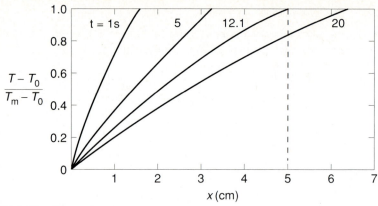

FIGURE 8.23. Normalized temperature profiles in solidifying aluminum after 1, 5, 12.1 and 20 s, calculated in Example 8.11.

Temperature Gradients in Both the Mold and the Casting

Between the extremes of negligible temperature gradients in the metal (sand casting) and negligible temperature gradients in the mold (cooled metal mold casting) is the case shown in Fig. 8.24, in which temperature gradients are developed in both the semi-infinitely thick mold wall and the solidified metal and, during solidification, the temperature at the mold–metal interface remains constant at T_s. From Eq. (8.35) the temperature profile in the mold is

$$\frac{T - T_0}{T_s - T_0} = \text{erfc} \left(\frac{-x}{2\sqrt{\alpha_m t}} \right) \tag{i}$$

and from Eq. (8.43), the temperature profile in the solid metal is

$$T = T_s + \frac{T_M - T_s}{\text{erf}(\phi)} \text{erf} \left(\frac{x}{2\sqrt{\alpha_s t}} \right) \tag{ii}$$

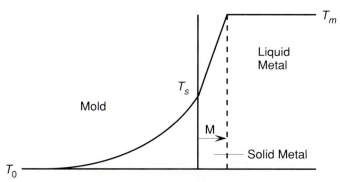

FIGURE 8.24. Temperature profile in a metal and a semi-infinite mold during solidification.

where, as with Eq. (8.44), ϕ is obtained from

$$\frac{T_M - T_s}{\sqrt{\pi}} \frac{C_{p,s}}{\Delta H_s} = \phi e^{\phi^2} \operatorname{erf}(\phi) \tag{iii}$$

In Fig. 8.24 the heat balance at the mold–metal interface requires that the flux to the interface from the solid equal the flux from the interface into the mold,

$$-k_m \frac{\partial T}{\partial x}\bigg|_{x=0(\text{mold})} = -k_s \frac{\partial T}{\partial x}\bigg|_{x=0(\text{metal})} \tag{iv}$$

From Eq. (i),

$$\frac{\dot{\partial}T}{\partial x}\bigg|_{x=0(\text{mold})} = \frac{T_s - T_0}{(\pi\alpha_m t)^{1/2}}$$

and from Eq. (ii),

$$\frac{\partial T}{\partial x}\bigg|_{x=0(\text{metal})} = \frac{T_M - T_s}{\operatorname{erf}(\phi)} \frac{1}{(\pi\alpha_s t)^{1/2}}$$

insertion of which into Eq. (iv) gives

$$k_m \frac{T_s - T_0}{(\pi\alpha_m t)^{1/2}} = k_s \frac{T_M - T_s}{(\pi\alpha_s t)^{1/2}} \frac{1}{\operatorname{erf}(\phi)}$$

or

$$(T_s - T_0)(k_m C_{p,m}\rho_m)^{1/2} = \frac{T_M - T_s}{\operatorname{erf}(\phi)} (k_s C_{p,s}\rho_s)^{1/2} \tag{v}$$

Substituting the expression for $(T_M - T_s)$ obtained from Eq. (iii) into Eq. (v) gives

$$T_s = T_0 + \phi e^{\phi^2} \frac{\sqrt{\pi}\, \Delta H_s}{C_{p,s}} \left(\frac{k_s C_{p,s}\rho_s}{k_m C_{p,m}\rho_m}\right)^{1/2} \tag{vi}$$

and substitution of the expression for T_s given by Eq. (vi) back into Eq. (iii) gives

$$T_M - T_0 - \phi e^{\phi^2} \frac{\sqrt{\pi}\, \Delta H_s}{C_{p,s}} \left(\frac{k_s C_{p,s}\rho_s}{k_m C_{p,m}\rho_m}\right)^{1/2} = \phi e^{\phi^2} \operatorname{erf}(\phi) \frac{\sqrt{\pi}\, \Delta H_s}{C_{p,s}}$$

$$\frac{T_M - T_0}{\sqrt{\pi}} \frac{C_{p,s}}{\Delta H_s} = \phi e^{\phi^2} \left[\operatorname{erf}(\phi) + \left(\frac{k_s C_{p,s}\rho_s}{k_m C_{p,m}\rho_m}\right)^{1/2}\right] \tag{vii}$$

Equation (vii) allows calculation of the value of ϕ that determines the temperature T_s in Eq. (vi), the temperature gradients in the mold and the solid metal in Eqs. (i) and (ii), and the rate of solidification given by

$$M = 2\phi\sqrt{\alpha_s t} \tag{viii}$$

EXAMPLE 8.12

Consider the casting of a 0.1-m-thick slab of iron at its melting temperature into a thick-walled copper mold, which is initially at 25°C. For iron,

$$T_M = 1535°C$$
$$\Delta H_s = 247{,}200 \text{ J/kg}$$
$$\rho = 7210 \text{ kg/m}^3$$
$$C_p = 786 \text{ J/kg·K}$$
$$k = 63 \text{ W/m·K}$$
$$\alpha = 1.1 \times 10^{-5} \text{ m}^2/\text{s}$$

For copper,

$$\rho = 8933 \text{ kg/m}^3$$
$$C_p = 385 \text{ J/kg·K}$$
$$k = 400 \text{ W/m·K}$$
$$\alpha = 11.6 \times 10^{-5} \text{ m}^2/\text{s}$$

Thus

$$\frac{T_M - T_0}{\sqrt{\pi}} \frac{C_{p,s}}{\Delta H_s} = \frac{1535 - 25}{\sqrt{\pi}} \times \frac{786}{247{,}200} = 2.708$$

and

$$\left(\frac{k_s C_{p,s} \rho_s}{k_m C_{p,m} \rho_m}\right)^{1/2} = \left(\frac{63 \times 786 \times 7210}{400 \times 385 \times 8933}\right)^{1/2} = 0.5094$$

Therefore, Eq. (vii) becomes

$$2.708 = \phi e^{\phi^2}[\text{erf}(\phi) + 0.5094] \tag{ix}$$

Using the approximation

$$\text{erf}(\phi) = \frac{2}{\sqrt{\pi}} \frac{3\phi}{\phi^2 + 3}$$

in Eq. (ix) gives the solution $\phi = 0.907$, which is within the range of applicability of the approximation, and thus

$$M = 2\phi\sqrt{\alpha_s t}$$
$$= 2 \times 0.907 \times (1.1 \times 10^{-5})^{1/2} t^{1/2}$$
$$= 6.016 \times 10^{-3} t^{1/2}$$

Therefore, for $L_c = M = 0.05$ m, $t_f = 69$ s.

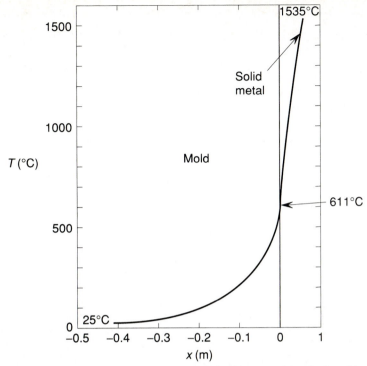

FIGURE 8.25. Temperature profile in the metal and the mold calculated in Example 8.12 when the solidification process is complete at $t = 69$ s.

From Eq. (vi),

$$T_s = T_1 + \phi e^{\phi^2} \frac{\sqrt{\pi}\,\Delta H_s}{C_{p,s}} \left(\frac{k_s C_{p,s}\rho_s}{k_m C_{p,m}\rho_m}\right)^{1/2}$$

$$= 25 + \frac{0.907 \exp(0.907^2)\sqrt{\pi} \times 247{,}200 \times 0.5094}{786}$$

$$= 25 + 586$$

$$= 611°C$$

From Eq. (ii) the temperature profile in the solid metal is

$$T = T_s + \frac{T_M - T_s}{\text{erf}(\phi)} \,\text{erf}\left(\frac{x}{2\sqrt{\alpha_s t}}\right)$$

which at $t = 69$ s gives

$$T = 611 + 1155 \,\text{erf}(18.15x) \tag{x}$$

and from Eq. (i), the temperature in the copper mold is

$$T = T_0 + (T_s - T_0) \,\text{erfc}\left(\frac{-x}{2\sqrt{\alpha_m t}}\right)$$

which at $t = 69$ s gives

$$T = 25 + 588 \text{ erfc}(-5.588x) \qquad \text{(xi)}$$

Equations (x) and (xi) are shown in Fig. 8.25.

8.7 Summary

Two thermal resistances operate during a cooling process: a resistance to heat transport by conduction in the solid and a resistance to heat transfer by convection at the surface. If the Biot number, which is defined as the ratio of the resistance to conduction to the resistance to convection, is less than 0.1, the temperature gradients that develop in the body during cooling are small enough to be ignored and the cooling rate can be calculated by equating the rate of heat transfer by convection from the surface with the rate of decrease in enthalpy of the body. This type of cooling is known as Newtonian cooling. When the Biot number is greater than 0.1, consideration of non-steady-state cooling requires integration of the general heat conduction equation. In finite systems the temperature transients that develop during cooling are strongly dependent on the shape and size of the body, and solutions to the general heat conduction equation can be arithmetically complicated. However, in semi-infinite systems which are sufficiently large that during the time period of interest, the temperature at one location is not influenced by temperature change at another location, integration of the general heat conduction equation produces relatively simple error functions. The complications of integration in finite systems are eliminated by using numerical techniques such as the finite-difference method.

The solidification of metal castings is determined by convection and conduction of heat from the bulk melt to the sold–liquid interface, the evolution of the latent heat of solidification at the interface, and the conduction of heat from the interface through the solidified metal and the mold wall. At one extreme, in which the thermal conductivity of the mold material is much smaller than that of the metal, the temperature gradients in the metal can be ignored. In this case, which is typical of sand casting, the solidification rate increases with increasing melting temperature of the metal, decreasing magnitude of the latent heat of solidification, and increasing heat diffusivity of the mold material. At the other extreme, which is exemplified by casting into a cooled metal mold, the temperature gradients that develop in the mold can be ignored. Between these extremes the temperature gradients in both the mold and the metal must be considered.

Problems

PROBLEM 8.1 _____

A heat transfer coefficient is to be determined by measuring the rate of decrease in the surface temperature of a warm sphere of pure copper by means of a thermocouple attached to its surface when it is immersed in a stream of cooler air. The initial uniform temperature of the surface of the 0.02-m-diameter sphere is 75°C, and after immersion for 60 s in the airstream, which has a temperature of 25°C, the surface temperature is 68°C. Calculate the value of the heat transfer coefficient.

PROBLEM 8.2

Steel rods of 0.05 m diameter are heat treated by being passed through a heat-treating furnace in which the atmosphere is at 700°C. If the furnace is 5 m in length and the rods enter at 50°C, at what speed must they pass through the furnace to be at 600°C when they leave the furnace? The heat transfer coefficient is 135 $W \cdot m^{-2} \cdot K^{-1}$.

PROBLEM 8.3

The temperature of a semi-infinite slab of steel, which is at the uniform temperature of 25°C, is raised instantaneously to 50°C. Calculate:

(a) The distance beneath the surface at which the local temperature is 30°C after 5 min
(b) The heat flux to the slab at $t = 5$ min
(c) The thermal energy that has been transferred to unit area of the slab during the first 5 min

PROBLEM 8.4

A thick oak wall at the uniform temperature 25°C is suddenly exposed to gaseous reaction products at 800°C. Calculate the time required to raise the temperature of the surface of the wall to its ignition temperature of 400°C. The heat transfer coefficient is 20 $W \cdot m^{-2} \cdot K^{-1}$.

PROBLEM 8.5

A thick slab of alumina at 600°C is exposed to a gas at 200°C and the heat transfer coefficient is 100 $W \cdot m^{-2} \cdot K^{-1}$. Calculate:

(a) The surface temperature of the slab
(b) The temperature at a point 0.01 m below the surface after 10 min of exposure

PROBLEM 8.6

Calculate the electrical current at which a B&S gauge (2.588 mm diameter) magnesium fuse operates when the heat transfer coefficient at its surface is 5 $W \cdot m^{-2} \cdot K^{-1}$ and the ambient temperature is 25°C. The specific resistivity of Mg is 4.6×10^{-6} $\Omega \cdot cm$.

PROBLEM 8.7

A plane wall of thickness 0.3 m at a uniform temperature of 100°C suddenly has both surfaces exposed to convection conditions with air at 20°C in which $h = 100$ $W \cdot m^{-2} \cdot K^{-1}$. Use the finite-difference technique to determine the temperature profile in the wall after 15 min of exposure. The properties of the material are $C_p = 800$ $J/(kg \cdot K)$, $\rho = 2500$ kg/m^3, and $k = 50$ $W/(m \cdot K)$.

PROBLEM 8.8

The plane wall at 100°C in Problem 8.7 is suddenly exposed to convection conditions with air at 20°C in which $h = 100$ W·m^{-2}·K^{-1} on one side of the wall and $h = 10$ W·m^{-2}·K^{-1} on the other side. Use the finite-difference technique to determine the temperature profile in the wall after 15 min of exposure.

PROBLEM 8.9

A plane wall of thickness 0.03 m is at a uniform temperature of 100°C. The temperature of one face of the wall is instantaneously deceased to 20°C, while the other face is perfectly insulated. Using the finite-difference technique, calculate the temperature profile in the wall after 2 min if the thermal diffusivity is 1.0×10^{-6} m^2/s.

PROBLEM 8.10

A steel pipeline of 1 m inner diameter and wall thickness 0.04 m is well insulated on its outer surface and, initially, is at a uniform temperature of 0°C. Oil at 60°C is allowed to flow through the pipeline, which causes a surface convection condition in which h is 500 W·m^{-2}·K^{-1}.

(a) Using the finite-difference technique, calculate the temperature profile in the wall of the pipe after the oil has been flowing for 5 min.
(b) Estimate the heat flux from the oil to the pipe at $t = 5$ min.
(c) Estimate the total thermal energy transferred to the pipe per meter of length after contact with the oil for 5 min.

As the diameter of the pipe is very much larger than the wall thickness, the heat flow can be considered to be one-dimensional.

PROBLEM 8.11

A nuclear fuel element in the shape of a plane slab of thickness 0.02 m generates heat at the rate $\dot{q} = 10^7$ W/m^3, which is transferred by convection to liquid sodium at 250°C under conditions of $h = 1100$ W·m^{-2}·K^{-1}.

(a) Calculate the temperature profile in the element when it is operating at steady state.

The uniform generation of heat suddenly increases to $\dot{q} = 20^8$ W/m^3. Using the finite-difference technique, determine:

(b) The temperature profile in the element 2 s after the increase in power
(c) How long the power surge can be endured before the sodium begins to boil.

For the nuclear fuel, $k = 30$ W·m^{-1}·K^{-1} and $\alpha = 10^{-5}$ m^2/s and sodium boils at 883°C.

PROBLEM 8.12

A steel billet of dimensions 0.4 m $\times$ 0.4 m, at an initially uniform temperature of

25°C, is placed in a reheating furnace at 800°C. Use the finite-difference technique to calculate the time required for the center of the billet to reach 600°C when the heat transfer coefficient is 50 W·m^{-2}·K^{-1}.

PROBLEM 8.13

Repeat the first part of Problem 8.3 using the finite-difference technique.

PROBLEM 8.14

Liquid steel at 1650°C is rapidly poured to a depth of 2.5 m into a ladle of diameter 3 m which has been preheated to 700°C. Calculate how much heat is transferred from the steel to the ladle by conduction during the first minute. For the refractory lining of the ladle,

$$k = 1.04 \text{ W·m}^{-1}\text{·K}^{-1}$$

$$\alpha = 3.1 \times 10^{-7} \text{m}^2/\text{s}$$

For the liquid steel;

$$\rho = 7050 \text{ kg/m}^3$$

$$C_p = 750 \text{ J·kg}^{-1}\text{·K}^{-1}$$

PROBLEM 8.15

Liquid steel at its freezing temperature of 1480°C is cast to form a slab of thickness 0.1 m. Calculate the times required for completion of the freezing process when the mold is (a) a thick sand mold at 25°C and (b) a water-cooled copper mold at 25°C. For the steel,

$$\alpha = 1.1 \times 10^{-5} \text{m}^2/\text{s}$$

$$C_p = 750 \text{ J·kg}^{-1}\text{·K}^{-1}$$

$$\Delta H = 2.56 \times 10^5 \text{ J·kg}^{-1}$$

$$\rho = 7100 \text{ kg/m}^3$$

For the sand,

$$\alpha = 2.5 \times 10^{-7} \text{ m}^2/\text{s}$$

$$k = 0.3 \text{ W·m}^{-1}\text{·K}^{-1}$$

PROBLEM 8.16

Is it feasible to cast liquid iron at its melting temperature into a thick-walled aluminum mold? For liquid iron,

$$\text{freezing temperature} = 1535°C$$

$$\Delta H = 247{,}200 \text{ J·kg}^{-1}$$

$$\rho = 7219 \text{ kg/m}^3$$

$$C_p = 786 \text{ J·kg}^{-1}\text{·K}^{-1}$$

$$k = 63 \text{ W·m}^{-1}\text{·K}^{-1}$$

$$\alpha = 1.1 \times 10^{-6} \text{ m}^2/\text{s}$$

For solid aluminum,

$$\text{melting temperature} = 660°\text{C}$$

$$\rho = 2700 \text{ kg/m}^3$$

$$C_p = 902 \text{ J·kg}^{-1}\text{·K}^{-1}$$

$$k = 236 \text{ W·m}^{-1}\text{·K}^{-1}$$

$$\alpha = 9.7 \times 10^{-5}\text{m}^2/\text{s}$$

PROBLEM 8.17

To produce a surface layer of martensite on the surface of a steel part, the temperature of the layer of steel that transforms from austenite to martensite as the result of a quench must be decreased from the soaking temperature of 900°C to 400°C in 2 s or less. If it is assumed that the quench instantaneously decreases the temperature of the surface from 900°C to 25°C, what thickness of martensite is produced? The thermal diffusivity of the steel is $1.2 \times 10^{-5} \text{ m}^2/\text{s}$.

9

Heat Transport by Thermal Radiation

9.1 Introduction

The transport of heat as thermal radiation is completely different from heat transport by conduction or convection. Thermal radiation is part of the electromagnetic spectrum shown in Fig. 9.1, and hence the propagation of energy by thermal radiation differs significantly from that by conduction and convection in that it does not require the presence of matter; the Earth is heated by thermal radiation from the Sun that is transported through the vacuum of space. Although the electromagnetic spectrum includes wavelengths in the range zero to infinity, significant thermal radiation is confined to wavelengths in the range 0.1 to 100 μm, which includes visible light in the range 0.4 to 0.7 μm and ultraviolet and infrared light at, respectively, lower and higher wavelengths. Electromagnetic waves are produced by oscillations of the electrical fields surrounding charges, and thus all matter at a finite temperature emits thermal radiation caused by atomic scale motions such as the vibration and rotation of molecules, the vibration of atoms, and certain electronic transitions. Although thermal emission can be thought of in terms of the propagation of discrete packets of energy, called quanta or photons, it is more conveniently dealt with as the propagation of electromagnetic waves of wavelength λ and frequency ν, which are related as

$$\lambda \times \nu = c \qquad (9.1)$$

where c is the velocity of light in the medium through which the radiation is passing. The speed of light in a vacuum, c_0, is 2.998×10^8 m/s.

A surface at a finite temperature emits thermal radiation over a range of wavelengths in all directions and the intensity and power of this radiation vary with the

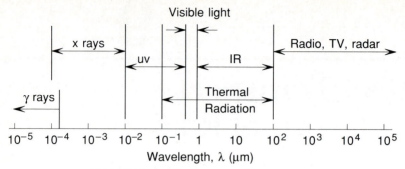

FIGURE 9.1. Electromagnetic spectrum.

wavelength of the radiation and with the direction of emission from the surface in which the radiation is emitted. The nature of the variation of the intensity of the radiation with wavelength is shown in Fig. 9.2(a) and the nature of its possible variation with direction is shown in Fig. 9.2(b). The situation in Fig. 9.2(b) is one in which the intensity of the emitted radiation has a maximum value in a direction normal to the surface, $\theta = 0$, and decreases with increasing θ. The variation shown in Fig. 9.2(a) is called the *spectral distribution* and the variation shown in Fig. 9.2(b) is called the *directional distribution*. These two differing types of distribution complicate the treatment of heat transport by radiation.

A surface can also receive thermal radiation from some other source, in which case the surface is said to be *irradiated*. If the matter being irradiated is opaque to the incoming radiation, the irradiation is reflected and/or absorbed by the surface, or if the irradiated matter is transparent to the incoming radiation, some of the

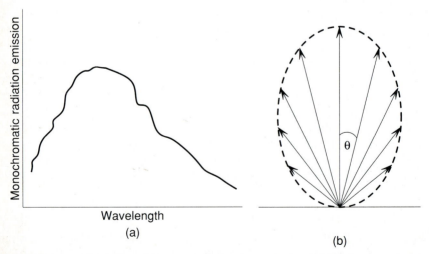

FIGURE 9.2. (a) Variation of the monochromatic radiation emission with wavelength of the radiation; (b) variation of radiation emission from a surface with the direction of the emission.

irradiation is transmitted through the matter. The fractions of the irradiation reflected, absorbed, and transmitted are called, respectively, the reflectivity, absorptivity, and transmissivity of the surface being irradiated. Thus three types of flux must be considered in the treatment of radiation: (1) the flux emanating from a surface by virtue of the finite temperature of the surface, which gives rise to the definition of the emissive power of the surface; (2) irradiation, which is the flux from another source of radiation incident on the surface; and (3) the radiosity of the surface, which is the sum of the intrinsic emission and the reflected irradiation.

9.2 Intensity and Emissive Power

In Fig. 9.3(a) radiation is being emitted in all directions and at all wavelengths by the surface 1. Consider the radiation emitted by the surface element dA_1, in the direction θ,ϕ, at the wavelength λ which passes thorugh the element of area dA_2, which is normal to the direction θ,ϕ and located at a distance r from the surface element dA_1. When viewed from dA_1, the area dA_2 subtends a solid angle $d\omega$ defined as

$$d\omega = \frac{dA_2}{r^2} \qquad (9.2)$$

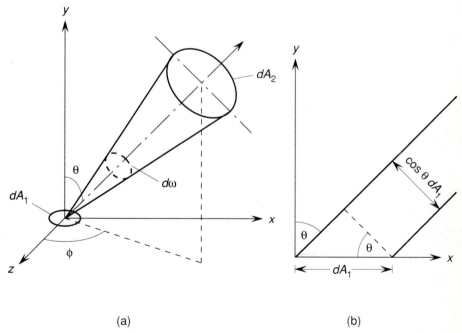

(a)   (b)

FIGURE 9.3. (a) Radiation of wavelength λ, emitted by the surface element dA_1 in the direction θ, ϕ, and intercepted by the area dA_2; (b) relationship between the area that dA_1 projects in the direction θ and the actual area dA_1.

in which the unit of ω is the steradian (sr). With reference to Fig. 3.11,

$$dA_2 = r^2 \sin \theta \ d\theta \ d\phi$$

and hence

$$d\omega = \sin \theta \ d\theta \ d\phi \qquad \text{(i)}$$

The solid angle subtended by a hemisphere placed over dA_1 is obtained by integrating Eq. (i) over the intervals $\theta = 0$ to $\pi/2$ and $\phi = 0$ to 2π:

$$\omega = \int_0^{2\pi} \int_0^{\pi/2} \sin \theta \ d\theta \ d\phi$$

$$= 2\pi [\cos \theta]_0^{\pi/2}$$

$$= 2\pi \quad \text{sr}$$

In considering the radiation emitted by the surface element dA_1, the *monochromatic* or *spectral intensity*, $I_{\lambda,e}$, of the emission is defined as

$I_{\lambda,e} =$ rate at which radiant energy is emitted

at wavelength λ

in direction θ, ϕ

per unit area of emitting surface normal to direction θ, ϕ

per unit solid angle about that direction

per unit wavelength interval $d\lambda$ about λ

With wavelength measured in micrometers (1 $\mu m = 10^{-6}$ m), spectral intensity has the units $W{\cdot}m^{-2}{\cdot}sr^{-1}{\cdot}\mu m^{-1}$. As shown in Fig. 9.3(b), the projected surface area normal to the direction θ, which is seen when viewed from area dA_2, is $\cos \theta \ dA_1$, and hence

$$I_{\lambda,e}(\lambda,\theta,\phi) = \frac{dq}{\cos \theta \ dA_1 \ d\omega \ d\lambda} \qquad (9.3)$$

In Eq. (9.3), the λ, θ, and ϕ in parentheses indicate that $I_{\lambda,e}$ is a function of λ, θ, and ϕ. The influences of wavelength and direction on the intensity of the radiation are separated by defining

$$dq_\lambda = \frac{dq}{d\lambda}$$

as the rate at which radiation of wavelength λ leaves dA_1 and passes through dA_2. Thus the spectral influence can be factored out and Eq. (9.2) can be written as

$$dq_\lambda = I_{\lambda,e}(\lambda,\theta,\phi) \cos \theta \ dA_1 \ d\omega \qquad (9.4)$$

or, from Eq. (i),

$$dq_\lambda' = \frac{dq_\lambda}{dA_1} = I_{\lambda,e}(\lambda,\theta,\phi) \cos \theta \sin \theta \ d\theta \ d\phi \qquad (9.5)$$

in which dq_λ' is the rate at which radiation of wavelength λ leaves dA_1 per unit area (not per unit projected area) and passes through dA_2. If $I_{\lambda,e}$ is known as a function of θ and ϕ, integration of Eq. (9.5) gives

$$q'_\lambda = \int_0^{2\pi} \int_0^{\pi/2} I_{\lambda,e}(\lambda,\theta,\phi) \cos \theta \sin \theta \; d\theta \; d\phi \tag{9.6}$$

and if $I_{\lambda,e}$ is known as a function of wavelength,

$$q' = \int_0^\infty q'_\lambda \; d\lambda \tag{9.7}$$

gives the total radiation heat flux emitted per unit area which is intercepted by the hemisphere above the surface 1.

Emissive Power

The *spectral, hemispherical emissive power, E_λ*, of radiation emitted by a surface is the rate at which radiation of wavelength λ is emitted in all directions per unit surface area per unit wavelength $d\lambda$ about λ. Thus, per unit area of surface, this quantity is obtained from Eq. (9.6) as

$$E_\lambda = q'_\lambda = \int_0^{2\pi} \int_0^{\pi/2} I_{\lambda,e}(\lambda,\theta,\phi) \cos \theta \sin \theta \; d\theta \; d\phi \tag{9.8}$$

and integrating Eq. (9.8) over all wavelengths gives the total emissive power of the radiation as

$$E = \int_0^\infty E_\lambda \; d\lambda \tag{9.9}$$

If the intensity of the emitted radiation is independent of the direction of emission, the surface is said to be a *diffuse emitter*, and for such a surface, Eq. (9.8) becomes

$$E_\lambda = I_{\lambda,e}(\lambda) \int_0^{2\pi} \int_0^{\pi/2} \cos \theta \sin \theta \; d\theta \; d\phi$$
$$= I_{\lambda,e}(\lambda) 2\pi [\tfrac{1}{2} \sin^2\theta]_0^{\pi/2}$$
$$= \pi I_{\lambda,e}(\lambda) \tag{9.10}$$

in which $I_{\lambda,e}$ has the units $\text{W·m}^{-2}\text{·sr}^{-1}$ and π has the units sr. Similarly, for a diffuse emitter, the total emissive power, obtained by averaging over all wavelengths, is

$$E = \pi I_e \tag{9.11}$$

in which I_e is the total intensity of the emitted radiation. Casual inspection of Eq. (9.11) might raise the question: Why is the total heat flux from a diffuse emitter not given by the radiant heat flux per steradian (the intensity) times the number of steradians in a hemisphere, 2π? The factor 2 does not occur in Eq. (9.11) because I_e is defined in terms of the area that dA_1 projects in the direction normal to the angle θ,ϕ, and E is defined in terms of the actual area dA_1.

Irradiation

In a manner that is directly analogous with the definition of the monochromatic or spectral intensity of radiation emitted by a surface, the monochromatic or spectral intensity of radiation from some other source which is intercepted by the surface $I_{\lambda,i}$ is defined as

$I_{\lambda,i}$ = rate at which energy is incident

at wavelength λ

from direction θ, ϕ

per unit area of intercepting surface normal to direction θ, ϕ

per unit solid angle about this direction

per unit wavelength interval $d\lambda$ about λ

The *spectral irradiation*, G_λ, in units of $W \cdot m^{-2} \cdot \mu m$ is then defined as the rate at which radiation from all directions is incident on a surface, per unit area of the surface, at the wavelength λ per unit wavelength $d\lambda$ about λ:

$$G_\lambda = \int_0^{2\pi} \int_0^{\pi/2} I_{\lambda,i}(\lambda,\theta,\phi) \cos\theta \sin\theta \, d\theta \, d\phi \tag{9.12}$$

and the total irradiation, G, is obtained as

$$G = \int_0^\infty G_\lambda \, d\lambda \tag{9.13}$$

If the incident radiation is diffuse,

$$G_\lambda = \pi I_{\lambda,i}(\lambda) \tag{9.14}$$

and hence the total irradiation is

$$G = \pi I_i \tag{9.15}$$

Radiosity

If some of the incident radiation on the surface is reflected by the surface, the radiosity of the surface is the sum of the direct emission and the reflected irradiation. The *spectral radiosity* J_λ is the rate at which radiation leaves unit area of the surface at the wavelength λ per unit wavelength interval $d\lambda$ in all directions. Thus

$$J_\lambda = \int_0^{2\pi} \int_0^{\pi/2} I_{\lambda,e+r}(\lambda,\theta,\phi) \cos\theta \sin\theta \, d\theta \, d\phi \tag{9.16}$$

and the total radiosity is

$$J = \int_0^\infty J_\lambda \, d\lambda \tag{9.17}$$

If the surface is both a diffuse emitter and a diffuse reflector,

$$J_\lambda = \pi I_{\lambda,e+r} \tag{9.18}$$

and

$$J = \pi I_{e+r}$$

9.3 Blackbody Radiation

The large variations in the radiative properties of real surfaces are dealt with by comparing the properties of real surfaces with those of an ideal surface called a *blackbody*. For any given temperature, radiation from a blackbody has the highest possible value of spectral intensity and hence the highest possible spectral emissive power. A blackbody is so-named because it absorbs all incident radiation of all wavelengths. It should not be confused with a surface that appears black to the human eye because the surface absorbs all incident radiation in the narrow band of visible light. Such a surface does not necessarily completely absorb radiation of all wavelengths. In addition to absorbing radiation of all wavelengths, a blackbody is a diffuse emitter. The spectral distribution of the intensity of radiation emitted by a blackbody has been determined by Planck to be

$$I_{\lambda,b}(\lambda,T) = \frac{2hc_0^2}{\lambda^5[\exp(hc_0/\lambda kT) - 1]} \tag{9.19}$$

in which h $(= 6.6256 \times 10^{-34}$ J·s) is Planck's constant, k $(= 1.3805 \times 10^{-23}$ J/K) is Boltzmann's constant, and c_0 $(= 2.998 \times 10^8$ m/s) is the speed of light in a vacuum. Since a blackbody is a diffuse emitter, its spectral emissive power is obtained from Eqs. (9.10) and (9.19) as

$$E_{\lambda,b}(\lambda,T) = \pi I_{\lambda,b}(\lambda,T) = \frac{C_1}{\lambda^5[\exp(C_2/\lambda T) - 1]} \quad \text{W·m}^{-2}\text{·}\mu\text{m} \tag{9.20}$$

in which C_1 $(= 3.742 \times 10^8$ W·μm^4·m^{-2}) and C_2 $(= 1.439 \times 10^4$ μm·K) are, respectively, the first and second radiation constants. Equation (9.21), which is known as *Planck's distribution*, is shown for several temperatures in Fig. 9.4. Several important features should be noted:

1. At any temperature, the spectral emissive power is a continuous function of wavelength.
2. At any wavelength, the spectral emissive power increases with increasing temperature.
3. With increasing temperature, the wavelength, at which the spectral emissive power has its maximum value, decreases.

Setting $dE_\lambda/d\lambda$ equal to zero gives the variation, with temperature, of the wavelength at which the spectral intensity has its maximum value, λ_{max}, as

$$\lambda_{max}T = C_3 \tag{9.21}$$

Equation (9.21) is Wien's displacement law and C_3 $(= 2897.6$ μm·K) is the third radiation constant. On the log-log plot of Fig. 9.4, Wien's displacement law is a straight line with a slope of -1.

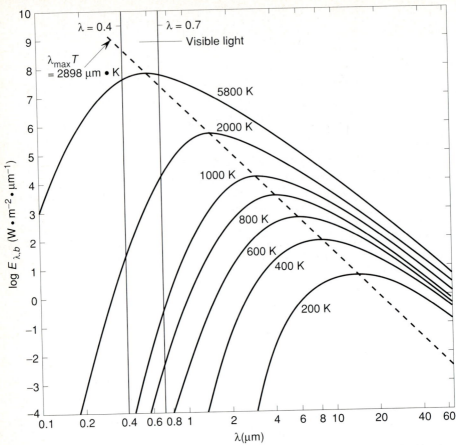

FIGURE 9.4. Planck's distribution of the spectral emissive power of a blackbody.

The total emissive power of a blackbody at any temperature T is the area under the curve in Fig. 9.4:

$$E_b = \int_0^\infty E_{\lambda,b}\, d\lambda = \sigma T^4 \tag{9.22}$$

Equation (9.22) is the *Stefan–Boltzmann law*, named after Stefan, who measured the dependence, on temperature, of the total emissive power, and after Boltzmann, who derived the relationship theoretically; σ ($= 5.670 \times 10^{-8}$ W·m^{-2}·K^{-4}) is the Stefan–Boltzmann constant. The Stefan–Boltzmann law shows that the rate at which radiation is emitted by a blackbody is a function only of temperature.

The emission from a blackbody in the interval of wavelengths from λ_1 to λ_2 is obtained from Eq. (9.22) as

$$\int_{\lambda_1}^{\lambda_2} E_{\lambda,b}\, d_\lambda$$

TABLE 9.1 Blackbody Radiation Functions

λT (μm·K)	$f_{(0\to\lambda)}$	λT (μm·K)	$f_{(0\to\lambda)}$
200	0.000000	6,200	0.754140
400	0.000000	6,400	0.769234
600	0.000000	6,600	0.783199
800	0.000016	6,800	0.796129
1,000	0.000321	7,000	0.808109
1,200	0.002134	7,200	0.819217
1,400	0.007790	7,400	0.829527
1,600	0.019718	7.600	0.839102
1,800	0.039341	7,800	0.848005
2,000	0.066728	8,000	0.856288
2,200	0.100888	8,500	0.874608
2,400	0.140256	9,000	0.890029
2,600	0.183120	9,500	0.903085
2,800	0.227897	10,000	0.914199
2,898	0.250108	10,500	0.923710
3,000	0.273232	11,000	0.931890
3,200	0.318102	11,500	0.939959
3,400	0.361735	12,000	0.945098
3,600	0.403607	13,000	0.955139
3,800	0.443382	14,000	0.962898
4,000	0.480877	15,000	0.969981
4,200	0.516014	16,000	0.973814
4,400	0.548796	18,000	0.980860
4,600	0.579280	20,000	0.985602
4,800	0.607559	25,000	0.992215
5,000	0.633747	30,000	0.995430
5,200	0.658970	40,000	0.997967
5,400	0.680360	50,000	0.998953
5,600	0.701046	75,000	0.999713
5,800	0.720158	100,000	0.999905
6,000	0.737818		

and hence the fraction of the total emission from a blackbody at the temperature T that occurs in this interval is

$$\frac{\int_{\lambda_1}^{\lambda_2} E_{\lambda,b}\, d\lambda}{\sigma T^4}$$

With the lower limit of $\lambda_1 = 0$, this fraction is defined as

$$f_{(0\to\lambda)} = \frac{\int_0^\lambda E_{\lambda,b}\, d\lambda}{\sigma T^4}$$

which, from Eq. (9.20), can be written as

$$f_{(0 \to \lambda)} = \frac{\displaystyle\int_0^{\lambda} \frac{C_1 \, d\lambda}{\lambda^5 [\exp(C_2/\lambda T) - 1]}}{\sigma T^4}$$

$$= \frac{C_1}{\sigma T^4} \int_0^{\lambda} \frac{d\lambda}{\lambda^5 [\exp(C_2/\lambda T) - 1]}$$

or, denoting $x = \lambda T$, then at constant T,

$$f_{(0 \to \lambda)} = \frac{C_1}{\sigma T^5} \int_0^{x} \frac{dx}{\lambda^5 [\exp(C_2/x) - 1]}$$

$$= \frac{C_1}{\sigma} \int_0^{x} \frac{dx}{x^5 [\exp(C_2/x) - 1]} \tag{9.23}$$

Values of $f_{(0 \to \lambda)}$ are listed for discrete values of λT in Table 9.1.

EXAMPLE 9.1 _____

The sun is a blackbody radiating at an effective temperature of 5800 K. Determine the fractions of the sun's emission that occur in (a) the ultraviolet range (between wavelengths of 0.01 and 0.4 μm), (b) the visible range (between wavelengths of 0.4 and 0.7 μm), and (c) the infrared range (between wavelengths of 0.7 and 1000 μm).

(a) With $T = 5800$ K, λT varies from 58 to 2320 μm·K in the ultraviolet range. Thus, from interpolation in Table 9.1, with $\lambda T = 58$ μm·K, $f_{(0 \to 0.01)} = 0$, and with $\lambda T = 2320$ μm·K, $f_{(0 \to 0.4)} = 0.125$. Therefore,

$$f_{(0.01 \to 0.04)} \text{ at } 5800 \text{ K} = 0.125 - 0 = 0.125$$

or 12.5% of the emission occurs in the ultaviolet range.

(b) With $T = 5800$ K, λT varies from 2320 to 4060 μm·K in the visible range. Thus

$$f_{(0 \to 0.40)} = 0.125 \quad \text{and} \quad f_{(0 \to 0.70)} = 0.491$$

Thus

$$f_{(0.40 \to 0.70)} \text{ at } 5800 \text{ K} = 0.491 - 0.125 = 0.366$$

or 36.6% occurs in the visible range.

(c) From Table 9.1, $f_{(0 \to 1000)}$ for $T = 5800$ K is essentially unity, and hence the fraction of the sun's radiation emitted in the infrared range is $1 - 0.125 - 0.366 = 0.509$.

9.4 Emissivity

Comparison of the radiation emitted by a real surface with that emitted by a black-body gives rise to the definition of the emissivity of the real surface, ϵ, as

$$\epsilon = \frac{\text{radiation emitted by a real surface at temperature } T}{\text{radiation emitted by a blackbody at the same temeprature } T}$$

However, as the radiation emitted by real surfaces varies with temperature, wave-length, and direction, several emissivities must be defined.

1. The *monochromatic, directional emissivity*, $\epsilon_{\lambda,\theta}(\lambda,\theta,\phi,T)$, defined as

$$\epsilon_{\lambda,\theta}(\lambda,\theta,\phi,T) = \frac{\begin{array}{c}\text{intensity of radiation of wavelength } \lambda \text{ from}\\ \text{a surface at temperature } T \text{ in direction } \theta,\phi\end{array}}{\begin{array}{c}\text{intensity of radiation of wavelength } \lambda \text{ emitted}\\ \text{by a blackbody at temperature } T\end{array}}$$

$$= \frac{I_{\lambda,e}(\lambda,\theta,\phi,T)}{I_{\lambda,b}(\lambda,T)} \tag{9.24}$$

2. The *total, directional emissivity*, $\epsilon_{\theta}(\theta,\phi,T)$, which is an average over all wave-lengths and is defined as

$$\epsilon_{\theta}(\theta,\phi,T) = \frac{\begin{array}{c}\text{intensity of radiation of all wavelengths from}\\ \text{a surface at temperature } T \text{ in direction } \theta, \phi\end{array}}{\begin{array}{c}\text{intensity of radiation of all wavelengths emitted}\\ \text{by a blackbody at temperature } T\end{array}}$$

$$= \frac{I_e(\theta,\phi,T)}{I_b(T)} \tag{9.25}$$

3. The *monochromatic, hemispherical* emissivity, ϵ_{λ}, defined as

$$\epsilon_{\lambda}(\lambda,T) = \frac{\begin{array}{c}\text{spectral hemispherical emissive power}\\ \text{of the surface at temperature } T\end{array}}{\begin{array}{c}\text{spectral hemispherical emissive}\\ \text{of a blackbody at temperature } T\end{array}}$$

$$= \frac{E_{\lambda}(\lambda,T)}{E_{\lambda,b}(\lambda,T)}$$

which, from Eq. (9.8), becomes

$$\epsilon_{\lambda}(\lambda,T) = \frac{\int_0^{2\pi}\int_0^{\pi/2} I_{\lambda,e}(\lambda,\theta,\phi,T)\cos\theta\sin\theta\,d\theta\,d\phi}{\int_0^{2\pi}\int_0^{\pi/2} I_{\lambda,b}(\lambda,T)\cos\theta\sin\theta\,d\theta\,d\phi}$$

Substitution from Eq. (9.24) gives

$$\epsilon_\lambda(\lambda,T) = \frac{\int_0^{2\pi} \int_0^{\pi/2} \epsilon_{\lambda,\theta}(\lambda,\theta,\phi,T) \cdot I_{\lambda,b}(\lambda,T) \cos\theta \sin\theta \, d\theta \, d\phi}{\int_0^{2\pi} \int_0^{\pi/2} I_{\lambda,b}(\lambda,T) \cos\theta \sin\theta \, d\theta \, d\phi}$$

or, as blackbody radiation is diffuse,

$$\epsilon_\lambda(\lambda,T) = \frac{\int_0^{2\pi} \int_0^{\pi/2} \epsilon_{\lambda,\theta}(\lambda,\theta,T) \cos\theta \sin\theta \, d\theta \, d\phi}{\int_0^{2\pi} \int_0^{\pi/2} \cos\theta \sin\theta \, d\theta \, d\phi}$$

$$= \frac{2\pi \int_0^{\pi/2} \epsilon_{\lambda,\theta}(\lambda,\theta,T) \cos\theta \sin\theta \, d\theta}{\pi}$$

$$= 2 \int_0^{\pi/2} \epsilon_{\lambda,\theta}(\lambda,\theta,T) \cos\theta \sin\theta \, d\theta \tag{9.26}$$

4. The *total, hemispherical* emissivity, $\epsilon(T)$, which on average over all wavelengths and directions, is defined as

$$\epsilon(T) = \frac{\text{total emissive power of radiation from a surface at temperature } T}{\text{total emissive power of blackbody radiation at temperature } T}$$

$$= \frac{E(T)}{E_b(T)}$$

$$= \frac{\int_0^\infty \epsilon_\lambda(\lambda,T) E_{\lambda,b}(\lambda,T) \, d\lambda}{E_b(T)} \tag{9.27}$$

Typical variations of ϵ_θ with θ for conducting materials and nonconducting materials are shown in Fig. 9.5(a). For conducting materials ϵ_θ is approximately constant for θ in the range 0 to 40°, after which it increases with increasing θ to a maximum value before decreasing to zero at 90°. In contrast, ϵ_θ for conducting materials is approximately constant up to $\theta = 70°$ and thereafter decreases rapidly. The variations, with θ, of ϵ_θ for highly polished surfaces of the nonconductors Al_2O_3 and CuO and the conductors Al and Cr are shown in Fig. 9.5(b). Figure 9.5 shows that the hemispherical emissivity, ϵ, is not significantly different from the normal emissivity, ϵ_n, corresponding to $\theta = 0$. Figure 9.6 shows the variation, with wavelength, of the spectral, normal emissivities of various materials, and Fig. 9.7 shows the corresponding variations, with temperature, of the total, normal emissivities.

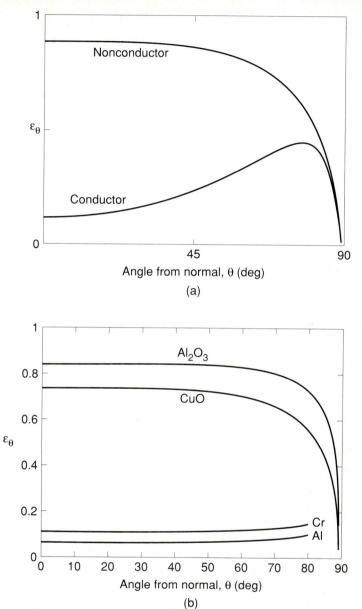

FIGURE 9.5. (a) Typical variations of the total directional emissivities, ϵ_θ, with angle of emission from the surfaces of a conductor and a nonconductor; (b) variations of ϵ_θ with angle of emission from the surfaces of two oxides (nonconductors) and two metals (conductors).

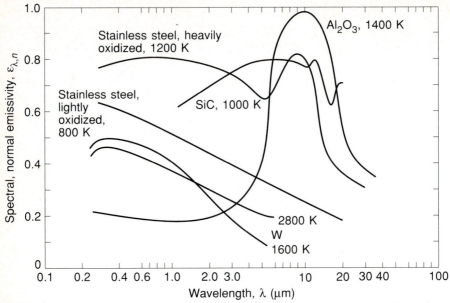

FIGURE 9.6. Variations, with wavelength, of the spectral normal emissivities of several substances.

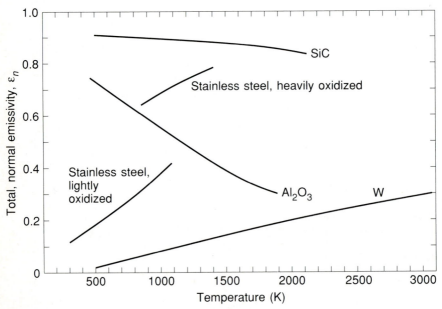

FIGURE 9.7. Variations, with temperature, of the total normal emissivities of several substances.

EXAMPLE 9.2

The spectral hemispherical emissivity of Al_2O_3 at 1400 K, shown in Fig. 9.6, can be approximated as shown in Fig. 9.8. Determine the total hemispherical emissivity of Al_2O_3 at 1400 K.

From Eq. (99.27),

$$\epsilon(T) = \frac{\int_0^\infty \epsilon_\lambda(\lambda)E_{\lambda,b}(\lambda,T)\,d\lambda}{E_b(T)}$$

which from Fig. 9.8 can be written as

$$\epsilon(T) = 0.2 \int_0^{4\mu m} \frac{E_{\lambda,b}(\lambda,T)}{E_b(T)}\,d\lambda + 0.95 \int_{4\mu m}^{11\mu m} \frac{E_{\lambda,b}(\lambda,T)}{E_b(T)}\,d\lambda$$

$$= 0.2 \int_0^{4\mu m} \frac{E_{\lambda,b}(\lambda,T)}{E_b(T)}\,d\lambda$$

$$+ 0.95 \left[\int_0^{11\mu m} \frac{E_{\lambda,b}(\lambda,T)}{E_b(T)}\,d\lambda - \int_0^{4\mu m} \frac{E_{\lambda,b}(\lambda,T)}{E_b(T)}\,d\lambda \right]$$

From Table 9.1,

$\lambda = 4\mu m$ and $T = 1400$ K; $\lambda T = 5600$ μm·K and $f_{(0 \to 4\mu m)} = 0.701$

$\lambda = 11$ μm   and $T = 1400$ K; $\lambda T = 15{,}400$ μm·K and $f_{(0 \to 11\mu m)} = 0.972$

Therefore,

$$\epsilon(1400\text{ K}) = 0.2 \times 0.701 + 0.95(0.972 - 0.701) = 0.398$$

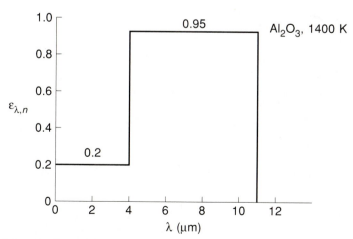

FIGURE 9.8. Idealized variation, with wavelength, of the spectral normal emissivity of Al_2O_3 used in Example 9.2.

9.5 Absorptivity, Reflectivity, and Transmissivity

Figure 9.9 shows the processes occurring at the surface of a body being irradiated. If the matter being irradiated is transparent to the irradiation, the irradiation can be reflected, absorbed, and transmitted, and the sum of the portions reflected, $G_{\lambda,ref}$, absorbed, $G_{\lambda,abs}$, and transmitted, $G_{\lambda,trans}$, equals G_λ:

$$G_\lambda = G_{\lambda,ref} + G_{\lambda,abs} + G_{\lambda,trans} \qquad (9.28)$$

If the medium is opaque to the irradiation (i.e., $G_{\lambda,trans} = 0$), treatment of phenomenon is greatly simplified and

$$G_\lambda = G_{\lambda,ref} + G_{\lambda,abs} \qquad (9.29)$$

The absorptivity of the surface is defined as the fraction of the irradiation that is absorbed. As was the case with emissivity, the absorptivity depends on both the wavelength of the irradiation and the direction of its incidence on the surface, and consequently, several types of absorptivity can be defined. However, integrating over all wavelengths and all directions gives the *total, hemispherical* absorptivity of the surface, α, as the fraction of the total irradiation that is absorbed,

$$\alpha = \frac{G_{abs}}{G} \qquad (9.30)$$

Similarly, the *total, hemispherical* reflectivity of the surface, ρ, is defined as the fraction of the total irradiation that is reflected,

$$\rho = \frac{G_{ref}}{G} \qquad (9.31)$$

and if the matter is transparent to the irradiation, the *total, hemispherical* transmis-

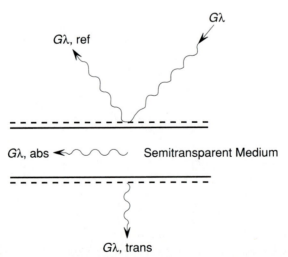

FIGURE 9.9. Processes that can occur at the surface of an irradiated body.

sivity is defined as the fraction of the total irradiation that is transmitted,

$$\tau = \frac{G_{\text{trans}}}{G} \qquad (9.32)$$

The sum of Eqs. (9.30), (9.31), and (9.32) gives

$$\alpha + \rho + \tau = 1$$

or for an opaque medium,

$$\alpha + \rho = 1 \qquad (9.33)$$

When the medium is opaque to the irradiation the absorption and reflection processes can be considered to be surface phenomena, which occur within fractions of a micrometer beneath the surface of the medium. The irradiation is thus reflected or absorbed by the surface and the two processes are influenced by the nature of the surface. Only the absorbed portion of the irradiation contributes to the thermal energy or enthalpy of the medium.

9.6 Kirchhoff's Law and the Hohlraum

Figure 9.10 shows an isothermal cavity, enclosed by a blackbody surface at the temperature T_s, which contains a small real body. The body emits radiation at the rate E and absorbs irradiation from the walls of the cavity at the rate $\alpha E_b(T_s)$ and the net rate of transfer of energy from the body is the difference between these two quantities. At thermal equilibrium the temperature of the body is also T_s, and hence the net rate of transfer of energy from the body is zero, in which case

$$E(T_s) = \alpha E_b(T_s) \qquad (9.34)$$

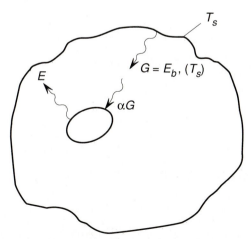

FIGURE 9.10. Isothermal cavity, surrounded by a blackbody surface, which contains a small real body.

or with ϵ being the total hemispherical emissivity of the body,

$$\epsilon E_b(T_s) = \alpha E_b(T_s)$$

or

$$\epsilon = \alpha \qquad (9.35)$$

that is, the total hemispherical emissivity of a surface within the enclosure is equal to its total hemispherical absorptivity. Equation (9.35) is known as *Kirchhoff's law*, and Eq. (9.32) shows that inasmuch that α is always either less than or equal to unity, $E(T_s)$ cannot be greater than $E_b(T_s)$, which indicates that a blackbody has the maximum possible emissive power at any temperature. Equation (9.35) is valid provided that either the radiation is black or that α_λ and ϵ_λ are independent of λ. A surface at which α_λ and ϵ_λ are independent of λ is said to be *gray*, and a surface for which $\alpha_{\lambda,\theta}$ and $\epsilon_{\lambda,\theta}$ are independent of λ and θ is said to be a *diffuse-gray* surface. The assumption of diffuse-gray behavior greatly simplifies the treatment of radiation exchange between surfaces.

Although a blackbody is hypothetical, its radiation field can be closely approximated by a hohlraum (heated cavity) shown in Fig. 9.11. In Fig. 9.11 the hole in the wall of a real isothermal cavity is made small enough that radiation entering the cavity through the hole undergoes multiple reflections from the walls, and only a small fraction of this radiation escapes from the cavity. As a result the radiation escaping from the cavity is representative of the radiation field within the cavity. Radiant energy emitted at some location on the cavity wall at the rate ϵE_b undergoes multiple reflections at other positions on the wall. At the first point of reflection a portion $\rho\epsilon E_b$ is reflected and the remaining portion $\alpha\epsilon E_b$ is absorbed. At the second point of reflection $\rho(\rho\epsilon E_b)$ is reflected and at the point of the nth reflection, $\rho^n\epsilon E_b$ is reflected. As the points of reflection are randomly located, the number of reflections that the radiation has undergone before emerging from the cavity through the

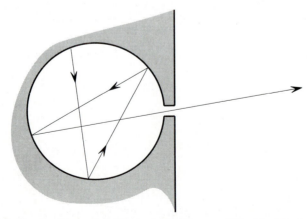

FIGURE 9.11. Isothermal cavity, or hohlraum, emitting blackbody radiation.

hole varies from 1 to infinity, and the emissive power of the emerging radiation is thus

$$E = \epsilon E_b(1 + \rho + \rho^2 + \rho^3 + \cdots) \tag{9.36}$$

Writing the sum in parentheses as S gives

$$S = 1 + \rho + \rho^2 + \rho^3 + \cdots$$

$$\frac{S}{\rho} = \frac{1}{\rho} + S$$

or

$$S = \frac{1}{1 - \rho}$$

However, in view of Eqs. (9.33) and (9.33),

$$S = \frac{1}{\alpha} = \frac{1}{\epsilon}$$

and hence Eq. (9.36) becomes

$$E = E_b$$

which shows that the hohlraum very closely approximates a blackbody.

9.7 Radiation Exchange Between Surfaces

In a system consisting of several surfaces at differing temperatures, the question of radiation exchange among the surfaces is of interest. For example, what fraction of the radiation emitted by one surface is intercepted by another surface, and as a result of the radiation exchange, which surfaces are gaining thermal energy and which are losing energy? The exchange of radiation between surfaces is facilitated by introduction of the *view factor*, which allows calculation of the fraction of the radiation emitted by one surface that is intercepted by another surface. Figure 9.12 shows the geometrical arrangement of two plane surfaces, one of area A_1 at the temperature T_1 and the other of area A_2, which is at the temperature T_2. The surface element dA_1 emits radiation in all directions, and some of this radiation is intercepted by the surface element dA_2. The distance between the two surface elements is R and the angles between the straight line joining the elements and the directions normal to the two surfaces are, respectively, θ_1 and θ_2. From Eq. (9.4), radiation emitted by dA_1 is intercepted by dA_2 at the rate

$$dq_{1 \to 2} = I_1 \cos \theta_1 \, dA_1 \, d\omega_{2-1} \tag{9.37}$$

in which I_1 is the intensity of the radiation leaving surface 1 and ω_{2-1} is the solid angle subtended by dA_2 when viewed from dA_1. The area projected by dA_2 normal

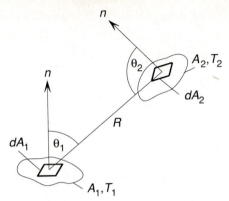

FIGURE 9.12.   Radiation emitted by surface element dA_1 that is intercepted by surface element dA_2.

to dA_1 is $\cos \theta_2 \, dA_2$ and hence, from Eq. (9.2)

$$d\omega = \frac{\cos \theta_2 \, dA_2}{R^2}$$

substitution of which into Eq. (9.37) gives

$$dq_{1 \to 2} = I \frac{\cos \theta_1 \, \cos \theta_2}{R^2} \, dA_1 \, dA_2 \tag{9.38}$$

If surface 1 emits and reflects diffusely, combination of Eq. (9.18) and Eq. (9.38) gives

$$dq_{1 \to 2} = J_1 \frac{\cos \theta_1 \, \cos \theta_2}{\pi R^2} \, dA_1 \, dA_2 \tag{9.39}$$

Assuming that the radiosity of surface 1, J_1, is uniform over the entire surface, the total rate at which radiation leaves the surface 1 and is intercepted by surface 2 is obtained by integrating Eq. (9.39) over both areas to get

$$q_{1 \to 2} = J_1 \int_{A_2} \int_{A_1} \frac{\cos \theta_1 \, \cos \theta_2}{\pi R^2} \, dA_1 \, dA_2 \tag{9.40}$$

The total radiation emitted by surface 1 is

$$A_1 J_1 \tag{9.41}$$

and hence the view factor F_{12}, which is the fraction of the total radiation emitted by surface 1 that is intercepted by surface 2, is obtained as the ratio of Eq. (9.40) to Eq. (9.41):

$$F_{12} = \frac{q_{1 \to 2}}{A_1 J_1} = \frac{1}{A_1} \int_{A_2} \int_{A_1} \frac{\cos \theta_1 \, \cos \theta_2}{\pi R^2} \, dA_1 \, dA_2 \tag{9.42}$$

Application of the derivation above to calculation of the fraction of the total radiation emitted by surface 2 that is intercepted by surface 1 gives the view factor F_{21} as

$$F_{21} = \frac{q_{2\to1}}{A_2 J_2} = \frac{1}{A_2} \int_{A_2} \int_{A_1} \frac{\cos\theta_1 \cos\theta_2}{\pi R^2} \, dA_1 \, dA_2 \tag{9.43}$$

and comparison of Eqs. (9.42) and (9.43) gives

$$A_1 F_{12} = A_2 F_{21} \tag{9.44}$$

Equation (9.44) is called the *reciprocity relation* and is applicable to any pair of radiating surfaces within an enclosure.

Within an enclosure formed by n radiating surfaces all of the radiation emitted by any one surface is intercepted by the other surfaces, and if the one surface is concave, it intercepts some of its own radiation. Thus, for such an enclosure containing n surfaces,

$$\sum_{j=1}^{n} F_{ij} = 1 \tag{9.45}$$

applies to each surface. Equation (9.45) is called the *summation rule*.

Figure 9.13 shows an enclosure containing three surfaces. In this enclosure $3^2 = 9$ view factors are required, with these being

$$F_{11} \quad F_{12} \quad F_{13}$$
$$F_{21} \quad F_{22} \quad F_{23}$$
$$F_{31} \quad F_{32} \quad F_{33}$$

Application of the reciprocity relations gives three equations,

$$A_1 J_{12} = A_2 J_{21}$$
$$A_1 J_{13} = A_3 J_{31}$$
$$A_2 J_{23} = A_3 J_{32}$$

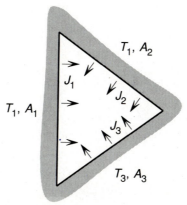

FIGURE 9.13.   Enclosure containing three surfaces.

and application of the summation rules gives three equations,

$$F_{11} + F_{12} + F_{13} = 1$$

$$F_{21} + F_{22} + F_{23} = 1$$

$$F_{31} + F_{32} + F_{33} = 1$$

Thus only $(9 - 6) = 3$ view factors need to be determined by equations of the form given by Eq. (9.43). In general, for an enclosure containing N surfaces, N^2 view factors are required, but as N view factors can be obtained from application of the summation rule and $N(N - 1)/2$ can be obtained from the reciprocity relations, only $N^2 - N - N(N - 1)/2 = N(N - 1)/2$ must be obtained directly.

Consider the simple enclosure shown in Fig. 9.14, which comprises a spherical body (surface 1) contained within a spherical cavity (surface 2). With two surfaces four view factors are required, but as surface 1 is convex, it does not intercept any of its own radiation and hence

$$F_{11} = 0 \tag{i}$$

and all of the radiation emitted by surface 1 is intercepted by surface 2, so that

$$F_{12} = 1 \tag{ii}$$

[Note that Eqs. (i) and (ii) obey the summation rule $F_{11} + F_{12} = 1$.] The reciprocity relation gives

$$A_1 F_{12} = A_2 F_{21} \tag{iii}$$

and hence from Eq. (ii),

$$F_{21} = \frac{A_1}{A_2} F_{12} = \frac{A_1}{A_2} \tag{iv}$$

The summation rule for surface 2 is

$$F_{22} + F_{21} = 1 \tag{v}$$

which, with Eq. (iv), gives

$$F_{22} = 1 - \frac{A_1}{A_2} \tag{vi}$$

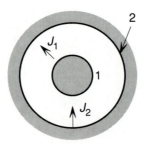

FIGURE 9.14. Spherical body contained within a spherical cavity.

The difficulty of evaluating the integrals in Eq. (9.42) has lead to presentation of view factors in graphical form, and the results for three three-dimensional geometries are shown in Figs. 9.15 to 9.17. Analytical expressions for view factors for several two- and three-dimensional geometries are shown, respectively, in Tables 9.2 and 9.3.

EXAMPLE 9.3

Calculate the view factors for the cylindrical enclosure of radius R and length $2R$ shown in Fig. 9.18.

With three surfaces, nine view factors are required, of which six can be obtained from the reciprocity relations and the summation rules. As surfaces 1 and 2 in the figure are plane, F_{11} and F_{22} are zero, and hence only one of the seven remaining view factors needs to be determined independently. With $L/R_i = 1$ and $R_j/L = 1$, this can be obtained from Fig. 9.16 as

$$F_{12} = 0.38$$

In the enclosure, $A_1 = A_2 = \pi R^2$ and $A_3 = 2\pi R^2$. From the summation rule,

$$F_{11} + F_{12} + F_{13} = 1$$

and thus

$$F_{13} = 1 - 0.38 = 0.62$$

From the reciprocity relation

$$A_1 F_{12} = A_2 F_{21}$$

or as $A_1 = A_2$,

$$F_{21} = F_{12} = 0.38$$

From the summation rule

$$F_{21} + F_{22} + F_{23} = 1$$

or

$$F_{23} = 1 - 0.38 = 0.62$$

From the reciprocity relation

$$A_1 F_{13} = A_3 F_{31}$$

$$F_{31} = \frac{A_1}{A_3} F_{13} = \frac{\pi R^2}{2\pi R^2} F_{13} = \frac{0.62}{2} = 0.31$$

From the summation rule

$$F_{31} + F_{32} + F_{33} = 1$$

$$F_{33} = 1 - 0.31 - 0.31 = 0.38$$

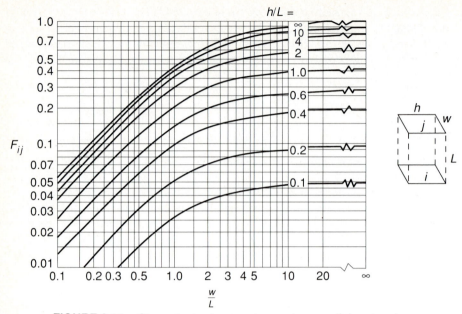

FIGURE 9.15.   Shape factor for equal opposing parallel rectangles.

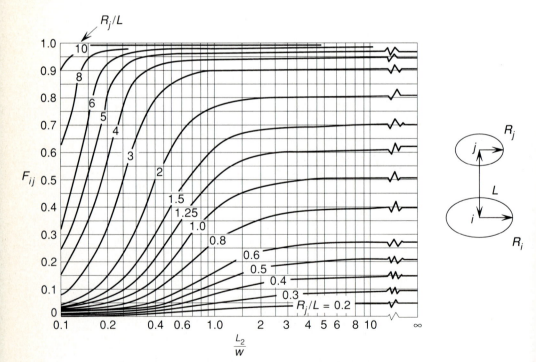

FIGURE 9.16.   Shape factor for two parallel coaxial disks.

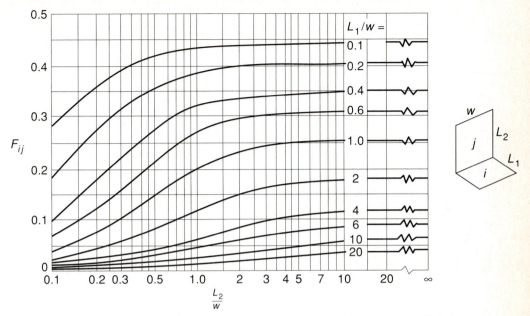

FIGURE 9.17. Shape factor for rectangles at 90° with a common edge.

TABLE 9.2 View factors for two-dimensional geometries

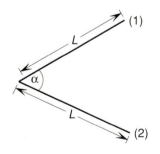

$$F_{12} = F_{21} = 1 - \sin \frac{\alpha}{2}$$

$$F_{12} = 0.5 \left[1 + \frac{h}{w} - \left(1 + \frac{h^2}{w^2} \right)^{1/2} \right]$$

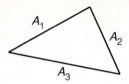

$$F_{12} = \frac{A_1 + A_2 - A_3}{2A_1}$$

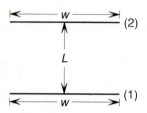

$$F_{12} = F_{21} = \left(1 + \frac{L^2}{w^2}\right)^{1/2} - \frac{L}{w}$$

TABLE 9.3   View factors for three-dimensional geometries

1.

2.

3.

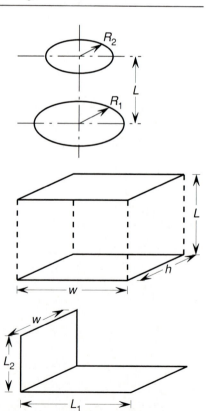

1.

$$F_{12} = 0.5 \left[x - \left(x^2 - \frac{4Z_2^2}{Z_1^2} \right)^{1/2} \right]$$

where

$$Z_1 = \frac{R_1}{L}, \quad Z_2 = \frac{R_2}{L}, \quad \text{and} \quad X = 1 + \frac{1 + Z_2^2}{Z_1^2}$$

2.

$$F_{12} = = \frac{2}{\pi X Y} \left\{ \ln \left[\frac{(1 + X^2)(1 + Y^2)}{1 + X^2 + Y^2} \right]^{1/2} + Y(1 + X^2)^{1/2} \tan^{-1} \frac{Y}{(1 + X^2)^{1/2}} \right.$$
$$\left. + X(1 + Y^2)^{1/2} \tan^{-1} \frac{X}{(1 + Y^2)^{1/2}} - X \tan^{-1} X - Y \tan^{-1} Y \right\}$$

where

$$X = \frac{w}{L} \quad \text{and} \quad Y = \frac{h}{L}$$

3.

$$F_{12} = \frac{1}{\pi W} \left(W \tan^{-1} \frac{1}{W} + H \tan^{-1} \frac{1}{H} - (H^2 + W^2)^{1/2} \tan^{-1} \frac{1}{(H^2 + W^2)^{1/2}} \right.$$
$$+ \frac{1}{4} \ln \left\{ \frac{(1 + W^2)(1 + H^2)}{1 + W^2 + H^2} \left[\frac{W^2(1 + W^2 + H^2)}{(1 + W^2)(W^2 + H^2)} \right]^{W^2} \right.$$
$$\left. \left. \times \left[\frac{H^2(1 + H^2 + W^2)}{(1 + H^2)(H^2 + W^2)} \right]^{H^2} \right\} \right)$$

where

$$H = \frac{L_2}{w} \quad \text{and} \quad W = \frac{L_1}{w}$$

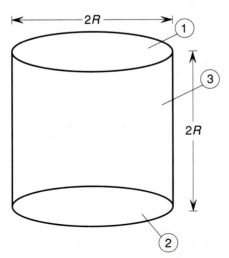

FIGURE 9.18. Cylindrical enclosure considered in Example 9.3.

From the symmetry of the geometry it is apparent that

$$F_{12} = F_{21}$$

$$F_{13} = F_{23}$$

$$F_{31} = F_{32}$$

which, as is seen, is the case. As the length of the cylinder is increased F_{33}, F_{13}, and F_{23} increase, and with the exception of F_{11} and F_{22}, which remain zero, the other four view factors decrease.

EXAMPLE 9.4

Calculate all of the view factors for the rectangular enclosure of dimensions 0.5 m × 1 m × 2 m shown in Fig. 9.19.

With six surfaces, 36 view factors are required. As all of the surfaces are planar,

$$F_{11} = F_{22} = F_{33} = F_{44} = F_{55} = F_{66} = 0$$

From Eq. (2) in Table 9.3 for surfaces 1 and 2, $w = 2$ m, $h = 1$ m, and $L = 0.5$ m. Thus $X = 2/0.5 = 4$ and $Y = 1/0.5 = 2$. Therefore,

$$F_{12} = F_{21} = 0.509$$

For surfaces 5 and 6, $w = 1$ m, $h = 0.5$ m, and $L = 2$ m. Thus $X = 1/2 = 0.5$ and $Y = 0.5/2 = 0.25$. Therefore,

$$F_{56} = F_{65} = 0.036$$

For surfaces 3 and 4, $w = 2$ m, $h = 0.5$ m, and $L = 1$ m. Thus $X = 2$ and $Y = 0.5$. Therefore,

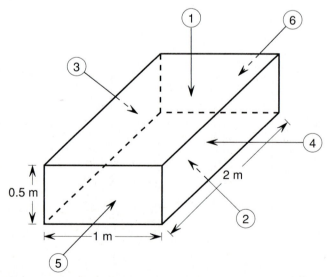

FIGURE 9.19. Rectangular enclosure considered in Example 9.4.

$$F_{34} = F_{43} = 0.165$$

From Eq. (3) in Table 9.3, for surfaces 2 and 3, $L_2 = 0.5$ m, $L_1 = 1$ m, and $w = 2$ m. Thus $H = 0.5/2 = 0.25$ and $W = 1/2 = 0.5$. Therefore,

$$F_{23} = 0.167$$

and by symmetry,

$$F_{24} = F_{13} = F_{14} = 0.167$$

For surfaces 2 and 5, $w = 1$ m, $L_1 = 2$ m, and $L_2 = 0.5$ m. Thus $H = 0.5$ and $W = 4$. Therefore,

$$F_{25} = 0.079$$

and by symmetry,

$$F_{45} = F_{46} = F_{36} = 0.079$$

For surfaces 3 and 5, $w = 0.5$ m, $L_1 = 2$ m, and $L_2 = 1$ m. Thus $H = 1/0.5 = 2$ and $W = 2/0.5 = 4$. Therefore,

$$F_{35} = 0.084$$

and by symmetry,

$$F_{45} = F_{46} = F_{36} = 0.084$$

From the reciprocity relation, with $A_1 = A_2 = 2$ m^2, $A_3 = A_4 = 1$ m^2, and $A_5 = A_6 = 0.5$ m^2,

$$F_{31} = \frac{A_1}{A_3} F_{13} = 2 \times 0.167 = 0.334$$

and by symmetry,

$$F_{32} = F_{41} = F_{42} = 0.334$$

From the reciprocity relation,

$$F_{51} = \frac{A_1}{A_5} F_{15} = 4 \times 0.079 = 0.316$$

and by symmetry,

$$F_{52} = F_{61} = F_{62} = 0.316$$

From the reciprocity relation,

$$F_{53} = \frac{A_3}{A_5} F_{35} = 2 \times 0.084 = 0.168$$

and by symmetry,

$$F_{54} = F_{63} = F_{64} = 0.168$$

These view factors, which are used in Problem 9.14, are listed in Table 9.4.

TABLE 9.4 F_{ij} Values

			j				
i	1	2	3	4	5	6	Σ
1	0	0.509	0.167	0.167	0.079	0.079	1.00
2	0.509	0	0.167	0.167	0.079	0.079	1.00
3	0.334	0.334	0	0.165	0.084	0.084	1.00
4	0.334	0.334	0.165	0	0.084	0.084	1.00
5	0.316	0.316	0.168	0.168	0	0.036	1.00
6	0.316	0.316	0.168	0.168	0.036	0	1.00

9.8 Radiation Exchange Between Blackbodies

The exchange of radiation among real surfaces is complicated by the fact that the radiation can be both absorbed and reflected. However, as a blackbody absorbs all incident radiation, in which case its radiosity, J, is simply the emissive power of a blackbody, E_b, the introduction to radiation exchange is simplified by first considering radiation exchange between blackbodies. In Fig. 9.20 the rate at which radiation *leaves* surface 1 and is *intercepted* by surface 2, $q_{1\rightarrow2}$, is obtained by rearranging Eq. (9.42) as

$$q_{1\rightarrow2} = (A_1 J_1)F_{12} \tag{9.46}$$

or since the surfaces are blackbody radiators,

$$q_{1\rightarrow2} = A_1 F_{12} E_{b,1} \tag{9.47}$$

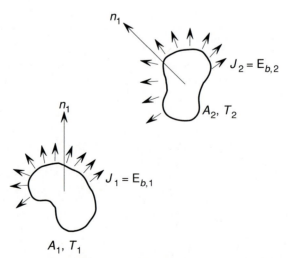

FIGURE 9.20. Blackbody radiation leaving surface 1 and intercepted by surface 2.

Similarly, the rate at which radiation *leaves* surface 2 and is *intercepted* by surface 1 is

$$q_{2\to1} = A_2 F_{21} E_{b,2} \qquad (9.48)$$

Thus the net exchange of radiation, q_{12}, is given by

$$q_{12} = q_{1-2} - q_{2-1} \qquad (9.49)$$

in which q_{1-2} is the rate at which radiation is *emitted* by 1 and *absorbed* by 2, and q_{2-1} is the rate at which radiation is *emitted* by 2 and *absorbed* by 1. The net radiation exchange, q_{12}, is thus the net rate at which radiation leaves surface 1 due to its interaction with surface 2, which is also the net rate at which surface 2 receives radiation due to its interaction with surface 1. As blackbodies absorb all incident radiation, $q_{1-2} = q_{1\to2}$, and hence

$$q_{12} = A_1 F_{12} E_{b,1} - A_2 F_{21} E_{b,2}$$

which, in view of the reciprocity relation given by Eq. (9.44), can be written as

$$q_{12} = A_1 F_{12}(E_{b,1} - E_{b,2})$$

or, from the Stefan–Boltzmann law given by Eq. (9.23),

$$q_{12} = A_1 F_{12} \sigma(T_1^4 - T_2^4) \qquad (9.50)$$

where T_1 and T_2 are the constant temperatures of the surfaces 1 and 2. With N surfaces in a blackbody enclosure maintained at different temperatures, the net transfer of radiation from surface 1 due to its interactions with the other surfaces is

$$q_1 = \sum_{j=2}^{N} A_1 F_{1j}(T_1^4 - T_j^4) \qquad (9.51)$$

EXAMPLE 9.5

Figure 9.21 shows a well-insulated electrically heated cylindrical crucible of diameter 12 cm and depth 20 cm which is open to the atmosphere at its upper end. If the inner walls of the crucible behave as blackbodies, calculate the power required to maintain the bottom surface (surface 1) at 1400°C and the vertical side walls (surface 2) at 1200°C. The ambient temperature outside the crucible is 25°C.

As the crucible is well insulated, the power requirement for the crucible equals the heat loss by radiation through the open mouth of the crucible and the mouth can be considered to be a hypothetical circular blackbody surface of diameter 12 cm at the temperature 25°C. The rate of heat loss from the crucible is thus

$$q = q_{13} + q_{23}$$

or, from Eq. (9.50),

$$q = A_1 F_{13} \sigma(T_1^4 - T_3^4) + A_2 F_{23} \sigma(T_2^4 - T_3^4) \qquad (i)$$

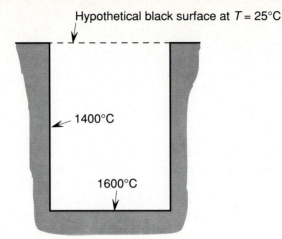

FIGURE 9.21. Well-insulated electrically heated crucible considered in Example 9.5.

From Fig. 9.16, with

$$\frac{L}{r_i} = \frac{20}{6} = 3.33$$

and

$$\frac{r_j}{L} = \frac{6}{20} = 0.3$$

$$F_{ij} = F_{13} = 0.075$$

As surface 1 is planar, $F_{11} = 0$, and hence from the summation rule,

$$F_{12} = 1 - F_{11} - F_{13} = 1 - 0.075 = 0.925$$

Then, from the reciprocity relation,

$$F_{21} = \frac{A_1}{A_2} F_{12} = \frac{\pi \times 6^2}{\pi \times 12 \times 20} \times 0.925 = 0.139$$

and from symmetry,

$$F_{21} = F_{23} = 0.139$$

Thus, in Eq. (i),

$$q = [\pi \times 0.06^2 \times 0.075 \times 5.67 \times 10^{-8} \times (1673^4 - 298^4)]$$
$$+ [\pi \times 0.12 \times 0.2 \times 0.139 \times 5.67 \times 10^{-8} \times (1473^4 - 298^4)]$$
$$= 376 + 2798$$
$$= 3175 \text{ W}$$

9.9 Radiation Exchange Between Diffuse-Gray Surfaces

Treatment of the net exchange of radiation among nonblack surfaces must take into account the fact that the radiation may experience multiple reflections from and absorption at all of the surfaces. The various interactions occurring at an opaque diffuse-gray surface are shown in Fig. 9.22. Radiation leaves the surface at the rate $J_1 A_1$ and the surface is irradiated at the rate $G_1 A_1$. If the radiosity is greater than the irradiation, maintenance of a constant surface temperature T_1 requires that thermal energy be supplied to the surface by some means other than radiation at the rate q_1 such that

$$q_1 = A_1(J_1 - G_1) \tag{9.52}$$

The radiosity of the surface is the sum of its emissive power and the reflected irradiation,

$$J_1 = E_1 + \rho_1 G_1 \tag{9.53}$$

insertion of which into Eq. (9.52) gives

$$q_1 = A_1[E_1 - (1 - \rho_1)G_1] \tag{9.54}$$

As the surface is opaque,

$$\alpha_1 + \rho_1 = 1$$

and hence

$$q_1 = A_1(E_1 - \alpha_1 G_1) \tag{9.55}$$

Alternatively, with a surface emissivity of ϵ_1,

$$E_1 = \epsilon_1 E_{b,1}$$

and from Kirchhoff's law, Eq. (9.35),

$$\epsilon_1 = \alpha_1 = 1 - \rho_1$$

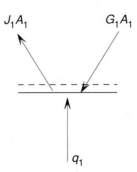

FIGURE 9.22.   Energy balance required for a constant temperature at an irradiated real surface.

Equation (9.53) can thus be expressed as

$$J_1 = \epsilon_1 E_{b,1} + (1 - \epsilon_1)G_1 \tag{9.56}$$

or

$$G_1 = \frac{J_1 - \epsilon_1 E_{b,1}}{1 - \epsilon_1} \tag{9.56}$$

substitution of which into Eq. (9.52) gives

$$q_1 = A_1 \left(J_1 - \frac{J_1 - \epsilon_1 E_{b,1}}{1 - \epsilon_1} \right)$$

$$= \frac{E_{b,1} - J_1}{(1 - \epsilon_1)/\epsilon_1 A_1} \tag{9.57}$$

In Eq. (9.57), q_1 is a flux or "current," $E_{b,1} - J_1$ is the "driving force" for the current, and hence from the electric analogy, the term

$$\frac{1 - \epsilon_1}{\epsilon_1 A_1}$$

is a resistance to the radiation flux leaving the surface. This resistance is called the *surface radiation resistance.*

The use of Eq. (9.57) requires that the radiosity J_1 be known, and this is influenced by the radiosities of the other surfaces in the enclosure. Figure 9.23 shows the exchange of radiation among three diffuse-gray surfaces in an enclosure. The irradiation received by surface 1 from surfaces 2 and 3 is

$$A_1 G_1 = A_2 F_{21} J_2 + A_3 F_{31} J_3$$

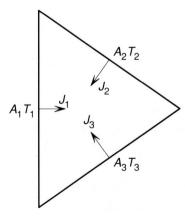

FIGURE 9.23. Exchange of radiation among three diffuse-gray surfaces in an enclosure.

which in view of the reciprocity relations can be written as

$$A_1 G_1 = A_1 F_{12} J_2 + A_1 F_{13} J_3$$

or

$$G_1 = F_{12} J_2 + F_{13} J_3$$

substitution of which into Eq. (9.52) gives

$$q_1 = A_1 (J_1 - F_{12} J_2 - F_{13} J_3) \tag{9.58}$$

As the surfaces in the enclosure are planar, the summation rule gives $F_{12} + F_{13} = 1$, and thus Eq. (9.58) can be expressed as

$$\begin{aligned} q_1 &= A_1 (F_{12} J_1 + F_{13} J_1 - F_{12} J_2 - F_{13} J_3) \\ &= A_1 [F_{12}(J_1 - J_2) + F_{13}(J_1 - J_3)] \end{aligned} \tag{9.59}$$

For an enclosure containing N surfaces, Eq. (9.59) becomes

$$q_1 = \sum_{j=2}^{N} A_1 F_{1j}(J_1 - J_j) \tag{9.60}$$

In Eq. (9.60), q_1 is again the flux or "current" and $J_1 - J_j$ is the "driving force" for the flux between surface 1 and surface j, and, hence from the electric analogy, the term

$$\frac{1}{A_1 F_{1j}}$$

is a resistance to the radiation flux between surface 1 and surface j, which is called the *view resistance*. Combination of Eqs. (9.57) and (9.60) gives

$$\frac{E_{b,1} - J_1}{(1 - \epsilon_1)/\epsilon_1 A_1} = \sum_{j=2}^{N} A_1 F_{1j}(J_1 - J_j) \tag{9.61}$$

With a large number, N, of surfaces in the enclosure, N equations are written from Eq. (9.61) in the form

$$a_{11} J_1 + a_{12} J_2 + \cdots + a_{1N} = C_1$$

$$a_{21} J_1 + a_{22} J_2 + \cdots + a_{2N} = C_2$$

$$\vdots$$

$$a_{N1} J_1 + a_{N2} J_2 + \cdots + a_{NN} = C_N$$

in which the coefficients a_{ij} and the constants C_i are determined by the blackbody emissive powers, the emissivities, and the areas of the N surfaces. In matrix form this is written as

$$[A][J] = [C]$$

which, on inversion, gives the radiosities as

$$[J] = [A]^{-1}[C]$$

EXAMPLE 9.6

Figure 9.24 shows the cross section of a long duct in which the inner wall conditions are $T_1 = 600$ K, $\epsilon_1 = 0.5$, $T_2 = 800$ K, $\epsilon_2 = 0.6$, and $T_3 = 1000$ K, $\epsilon_3 = 0.7$. Calculate the radiosities of the inner surfaces of the duct and the net heat fluxes at the walls.

Equation (9.61) is written for each of the three surfaces as

$$\frac{E_{b,1} - J_1}{(1 - \epsilon_1)/\epsilon_1 A_1} = A_1 F_{12}(J_1 - J_2) + A_1 F_{13}(J_1 - J_3) \qquad \text{i(a)}$$

$$\frac{E_{b,2} - J_2}{(1 - \epsilon_2)/\epsilon_2 A_2} = A_2 F_{21}(J_2 - J_1) + A_2 F_{23}(J_2 - J_3) \qquad \text{i(b)}$$

$$\frac{E_{b,3} - J_3}{(1 - \epsilon_3)/\epsilon_3 A_3} = A_3 F_{31}(J_3 - J_1) + A_3 F_{32}(J_3 - J_2) \qquad \text{i(c)}$$

The areas, per meter length of the duct, are

$$A_1 = 3 \text{ m}^2$$
$$A_2 = 4 \text{ m}^2$$
$$A_3 = 5 \text{ m}^2$$

From the Stefan–Boltzmann law, Eq. (9.23),

$$E_{b,1} = 5.67 \times 10^{-8} \times 600^4 = 7348 \text{ W/m}^2$$
$$E_{b,2} = 5.67 \times 10^{-8} \times 800^4 = 23,224 \text{ W/m}^2$$
$$E_{b,3} = 5.67 \times 10^{-8} \times 1000^4 = 56,700 \text{ W/m}^2$$

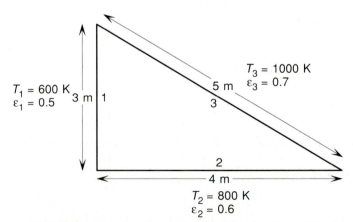

FIGURE 9.24. Enclosure considered in Example 9.6.

From Table 9.2 the view factors per unit length of the duct are

$$F_{12} = \frac{A_1 + A_2 - A_3}{2A_1} = \frac{3 + 4 - 5}{2 \times 3} = 0.333$$

$$F_{13} = \frac{A_1 + A_3 - A_2}{2A_1} = \frac{3 + 5 - 4}{2 \times 3} = 0.667$$

$$F_{23} = \frac{A_2 + A_3 - A_1}{2A_2} = \frac{4 + 5 - 3}{2 \times 4} = 0.75$$

and from the reciprocity relations,

$$A_1 F_{12} = A_2 F_{21} = 3 \times 0.333 = 1.0$$

$$A_1 F_{13} = A_3 F_{31} = 3 \times 0.667 = 2$$

$$A_2 F_{23} = A_3 F_{32} = 4 \times 0.75 = 3$$

Substitution into Eqs. i(a) to i(c) gives

$$\frac{7348 - J_1}{0.5/(0.5 \times 3)} = (J_1 - J_2) + 2(J_1 - J_3) \qquad \text{ii(a)}$$

$$\frac{23,224 - J_2}{0.4/(0.6 \times 4)} = (J_2 - J_1) + 3(J_2 - J_3) \qquad \text{ii(b)}$$

$$\frac{56,700 - J_3}{0.3/(0.7 \times 5)} = 2(J_3 - J_1) + 3(J_3 - J_2) \qquad \text{ii(c)}$$

which, on rearrangement, gives

$$6J_1 - J_2 - 2J_3 = 22{,}044 \qquad \text{iii(a)}$$

$$-J_1 + 10J_2 - 3J_3 = 139{,}344 \qquad \text{iii(b)}$$

$$-2J_1 - 3J_2 + 16.667J_3 = 661{,}500 \qquad \text{iii(c)}$$

Solution of Eqs. iii(a) to iii(c) gives the radiosities as

$$J_1 = 24{,}900 \text{ W/meter length}$$

$$J_2 = 30{,}900 \text{ W/meter length}$$

and

$$J_3 = 48{,}240 \text{ W/meter length}$$

and from Eq. (9.57), the net fluxes at the walls are

$$q_1 = \frac{7348 - 24{,}900}{0.5/(0.5 \times 3)} = -52{,}670 \text{ W/m}$$

$$q_2 = \frac{23{,}224 - 30{,}900}{0.4/(0.6 \times 4)} = -46{,}030 \text{ W/m}$$

and

$$q_3 = \frac{56,700 - 48,240}{0.3/(0.7 \times 5)} = +98,700 \text{ W/m}$$

The sum of the net fluxes at the walls is zero.

9.10 Electric Analogy

The "circuit diagram" arising from application of the electric analogy to the exchange of radiation between two planar diffuse-gray surfaces is shown in Fig. 9.25. The diagram contains nodes representing the potentials of the blackbody emissive powers of the surfaces and nodes representing the potentials of the radiosities of the surfaces, and $E_{b,1} - E_{b,2}$ is the driving force for the transfer of heat from surface 1 to surface 2. The surface resistances occur between the E_b nodes and the J nodes and the view resistances occur between the J nodes. The resistances occur in series and hence the total resistance to q_{12} is the sum of the individual resistances. The "current" flowing in the circuit is determined by analogy with Ohm's law as

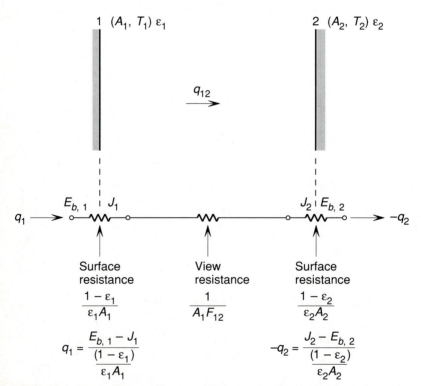

FIGURE 9.25. Circuit diagram for the exchange of radiation between two planar diffuse-gray surfaces.

$$\text{current} = \frac{\text{voltage}}{\text{resistance}}$$

or

$$q_{12} = \frac{E_{b,1} - E_{b,2}}{\dfrac{1 - \epsilon_1}{\epsilon_1 A_1} + \dfrac{1}{A_2 F_{12}} + \dfrac{1 - \epsilon_2}{\epsilon_2 A_2}} \tag{9.62}$$

For large (infinite) parallel plates in which $A_1 = A_2 = A$ and $F_{12} = 1$, Eq. (9.62) becomes

$$q_{12} = \frac{A(E_{b,1} - E_{b,2})}{1/\epsilon_1 + 1/\epsilon_2 - 1} \tag{9.63}$$

For the long (infinite) concentric cylinders shown in Fig. 9.26(a), $F_{12} = 1$, $A_1/A_2 = R_1/R_2$, and Eq. (9.62) becomes

$$q_{12} = \frac{A_1(E_{b,1} - E_{b,2})}{\dfrac{1}{\epsilon_1} + \dfrac{1 - \epsilon_2}{\epsilon_2}\dfrac{R_1}{R_2}}$$

Similarly, for the concentric spheres shown in Fig. 9.26(b), with $F_{12} = 1$ and $A_1/A_2 = R_1^2/R_2^2$,

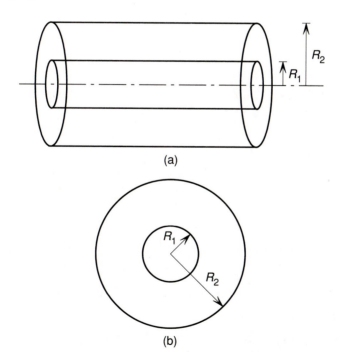

(a)

(b)

FIGURE 9.26. Exchange of radiation between (a) two diffuse-gray concentric cylinders, and (b) two diffuse-gray concentric spheres.

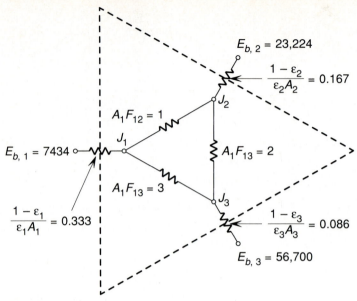

FIGURE 9.27.   Circuit diagram for the duct shown in Fig. 9.24.

$$q_{12} = \frac{A_1(E_{b,1} - E_{b,2})}{\dfrac{1}{\epsilon_1} + \dfrac{1 - \epsilon_2}{\epsilon_2}\left(\dfrac{R_1}{R_2}\right)^2}$$

Figure 9.27 shows the circuit diagram for the duct considered in Fig. 9.24. The net current entering each J node is zero, and thus for node J_1,

$$\frac{7348 - J_1}{0.333} + \frac{J_2 - J_1}{1} + \frac{J_3 - J_1}{0.5} = 0$$

for node J_2,

$$\frac{23,224 - J_2}{0.167} + \frac{J_1 - J_2}{1} + \frac{J_3 - J_2}{0.333} = 0$$

and for node J_3,

$$\frac{56,700 - J_3}{0.086} + \frac{J_1 - J_3}{0.5} + \frac{J_2 - J_3}{0.333} = 0$$

which, on simplification, yield Eqs. iii(a) to iii(c).

9.11 Radiation Shields

Radiation shields are sheets of material of low emissivity which when placed between two radiating surfaces, decrease the net radiation transfer between the surfaces.

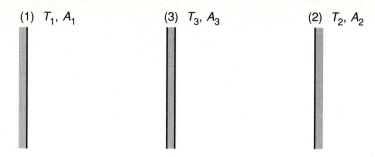

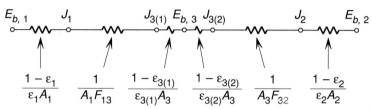

FIGURE 9.28. Radiation shield placed between two radiating surfaces and the corresponding circuit diagram.

The circuit diagram for a radiation shield placed between surfaces 1 and 2 in Fig. 9.25 is shown in Fig. 9.28. As the emissivities of the two surfaces of the radiation shield are not necessarily equal, the emissivity of the surface facing surface 1 is designated $\epsilon_{3(1)}$ and the emissivity of the surface facing surface 2 is designated $\epsilon_{3(2)}$. The presence of the radiation shield increases the resistance to radiation exchange from

$$\frac{1 - \epsilon_1}{\epsilon_1 A_1} + \frac{1}{A_1 F_{12}} + \frac{1 - \epsilon_2}{\epsilon_2 A_2}$$

in Eq. (9.63) to

$$\frac{1 - \epsilon_1}{\epsilon_1 A_1} + \frac{1}{A_2 F_{13}} + \frac{1 - \epsilon_{3(1)}}{\epsilon_{3(1)} A_3} + \frac{1 - \epsilon_{3(2)}}{\epsilon_{3(2)} A_3} + \frac{1}{A_2 F_{32}} + \frac{1 - \epsilon_2}{\epsilon_2 A_2}$$

As $F_{13} = F_{32} = 1$, the radiation transfer is decreased from that given by Eq. (9.63) to

$$q_{12} = \frac{A_1 \sigma (T_1^4 - T_1^4)}{\dfrac{1}{\epsilon_1} + \dfrac{1}{\epsilon_2} + \dfrac{1 - \epsilon_{3(1)}}{\epsilon_{3(1)}} + \dfrac{1 - \epsilon_{3(2)}}{\epsilon_{3(2)}}} \qquad (9.64)$$

With small values of $\epsilon_{3(2)}$ and $\epsilon_{3(2)}$ the resistance introduced by the radiation shield can be increased significantly.

EXAMPLE 9.7

Determine the influence of inserting a radiation shield of $\epsilon_{3(1)} = \epsilon_{3(2)} = \epsilon_3 = 0.1$ between the surfaces 1 and 2 for which $T_1 = 1000$ K, $\epsilon_2 = 0.5$ and $T_2 = 300$ K, $\epsilon_2 = 0.5$.

Without the radiation shield Eq. (9.63) gives

$$\frac{q_{12}}{A} = \frac{\sigma(T_1^4 - T_2^4)}{1/\epsilon_1 + 1/\epsilon_2 - 1}$$

$$= \frac{5.67 \times 10^{-8}(1000^4 - 300^4)}{2 + 2 - 1}$$

$$= 18{,}745 \text{ W/m}^2$$

With the radiation shield in place, Eq. (9.64) gives

$$\frac{q_{12}}{A_1} = \frac{\sigma(T_1^4 - T_2^4)}{\dfrac{1}{\epsilon_1} + \dfrac{1}{\epsilon_2} + \dfrac{1 - \epsilon_3}{\epsilon_3} + \dfrac{1 - \epsilon_3}{\epsilon_3}}$$

$$= \frac{5.67 \times 10^{-8}(1000^4 - 300^4)}{2 + 2 + 0.9/0.1 + 0.9/0.1}$$

$$= 2556 \text{ W/m}^2$$

Insertion of the radiation shield thus increases the resistance per unit area from 3 to 22 and hence decreases q_{12} by the factor $22/3 = 7.33$. The temperature of the radiation shield is obtained by considering the radiation exchange between 1 and 3 or between 3 and 1. For exchange between surface 1 and the shield

$$\frac{q_{12}}{A} = \frac{\sigma(T_1^4 - T_3^4)}{1/\epsilon_1 + 1/\epsilon_3 - 1}$$

$$2556 = \frac{5.67 \times 10^{-8}(1000^4 - T_3^4)}{1/0.5 + 1/0.1 - 1}$$

which gives $T_3 = 842$ K, or, alternatively, for exchange between the shield and surface 2,

$$2556 = \frac{5.67 \times 10^{-8}(T_3^4 - 300^4)}{1/0.1 + 1/0.5 - 1}$$

which also gives $T_3 = 842$ K. If two shields (surfaces 3 and 4) are placed between surfaces 1 and 2, the resistance per unit area is increased to

$$\frac{1}{\epsilon_1} + \frac{1}{\epsilon_2} + \frac{1 - \epsilon_{3(1)}}{\epsilon_{3(1)}} + \frac{1 - \epsilon_{3(4)}}{\epsilon_{3(4)}} + \frac{1 - \epsilon_{4(3)}}{\epsilon_{4(3)}} + \frac{1 - \epsilon_{4(2)}}{\epsilon_{4(2)}} + 1$$

which, with $\epsilon_1 + \epsilon_2 = 0.5$ and $\epsilon_{3(1)} = \epsilon_{3(4)} = \epsilon_{4(3)} = \epsilon_{4(2)} = 0.1$, gives a resistance per unit area of 41, in which case q_{12}/A is decreased to

$$18{,}746 \times \frac{3}{41} = 1371 \text{ W/m}^2$$

The temperature of radiation shield 3 is obtained from

$$\frac{q_{12}}{A} = 1371 = \frac{5.67 \times 10^{-8}(1000^4 - T_3^4)}{1/0.5 + 1/0.1 - 1}$$

as $T_3 = 925$ K and the temperature of radiation shield 4 is obtained from

$$\frac{q_{12}}{A} = 1371 = \frac{5.67 \times 10^{-8}(T_4^4 - 300^4)}{1/0.5 + 1/0.1 - 1}$$

as $T_4 = 723$ K.

In general, if all of the emissivities are equal, the resistance R_i per unit area is with no shields,

$$R_0 = \frac{1}{\epsilon} + \frac{1}{\epsilon} - 1 = \frac{2 - \epsilon}{\epsilon}$$

with one shield,

$$R_1 = \frac{2}{\epsilon} + \frac{2(1 - \epsilon)}{\epsilon} = \frac{2(2 - \epsilon)}{\epsilon} = 2R_0$$

with two shields,

$$R_2 = \frac{2}{\epsilon} + \frac{4(1 - \epsilon)}{\epsilon} + 1 = \frac{3(2 - \epsilon)}{\epsilon} = 3R_0$$

with N shields,

$$R_N = (N + 1)R_0$$

and hence the radiation transfer rate with N shields, $(q_{12})_N$, is

$$(q_{12})_N = \frac{1}{N + 1} (q_{12})_0$$

9.12 Reradiating Surface

A reradiating surface in a radiation enclosure is simply one that is sufficiently well insulated that it can be considered to be an adiabatic wall at which $q = 0$. This has the consequence that if at the surface i, $q_i = 0$, then from Eq. (9.52),

$$J_i = G_i \tag{9.65}$$

and from Eq. (9.57),

$$E_{b,i} = J_i$$

Thus, in an enclosure, the equilibrium temperature of a reradiating surface is determined only by its interaction with the other surfaces in the enclosure and is independent of its emissivity. Figure 9.29(a) shows a three-surface enclosure in which surface 3 is reradiating, and Fig. 9.29(b) shows the corresponding circuit diagram. As $q_R = 0$, the heat transfer from surface 1 equals the heat transfer to surface 2:

$$q_1 = -q_2 = \frac{E_{b,1} - E_{b,2}}{\dfrac{1 - \epsilon_1}{\epsilon_1 A_1} + R + \dfrac{1 - \epsilon_2}{\epsilon_2 A_2}}$$

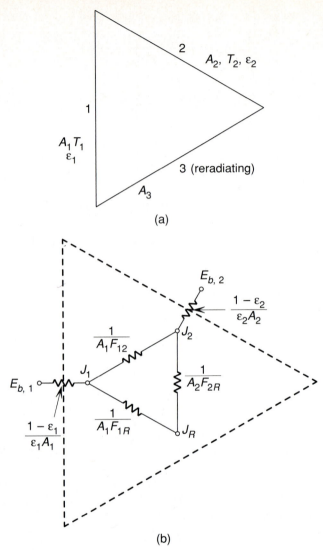

FIGURE 9.29. (a) Enclosure containing a reradiating surface; (b) circuit diagram for (a).

where R is the resistance in the circuit between node J_1 and node J_2. With

$$R_1 = \frac{1}{A_1 F_{12}}$$

$$R_2 = \frac{1}{A_1 F_{1R}}$$

and

$$R_3 = \frac{1}{A_2 F_{2R}}$$

$$\frac{1}{R} = \frac{1}{R_1} + \frac{1}{R_2 + R_3}$$

or

$$R = \frac{1}{A_1 F_{12} + (1/A_1 F_{1R} + 1/A_2 F_{2R})^{-1}}$$

and therefore,

$$q_1 = -q_2 = \frac{E_{b,1} - E_{b,2}}{\dfrac{1 - \epsilon_1}{\epsilon_1 A_1} + \dfrac{1}{A_1 F_{12} + (1/A_1 F_{1R} + 1/(A_2 F_{2R})^{-1}} + \dfrac{1 - \epsilon_2}{\epsilon_2 A_2}} \tag{9.66}$$

Thus if T_1 and T_2 are known, q_1 and q_2, obtained from Eq. (9.66), allow J_1 and J_2 to be obtained from Eq. (9.57) as

$$q_1 = \frac{E_{b,1} - J_1}{(1 - \epsilon_1)/\epsilon_1 A_1}$$

and

$$q_2 = \frac{E_{b,2} - J_2}{(1 - \epsilon_2)\epsilon_2 A_2}$$

Then, as $q_R = 0$, the current from J_1 to J_R equals the current from J_R to J_3:

$$\frac{J_1 - J_R}{1/A_1 F_{1R}} = \frac{J_R - J_2}{1/A_2 F_{2R}} \tag{9.67}$$

which allows J_R to be determined. The temperature of the reradiating surface is then determined from Eq. (9.65) as

$$J_R = E_{b,R} = \sigma T_R^4$$

EXAMPLE 9.8 _____

Consider that surface 3 in the duct shown in Fig. 9.27 is reradiating and that the conditions of the other surfaces are $T_1 = 600$ K, $\epsilon_1 = 0.5$ and $T_2 = 800$ K, $\epsilon_2 = 0.6$. Calculate the radiosities of the three surfaces and the equilibrium temperature of the reradiating surface 3.

As in Example 9.6,

$$A_1 F_{12} = 1, \quad A_1 F_{13} = 2, \quad \text{and} \quad A_2 F_{23} = 3$$

From Eq. (9.66),

$$q_1 = -q_2 = \cfrac{E_{b,1} - E_{b,2}}{\cfrac{1 - \epsilon_1}{\epsilon_1 A_1} + \cfrac{1}{A_1 F_{12} + (1/A_1 F_{1R} + 1/A_2 F_{2R})^{-1}} + \cfrac{1 - \epsilon_2}{\epsilon_2 A_2}}$$

$$= \cfrac{5.67 \times 10^{-8}(600^4 - 800^4)}{\cfrac{0.5}{0.5 \times 3} + \cfrac{1}{1 + 1/(0.5 + 0.333)} + \cfrac{0.4}{0.6 \times 4}}$$

$$= -16{,}634 \text{ W/meter length}$$

$$= -q_2$$

From Eq. (9.57),

$$q_1 = \frac{E_{b,1} - J_1}{(1 - \epsilon_1)/\epsilon_1 A_1} = -16{,}634 = \frac{5.67 \times 10^{-8} \times 600^4 - J_1}{0.5/(0.5 \times 3)}$$

which gives

$$J_1 = 12{,}892 \text{ W/meter length}$$

and

$$q_2 = \frac{E_{b,2} - J_2}{(1 - \epsilon_2)/\epsilon_2 A_2} = 16{,}634 = \frac{5.67 \times 10^{-8} \times 800^4 - J_2}{0.4/(0.6 \times 4)}$$

which gives

$$J_2 = 26{,}452 \text{ W/meter length}$$

Then, from Eq. (9.67),

$$\frac{J_1 - J_3}{1/A_1 F_{13}} - \frac{J_3 - J_2}{1/A_2 F_{23}} = 0$$

$$\frac{12{,}892 - J_3}{0.5} - \frac{J_3 - 20{,}452}{0.333} = 0$$

which gives

$$J_3 = 17{,}430 = 5.67 \times 10^{-8} T_3^4 \text{ W/meter length}$$

and

$$T_3 = 745 \text{ K}$$

9.13 Heat Transfer from a Surface by Convection and Radiation

Equation (6.2) gives the rate of heat transfer by convection from a surface at the temperature T_s to it surroundings at the ambient temperature T_∞ as

$$q'_{conv} = h(T_s - T_\infty) \tag{9.68}$$

where for heat transfer to a gaseous atmosphere by free convection, h can vary in the range 5 to 25 W/m²·K, and for heat transfer by forced convection, h can vary in the range 25 to 250 W/m²·K. For heat transfer by radiation from a surface of emissivity ϵ at the temperature T_s to surroundings at the temperature T_∞,

$$q'_{rad} = \epsilon\sigma(T_s^4 - T_\infty^4) \tag{9.69}$$

Equation (9.69) can be rewritten as

$$
\begin{aligned}
q'_{rad} &= \epsilon\sigma(T_s^2 + T_\infty^2)(T_s^2 - T_\infty^2) \\
&= \epsilon\sigma(T_s^2 + T_\infty^2)(T_s + T_\infty)(T_s - T_\infty)
\end{aligned}
$$

which, in comparison with Eq. (9.68), shows that a *radiation heat transfer coefficient*, h_r, can be defined as

$$h_r = \epsilon\sigma(T_s^2 + T_\infty^2)(T_s + T_\infty) \tag{9.70}$$

such that, in the convention of Eq. (9.68),

$$q'_{rad} = h_r(T_s - T_\infty) \tag{9.71}$$

The total rate of heat transfer from the surface by convection and by radiation is thus

$$q'_{total} = q'_{conv} + q'_{rad}$$

and the relative contributions of convection and radiation to the total heat transfer rate are determined by the relative magnitudes of h and h_r. In general, h is a weak function of temperature, whereas from Eq. (9.70), h_r is a strong function.

Figure 9.30, which shows the variation of h_r with ϵ and T_∞ for radiation from a surface to surroundings at 298 K, gives an indication of the conditions under which the contributions of convection and radiation are comparable. Figure 9.31 shows the contributions of forced convection (with $h = 150$ W/m²·K) and radiation from a surface of $\epsilon = 0.5$ to its surroundings at $T_\infty = 298$ K as a function of surface temperature in the range 300 K to 1500 K. Within this range the contribution due to forced convection is always larger than that due to radiation, although the contribution of the latter to the total increases from 5% at $T_s = 500$ K to 44% at $T_s = 1500$ K. On the other hand, if the convection contribution is by free convection at $h = 10$ W/m²·K, q'_{conv} at 1500 K is 12,020 W/m², in comparison with a q'_{rad} of 143,300 W/m² (i.e., the convection contribution is only 8.4% of the total).

EXAMPLE 9.9

After being heated to a uniform temperature of 1173 K in a furnace, a strip of steel of thickness 0.03 m is removed from the furnace and allowed to cool in still air at 300 K under conditions in which the convection heat transfer coefficient is 20 W·m⁻²·K⁻¹. Calculate the temperature of the strip after 5 min of cooling. The data are

$$\rho = 7680 \text{ kg/m}^3$$

$$C_p = 470 \text{ J·kg}^{-1}\text{·K}^{-1}$$

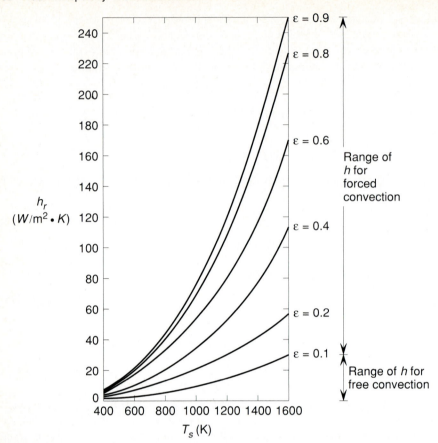

FIGURE 9.30. Variation, with surface temperature and emissivity, of the radiation heat transfer coefficient for emission of radiation to surroundings at 298 K.

$$k = 28 \ \text{W·m}^{-1}\text{·K}^{-1}$$

$$\epsilon = 0.6$$

First assume that heat transfer by radiation is the sole means of cooling. As all of the radiation emitted by *both* sides of the strip is intercepted by the surroundings, the view factor $F_{\text{strip}-\text{surr}}$ is unity and hence

$$q'_{\text{rad}} = 2\epsilon\sigma(T_s^4 - 300^4)$$

If the cooling of the strip is Newtonian, the rate of transfer of heat from the strip equals its rate of decrease in enthalpy. Thus per unit surface area (both sides) of the strip of thickness Δz,

$$2\epsilon\sigma(T_s^4 - 300^4) = -\rho C_p \, \Delta z \, \frac{dT_s}{dt}$$

$$2 \times 0.6 \times 5.67 \times 10^{-8}(T_s^4 - 300^4) = -7680 \times 470 \times 0.03 \, \frac{dT_s}{dt}$$

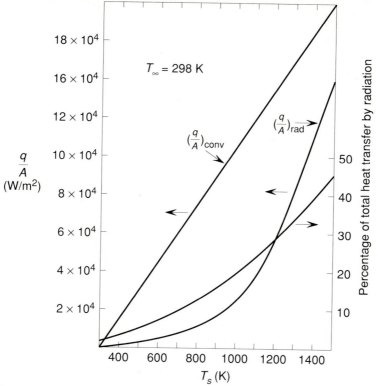

FIGURE 9.31. Contributions of forced convection (with $h = 150$ W/m² · K⁻¹) and radiation from a surface of $\epsilon = 0.5$ to its surroundings at $T_\infty = 298$ K as a function of surface temperature.

or

$$\frac{dT_s}{T_s^4 - 300^4} = -6.283 \times 10^{-13} \, dt \qquad \text{(i)}$$

From the identity

$$\int \frac{dx}{x^4 - a^4} = \frac{1}{4a^3} \ln \frac{a - x}{a + x} - \frac{1}{2a^3} \arctan \frac{x}{a}$$

the integral in Eq. (i) from $T_s = T_i$ at $t = 0$ to T_s and T_f at $t = t$ gives

$$\left[\frac{1}{4 \times 300^3} \ln \frac{300 + T_s}{300 - T_s} + \frac{1}{2 \times 300^3} \arctan \frac{T_s}{300} \right]_{T_i}^{T_f} = -6.283 \times 10^{-13} t$$

which for $t = 300$ s and $T_i = 1173$ K gives

$$\ln \frac{T_f - 300}{T_f + 300} - 2 \arctan \frac{T_f}{300} + 3.1639 = -6.786 \times 10^{-5} \times 300$$

which has the solution

$$T_f = 946 \text{ K}$$

after cooling for 300 s.

If it is assumed that heat transfer by convection is the sole means of cooling, then

$$q'_{\text{conv}} = 2h(T_s - 300) = -\rho C_p \, \Delta z \frac{dT_s}{dt}$$

which with $h = 20 \text{ W·m}^{-2}\text{·K}^{-1}$ gives

$$\frac{dT_s}{T_s - 300} = -\frac{2 \times 20}{7680 \times 470 \times 0.03} \, dt = -3.694 \times 10^{-4} \, dt$$

Thus, after 300 s,

$$\ln \frac{T_f - 300}{1173 - 300} = -3.694 \times 10^{-4} \times 300$$

which gives

$$T_f = 1081 \text{ K}$$

When both radiation and convection are considered,

$$q'_{\text{rad}} + q'_{\text{conv}} = -\rho C_p \, \Delta z \frac{dT_s}{dt}$$

or

$$[2 \times 0.6 \times 5.67 \times 10^{-8}(T_s^4 - 300^4)] + [2 \times 20(T_s - 300)]$$
$$= -7680 \times 470 \times 0.03 \frac{dT_s}{dt}$$

which gives

$$\frac{dT_s}{dt} = 0.1159 - 6.283 \times 10^{-13}T_s^4 - 3.694 \times 10^{-4}T_s$$

and numerical integration from 0 to 300 s gives

$$T_f = 891 \text{ K}$$

From Eq. (9.70), at 1173 K the radiation heat transfer coefficient is

$$\begin{aligned}
h_r &= \epsilon\sigma(T_s^2 + T_\infty^2)(T_s + T_\infty) \\
&= 0.6 \times 5.67 \times 10^{-8}(1173^2 + 300^2)(1173 + 300) \\
&= 73.5 \text{ W·m}^{-2}\text{·K}^{-1}
\end{aligned}$$

With $L_c = \Delta z/2$, the Biot number is

$$\text{Bi} = \frac{h_r L_c}{k} = \frac{73.5 \times 0.015}{28} = 0.039$$

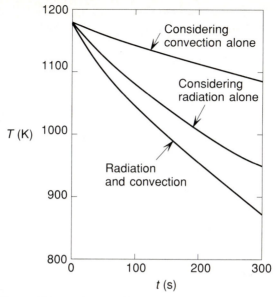

FIGURE 9.32. Rate of Newtonian cooling of the steel strip, considered in Example 9.9, assuming heat transfer by radiation alone, heat transfer by convection alone, and heat transfer by both radiation and convection.

which, being less than 0.1, indicates that the assumption of Newtonian cooling is valid. The cooling rates of the strip, assuming heat transfer by radiation only, heat transfer by convection only, and heat transfer by both radiation and convection, are shown in Fig. 9.32.

9.14 Summary

Thermal radiation, which is part of the electromagnetic spectrum, is emitted over a range of wavelengths in all directions from the surfaces of all bodies at finite temperatures. The intensity of the radiation varies with wavelength and direction of emission, and hence the spectral and directional distributions of the intensity have to be taken into consideration. The total intensity and total emissive power are obtained by integration over all wavelengths and all directions. A surface can receive radiation from another source, in which case it is said to be irradiated, and the irradiation can be reflected or absorbed or if the irradiated body is transparent to the irradiation, can be transmitted through the body. The absorptivity, reflectivity, and transmissivity of the irradiated body are, respectively, the fractions of the irradiation that are absorbed, reflected, and transmitted. Only the absorbed irradiation contributes to the enthalpy of the body. The radiosity of the surface of the body is the sum of its emissive power and the reflected irradiation.

A blackbody has an ideal surface that absorbs all incident irradiation and has the highest possible values of spectral intensity (intensity at a given wavelength) and spectral emissive power. The spectral distribution of the intensity of the radiation emitted by a blackbody is given by Planck's equation, integration of which over all wavelengths gives the Stefan–Boltzmann equation, which states that the total emissive power of a blackbody is proportional to the fourth power of its temperature.

The emissivity of a real surface is the ratio of the radiation emitted by the real surface to that emitted by a blackbody at the same temperature. The emissivity of a real surface varies with wavelength and direction of emission and the total hemispherical emissivity is obtained by integrating over all wavelengths and all directions. A surface at which the emissivity and absorptivity are independent of wavelength is said to be gray, and a surface at which the emissivity and absorptivity are independent of the direction of the radiation are said to be diffuse. A surface with both properties is a diffuse-gray surface.

Consideration of the exchange of radiation among several surfaces at different temperatures requires introduction of the view factor, which is the fraction of the radiation emitted by one surface that is intercepted by another surface. View factors in an enclosure formed by several surfaces, which are obtained by integration, are related to one another by the reciprocity relation and by the summation rule, and hence in an enclosure formed by n surfaces only $n(n-1)/2$ view factors need be obtained directly.

The treatment of radiation exchange among nonblackbody surfaces differs from the treatment of radiation exchange among blackbody surfaces in that the former may experience multiple reflections from and absorptions at all of the surfaces, in which case the radiosity of each surface is influenced by the radiosities of all of the other surfaces in the enclosure. This treatment is facilitated by the electric analogy in which surface and view resistances to radiation are defined. Treatment of the exchange of radiation among blackbody surfaces in an enclosure is simplified by the fact that blackbody surfaces absorb all incident irradiation and hence the radiosity of the surface is simply its emissive power.

Problems

PROBLEM 9.1 ⸻⸻⸻⸻⸻⸻⸻⸻⸻⸻⸻⸻⸻⸻⸻⸻⸻⸻⸻⸻

Calculate:

(a) The emissive power of the radiation emerging from a small aperture on a large isothermal enclosure that is at the uniform temperature of 1500 K
(b) The wavelength below which 50% of the emission occurs
(c) The maximum spectral emissive power and the wavelength at which the maximum spectral emissive power occurs

PROBLEM 9.2 ⸻⸻⸻⸻⸻⸻⸻⸻⸻⸻⸻⸻⸻⸻⸻⸻⸻⸻⸻⸻

A shallow pan of water with bottom and sides insulated is placed outside on a clear night when the effective sky radiation temperature is 230 K. If the emissivity of the surface of the water is 0.9 and the convection heat transfer coefficient is 25

$W{\cdot}m^{-2}{\cdot}K^{-1}$, what is the minimum air temperature that can be tolerated without freezing the water?

PROBLEM 9.3

An opaque horizontal plate, which is insulated on its backside, receives irradiation at the rate 3000 W/m^2, of which 500 W/m^2 is reflected. The plate is at a temperature of 200°C and has an emissive power of 500 W/m^2. Air at 25°C flows over the plate and the convection heat transfer coefficient is 20 $W{\cdot}m^{-2}{\cdot}K^{-1}$. Calculate:

(a) The emissivity, absorptivity, and radiosity of the plate
(b) The net heat transfer rate per unit area

PROBLEM 9.4

Solar irradiation of 1000 W/m^2 is incident on the top side of a plate the surface of which has a solar absorptivity of of 0.9 and an emissivity of 0.1. The surrounding air is at 20°C and the heat transfer coefficient is 15 $W{\cdot}m^{-2}{\cdot}K^{-1}$. If the backside of the plate is insulated, what is the steady-state temperature of the plate?

PROBLEM 9.5

A horizontal opaque surface at a steady-state temperature of 50°C is exposed to a flow of air at 23°C with a convection heat transfer coefficient of 20 $W{\cdot}m^{-2}{\cdot}K^{-1}$. The emissive power of the surface is 60 W/m^2, the irradiation incident on the surface is 1000 W/m^2, and the reflectivity of the surface is 0.4. Calculate:

(a) The absorptivity of the surface
(b) The net heat transfer by radiation, and
(c) The total heat transfer rate to or from the surface

PROBLEM 9.6

The inner top surface and the vertical side walls of a cubical muffle furnace with black walls and sides of length 2 m are at 500°C and the bottom inner surface is at 1000°C. Calculate the rate of heat transfer by radiation from (a) the bottom to the top inner surface and (b) from the bottom to the side walls.

PROBLEM 9.7

The flat bottom of a cylindrical crucible, which is fitted with a tight lid, is at 500°C and the inner surface of the lid and the vertical walls are at 400°C. The crucible has an inner diameter of 4 cm and is 4 cm in height. If the surfaces are black, what is the heat flux by radiation from the the bottom surface to (a) the inner surface of the lid and (b) the vertical walls?

PROBLEM 9.8

Air flows in a duct, the inner wall temperature of which is 200°C. A thermocouple placed in the airflow measures a temperature of 113°C. Calculate the temperature of the gas if the surface emissivities of the duct wall and the surface of the thermocouple are, respectively, 0.85 and 0.6 and the heat transfer coefficient for convection at the surface of the thermocouple is $125 \ W \cdot m^{-2} \cdot K^{-1}$.

In the following problems the surfaces can be considered to be diffuse-gray.

PROBLEM 9.9

A long heat-treating furnace has a rectangular cross section 3 m in height and 4 m in width. The surface of the roof is at 1000°C, the side walls are at 800°C, the floor is at 400°C, and all the surfaces have an emissivity of 0.8. Calculate the radiant heat transfer to the floor of the furnace per meter of length.

PROBLEM 9.10

An electrically heated wire of radius 1 mm, emissivity 0.5, and surface temperature 500°C is surrounded by a thin cylinder of radius 1 cm that is at 25°C and has an emissivity of 0.6.

(a) Calculate the radiation heat flux from the wire to the outer cylinder per unit length.

(b) A thin cylindrical radiation shield of radius 5 mm and emissivity 0.1 is placed between the wire and the outer cylinder. Calculate the new radiation heat flux from the wire to the outer cylinder.

Is the radiation heat transfer from the wire to the outer cylinder influenced by:

(c) The radius of the radiation shield?

(d) The degree of eccentricity of the shield?

(e) What radius of radiation shield is required to decrease the radiation heat flux occurring without a shield by a factor of 4?

(f) What is the steady-state temperature of a shield of this radius?

(g) At what radius is the temperature of the shield 600 K?

Consider all surfaces to be gray.

PROBLEM 9.11

An electrically heated copper wire of 2 cm diameter is surrounded by a thin cylinder of diameter 5 cm. The annular space between the wire and the cylinder is evacuated, the wire dissipates 50 W per meter of length, the surface emissivities of the wire and the cylinder are 0.8, and the steady-state temperature of the cylinder is 300 K. Calculate the steady-state temperature of the surface of the wire.

PROBLEM 9.12

A long duct with an equilateral triangular cross section and sides of width 1 m has the following surface conditions:

(1) Side 1 at 600°C with $\epsilon_1 = 0.6$

(2) Side 2 at 400°C with $\epsilon_2 = 0.4$

(3) Side 3 at 200°C with $\epsilon_3 = 0.2$

Calculate the radiation heat fluxes at the three walls.

PROBLEM 9.13 _____

If side 3 in Problem 9.12 were reradiating and the other two sides had the same surface conditions as in Problem 9.12, calculate:

(a) The radiation flux between sides 1 and 2
(b) The temperature of the reradiating side

PROBLEM 9.14 _____

An electrically heated furnace enclosure is 2 m long, 1 m wide, and 0.5 m in height. A 30-kW electric heater embedded in the roof produces a temperature of 700°C at the inner surface of the roof. The emissivity of the roof is 0.7 and the emissivity of the floor surface is 0.9. The four vertical walls are reradiating. Calculate the steady-state temperature of the surface of the floor. (Ignore any heat transport by convection.)

PROBLEM 9.15 _____

An unlagged horizontal steam pipe of diameter 0.05 m and surface emissivity 0.9 losses heat to its surroundings by radiation and by natural convection. If the surrounding air is at 10°C, calculate the surface temperature of the pipe at which the rate of heat loss due to radiation is twice the rate of heat loss by natural convection.

10

Mass Transport by Diffusion in the Solid State

10.1 Introduction

The physical and mechanical properties of all solid materials depend to some extent on structure, and development of desirable engineering properties requires the ability to manipulate such aspects of structure as crystal structure, phase morphology, and grain size. Although the relative stabilities of phases are determined by thermodynamic considerations, the rates at which phase changes can occur are determined by the rates at which atoms or molecules or ions can be transported in the solid state. An understanding of the kinetics of phase change in the solid state thus requires an appreciation of solid-state diffusion.

10.2 Atomic Diffusion as a Random-Walk Process

Diffusion in the solid state is a manifestation of the random motion of atoms. Consider Fig. 10.1, which shows a unit cell of face-centered cubic iron and the positions in the unit cell that can be occupied by interstitial carbon atoms dissolved in the iron. At any finite temperature the atoms are vibrating about their mean positions in the crystal lattice, and every now and then, when conditions are favorable, a carbon atom can jump from one interstitial site to a neighboring site. Thus a carbon atom located at the center of the unit cell in Fig. 10.1 can jump to any of the 12 neighboring sites located at the midpoints of the edges of the cube. As the jumping of the carbon atom is a random process, the probabilities that the carbon will jump to each of the 12 neighboring vacant sites are equal. The question is thus, how far will the carbon atom move in a given time? The problem can be addressed by considering the case

476

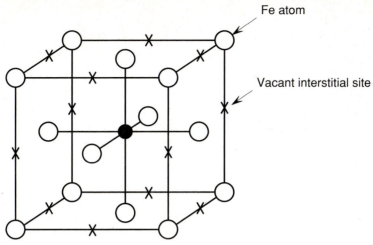

Fe atom

Vacant interstitial site

FIGURE 10.1. Face-centered cubic crystal structure of Fe showing the locations of the octahedral interstitial sites.

of the drunken sailor leaving a bar and attempting to return to his ship. Although the sailor is capable of walking, he does not have control of the direction of his steps and thus his motion is that of a random walk. To simplify the analysis, consider that the random motion of the sailor is in one dimension, in which a step is made either to the left or to the right. The sailor begins at the origin, $x = 0$, and takes n steps; he travels a distance x_1 on the first step, x_2 on the second step, and so on. After n steps the average distance $<x>$ that he has traveled from the origin is

$$<x> = \frac{x_1 + x_2 + \cdots x_n}{n} \tag{10.1}$$

As steps taken to the left and the right are equally likely, when a sufficiently large number of steps has been taken, the number of steps made to the left equals the number made to the right and hence the *mean* progress from the origin, $<x>$, is zero. The influence of the negative values of x_i that occur in Eq. (10.1) is eliminated by considering the sum of the squares of the values of i to give the *mean square distance*, $<x^2>$, traveled by the sailor in n steps as

$$<x^2> = \frac{x_1^2 + x_2^2 + \cdots + x_n^2}{n}$$

As the individual squares are positive, the mean square distance is a positive quantity and the dependence of $<x^2>$ on n is determined as follows. After having taken $n - 1$ steps the sailor has traveled the distance x_{n-1}. If he takes one more step of length l, his distance from the origin will be either

$$x_n = x_{n-1} + l \tag{10.2}$$

or

$$x_n = x_{n-1} - l \tag{10.3}$$

Squaring both sides of Eqs. (10.2) and (10.3) gives

$$x_n^2 = x_{n-1}^2 + 2lx_{n-1} + l^2$$

and

$$x_n^2 = x_{n-1}^2 - 2lx_{n-1} + l^2$$

The average of the two possibilities is thus

$$x_n^2 = x_{n-1}^2 + l^2 \tag{10.4}$$

which is the result for x_n^2 when the distance traveled after $n - 1$ steps is exactly x_{n-1}. For an averaged value of $<x_{n-1}^2>$, Eq. (10.4) becomes

$$<x_n^2> = <x_{n-1}^2> + l^2$$

Thus, at the start of the random walk before any steps have been taken,

$$<x_0^2> = 0$$

After the first step

$$<x_1^2> = <x_0^2> + l^2 = l^2$$

After the second step

$$<x_2^2> = <x_1^2> + l^2 = 2l^2$$

and so on, until after the nth step,

$$<x_n^2> = <x_{n-1}^2> + l^2 = nl^2$$

If the n steps are made at the rate of r per second, the time taken to make the n steps is n/r (i.e., $n = rt$) and hence

$$<x_n^2> = rtl^2$$

or

$$<x_n^2> \propto t \tag{10.5}$$

Thus the main characteristic of a random walking process is that the mean square distance that the sailor, or atom, has traveled, increases linearly with time.

Figure 10.2 shows two rectangular volumes each of length $<x^2>^{1/2}$ in the x-direction and of unit cross-sectional area. The volume on the left contains particles at the concentration C_L particles per unit volume and the volume on the right contains particles at the concentration C_R particles per unit volume. In time t a randomly walking particle travels a mean square distance of $<x^2>$ or a mean distance of $<x^2>^{1/2}$, and hence in time t, all of the particles in the volume on the left have either walked out of the volume in the $-x$ direction or have walked out of the volume in the $+x$ direction. As motions to the right and left are equally probable, the number of particles moving across the transit plan from the left volume to the right volume in time t is $0.5 \times$ (concentration) $\times$ (volume), or $0.5C_L<x^2>^{1/2}$, or the flux of particles (the number of particles transported through unit area in unit time), from left to right, $J_{L \rightarrow R}$, is

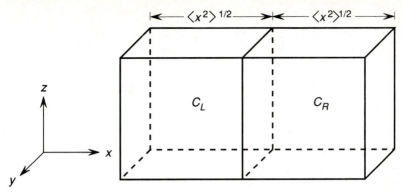

FIGURE 10.2. Two rectangular volumes containing randomly walking particles at the concentration C_L in the left volume and C_R in the right volume.

$$J_{L \to R} = \frac{0.5 C_L <x^2>^{1/2}}{t}$$

Similarly, the flux of particles from the volume on the right to the volume on the left is

$$J_{R \to L} = \frac{0.5 C_R <x^2>^{1/2}}{t}$$

The *net* flux of particles across the transit plane in the x direction is thus

$$J = J_{L \to R} - J_{R \to L} = \frac{0.5 <x^2>^{1/2}(C_L - C_R)}{t} \tag{10.6}$$

In Fig. 10.2 the concentration gradient in the x-direction is

$$\frac{dC}{dx} = \frac{C_R - C_L}{<x^2>^{1/2}}$$

substitution of which into Eq. (10.6) gives

$$J = -\frac{0.5 <x^2>}{t} \frac{dC}{dx} \tag{10.7}$$

Equation (10.7) shows that diffusion is caused by the existence of a concentration gradient, and that the flux of particles by diffusion is down the concentration gradient. Furthermore, from Eq. (10.5), as $<x_n^2>$ is proportional to t,

$$J \propto -\frac{dC}{dx} \tag{10.8}$$

(i.e., the diffusion flux is proportional to the concentration gradient).

10.3 Fick's First Law of Diffusion

The relationship given by Eq. (10.8) was first enunciated in 1855 by Fick as

$$J_i = -D \frac{dC_i}{dx} \tag{10.9}$$

which is known as *Fick's first law of diffusion*. The proportionality constant, D, is called the diffusion coefficient. With the diffusion flux of species i in $mol \cdot m^{-2} \cdot s^{-1}$ and the concentration gradient in $mol \cdot m^{-3} \cdot m^{-1}$, the diffusion coefficient has the units $m^2 \cdot s^{-1}$, which are the same as those of kinematic viscosity, v, and thermal diffusivity, α. Newton's law of viscosity, Eq. (2.4), Fourier's law of heat conduction, Eq. (6.1), and Fick's first law of diffusion can be compared as follows. Newton's law of viscosity,

$$\tau_{yx} = -\eta \frac{dv_x}{dy} \tag{2.4}$$

recast for a fluid of constant density as

$$\tau_{yx} = -v \frac{d}{dy} (\rho v_x)$$

shows that the transport of momentum is caused by the existence of a momentum gradient. Similarly, Fourier's law of heat conduction,

$$q'_x = -k \frac{dT}{dx} \tag{6.1}$$

recast for a medium of constant density and heat capacity as

$$q'_x = -\alpha \frac{d}{dx} (\rho C_p T)$$

shows that the transport of thermal energy is caused by a thermal energy gradient, and Fick's first law of diffusion shows that transport of mass by diffusion is caused by a gradient in mass concentration.

EXAMPLE 10.1 _____

Diffusion of Hydrogen Through a Planar Iron Wall

Consider the iron film of thickness Δx shown in Fig. 10.3, which separates two volumes of hydrogen gas. The gas on the left side of the film is at the pressure P_1 and the gas on the right is at the lower pressure P_2. For constant values of P_1 and P_2 the diffusion flux of hydrogen through the iron film from the high-pressure side to the low-pressure side is obtained from Fick's first law as

$$J_H = -D_H \frac{dC}{dx} = \frac{D_H(C_{H(P_1)} - C_{H(P_2)})}{\Delta x} \tag{i}$$

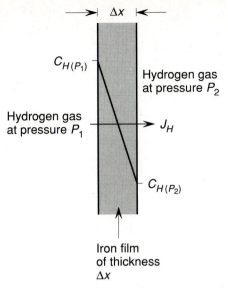

FIGURE 10.3. Diffusion of hydrogen through a plane wall.

where D_H is the diffusion coefficient for hydrogen in iron, $C_{H(P_1)}$ is the concentration of hydrogen in iron in equilibrium with hydrogen gas at the pressure P_1, and $C_{H(P_2)}$ is the corresponding value at the pressure P_2. Hydrogen dissolved in iron obeys *Sievert's law*; that is, for the equilibrium

$$\tfrac{1}{2}H_{2(g)} = [H]_{\text{dissolved in iron}}$$

the equilibrium constant is

$$K = \frac{[\text{ppm H}]}{P_{H_2}^{1/2}}$$

where [ppm H] is the concentration of atomic hydrogen in the iron, in parts per million by weight, in equilibrium with hydrogen gas at the pressure P_{H_2}. At 400°C and a pressure of hydrogen of 1.013×10^5 Pa (1 atm), the solubility of hydrogen in iron is 3 ppm by weight. Thus at 400°C,

$$K = \frac{3}{(1.013 \times 10^5)^{1/2}} = 9.43 \times 10^{-3}$$

or

$$[\text{ppm H}] = 9.43 \times 10^{-3} P_{H_2}^{1/2} \qquad \text{(ii)}$$

A concentration of [ppm H] corresponds to [ppm H] kg of hydrogen per 10^6 kg of iron. The density of iron at 400°C is 7730 kg/m³, and thus a concentration of [ppm H] by weight is

$$C_H = \frac{[\text{ppm H}] \times 7730}{10^6} = \frac{[\text{ppm H}]}{129.4} \frac{\text{kg H}}{\text{m}^3}$$

which, in combination with Eq. (ii), gives

$$C_H = \frac{9.43 \times 10^{-3}}{129.4} P_{H_2}^{1/2} = 7.29 \times 10^{-5} P_{H_2}^{1/2} \qquad \text{(iii)}$$

Substitution of Eq. (iii) into Eq. (i) then gives

$$J_H = \frac{7.29 \times 10^{-5} D_H (P_1^{1/2} - P_2^{1/2})}{\Delta x} \qquad \text{(iv)}$$

At 400°C, $D_H = 10^{-8}$ m²/s, and with $\Delta x = 0.001$ m, $P_1 = 1.013 \times 10^5$ Pa, and $P_2 = 0$,

$$J_H = \frac{7.29 \times 10^{-5} \times 10^{-8} \times (1.013 \times 10^5)^{1/2}}{0.001} = 2.32 \times 10^{-7} \text{ kg·m}^{-2}\text{·s}$$

Consider a cubical iron tank of wall thickness 0.001 m and volume 0.1 m³ that contains hydrogen gas at a pressure of 1.013×10^5 Pa at 400°C. If the pressure outside the tank is zero, after what time has the diffusion of hydrogen through the walls of the tank occurred to the extent that the pressure in the tank has decreased to half of its original value? The mass balance gives

rate of loss of mass from tank = mass diffusion flux through walls

Assuming that hydrogen obeys the ideal gas law, the rate of loss of mass of hydrogen from the tank, dm/dt, is related to the rate of decrease in pressure as

$$\frac{dm}{dt} = \frac{MV}{RT} \frac{dP}{dt}$$

where M is the molecular weight of hydrogen, T is temperature, V is the volume of the tank, and R is the gas constant. Therefore,

$$\frac{dm}{dt} = \frac{2 \times 10^{-3} \times 0.1}{8.3144 \times 673} \frac{dP}{dt} = 3.574 \times 10^{-8} \frac{dP}{dt} \text{ kg·s}^{-1}$$

The cubical tank of volume 0.1 m³ has sides of length 0.464 m and hence a surface area of $6 \times 0.464^2 = 1.29$ m², and thus the diffusion flux through the walls is the product of the flux per unit area given by Eq. (iv) and the area:

$$J = \frac{1.29 \times 7.29 \times 10^{-5} \times 10^{-8}}{0.001} \sqrt{P} = 9.40 \times 10^{-10} \sqrt{P} \text{ kg·s}^{-1}$$

Therefore,

$$-3.574 \times 10^{-8} \frac{dP}{dt} = 9.40 \times 10^{-10} \sqrt{P}$$

or

$$\frac{dP}{\sqrt{P}} = -2.631 \times 10^{-2} dt$$

Integrating between P_0 at $t = 0$ and P at $t = t$ gives

$$P_0^{1/2} - P^{1/2} = 1.316 \times 10^{-2}t$$

which with $P_0 = 1.013 \times 10^5$ Pa and $P = 0.507 \times 10^5$ gives

$$t = 7075 \text{ s} = 1.97 \text{ h}$$

as the time required for the pressure to decrease to half of its original value.

EXAMPLE 10.2 ————————————————————————————

Diffusion of Carbon Through the Wall of a Hollow Iron Cylinder

Consider a hollow iron cylinder of length L, the inner and outer surfaces of which are in contact with carburizing gases (mixtures of CO and CO_2) of different carburizing potentials. Local thermodynamic equilibrium is established between the gases and the surfaces and thus the carbon contents of the surfaces are fixed by the activy of carbon in the gases. As the concentrations of carbon at the two surfaces are different and are maintained at constant values by the gases, at steady state there is a constant flux of carbon by diffusion through the wall of the cylinder given by

$$J \left(\frac{\text{kg}}{\text{s}} \right) = -A(\text{m}^2)D_C \left(\frac{\text{m}^2}{\text{s}} \right) \frac{dC}{dr} \left(\frac{\text{kg}}{\text{m}^3 \cdot \text{m}} \right)$$

where the area through which the carbon is diffusing is $A = 2\pi r L$. Thus

$$J = -2\pi r L D_C \frac{dC}{dr} = -2\pi L D_C \frac{dC}{d \ln r} \qquad (10.10)$$

Equation (10.10) indicates that at steady state and with a constant D_C, $dC/d \ln r$ is a constant (i.e., the concentration of carbon in the iron is a linear function of the logarithm of the radial position in the wall). Experiments of this type conducted by Smith [R. P. Smith, *Acta. Met* (1953), vol. 1, p. 578] gave the results shown in Fig. 10.4, in which the gradient $dC/d \ln r$ increases in moving through the cylinder wall from the high-carbon outer surface to the low-carbon inner surface. As Eq. (10.10) requires that the term $D_C dC/d \ln r$ be constant, Fig. 10.4 shows that the diffusion coefficient of carbon in iron increases with increasing carbon content. This variation, as determined by measuring the flux of carbon through the wall of the cylinder and the slopes in Fig. 10.4, is shown in Fig. 10.5.

10.4 One-Dimensional Non-Steady-State Diffusion in a Solid; Fick's Second Law of Diffusion

Consider a mass balance on the control volume of unit cross-sectional area and length Δx shown in Fig. 10.6. For one-dimensional diffusion of a solute, i, in the x-direction, the mass balance is

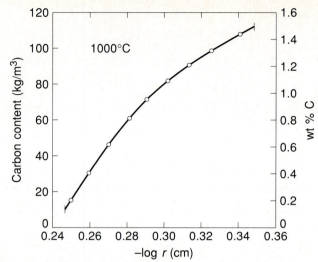

FIGURE 10.4. Concentration profile of carbon in the wall of a hollow steel cylinder under conditions of steady-state diffusion.

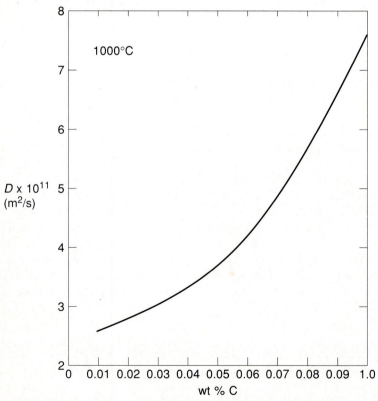

FIGURE 10.5. Dependence, on composition, of the diffusion coefficient of carbon in iron at 1000°C.

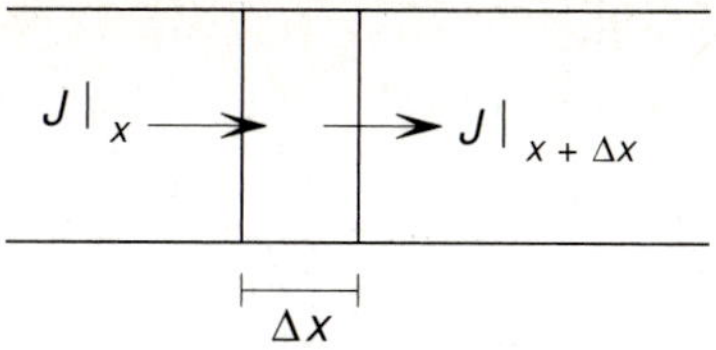

FIGURE 10.6. Mass balance in the control volume of unit cross-sectional area and length Δx.

rate of accumulation of i in control volume
= (rate at which i is transported into control volume at x by diffusion)
− (rate at which i is transported out of control volume at $x + \Delta x$ by diffusion)

that is,

$$\frac{\partial C_i}{\partial t} \Delta x = J_i|_x - \left(J_i|_x + \frac{\partial J_i}{\partial x} \Delta x \right)$$

or

$$\frac{\partial C_i}{\partial t} = -\frac{\partial J_i}{\partial x} \tag{10.11}$$

Substitution of Eq. (10.11) into Eq. (10.9) gives

$$\frac{\partial C_i}{\partial t} = \frac{\partial}{\partial x} \left(D \frac{\partial C_i}{\partial x} \right) \tag{10.12}$$

or for a constant D,

$$\frac{\partial C_i}{\partial t} = D \frac{\partial^2 C_i}{\partial x^2} \tag{10.13}$$

Equations (10.12) and (10.13) are called *Fick's second law of diffusion*, which with a constant D is analogous to Eq. (8.13) for one-dimensional heat flow by conduction,

$$\frac{\partial T}{\partial t} = \alpha \frac{\partial^2 T}{\partial x^2} \tag{8.13}$$

In Section 8.3 an examination was made of the situation in which one end of a semi-infinite rod, along which the temperature is uniformly T_i, is instantaneously raised to the higher temperature T_0, and heat flows into the rod by conduction. The diffusion analog is that in which one end of a semi-infinite rod, in which the concentration of the species i is initially uniform at the value C_i^0, is instantaneously raised to the concentration C_i^s, and the species i diffuses into the rod. In the case of heat flow by conduction, introduction of the function

$$\eta = \frac{x}{2\sqrt{\alpha t}} \tag{8.16}$$

allowed Eq. (8.13) to be converted from a partial differential equation to an ordinary differential equation. In a similar manner, introduction of the variable

$$\eta = \frac{x}{2\sqrt{Dt}}$$

allows Eq. (10.13) to be converted to the ordinary differential equation

$$\frac{dC}{d\eta} = -\frac{1}{2\eta}\frac{d^2C}{d\eta^2} \tag{10.14}$$

Solution of Eq. (8.13) for the boundary conditions

$$T = \begin{cases} T_0 & \text{at } x = 0 \quad \text{and} \quad t > 0 \\ T_i & \text{at } t = 0 \quad \text{and} \quad x > 0 \end{cases}$$

gave the temperature profile in the rod as

$$\frac{T - T_i}{T_0 - T_i} = \text{erfc}(\eta) = \text{erfc}\left(\frac{x}{2\sqrt{\alpha t}}\right) \tag{8.17}$$

In an identical manner solution of Eq. (10.14), with the boundary conditions

$$C_i = \begin{cases} C_i^s & \text{at } x = 0 \quad \text{and} \quad t > 0 \\ C_i^0 & \text{at } t = 0 \quad \text{and} \quad x > 0 \end{cases}$$

gives

$$\frac{C_i - C_i^0}{C_i^s - C_i^0} = \text{erfc}(\eta) = \text{erfc}\left(\frac{x}{2\sqrt{Dt}}\right) \tag{10.15}$$

which is the complementary error function shown in Fig. 8.4. Alternatively,

$$\frac{C_i - C_i^s}{C_i^0 - C_i^s} = 1 - \frac{C_i - C_i^0}{C_i^s - C_i^0} = 1 - \text{erfc}(\eta)$$
$$= \text{erf}(\eta) \tag{10.16}$$

which is the error function shown in Fig. 8.4. If C_i^0 is zero, Eq. (10.15) reduces to

$$\frac{C_i}{C_i^s} = \text{erfc}\left(\frac{x}{2\sqrt{Dt}}\right) \tag{10.17}$$

which is shown in Fig. 10.7.

EXAMPLE 10.3 ───

Carburization of Iron

Consider the unidirectional carburization of iron at 1000°C in which the surface of the initially pure iron is brought into contact with a carburizing gas at time $t = 0$, which instantaneously raises the concentration of carbon in the iron at the surface to C_s. For an average diffusion coefficient of 3.0×10^{-11} m²/s, the

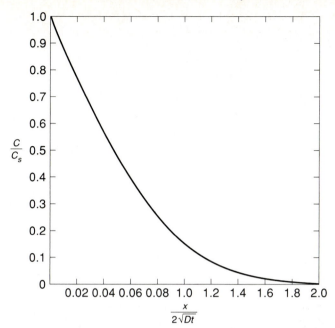

FIGURE 10.7. Complementary error function dependence of the normalized concentration of a species diffusing in a semi-infinite medium.

concentration profiles obtained from Eq. (10.16) at $t = 1, 6, 12$, and 24 h are shown in Fig. 10.8. (a) What is the value of C/C_s at $x = 0.5$ mm after 1 h of carburization? (b) After what time does $C/C_s = 0.5$ at $x = 8$ mm?

(a) With $x = 5 \times 10^{-4}$ m and $t = 3600$ s,

$$\eta = \frac{x}{2\sqrt{Dt}} = \frac{5 \times 10^{-4}}{2(3 \times 10^{-11} \times 3600)^{1/2}} = 0.761$$

and

$$\frac{C}{C_s} = \text{erfc}(0.761) = 0.282$$

which is point A in Fig. 10.8.

(b) From Fig. 10.7, $C/C_s = 0.5$ at $x/2\sqrt{Dt} = 0.48$. Thus, at $x = 8 \times 10^{-4}$ m,

$$t = \left(\frac{8 \times 10^{-4}}{0.48 \times 2}\right)^2 \times \frac{1}{3 \times 10^{-11}} = 2.31 \times 10^4 \text{ s} = 6.43 \text{ h}$$

A *diffusion distance* is commonly defined as $x/\sqrt{Dt} = 1$ or

$$x = \sqrt{Dt}$$

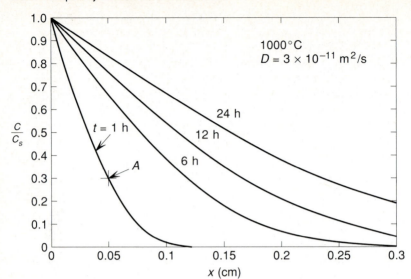

FIGURE 10.8. Variation, with x, of the normalized concentration of carbon diffusing into a semi-infinite rod of iron, considered in Example 10.3.

With $x = \sqrt{Dt}$ or $\eta = 0.5$,

$$\frac{C}{C_s} = \text{erfc}(0.5) = 0.48$$

which is very close to 0.5. Thus, in time t, the diffusion distance is the value of x at which $C/C_s \sim 0.5$. If it were required that a diffusion distance of 0.1 cm be achieved by carburizing at 1000°C, C/C_s, would have the value of approximately 0.5 after

$$t = \frac{x^2}{D} = \frac{(1 \times 10^{-3})^2}{3 \times 10^{-11}} = 3333 \text{ s} = 9.26 \text{ h}$$

Combination of Eqs. (10.7) and (10.9) yields

$$<x^2> = 2Dt \qquad\qquad (10.18)$$

which is known as the *Einstein–Smoluchowski equation*, in which $<x^2>$ is the mean square distance through which the diffusing particles have traveled. The root mean square distance, $<x^2>^{1/2}$, corresponds to

$$\eta = \frac{<x^2>^{1/2}}{2\sqrt{Dt}} = \frac{1}{\sqrt{2}} = 0.707$$

and

$$\text{erfc}(0.707) = \frac{C}{C_s} = 0.32$$

Thus the root mean square distance is that at which, after diffusion for time t, C/C_s is approximately one-third, and by comparing the areas under the curve in Fig. 10.7 from $\eta = 0$ to $\eta = 0.707$ and from $\eta = 0.707$ to $\eta = \infty$, after time t approximately 79% of the atoms that have entered the solid are in the region between $x = 0$ and $x = \langle x^2 \rangle^{1/2}$ and approximately 21% of the atoms have diffused farther than the root-mean-square distance. From Fick's first law, the flux of carbon atoms across the surface at $x = 0$ is

$$J\big|_{x=0} = -D \frac{\partial C}{\partial x}\bigg|_{x=0}$$

With

$$\frac{C}{C_s} = \operatorname{erfc}(\eta) = 1 - \frac{2}{\sqrt{\pi}} \int_0^z e^{-\eta^2} \, d\eta = 1 - \frac{2}{\sqrt{\pi}} \int_0^z e^{-x^2/4Dt} \, d\frac{x}{2\sqrt{Dt}}$$

$$\frac{\partial C}{\partial x}\bigg|_{x=0} = \frac{-2C_s}{\sqrt{\pi}} e^0 \frac{1}{2\sqrt{Dt}}$$

and thus

$$J\big|_{x=0} = C_s \sqrt{\frac{D}{\pi t}} \tag{10.19}$$

The mass of carbon entering the steel in time t is obtained from Eq. (10.19) as

$$\int_0^t J \, dt = C_s \sqrt{\frac{D}{\pi}} \int_0^t t^{-1/2} \, dt = 2C_s \sqrt{\frac{Dt}{\pi}} \tag{10.20}$$

For decarburization of a steel of initially uniform composition C_0, in which the surface concentration is instantaneously decreased to a lower constant value, C_s, at $t = 0$, the concentration profile is

$$\frac{C - C_s}{C_0 - C_s} = \operatorname{erf}\left(\frac{x}{2\sqrt{Dt}}\right) \tag{10.21}$$

10.5 Infinite Diffusion Couple

When one end of a semi-infinite rod of steel of uniform carbon content C_0, is brought into intimate contact with one end of a semi-infinite rod of pure iron the assembly is called an infinite diffusion couple and carbon diffuses from the steel rod into the iron rod. The diffusion couple is infinite in that, within the period of time when diffusion is occurring, no changes in concentration occur near the two ends of the couple. In this case the boundary conditions are

$$C = \begin{cases} 0 & \text{at } x < 0 \quad \text{and} \quad t = 0 \\ C_0 & \text{at } x > 0 \quad \text{and} \quad t = 0 \end{cases}$$

and the solution to Eq. (10.12) is

$$\frac{C}{C_0} = \frac{1}{2}\left[1 + \mathrm{erf}\left(\frac{x}{2\sqrt{Dt}}\right)\right] \tag{10.22}$$

Equation (10.22) is shown in Fig. 10.9. At $x > 0$,

$$\frac{C}{C_0} = 0.5 + \frac{1}{2}\,\mathrm{erf}\left(\frac{x}{2\sqrt{Dt}}\right)$$

and with $x < 0$, as $\mathrm{erf}\,(-x) = -\mathrm{erf}(x)$,

$$\frac{C}{C_0} = 0.5 - \frac{1}{2}\,\mathrm{erf}\left(\frac{x}{2\sqrt{Dt}}\right)$$

and $C/C_0 = 0.5$ at $x = 0$ and $t > 0$. Each value of C/C_0 is associated with a particular value of $\eta = x/2\sqrt{Dt}$. For example, for $\eta = 1$, $\mathrm{erf}\,(1) = 0.845$ and hence $C/C_0 = 0.921$. Thus the value of x at which $C/C_0 = 0.921$ varies with time as

$$x = 2\sqrt{Dt}$$

The flux of carbon across the plane at $x = 0$ is obtained as

$$J = -D\,\frac{\partial C}{\partial x}\bigg|_{x=0}$$

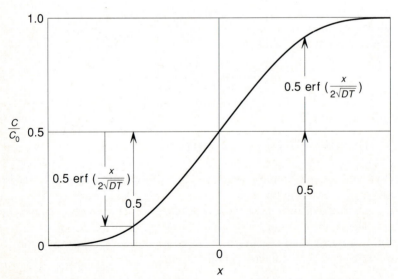

FIGURE 10.9. Normalized concentration profile produced by diffusion in an infinite diffusion couple.

From Eq. (10.22),

$$\frac{\partial C}{\partial x} = \frac{C_0}{2} \frac{2}{\sqrt{\pi}} \frac{1}{2\sqrt{Dt}} e^{-x^2/4Dt}$$

and thus

$$J = -D \frac{\partial C}{\partial x}\bigg|_{x=0} = -\frac{C_0}{2} \sqrt{\frac{D}{\pi t}} \tag{10.23}$$

and the mass of carbon which has crossed the plane at $x = 0$ in time t is

$$M = \int_0^t J \, dt = -C_0 \sqrt{\frac{Dt}{\pi}} \tag{10.24}$$

The concentration profiles in a diffusion couple at 1000°C after times of 1, 5, and 10 h are shown in Fig. 10.10.

10.6 One-Dimensional Diffusion in a Semi-infinite System Involving a Change of Phase

The portion of the iron–carbon phase diagram shown in Fig. 10.11 shows the phase relationship between austenite (γ) and ferrite (α). Thus if at a temperature in the range 723 to 910°C, one end of a semi-infinite rod of austenite is brought into contact with a decarburizing gas in which the chemical potential of carbon corresponds to a carbon content in the ferrite phase field, ferrite is nucleated on the surface of the austenite and grows into the rod; that is, decarburization causes the movement of

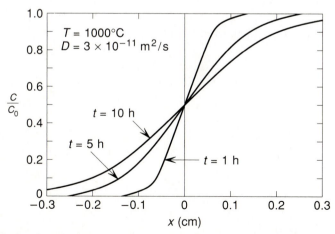

FIGURE 10.10. Normalized concentration profiles for carbon diffusing in an infinite diffusion couple.

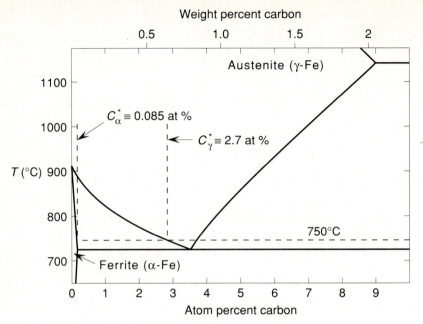

FIGURE 10.11. Part of the Fe–C phase diagram.

the austenite–ferrite phase boundary away from the surface of the rod, and at any finite time t, the concentration profile in the semi-infinite rod is as shown in Fig. 10.12. As equilibrium is maintained at the phase boundary, the carbon contents of the austenite at the boundary, C_γ^*, and of ferrite at the boundary, C_α^*, have constant values determined by the phase diagram. The question of interest is: At what rate does the phase boundary at $x = M$ move into the rod?

Diffusion in both phases is governed by Eq. (10.13), which is converted to the ordinary differential equation given by Eq. (10.14):

$$\frac{dC}{d\eta} = -\frac{1}{2\eta}\frac{d^2C}{d\eta^2} \tag{10.14}$$

Setting

$$y = \frac{dC}{d\eta}$$

Eq. (10.14) can be written as

$$\frac{dy}{y} = -2\eta \, d\eta$$

integration of which gives

$$\ln y = -\eta^2 + \ln A'$$

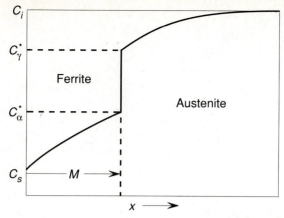

FIGURE 10.12.   Schematic representation of the concentration profile of carbon in iron when decarburization of austenite causes nucleation and growth of ferrite at the surface.

where $\ln A'$ is the integration constant, or

$$y = \frac{dC}{d\eta} = A'e^{-\eta^2}$$

A second integration gives

$$\int dC = A' \int e^{-\eta^2}\, d\eta \tag{i}$$

The initial condition is

$$C_\gamma = C_i \quad \text{at } t = 0 \ (\eta = \infty) \quad \text{and} \quad x \geq 0$$

which when applied to Eq. (i) gives

$$\int_{C_\gamma}^{C_i} dC = A' \int_{\eta}^{\infty} e^{-\eta^2}\, d\eta$$

or

$$C_i = C_\gamma - A' \frac{2}{\sqrt{\pi}} \operatorname{erfc}(\eta)$$

or

$$C_\gamma = C_i - A \operatorname{erfc}\left(\frac{x}{2\sqrt{D_\gamma t}}\right) \tag{ii}$$

where D_γ is the diffusion coefficient of carbon in austenite. The boundary condition is

$$C_\alpha = C_s \quad \text{at } x = 0 \ (\eta = 0) \quad \text{and} \quad t > 0$$

which, when applied to Eq. (i) gives

$$\int_{C_s}^{C_\alpha} dC = A' \int_0^\eta e^{-\eta^2} d\eta$$

or

$$C_\alpha - C_s = B \text{ erf}\left(\frac{x}{2\sqrt{D_\alpha t}}\right) \tag{iii}$$

where D_α is the diffusion coefficient of carbon in ferrite. At $x = M$ Eq. (ii) becomes

$$C_\gamma^* = C_i - A \text{ erfc}\left(\frac{M}{2\sqrt{D_\gamma t}}\right) \tag{iv}$$

which shows that as C_γ^*, C_i and A are constants,

$$\frac{M}{2\sqrt{D_\gamma t}} = \text{constant} = \phi$$

or

$$M = 2\phi\sqrt{D_\gamma t} \tag{v}$$

Similarly, at $x = M$, Eq. (iii) becomes

$$C_\alpha^* = C_s + B \text{ erf}\left(\frac{M}{2\sqrt{D_\alpha t}}\right)$$

$$= C_s + B \text{ erf}(\phi\alpha) \tag{vi}$$

where

$$\alpha = \sqrt{\frac{D_\gamma}{D_\alpha}}$$

Thus from Eq. (iv),

$$A = \frac{C_i - C_\gamma^*}{\text{erfc}(\phi)}$$

and from Eq. (vi),

$$B = \frac{C_\alpha^* - C_s}{\text{erf}(\phi\alpha)}$$

Eq. (ii) thus becomes

$$C_\gamma = C_i - \frac{C_i - C_\gamma^*}{\text{erfc}(\phi)} \text{ erfc}\left(\frac{x}{2\sqrt{D_\gamma t}}\right) \tag{vii}$$

and Eq. (iii) becomes

$$C_\alpha = C_s + \frac{C_\alpha^* - C_s}{\mathrm{erf}(\phi\alpha)}\,\mathrm{erf}\left(\frac{x}{2\sqrt{D_\alpha t}}\right) \tag{viii}$$

Mass balance at the phase boundary requires that the diffusion flux in the austenite to the boundary at $x = M$ minus the diffusion flux in the ferrite away from the boundary at $x = M$ equals the rate at which carbon is added to the α-phase:

$$-D_\gamma \frac{\partial C_\gamma}{\partial x}\bigg|_{x=M} + D_\alpha \frac{\partial C_\alpha}{\partial x}\bigg|_{x=M} = (C_\gamma^* - C_\alpha^*)\frac{dM}{dt} \tag{ix}$$

Considering each of the terms in Eq. (ix) one at a time, From Eq. (vii),

$$-D_\gamma \frac{\partial C_\gamma}{\partial x}\bigg|_{x=M} = \frac{D_\gamma(C_\gamma^* - C_i)}{\mathrm{erfc}(\phi)}\frac{1}{2\sqrt{D_\gamma t}}\frac{2}{\sqrt{\pi}}e^{-\phi^2}$$

From Eq. (viii),

$$D_\alpha \frac{\partial C_\alpha}{\partial x}\bigg|_{x=M} = \frac{D_\alpha(C_\alpha^* - C_s)}{\mathrm{erf}(\phi\alpha)}\frac{1}{2\sqrt{D_\alpha t}}\frac{2}{\sqrt{\pi}}e^{-(\phi\alpha)^2}$$

and from Eq. (v)

$$(C_\gamma^* - C_\alpha^*)\frac{dM}{dt} = (C_\gamma^* - C_\alpha^*)\phi\sqrt{\frac{D_\gamma}{t}}$$

Inserting into Eq. (ix) and simplifying gives

$$C_\gamma^* - C_\alpha^* = \frac{C_\gamma^* - C_i}{\sqrt{\pi}\,\phi e^{\phi^2}\,\mathrm{erfc}(\phi)} + \frac{C_\alpha^* - C_s}{\sqrt{\pi}\,\phi\alpha e^{(\phi\alpha)^2}\,\mathrm{erf}(\phi\alpha)} \tag{x}$$

from which the value of ϕ is obtained. This expression is the mass transport analog of Eq. (8.44). For diffusion of carbon into ferrite, which causes nucleation of austenite at the surface and movement of the austenite-ferrite interface into the steel the expression is

$$C_\alpha^* - C_\gamma^* = \frac{C_\alpha^* - C_i}{\sqrt{\pi}\,\alpha\phi e^{(\alpha\phi)^2}\,\mathrm{erfc}(\phi)} + \frac{C_\gamma^* - C_s}{\sqrt{\pi}\,\phi e^{\phi^2}\,\mathrm{eft}(\phi)} \tag{ix}$$

in which C_i is the initial concentration of carbon in the ferrite and C_s is the constant concentration of carbon in the austenite at $x = 0$.

EXAMPLE 10.4

Consider the decarburization of a steel with an initial carbon content of 3 atom percent at 750°C by a decarburizing gas that maintains a carbon content of 0.05 atom percent at the surface. From Fig. 10.11, the carbon contents of the ferrite and austenite phases at the phase boundary are, respectively, 0.09 and 2.7 at %. In dilute solutions, as concentration in atom percent is proportional to concentration in g mol/m³ and Eq. (x) contains only expressions for differences in concentration, the concentrations can be selected as

$$C_i = 3.0 \text{ at } \%$$

$$C_\gamma^* = 2.7 \text{ at } \%$$

$$C_\alpha^* = 0.09 \text{ at } \%$$

$$C_s = 0.05 \text{ at } \%$$

At 750°C, $D_\gamma = 1.1 \times 10^{-12}$ m²/s and $D_\alpha = 5 \times 10^{-11}$ m²/s. Thus

$$\alpha = \left(\frac{1.1 \times 10^{-12}}{5 \times 10^{-11}}\right)^{1/2} = 0.148$$

and Eq. (x) becomes

$$\sqrt{\pi}\,(2.7 - 0.09) = \frac{2.7 - 3.0}{\phi e^{\phi^2} \text{erfc}(\phi)} + \frac{0.09 - 0.05}{0.148\phi \exp(0.148\phi)^2 \text{erf}(0.148\phi)}$$

or

$$4.63 + \frac{0.3}{\phi e^{\phi^2} \text{erfc}(\phi)} - \frac{0.04}{0.148\phi \exp(0.148\phi)^2 \text{erf}(0.148\phi)} = 0$$

A plot of the left side of this equation as a function of ϕ is shown in Fig. 10.13, from which the solution is obtained as $\phi = 0.54$. Therefore, from Eq. (v),

$$M = 2 \times 0.54\sqrt{1.1 \times 10^{-12}t}$$
$$= 1.13 \times 10^{-6}\sqrt{t} \text{ m}$$

and the rate of advance of the interface is

$$\frac{dM}{dt} = \frac{5.66 \times 10^{-7}}{\sqrt{t}} \quad \text{m/s}$$

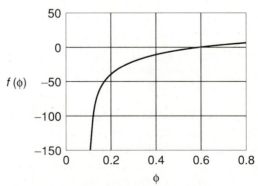

FIGURE 10.13. Variation of $f(\phi)$ with ϕ.

The time required for decarburization to produce a 0.1-mm-thick ferrite layer is

$$\left(\frac{10^{-4}}{1.13 \times 10^{-6}}\right)^2 = 7830 \text{ s} = 2.2 \text{ h}$$

and at this time the concentration profiles are given by Eq. (vii) as

$$(\text{at } \% \ C)_\gamma = 3.0 - \frac{3 - 2.7}{\text{erfc}(0.54)} \text{erfc}\left(\frac{x}{2\sqrt{1.1 \times 10^{-12} \times 7830}}\right)$$

$$= 3.0 - 0.67 \text{ erfc}(5387x)$$

and by Eq. (viii) as

$$(\text{at } \% \ C)_\alpha = 0.05 + \frac{0.09 - 0.05}{\text{erf}(0.08)} \text{erf}\left(\frac{x}{2\sqrt{5 \times 10^{-11} \times 7830}}\right)$$

$$= 0.05 + 0.44 \text{ erf}(799x)$$

These concentration profiles are shown in Fig. 10.14.

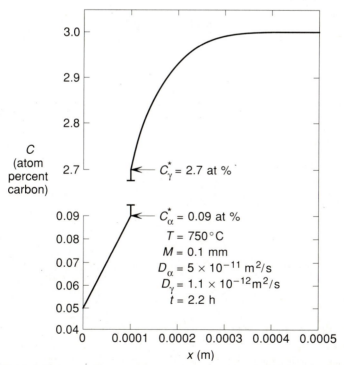

FIGURE 10.14. Concentration profile of carbon in ferrite and austenite calculated in Example 10.4.

10.7 Steady-State Diffusion Through a Composite Wall

Consider Fig. 10.15, which shows the diffusion of hydrogen through a plane composite wall consisting of sheets of the metals A and B each of thickness Δx. The hydrogen gas in contact with metal A is at the pressure P_1 and the hydrogen in contact with B is at the pressure P_2, and the chemical diffusion coefficients of hydrogen in A and B are, respectively, D_{H-A} and D_{H-B}. At steady state

$$J_H(\text{in } A) = J_H(\text{in } B)$$

or

$$D_{H-A}\frac{C_1 - C_1^*}{\Delta x} = D_{H-B}\frac{C_2^* - C_2}{\Delta x} \tag{i}$$

in which C_1 is the solubility of hydrogen in A at the pressure P_1, C_2 is the solubility of hydrogen in B at the pressure P_2, and C_1^* and C_2^* are the concentrations of hydrogen in A and B, respectively, at the interface between A and B. The concentrations C_1^* and C_2^* are determined by the solubilities and diffusivities of hydrogen in the metals A and B, and if the solubilities differ from one another, C_1^* and C_2^* do not have the same value. However, the thermodynamic activity of hydrogen has a single value at the interface, and hence treatment of the diffusion flux is simplified by considering activity gradients in the composite wall rather than concentration gradients. For the equilibrium

$$\tfrac{1}{2}H_{2(gas)} = [H]_{\text{dissolved in the metal}}$$

if hydrogen in solution obeys Sievert's law, the equilibrium constant can be written as

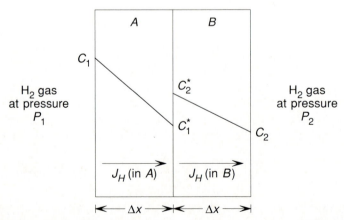

FIGURE 10.15.   Schematic representation of the concentration profiles of hydrogen diffusing through a composite plain wall.

$$K = \frac{C_H}{P_{H_2}^{1/2}}$$

or

$$C_H = KP_{H_2}^{1/2}$$

in which K is numerically equal to the concentration of hydrogen in equilibrium with hydrogen gas at a pressure of 1 atm. The activity of hydrogen is equal to the square root of the partial pressure of its pressure and thus

$$a_H = P_{H_2}^{1/2} = \frac{C_H}{K} \tag{ii}$$

substitution of which into Eq. (i) gives

$$J_H = D_{H-A}K_A \frac{P_1^{1/2} - a_H^*}{\Delta x} = D_{H-B} \frac{K_B(a_H^* - P_2^{1/2})}{\Delta x} \tag{iii}$$

a_H^*, the activity of hydrogen at the interface between A and B, is thus obtained as

$$a_H^* = \frac{D_{H-A}K_A P_1^{1/2} + D_{H-B}K_B P_2^{1/2}}{D_{H-A}K_A + D_{H-B}K_B} \tag{iv}$$

and

$$C_1^* = K_A a^* \tag{v}$$

$$C_2^* = K_B a^* \tag{vi}$$

EXAMPLE 10.5

Consider the diffusion of hydrogen through a composite wall comprising 1 mm thickness of Ni and 1 mm thickness of Pd at 400°C. The hydrogen gas in contact with the Ni is at 1 atm and the hydrogen in contact with the Pd is at 0.1 atm. The required data are

	Ni	**Pd**
Atomic weight g/g mol	58.71	106.4
Density at 400°C (kg/m³)	8770	11,850
Diffusivity of H at 400°C (m²/s)	4.9×10^{-10}	6.4×10^{-9}
Solubility of H at 400°C and		
$\quad P_{H_2} = 1$ atm (atomic ppm)	250	10,000

A concentration of 250 atomic ppm of H in Ni corresponds to

$$1 \text{ kg H per } \frac{58.71 \times 10^6}{250} \text{ kg Ni}$$

$$1 \text{ kg H per } \frac{58.71 \times 10^6}{250 \times 8770} \text{ m}^3 \text{ Ni}$$

or

$$3.73 \times 10^{-2} \text{ kg H/m}^3$$

and similarly, 10,000 atomic ppm of H in Pd corresponds to 1.11 kg H/m^3. Therefore,

$$K_{Ni} = 3.73 \times 10^{-2}$$

and

$$K_{Pd} = 1.11$$

Thus from Eq. (iv),

$$a_H^* = \frac{D_{H-Ni}K_{Ni}P_1^{1/2} + D_{H-Pd}K_{Pd}P_2^{1/2}}{D_{H-Ni}K_{Ni} + D_{H-Pd}K_{Pd}}$$

$$= \frac{(4.9 \times 10^{-10} \times 3.73 \times 10^{-2} \times 1) +}{(6.4 \times 10^{-9} \times 1.11 \times 0.1^{1/2})}$$
$$\overline{(4.9 \times 10^{-10} \times 3.73 \times 10^{-2}) + (6.4 \times 10^{-9} \times 1.11)}$$

$$= 0.318$$

and from Eqs. (v) and (vi),

$$C_1^* = K_{Ni}a^* = 3.73 \times 10^{-2} \times 0.318 = 1.19 \times 10^{-2} \text{ kg/m}^3$$
$$C_2^* = K_{Pd}a^* = 1.11 \times 0.318 = 0.353 \text{ kg/m}^3$$

and

$$C_2 = K_{Pd}P_2^{1/2} = 1.11 \times 0.1^{1/2} = 0.351 \text{ kg/m}^3$$

Thus from Eq. (i),

$$J_H \text{ (in Ni)} = \frac{4.9 \times 10^{-10} \times (3.73 \times 10^{-2} - 1.19 \times 10^{-2})}{0.001}$$

$$= 1.24 \times 10^{-8} \text{ kg H} \cdot \text{m}^{-2} \cdot \text{s}^{-1}$$
$$= J_H \text{ (in Pd)}$$
$$= \frac{6.4 \times 10^{-10} \times (0.353 - 0.351)}{0.001}$$
$$= 1.24 \times 10^{-8} \text{ kg H} \cdot \text{m}^{-2} \cdot \text{s}^{-1}$$

The concentration and activity profiles are shown in Fig. 10.16. The activity profile is the mass transport analog of the temperature profile through a composite wall shown in Fig. 6.9. The greater rate of decrease of activity through the Ni layer arises because Ni has a greater resistance to transport of hydrogen (the analog of a smaller thermal conductivity in heat transport) than has Pd, which is reflected in D_{H-Ni} being smaller than D_{H-Pd}. If the Pd were on the high-hydrogen-pressure side of the composite wall, then

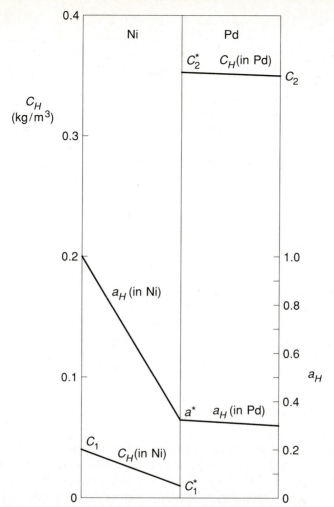

FIGURE 10.16. Concentration and activity profiles of hydrogen diffusing through a composite plane wall of Ni and Pd, considered in Example 10.5.

$$a^* = \frac{(6.4 \times 10^{-9} \times 1.11 \times 1) + (4.9 \times 10^{-10} \times 3.73 \times 10^{-2} \times 0.1^{1/2})}{(6.4 \times 10^{-9} \times 1.11) + (4.9 \times 10^{-10} \times 3.73 \times 10^{-2})}$$

$$= 0.998$$

and

$$C_1^* = K_{Pd}a^* = 1.11 \times 0.998 = 1.11 \text{ kg H/m}^3$$

$$C_2^* = K_{Ni}a^* = 3.73 \times 10^{-2} \times 0.998 = 3.72 \times 10^{-2} \text{ kg H/m}^3$$

$$C_2 = K_{Ni}P_2^{1/2} = 3.73 \times 10^{-2} \times 0.316 = 1.18 \times 10^{-2} \text{ kg H/m}^3$$

10.8 Diffusion in Substitutional Solid Solutions

In the preceding discussion the diffusion of interstitial solutes, such as carbon and hydrogen in iron was considered, in which the diffusing atoms jump from one interstitial site to the next in the crystal lattice of the solvent. The solute atoms are like small animals wandering through a dense forest, in which the trees are unaware of the motion of the animals, and thus in considering the diffusion of carbon in an infinite diffusion couple comprising steels of differing carbon contents, no consideration was given to the possible movements of the individual iron atoms. Consider now a diffusion couple comprising metals of the same crystal structure and similar atomic radii, such as silver and gold. At what rates do the silver and gold diffuse into each other, and how is the diffusion process described? The answer to the first question was provided by the experiments of Smigelskas and Kirkendall [A. Smigelskas and E. Kirkendall, *Trans. AIME* (1947), vol. 171, p. 130] and the answer to the second question was provided by Darken [L. S. Darken, *Trans. AIME* (1948), vol. 174, p. 184]. In Smigelskas and Kirkendall's experiment a rectangular bar of 70–30 brass, shown in Fig. 10.17, was wound with fine molybdenum wire and then electroplated with a layer of pure copper. Molybdenum is insoluble in both copper and brass and the wire served as a marker of the position of the interface. During annealing it was observed that the distance d decreased monotonically with time, which indicated that the flux of zinc atoms from the brass outward past the markers is greater than the flux of copper atoms inward past the markers. The phenomenon is known as the *Kirkendall effect*.

10.9 Darken's Analysis

Consider a train traveling due north at 60 mph in which passenger A is walking toward the front of the train at 4 mph and passenger B is walking toward the rear of the train at 5 mph. Are the passengers traveling in northerly or southerly directions? The answer depends on the frame of reference from which observation of the pas-

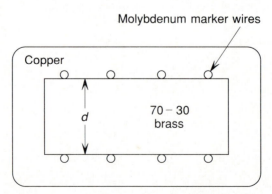

FIGURE 10.17. Diffusion couple used by Smigelskas and Kirkendall to observe the Kirkendall effect.

sengers is made. A stationary observer standing beside the track sees the train traveling in a northerly direction at 60 mph, passenger A traveling in a northerly direction at 64 mph and passenger B traveling in a northerly direction at 55 mph. However, a passenger seated on the train sees passenger A traveling in a northerly direction at 4 mph and passenger B traveling in a southerly direction at 5 mph. In this analogy the seated passengers are being transported by the motion of the train and the walking passengers, who are diffusing in the bulk flow, are being transported by both bulk flow and diffusion. In Kirkendall's experiment involving an infinite diffusion couple of the metals A and B in which the original interface is identified by inert markers, an observer seated on a marker sees only the movement of A and B across the marker plane that is caused by diffusion down a concentration gradient given by

$$J_A = -D_A \frac{\partial C_A}{\partial x}$$

and

$$J_B = -D_B \frac{\partial C_B}{\partial x}$$

However, an observer near an end of the diffusion couple, in a region where no concentration gradients occur, sees the movement of A and B at the position of the marker as being the sum of that due to diffusion across the plane of the marker and the motion of the marker. Thus if the marker is moving with the velocity v relative to the observer near one end of the diffusion couple and the concentration of A at the position of the marker is C_A, the observer sees the flux of A as

$$\dot{N}_A = J_A + C_A v = -D_A \frac{\partial C_A}{\partial x} + C_A v \tag{i}$$

in which the first term on the right-hand side is the rate of transport of A across the plane of the marker by diffusion, and the second term is the rate of movement of the marker plane relative to the observer. For a system in which both diffusion and bulk movement are occurring, Eq. (10.11) becomes

$$\frac{\partial C_A}{\partial t} = -\frac{\partial \dot{N}_A}{\partial x} \quad \text{and} \quad \frac{\partial C_B}{\partial t} = -\frac{\partial \dot{N}_B}{\partial x} \tag{ii}$$

and as the total number of moles per unit volume is

$$C = C_A + C_B$$

Equations (i) and (ii) give

$$\frac{\partial C}{\partial t} = \frac{\partial C_A}{\partial t} + \frac{\partial C_B}{\partial t} = \frac{\partial}{\partial x}\left(D_A \frac{\partial C_A}{\partial x} + D_B \frac{\partial C_B}{\partial x} - Cv\right) \tag{iii}$$

If the molar volume of alloys in the system $A-B$ is independent of composition (i.e., is constant), $\partial C/\partial t = 0$, in which case integration of Eq. (iii) gives

$$D_A \frac{\partial C_A}{\partial x} + D_B \frac{\partial C_B}{\partial x} - Cv = I \tag{iv}$$

in which I is the integration constant. Equation (iv) is valid along the entire length of the diffusion couple and v is the velocity of a marker placed at any position in the couple relative to the observer near one of its ends. By the definition of an infinite diffusion couple, no concentration gradients occur near the ends, in which case Eq. (iv) gives

$$I = Cv$$

But with no concentration gradients, and hence no diffusion occurring near the ends, $v = 0$ and hence $I = 0$. Thus Eq. (iv) becomes

$$v = \frac{1}{C}\left(D_A \frac{\partial C_A}{\partial x} + D_B \frac{\partial C_B}{\partial x}\right) \tag{v}$$

which is the velocity of the marker plane at which the concentration gradients are $\partial C_A/\partial x$ and $\partial C_B/\partial x$. Equation (v) is then substituted back with Eq. (i) into Eq. (iii) for A to obtain

$$\frac{\partial C_A}{\partial t} = -\frac{\partial \dot{N}_A}{\partial x}$$

$$= \frac{\partial}{\partial x}\left(D_A \frac{\partial C_A}{\partial x} - C_A v\right)$$

$$= \frac{\partial}{\partial x}\left(D_A \frac{\partial C_A}{\partial x} - \frac{C_A}{C}D_A\frac{\partial C_A}{\partial x} - \frac{C_A}{C}D_B\frac{\partial C_B}{\partial x}\right) \tag{vi}$$

As C is constant, $\partial C_A/\partial x = -\partial C_B/\partial x$ and hence Eq. (vi) becomes

$$\frac{\partial C_A}{\partial t} = \frac{\partial}{\partial x}\left(\frac{C_A + C_B}{C}D_A\frac{\partial C_A}{\partial x} - \frac{C_A}{C}D_A\frac{\partial C_A}{\partial x} + \frac{C_A}{C}D_B\frac{\partial C_A}{\partial x}\right)$$

$$= \frac{\partial}{\partial x}\left(\frac{C_B D_A + C_A D_B}{C}\frac{\partial C_A}{\partial x}\right)$$

or as $C_A/C = X_A$, the mole fraction of A, and $C_B/C = X_B$, the mole fraction of B,

$$\frac{\partial C_A}{\partial t} = \frac{\partial}{\partial x}\left[(X_B D_A + X_A D_B)\frac{\partial C_A}{\partial x}\right] \tag{10.25}$$

which is identical with Fick's second law of diffusion, given by Eq. (10.13), if

$$D = X_B D_A + X_A D_B \tag{10.26}$$

In Eq. (10.26) D is the interdiffusion coefficient for the system $A-B$ and D_A and D_B are the chemical diffusion coefficients for A and B. Also, rewriting Eq. (v) as

$$v = \frac{1}{C}\left(D_A \frac{\partial C_A}{\partial x} - D_B \frac{\partial C_A}{\partial x}\right)$$

$$= (D_A - D_B)\frac{\partial X_A}{\partial x} \tag{10.27}$$

gives the velocity of the marker plane at which the gradient of the mole fraction of A is $\partial X_A/\partial x$. The marker planes in the diffusion couple are stationary only when $D_A = D_B$.

Equation (10.26) shows that the interdiffusion coefficient, D, in Fick's law for the system $A-B$ is not a constant. The question now arises: Are the chemical diffusion coefficients D_A and D_B constant?

Atoms move as a result of the application of some kind of force, and formally, a force is the negative of the gradient of some type of potential. For example, the gravitational force is the negative of the gravitational potential, and when applied to a mass in a gravitational field, it causes movement of the mass. Similarly, in an electric field the electric force acting on unit electrical charge is the negative of the gradient of the electrostatic potential, and the application of this force to an electrical charge in the electric field causes movement of the charge. In a system $A-B$ in which concentration gradients occur, the potentials of interest are the chemical potentials of A and B. The chemical potential of a gram mole of the species i in an isothermal, isobaric system is defined as

$$\mu_i = \mu_i^\circ + RT \ln a_i \tag{vii}$$

in which a_i is the thermodynamic activity of i relative to some standard state (usually pure i) in which state the chemical potential is μ_i° and the activity of i is, by definition, unity. Alternatively, the chemical potential per atom of i is

$$\mu_i' = \frac{\mu_i}{N_0} = \frac{\mu_i^\circ}{N_0} + kT \ln a_i$$

in which N_0 is Avogadro's number. The existence of concentration gradients in a system gives rise to gradients in the chemical potentials of the species in the system, and the force, F, acting on an atom can be identified as being the negative of the gradient of the chemical potential; for example, for an atom of A,

$$F = -\frac{1}{N_0}\frac{d\mu_i}{dx}$$

and this force causes atomic movement in that direction which decreases the chemical potential of A. When acted on by this force, the atom acquires a mean velocity v and the mobility, B, of the atom is defined as its velocity per unit applied force:

$$B_A = \frac{v}{F}$$

The flux of A atoms (atoms·m^{-2}·s^{-1}) caused by the force is thus

$$J_A' = N_0 C_A v = N_0 C_A B_A F = -N_0 C_A B_A \frac{1}{N_0}\frac{d\mu_i}{dx} \tag{viii}$$

The thermodynamic activity of A, defined in Eq. (vii), is the product of an activity coefficient γ_A and the mole fraction of A, $X_A = C_A/C$, and thus Eq. (vii) gives

$a_A = \gamma_A \cdot X_A$

$$\frac{d\mu_A}{dx} = RT\left(\frac{d \ln X_A}{dx} + \frac{d \ln \gamma_A}{dx}\right) \tag{ix}$$

substitution of which into Eq. (viii) gives the flux of A in g mol·m^{-2}·s^{-1} as

$$J_A = \frac{J_A'}{N_0} = -C_A B_A \hbar T \left(\frac{d \ln X_A}{dx} + \frac{d \ln \gamma_A}{dx} \right)$$

$$= -C_A B_A \hbar T \left(\frac{d \ln C_A}{dx} + \frac{d \ln \gamma_A}{dx} \right)$$

$$= -C_A B_A \hbar T \left(\frac{1}{C_A} \frac{dC_A}{dx} + \frac{d \ln \gamma_A}{dx} \right)$$

$$= -B_A \hbar T \frac{dC_A}{dx} \left(1 + \frac{d \ln \gamma_A}{d \ln C_A} \right)$$

$$= -B_A \hbar T \left(1 + \frac{d \ln \gamma_A}{d \ln X_A} \right) \frac{dC_A}{dx} \tag{10.28}$$

which on comparison with Fick's first law gives the chemical diffusion coefficient for A as

$$D_A = B_A \hbar T \left(1 + \frac{d \ln \gamma_A}{d \ln X_A} \right) \tag{10.29}$$

If the system $A–B$ obeys Raoult's law ($\gamma_A = 1$) or Henry's law ($\gamma_A = $ constant) over some range of composition, Eq. (10.29) becomes

$$D_A = B_A \hbar T \tag{10.30}$$

and Eq. (10.30) is known as the *Nernst–Einstein equation*. The thermodynamic solution behavior of a solute in a binary system approaches Henry's law, and the behavior of the solvent approaches Raoult's law as the concentration of the solute approaches zero. However, in many solutions it can be considered that Henry's law is obeyed by the solute over some finite range of dilute solution, in which case Raoult's law is obeyed by the solvent in the same range of composition. Within these ranges of composition the chemical diffusion coefficients of the species are constant, but in concentrated solutions they vary with composition in accordance with the nonideal thermodynamic solution properties of the system.

10.10 Self-Diffusion Coefficient

If a thin film of the radioisotope A^* of the metal A, containing α gram moles of A^* per unit cross-sectional area, is plated on to one end of a semi-infinite rod of the stable isotope A, and the rod is annealed, the concentration profile that develops is given by

$$C_A^* = \frac{\alpha}{\sqrt{\pi D_A^* t}} \exp \left(\frac{-x^2}{4 D_A^* t} \right) \tag{10.31}$$

in which D_A^* is self-diffusion coefficient of A. Equation (10.31) is shown for several values of D^*t in Fig. 10.18, which shows that with increasing time, the concentration profile flattens as the A^* spreads through the rod. From Eq. (10.31) the concentration of C_A^* at $x = 0$ is proportional to $1/t^{1/2}$ and as

$$\int_0^\infty e^{-\eta^2}\, d\eta = \frac{\sqrt{\pi}}{2}$$

then

$$\int_0^\infty C_A^*\, dx = \alpha$$

and hence the area under each of the curves in Fig. 10.18 is unity. A plot of the logarithm of the experimentally measured C_A^* against x^2 gives a straight line with a slope of $-1/4D_A^*t$, which allows determination of the self-diffusion coefficient, D_A^*.

In the rod, because A and A^* are chemically identical, there are no chemical concentration gradients and hence no gradients in chemical potential. Thus in the absence of a driving force for diffusion, the concentration profiles shown in Fig. 10.18 are developed by a random walking process of A atoms and the self-diffusion coefficient is a quantitative measure of this process. The probability that an A^* atom has randomly walked to a region between x and $x + dx$, $p(x)\, dx$, is the number of A^* atoms between x and $x + dx$ divided by the total number of A^* atoms,

$$p(x)\, dx = \frac{C_A^*\, dx}{\alpha} = \frac{1}{\sqrt{\pi D_A^* t}} \exp\left(\frac{-x^2}{4D_A^* t}\right) dx$$

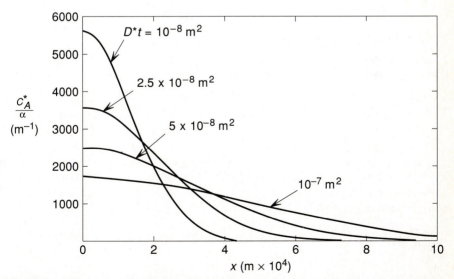

FIGURE 10.18. Concentration profiles of a radioisotope A^* diffusing from an initial plane source into a semi-infinite rod of A.

and the mean square distance, $<x^2>$, through which an atom of $A*$ has diffused in time t is thus

$$<x^2> = \int_0^{\infty} x^2 p(x) \, dx$$

$$= \frac{1}{\sqrt{\pi D_A^* t}} \int_0^{\infty} x^2 \exp\left(-\frac{x^2}{4D_A^* t}\right) dx$$

$$= \frac{1}{\sqrt{\pi D_A^* t}} \left[2\sqrt{\pi}(D_A^* t)^{3/2} \, \text{erf}\left(\frac{x}{2\sqrt{D_A^* t}}\right) - 2D_A^* tx \exp\left(\frac{-x^2}{4D_A^* t}\right)\right]_0^{\infty}$$

$$= 2D_A^* t$$

which is the Einstein–Smolochowski equation derived as Eq. (10.18).

Is the measured self-diffusion coefficient of the radioactive atoms of A, D_A^*, equal to the chemical diffusion coefficient of A in the A–B alloy? Consider an alloy diffusion couple in which C_B has the same value in both halves of the couple and the initial value of C_A in one half equals the initial value of the sum of C_A and C_A^* in the other half. From Eq. (10.28) the self-diffusion coefficient is

$$D_A^* = B_A^* kT \left(1 + \frac{d \ln \gamma_A^*}{d \ln X_A^*}\right) \tag{10.32}$$

However, as the stable and radioactive isotopes are chemically identical, γ_A depends only on the chemical composition of the alloy (i.e., on the sum of C_A and C_A^*) and is independent of the ratio C_A^*/C_A, and thus for a constant $C_A + C_A^*$, the term $d \ln \gamma_A^*/d \ln X_A^*$ in Eq. (10.32) is zero. Furthermore, the mobilities of the two isotopes of A are the same, and thus

$$D_A^* = B_A^* kT = B_A kT$$

which from Eq. (10.28) leads to

$$D_A = D_A^* \left(1 + \frac{d \ln \gamma_A}{d \ln X_A}\right) \tag{10.33}$$

Thus the chemical diffusion coefficient of A, D_A, in a system containing concentration gradients is only equal to the self-diffusion coefficient of A, D_A^*, if the system obeys Raoultian (and hence Henrian) ideal behavior.

As the variations of D_A and D_B with composition are caused by nonideal thermodynamic behavior in the system, the interdiffusion coefficient of the system A–B can be related to the self-diffusion coefficients by means of the Gibbs–Duhem equation for the binary system

$$X_A \, d\mu_A + X_B \, d\mu_B = 0$$

or

$$X_A RT \, d \ln X_A + X_A RT \, d \ln \gamma_A + X_B RT \, d \ln X_B + X_B RT \, d \ln \gamma_B = 0 \tag{x}$$

However, as

$$X_A + X_B = 1$$

$$dX_A + dX_B = 0$$

$$\frac{X_A}{X_A} dX_A + \frac{X_B}{X_B} dX_B = 0$$

$$X_A d \ln X_A + X_B d \ln X_B = 0$$

the sum of the first and third terms in Eq. (x) is zero, and hence

$$X_A RT \, d \ln \gamma_A + X_B RT \, d \ln \gamma_B = 0$$

or

$$X_A RT \frac{d \ln \gamma_A}{dX_A} = X_B RT \frac{d \ln \gamma_B}{dX_B}$$

or

$$RT \frac{d \ln \gamma_A}{d \ln X_A} = RT \frac{d \ln \gamma_B}{d \ln X_A} \qquad (xi)$$

Thus combination of Eqs. (10.26), (10.33) and (x) gives

$$D = X_B D_A + X_A D_B$$

$$= X_B D_A^* \left(1 + \frac{d \ln \gamma_A}{d \ln X_A}\right) + X_A D_B^* \left(1 + \frac{d \ln \gamma_B}{d \ln X_B}\right)$$

$$= (X_B D_A^* + X_A D_B^*) \left(1 + \frac{d \ln \gamma_A}{d \ln X_A}\right) \qquad (10.34)$$

EXAMPLE 10.6 _____

In the doping of silicon with phosphorus to produce a semiconductor, a thin layer of phosphorus, placed on the surface of the silicon by vapor deposition, is allowed to diffuse into the silicon. If a film of thickness 10^{-6} m is vapor deposited and diffusion is allowed to occur at 1000°C for 5 h, at what penetration depth does the mole fraction of phosphorus have a value of 10^{-3}?

At 1273 K,

$$D_P = 1.2 \times 10^{-17} \text{ m}^2/\text{s}$$

$$\rho_P = 2000 \text{ kg/m}^3$$

$$\rho_{Si} = 2300 \text{ kg/m}^3$$

$$MW_P = 0.03097 \text{ kg/g mol}$$

$$MW_{Si} = 0.02809 \text{ kg/g mol}$$

The concentration profile developed from a plane source is given by

$$C_P = \frac{\alpha}{\sqrt{\pi D_p t}} \exp\left(\frac{-x^2}{4 D_p t}\right) \tag{10.31}$$

For a vapor-deposited layer of 10^{-6} m, the quantity α, per unit area of surface, is

$$\frac{2000 \times 10^{-6}}{0.03097} = 6.46 \times 10^{-2} \text{ g mol/m}^2$$

A mole fraction of 10^{-3} of P in Si corresponds to a concentration of

$$10^{-3} \text{ g mol of P per } \frac{0.02809}{2300} \text{ m}^3$$

or 81.9 g mol of P per cubic meter. Therefore, in Eq. (10.31), with $t = 5$ h $= 18,000$ s,

$$81.9 = \frac{6.46 \times 10^{-2}}{\sqrt{\pi \times 1.2 \times 10^{-17} \times 18,000}} \exp\left(\frac{-x^2}{4 \times 1.2 \times 10^{-17} \times 18,000}\right)$$

which gives the penetration depth as

$$x = 2.44 \times 10^{-6} \text{ m}$$

10.11 Measurement of the Interdiffusion Coefficient; Boltzmann–Matano Analysis

When the interdiffusion coefficient varies with composition, the Boltzmann–Matano analysis allows D to be obtained from an experimentally measured concentration profile in an infinite diffusion couple. The function λ, defined as

$$\lambda = \frac{x}{\sqrt{t}} \tag{i}$$

is introduced into Eq. (10.12):

$$\frac{\partial C}{\partial t} = \frac{\partial}{\partial x}\left(D \frac{\partial C}{\partial x}\right) \tag{10.12}$$

From Eq. (i),

$$\frac{\partial C}{\partial t} = \frac{dC}{d\lambda} \frac{\partial \lambda}{\partial t} = -\frac{1}{2} \frac{x}{t^{3/2}} \frac{dC}{d\lambda}$$

and

$$\frac{\partial C}{\partial x} = \frac{dC}{d\lambda} \frac{\partial \lambda}{\partial x} = \frac{1}{t^{1/2}} \frac{dC}{d\lambda}$$

substitution of which into Eq. (10.12) gives

$$-\frac{1}{2}\frac{\lambda}{t}\frac{dC}{d\lambda} = \frac{1}{t^{1/2}}\frac{\partial}{\partial x}\left(D\frac{dC}{d\lambda}\right) = \frac{1}{t}\frac{d}{d\lambda}\left(D\frac{dC}{d\lambda}\right)$$

or

$$-\frac{1}{2}\lambda\frac{dC}{d\lambda} = \frac{d}{d\lambda}\left(D\frac{dC}{d\lambda}\right)$$

which, being an ordinary differential equation, can be written as

$$-\frac{1}{2}\lambda\, dC = d\left(D\frac{dC}{d\lambda}\right) \tag{ii}$$

The infinite binary diffusion couple is made from pure A and pure B and has the initial conditions

$$C = \begin{cases} C_0 & \text{for } x < 0 \quad \text{at } t = 0 \quad (\lambda = -\infty) \\ 0 & \text{for } x > 0 \quad \text{at } t = 0 \quad (\lambda = \infty) \end{cases}$$

and integration of Eq. (ii) from $C = C$ to $C = 0$ gives

$$-\frac{1}{2}\int_{C=0}^{C}\lambda\, dC = \left[D\frac{dC}{d\lambda}\right]_{C=0}^{C}$$

or as the concentration profile is always measured after diffusion has occurred for some diffusion time, t,

$$-\frac{1}{2}\int_{C=0}^{C}x\, dC = Dt\left[\frac{dC}{dx}\right]_{C=0}^{C}$$

At a position in the couple where $C = 0$, $dC/dx = 0$ and hence

$$-\frac{1}{2}\int_{C=0}^{C}x\, dC = Dt\left(\frac{\partial C}{\partial x}\right)\bigg|_{C} \tag{iii}$$

Similarly, at a position in the couple where $C = C_0$, dC/dx is also zero, in which case Eq. (iii) gives

$$\int_{0}^{C_0} x\, dC = 0$$

which defines the plane in the diffusion couple at which $x = 0$. This plane is known as the *Matano interface*. The plane at which $x = 0$ is shown in Fig. 10.19 as that which makes the light gray regions of equal area. Equation (iii) then gives

$$D = -\frac{1}{2t}\left(\frac{dx}{dC}\right)\bigg|_{C}\int_{0}^{C}x\, dC \tag{10.35}$$

as the value of the interdiffusion coefficient for the composition C. In Fig. 10.19 the

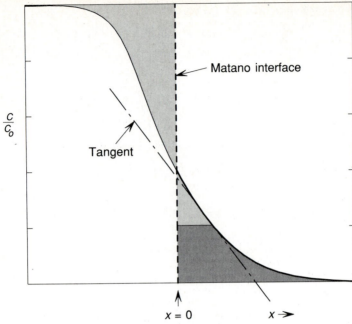

FIGURE 10.19.   Concentration profile in an infinite diffusion couple used in the Boltzmann–Matano analysis for determination of the interdiffusion coefficient.

value of

$$\int_0^{C=0.2C_0} x \, dC$$

is the dark gray area under the concentration profile between $C = 0.2C_0$ and $C = 0$, and $(dx/dC)|_{C=0.2C_0}$ is the slope of the tangent to the concentration profile at $C = 0.2C_0$.

The measured self-diffusion coefficients of nickel and gold in the system Ni–Au at 900°C are shown in Fig. 10.20, and the measured interdiffusion coefficients are shown in comparison with values calculated using Eq. (10.34) in Fig. 10.21 [J. E. Reynolds, B. L. Averbach, and M. Cohen, *Acta Metall.* (1957), vol. 5, p. 29]. The thermodynamic activities of nickel and gold and the thermodynamic factor in Eq. (10.34),

$$1 + \frac{d \ln \gamma_{Ni}}{d \ln X_{Ni}}$$

are shown in Fig. 10.22. The system Au–Ni has a miscibility gap in the solid state that begins at $X_{Ni} = 0.8$ and $T = 820°C$, and the minimum in the thermodynamic factor at this composition at 900°C reflects the tendency of the system toward immiscibility. The minimum in the interdiffusion coefficient in Fig. 10.21 is caused by the thermodynamic factor and not by any peculiarity in the composition dependence

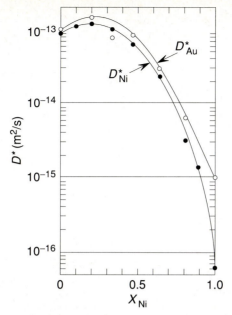

FIGURE 10.20.   Variations, with composition, of the self-diffusion coefficients of Ni and Au in the system Ni–Au at 900°C.

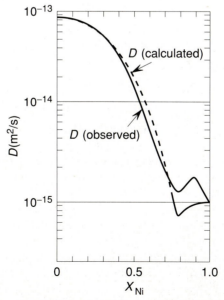

FIGURE 10.21.   Comparison of the measured and calculated interdiffusion coefficients in the system Ni–Au at 900°C.

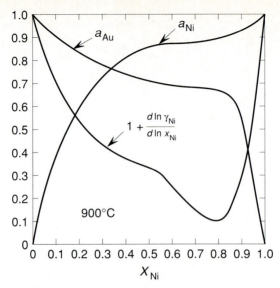

FIGURE 10.22. Variations of the thermodynamic activities and the thermodynamic factor with composition in the system Ni–Au at 900°C.

of the term

$$X_{Ni}D_{Au}^* + X_{Au}D_{Ni}^*$$

in Eq. (10.34). The very low interdiffusion coefficients in the composition range over the peak of the immiscibility gap are thus caused by a low thermodynamic driving force in this range and not by any anomaly in the atomic mobilities. The variations of D with composition in four other binary metal systems are shown in Fig. 10.23.

10.12 Influence of Temperature on the Diffusion Coefficient

Figure 10.24(a) shows a different perspective of the unit cell of face-centered cubic iron shown in Fig. 10.1 and identifies three interstitial sites along the new x-axis available for occupancy by a small impurity atom such as carbon. At 950°C the diameter of the Fe atom in the face-centered cubic structure is 2.58 Å and thus the diameter of the interstitial site is 1.07 Å and the distance between atoms a and b in Fig. 10.24(a) is 1.04 Å. As the carbon atom has a diameter of 1.54 Å, the presence of a carbon atom on an interstitial site causes local distortion of the iron lattice, which is why the solubility of carbon in iron is relatively small. An interstitial atom on an interstitial site such as site 2 vibrates in all directions about the center of the site with a frequency of vibration, ν, which, typically, is on the order of 10^{13} s^{-1}. For most of the time the vibrating atom is contained in an "energy well" that arises

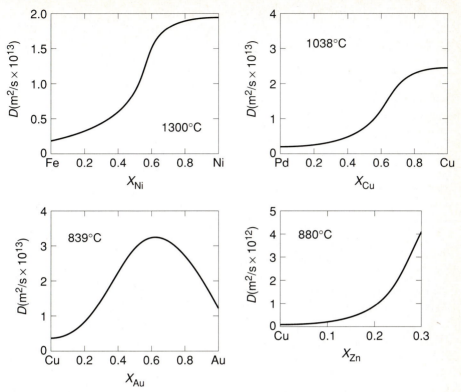

FIGURE 10.23.   Composition dependencies of the interdiffusion coefficients in 4 binary systems.

from the presence of the six neighboring Fe atoms at locations a to f (i.e., if the interstitial moves in any direction from the center of the well it experiences a repulsive force exerted by the neighboring Fe atoms). The variation, with position along the x-axis, of the energy of an interstitial atom, shown in Fig. 10.24(b), shows that if, as a result of its vibration, an interstitial atom is to squeeze successfully between the atoms A and B and jump to the interstitial site 1, it must have at least the energy E_a, the activation energy required for the jump. The question now arises: At what rate do interstitial atoms jump successfully from one interstitial site to another?

In a large population of n atoms the number of atoms, n_i, which have energies greater or equal to E_i is $\exp(-E_i/kT)$. Thus the fraction of interstitial atoms in solution in a host lattice that have an energy greater than or equal to the required activation energy for a successful jump from one site to another is $\exp(-E_a/kT)$, or equivalently, the probability that any one atom is sufficiently energetic for a successful jump is $\exp(-E_a/kT)$. In Fig. 10.1 the interstitial atom at the center of the unit cell can jump to any one of 12 neighboring interstitial sites, and thus the frequency, Γ, with which an interstitial atom at a site such as 2 in Fig. 10.24(a)

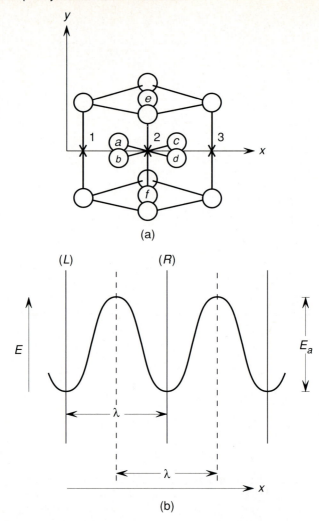

(a)

(b)

FIGURE 10.24.   (a) Interstitial sites in the face-centered cubic Fe lattice; (b) variation of energy with position between interstitial lattice sites.

jumps to site 1 is

> (frequency of vibration)
> × (probability that, during any vibration, the atom is
> sufficiently energetic to make a successful jump)
> × (probability that jump is in direction 1 → 2)

that is,

$$\Gamma = \nu e^{-E_a/kT} \times \tfrac{1}{12}$$

The activation energy for diffusion of carbon in face-centered cubic iron is 2.223 × 10^{-19} J per atom or 133,900 J/g mol and thus at 950°C,

$$\Gamma = 10^{13} \times \exp\left(\frac{-2.223 \times 10^{-19}}{1.3805 \times 10^{-23} \times 1223}\right) \times \frac{1}{12}$$

$$= 10^{13} \times 1.91 \times 10^{-6} \times \frac{1}{12}$$

$$= 1.59 \times 10^{6} \text{ s}^{-1}$$

(i.e., a carbon atom makes 1.59×10^{6} jumps per second).

If the number if interstitials per unit area on the plane R in Fig. 10.24(b) is N_R, the flux of interstitials from the plane R to the plane L in the $-x$-direction is

$$J_{R \to L} = \Gamma N_R$$

and if the number of interstitials per unit area on the plane L in Fig. 10.24(b) is N_L, the flux of interstitials from the plane L to the plane R in the x-direction is

$$J_{L \to R} = \Gamma N_L$$

The net flux of interstitials in the x-direction is thus

$$J = J_{L \to R} - J_{R \to L} = \Gamma(N_L - N_R) \tag{i}$$

With the planes L and R being separated by the jump distance λ the concentration of interstitials on a plane, C, is

$$C\left(\frac{\text{atoms}}{\text{m}^3}\right) = N\left(\frac{\text{atoms}}{\text{m}^2}\right) \times \frac{1}{\lambda}\left(\frac{1}{\text{m}}\right) \tag{ii}$$

substitution of which into Eq. (i) gives

$$J = \Gamma\lambda(C_L - C_R)$$

$$= -\Gamma\lambda^2 \left(\frac{C_R - C_L}{\lambda}\right) \tag{iii}$$

In Eq. (iii) the term in parentheses is the concentration gradient of the interstitials in the x direction and hence Eq. (iii) can be written as

$$J = -\Gamma\lambda^2 \frac{\partial C}{\partial x} \tag{iv}$$

which, on comparison with Fick's first law of diffusion,

$$J = -D \frac{\partial C}{\partial x}$$

shows that

$$D = \Gamma\lambda^2$$

or, from Eq. (10.36),

$$D = \tfrac{1}{12} \nu\lambda^2 e^{-E_a/\hbar T} \tag{10.37}$$

The preexponential terms in Eq. (10.37) can be lumped together as D_0 to give the exponential dependence of the diffusion coefficient on temperature as

$$D = D_0 e^{-E_a/\tilde{k}T} \tag{10.38}$$

In Eq. (10.38), E_a is the activation energy for diffusion per atom, and multiplying this quantity by Avogadro's number, N_0, gives the activation energy for diffusion per gram mole of diffusing species, in which case the diffusion coefficient is given by

$$D = D_0 e^{-E_a/RT} \tag{10.39}$$

in which $R = N_0 \tilde{k}$ is the universal gas constant. Whether or not the activation energy is in units of energy per atom or energy per gram mole can be determined by observing whether the expression contains Boltzmann's constant, $\tilde{k}$, or the gas constant, R.

The variation of the logarithm of the diffusion coefficient of carbon in body-centered cubic iron with inverse temperature over the range -38 to $800°C$ is shown in Fig. 10.25. The variation is linear with the equation

$$\log_{10} = \frac{-4400}{T} - 5.70$$

From the slope of the line the activation energy for diffusion is

$$4400 \times 2.303 \times 8.3144 = 84,300 \text{ J/g mol}$$

10.13 Summary

Diffusion in the solid state occurs by the random motions of atoms, and the mean square distance through which an atom moves by random motion is proportional to the time for which the diffusion has occurred. In binary systems diffusion occurs down concentration gradients, and Fick's first law states that the diffusion flux is proportional to the concentration gradient, with the proportionality constant being called the diffusion coefficient of the diffusing species. Fick's first law is thus analogous to Fourier's law of heat transport by conduction, and Fick's second law, which states that the rate of change of concentration of the diffusing species at any location is equal to the rate of change of the diffusion flux with distance (which, itself, is proportional to the second derivative of concentration with respect to distance), is analogous to the general heat conduction equation. Thus although the phenomenon of mass transport by diffusion is intrinsically different from the phenomenon of heat transport by conduction, the similarity of the laws governing the phenomena makes the mathematical treatment of the diffusion of interstitials in a solid identical with the mathematical treatment of heat transport by conduction.

Observation of diffusion in substitutional solid solution shows that the diffusion coefficient, as defined by Fick's first law, can vary significantly with composition. This diffusion coefficient, which is called the interdiffusion coefficient, is thus only a measure of the rate at which concentration gradients in the system are eliminated by diffusion processes, and it is not an intrinsic property. It is a function of the chemical diffusion coefficients of the component species and is a function of composition. Fick's laws hold only for thermodynamically ideal solutions in which the

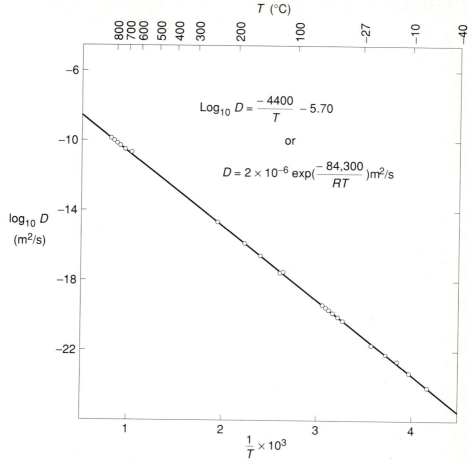

FIGURE 10.25. Variation, with temperature, of the diffusion coefficient for C in body-centered cubic Fe.

thermodynamic activity of the solute is proportional to its concentration (i.e., the solute obeys Henry's law and the solvent obeys Raoult's law). In general, mass transport by diffusion occurs down gradients of chemical potential (thermodynamic activity gradients), with the consequence that the chemical diffusion coefficients of the diffusing species are only independent of composition if the system is thermo-dynamically ideal. In such systems the chemical diffusion coefficient of a species is proportional to its mobility, where the mobility is defined as the velocity of the diffusing atom per unit force acting on it, and the force arises from the gradient in chemical potential. The chemical diffusion coefficient of a diffusing species in a multicomponent system is also only equal to its self-diffusion coefficient (which is determined by measuring isotope diffusion in a one-component system) if the multi-component system is thermodynamically ideal. In systems in which the chemical diffusivities of the components have different values, the diffusion process causes movement of the positions of lattice planes relative to a fixed reference and hence mass transport in the system is the sum of motion by diffusion down a concentration gradient and motion of the lattice planes.

Diffusion is a thermally activated process and hence the diffusion coefficient increases exponentially with increasing temperature, with the rate of increase being determined by the activation energy required for an atom to jump from one site to another. This phenomenon shows that there is not a physical analogy between mass transport by diffusion and heat transport by conduction. Conduction is not a thermally activated process, and the temperature dependence of the thermal diffusivity is determined by completely different phenomena.

Problems

PROBLEM 10.1 _____

A carbon steel containing 0.1 wt % C is carburized at 950°C by a gas in which the carbon activity is equivalent to 1 wt % C in iron.

(a) Calculate the time required to obtain a carbon content of 0.5 wt % at a depth of penetration of 0.05 cm
(b) What carburizing temperature is required to get a concentration of 0.5 wt % at a depth of penetration of 0.1 cm in the same time that 0.5 wt % was attained at 0.05 cm at 950°C?

Assume that the carbon activity in the carburizing gas at the higher temperature still corresponds to 1 wt % C in iron. The diffusion coefficient for carbon in austenitic iron is

$$D = 7 \times 10^{-6} \exp\left(\frac{-133,900}{RT}\right) \quad m^2/s$$

PROBLEM 10.2 _____

Hydrogen is stored at 400°C in a long cylindrical steel vessel of inner diameter 0.1 m. Calculate the wall thickness of the vessel at which the hydrogen leaks from the vessel by diffusion through the wall at the rate of 0.1 cm³ (STP) per second per meter length of the vessel. The pressure of hydrogen inside the vessel is maintained at 101.3 kPa and the hydrogen pressure outside the vessel is negligibly small. At 400°C the diffusion coefficient for hydrogen in iron is 10^{-8} m²/s, and from Sievert's law,

$$C_H(kg/m^3) = 7.29 \times 10^{-5} P^{1/2}(Pa)$$

PROBLEM 10.3 _____

A 1-mm-thick sheet of steel at 750°C is subjected to conditions that maintain the carbon content at zero on one surface and at 3 at % carbon on the other surface. At

750°C the carbon contents of ferrite and austenite which are in equilibrium with each other are, respectively. 0.09 and 2.7 at %.

(a) Calculate the flux of carbon through the sheet at steady state.
(b) Determine the position of the ferrite–austenite interface in the sheet.

The diffusion coefficients of carbon in ferrite and austenite at 750°C are, respectively, 5.5×10^{-11} and 1.1×10^{-12} m²/s and the density of the steel is 7680 kg/m³.

PROBLEM 10.4

Hydrogen in solution in a thick slab of nickel is removed by subjecting one face of the slab to a vacuum at 600°C.

(a) If the initial concentration of H in the nickel is 8 g mol/m³, after what time is the local concentration of H at a depth of 10^{-3} m below the surface equal to 4 g mol/m³?
(b) After this diffusion time, what is the flux of hydrogen from the surface?
(c) How much hydrogen has been removed from the slab per unit area?

For hydrogen in nickel,

$$D_H = 7.77 \times 10^{-7} \exp\left(\frac{-41{,}200}{RT}\right) \quad \text{m}^2/\text{s}$$

PROBLEM 10.5

The surface of a piece of pure iron is placed in contact with a carburizing gas at 800°C in which the activity of carbon is equivalent to a concentration of 0.8 wt % carbon in iron. Calculate the position of the austenite–ferrite phase boundary after 2 h. At 800°C the austenite and ferrite compositions in equilibrium with one another contain, respectively, 0.24 and 0.02 wt % carbon, the diffusion coefficient for carbon in ferrite is given by

$$D_\alpha = 2 \times 10^{-6} \exp\left(\frac{-84{,}100}{RT}\right) \quad \text{m}^2/\text{s}$$

and the diffusion coefficient for carbon in austenite is given by

$$D_\gamma = 7 \times 10^{-6} \exp\left(\frac{-133{,}900}{RT}\right) \quad \text{m}^2/\text{s}$$

11

Mass Transport in Fluids

11.1 Introduction

Mass transport in flowing fluids occurs by diffusion down concentration gradients and by bulk flow of the fluid, and in considering mass transport from a solid surface to a flowing fluid with which the solid is in contact, a heavy reliance is made on the analogy with heat transfer from a solid to a flowing fluid. Unfortunately, quantitative knowledge of the diffusion coefficients in systems in which mass transport is occurring is significantly less than quantitative knowledge of the thermal diffusivities in systems in which heat transfer is occurring, and thus although the mathematics of mass transport in a large number of situations can be developed by analogy with heat transfer, numerical calculation is limited to systems for which diffusion data are available.

11.2 Mass and Molar Fluxes in a Fluid

If a fluid of uniform composition containing the species i at the mass concentration ρ_i kg/m^3 is flowing with the velocity $\mathbf{v}$ m/s, the rate of mass transport of i due to bulk motion of the fluid, as observed by a stationary observer, is $\rho_i \mathbf{v}$ kg·m^{-2}·s^{-1}. However, to an observer who is also traveling at the velocity $\mathbf{v}$, the rate of mass transport of i is zero. If the concentration of i in the flowing fluid is not uniform, i is being transported through the fluid by diffusion in the x-direction at the rate

$$ j_{i,x} = -D \frac{\partial \rho_i}{\partial x} \qquad \text{kg·m}^{-2}\text{·s}^{-1} $$

and this is the rate of mass transport of i in the x-direction observed by an observer traveling at the velocity $\mathbf{v}$. However, in this case, the stationary observer sees the transport of i due to the bulk motion of the fluid and the transport of i due to diffusion within the fluid as the sum of j_i and $\rho_i \mathbf{v}$.

The local mass average velocity, $\mathbf{v}$, of a binary fluid of density ρ containing components A and B is defined by

$$\rho \mathbf{v} = \rho_A \mathbf{v}_A + \rho_B \mathbf{v}_B \qquad (11.1)$$

In this expression $\rho \mathbf{v}$ is the rate at which mass passes through unit cross-sectional area perpendicular to the direction of flow, and $\mathbf{v}_A$ and $\mathbf{v}_B$ are, respectively, the averaged values of the absolute velocities of the individual particles of A and B as observed by a stationary observer. The local mass average velocity is thus the absolute velocity of the center of mass of a volume element of the fluid. Alternatively, Eq. (11.1) can be written as

$$\mathbf{v} = \omega_A \mathbf{v}_A + \omega_B \mathbf{v}_B \qquad (11.2)$$

in which $\omega_A = \rho_A/\rho$ and $\omega_B = \rho_B/\rho$ are the mass fractions of A and B in the mixture. The existence of concentration gradients in the mixtures gives rise to diffusion fluxes, which for diffusion in the x-direction are

$$j_{A,x} = -D \frac{\partial \rho_A}{\partial x}$$

and

$$j_{B,x} = -D \frac{\partial \rho_B}{\partial x}$$

Thus the absolute mass fluxes in the x-direction are for A,

$$\dot{n}_{A,x} = \rho_A v_{A,x} = j_{A,x} + \rho_A v_x \qquad \text{kg·m}^{-2}\text{·s}^{-1} \qquad (11.3)$$

and for B,

$$\dot{n}_{B,x} = \rho_B v_{B,x} = j_{B,x} + \rho_B v_x \qquad \text{kg·m}^{-2}\text{·s}^{-1} \qquad (11.4)$$

where v_x is the x-component of the local mass average velocity $\mathbf{v}$ and $v_{A,x}$ and $v_{B,x}$ are the x-components of $\mathbf{v}_A$ and $\mathbf{v}_B$. The sum of the absolute mass fluxes is thus

$$\dot{n}_{A,x} + \dot{n}_{B,x} = \rho_A v_{A,x} + \rho_B v_{B,x} \qquad (11.5)$$

which, from Eq. (11.1), written for the x-component of the local mass average velocity, gives

$$\dot{n}_{A,x} + \dot{n}_{B,x} = \rho v_x \qquad (11.6)$$

From Eqs. (11.3) and (11.4),

$$j_{A,x} = \rho_A(v_{A,x} - v_x) \qquad (11.7)$$

and

$$j_{B,x} = \rho_B(v_{B,x} - v_x) \qquad (11.8)$$

the sum of which is

$$j_{A,x} + j_{B,x} = \rho_A v_{A,x} + \rho_B v_{B,x} - \rho v_x = 0 \tag{11.9}$$

Mass transport can also be considered on a molar basis, in which case the local molar average velocity, $\mathbf{v}$, is defined by

$$C\mathbf{v} = C_A \mathbf{v}_A + C_B \mathbf{v}_B \tag{11.10}$$

where C_A and C_B are the molar concentrations of A and B in the fluid, and $C = C_A + C_B$. In this case $C\mathbf{v}$ is the rate at which matter passes through unit cross section perpendicular to the direction of the flow in units of kg mol·m^{-2}·s^{-1}. Although the velocities $\mathbf{v}_A$ and $\mathbf{v}_B$ are identical in Eqs. (11.1) and (11.10), the local molar average velocity $\mathbf{v}$ has the same value as the local mass average velocity $\mathbf{v}$ only when the molecular weight of A is equal to the molecular weight of B. Alternatively, Eq. (11.10) can be written as

$$\mathbf{v} = X_A \mathbf{v}_A + X_B \mathbf{v}_B \tag{11.11}$$

in which X_A and X_B are the mole fractions of A and B in the fluid. In this case concentration gradients in the fluid give rise to molar diffusion fluxes in the x-direction of

$$J_{A,x}^* = -D \frac{\partial C_A}{\partial x} \tag{11.12}$$

and

$$J_{B,x}^* = -D \frac{\partial C_B}{\partial x}$$

The absolute molar fluxes in the x-direction are thus

$$\dot{N}_{A,x} = C_A v_{A,x} = J_{A,x}^* + C_A v_x^* \tag{11.13}$$

and

$$\dot{N}_{B,x} = C_B v_{B,x} = J_{B,x}^* + C_B v_x^* \tag{11.14}$$

and the total absolute molar flux is

$$\dot{N}_{A,x} + \dot{N}_{B,x} = C_A v_{A,x} + C_B v_{B,x} = C v_x^* \tag{11.15}$$

and

$$J_{A,x}^* + J_{B,x}^* = C_A v_{A,x} + C_B v_{B,x} - C v_x^* = 0 \tag{11.16}$$

11.3 Equations of Diffusion with Convection in a Binary Mixture A–B

Figure 11.1 shows a control volume of dimensions Δx, Δy, and Δz that is fixed in space relative to a stationary observer and through which a binary mixture of A and

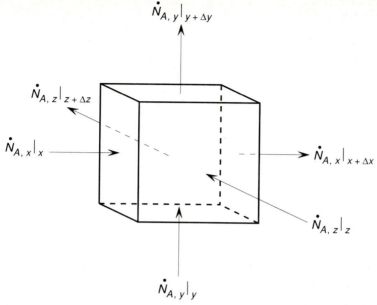

FIGURE 11.1. Control volume, fixed in space relative to a stationary observer, through which a binary mixture of A and B is flowing with the molar average velocity v.

B is flowing with the molar average velocity **v**. A mass balance for A requires that

rate of change of concentration of A in the control volume
= net rate at which A enters the control volume

The rate of change of the concentration of A in the control volume is

$$\frac{\partial C_A}{\partial t} \, \Delta x \, \Delta y \, \Delta z$$

and the net rate of transport of A into the control volume in the x, y, and z directions is

$$-\frac{\partial}{\partial x}(C_A v_{A,x}) \, \Delta x (\Delta y \, \Delta z) - \frac{\partial}{\partial y}(C_A v_{A,y}) \, \Delta y \, (\Delta x \, \Delta z) - \frac{\partial}{\partial z}(C_A v_{A,z}) \, \Delta z \, (\Delta x \, \Delta y)$$

The mass balance for A is thus

$$\frac{\partial C_A}{\partial t} = -\frac{\partial}{\partial x}(C_A v_{A,x}) - \frac{\partial}{\partial y}(C_A v_{A,y}) - \frac{\partial}{\partial z}(C_A v_{A,z}) \qquad (11.17)$$

and similarly, the mass balance for B is

$$\frac{\partial C_B}{\partial t} = -\frac{\partial}{\partial x}(C_B v_{B,x}) - \frac{\partial}{\partial y}(C_B v_{B,y}) - \frac{\partial}{\partial z}(C_B v_{B,z}) \qquad (11.18)$$

The sum of Eqs. (11.17) and (11.18) is

$$\frac{\partial(C_A + C_B)}{\partial t} = -\frac{\partial}{\partial x}(C_A v_{A,x} + C_B v_{B,x}) - \frac{\partial}{\partial y}(C_A v_{A,y} + C_B v_{B,y})$$
$$- \frac{\partial}{\partial z}(C_A v_{A,z} + C_B v_{B,z})$$

which, from Eq. (11.10), gives

$$\frac{\partial C}{\partial t} = -\frac{\partial}{\partial x}(C v_x^*) - \frac{\partial}{\partial y}(C v_y^*) - \frac{\partial}{\partial z}(C v_z^*) \tag{11.19}$$

Equation (11.19) is thus the rate of change, with time, of the molar concentration in the stationary control volume.

The molar concentration of A in the flowing fluid is a function of time and position,

$$C_A = C_A(t,x,y,z)$$

and therefore

$$dC_A = \left(\frac{\partial C_A}{\partial t}\right)dt + \left(\frac{\partial C_A}{\partial x}\right)dx + \left(\frac{\partial C_A}{\partial y}\right)dy + \left(\frac{\partial C_A}{\partial z}\right)dz$$

or

$$\frac{dC_A}{dt} = \frac{\partial C_A}{\partial t} + \frac{\partial C_A}{\partial x}\frac{dx}{dt} + \frac{\partial C_A}{\partial y}\frac{dy}{dt} + \frac{\partial C_A}{\partial z}\frac{dz}{dt} \tag{11.20}$$

which is the substantial derivative of C_A,

$$\frac{DC_A}{Dt} = \frac{\partial C_A}{\partial t} + v_x\frac{\partial C_A}{\partial x} + v_y\frac{\partial C_A}{\partial y} + v_z\frac{\partial C_A}{\partial z} \tag{11.21}$$

If, in Eq. (11.21), $v_x = v_x^*$, $v_y = v_y^*$, and v_z and v_z^*, then

$$\frac{DC_A}{Dt} = \frac{\partial C_A}{\partial t} + v_x^*\frac{\partial C_A}{\partial x} + v_y^*\frac{\partial C_A}{\partial y} + v_z^*\frac{\partial C_A}{\partial z} \tag{11.22}$$

is the rate of change, with time, of the molar concentration of A in the fluid as observed by an observer traveling at the velocity **v**. Substitution of Eq. (11.13) into Eq. (11.17) gives

$$\frac{\partial C_A}{\partial t} = -\frac{\partial}{\partial x}(J_{A,x}^* + C_A v_x^*) - \frac{\partial}{\partial y}(J_{A,y}^* + C_A v_y^*) - \frac{\partial}{\partial z}(J_{A,z}^* + C_A v_z^*)$$

$$= -C_A\left(\frac{\partial v_x^*}{\partial x} + \frac{\partial v_y^*}{\partial y} + \frac{\partial v_z^*}{\partial z}\right) - \frac{\partial J_{A,x}^*}{\partial x} - \frac{\partial J_{A,y}^*}{\partial y} - \frac{\partial J_{A,z}^*}{\partial z}$$

$$- v_x^*\frac{\partial C_A}{\partial x} - v_y^*\frac{\partial C_A}{\partial y} - v_z^*\frac{\partial C_A}{\partial z} \tag{11.23}$$

and substitution of Eq. (11.23) into Eq. (11.22) gives

$$\frac{DC_A}{Dt} = -C_A \left(\frac{\partial v_x^*}{\partial x} + \frac{\partial v_y^*}{\partial y} + \frac{\partial v_z^*}{\partial z} \right) - \left(\frac{\partial J_{A,x}^*}{\partial x} + \frac{\partial J_{A,y}^*}{\partial y} + \frac{\partial J_{A,z}^*}{\partial z} \right) \quad (11.24)$$

Similarly,

$$\frac{DC_B}{Dt} = -C_B \left(\frac{\partial v_x^*}{\partial x} + \frac{\partial v_y^*}{\partial y} + \frac{\partial v_z^*}{\partial z} \right) - \left(\frac{\partial J_{B,x}^*}{\partial x} + \frac{\partial J_{B,y}^*}{\partial y} + \frac{\partial J_{B,z}^*}{\partial z} \right) \quad (11.25)$$

From Eq. (11.16),

$$(J_{A,x}^* + J_{B,x}^*) = (J_{A,y}^* + J_{B,y}^*) = (J_{A,z}^* + J_{B,z}^*) = 0$$

and thus the sum of Eqs. (11.24) and (11.25) gives

$$\frac{DC}{Dt} = -C \left(\frac{\partial v_x^*}{\partial x} + \frac{\partial v_y^*}{\partial y} + \frac{\partial v_z^*}{\partial z} \right) \quad (11.26)$$

For a fluid of constant density, the equation of continuity, Eq. (3.3), gives

$$\frac{\partial v_x^*}{\partial x} + \frac{\partial v_y^*}{\partial y} + \frac{\partial v_z^*}{\partial z} = 0 \quad (3.3)$$

and hence, for such a fluid, in which D is constant, Eq. (11.24) becomes

$$\frac{DC_A}{Dt} = - \left(\frac{\partial J_{A,x}^*}{\partial x} + \frac{\partial J_{A,y}^*}{\partial y} + \frac{\partial J_{A,z}^*}{\partial z} \right)$$

$$= D \left(\frac{\partial^2 C_A}{\partial x^2} + \frac{\partial^2 C_A}{\partial y^2} + \frac{\partial^2 C_A}{\partial z^2} \right) \quad (11.27)$$

11.4 One-Dimensional Transport in a Binary Mixture of Ideal Gases

In a mixture of ideal gases the sum of the partial pressures of the component gases, p_i, equals the total pressure of the gas, P,

$$\sum_i p_i = P \quad (11.28)$$

and from the ideal gas law,

$$p_i V = n_i RT$$

and thus the molar concentrations of the component gases are

$$C_i = \frac{n_i}{V} = \frac{p_i}{RT} \quad (11.29)$$

Consider steady-state, one-dimensional diffusion in the ideal gas mixture at the constant total pressure P, in which D is constant.

11.5 Equimolar Counterdiffusion

For steady-state one-dimensional counterdiffusion of A and B

$$C_A v_{A,x} = -C_B v_{B,x}$$

and thus from Eq. (11.16), $\mathbf{v}^* = 0$. Also, at steady state $\partial C_A/\partial t = 0$ and for one-dimensional diffusion $\partial C_A/\partial y$ and $\partial C_A/\partial z$ are zero. Thus Eq. (11.27) gives

$$\frac{d^2 C_A}{dx^2} = 0$$

or from Eq. (11.29),

$$\frac{d^2 p_A}{dx^2} = 0 \tag{11.30}$$

The first integration of Eq. (11.30) gives

$$\frac{dp_A}{dx} = C_1$$

and the second integration gives

$$p_A = C_1 x + C_2 \tag{11.31}$$

[i.e., the partial pressure of A (and hence of B) is a linear function of x]. With the boundary conditions

$$p_A = \begin{cases} p_{A(0)} & \text{at } x = 0 \\ p_{A(L)} & \text{at } x = L \end{cases}$$

$$C_2 = p_{A(0)}$$

and

$$C_1 = \frac{p_{A(L)} - p_{A(0)}}{L}$$

and Eq. (11.31) becomes

$$p_A = (p_{A(L)} - p_{A(0)}) \frac{x}{L} + p_{A(0)} \tag{11.32}$$

The gradient of the partial pressure of A is thus

$$\frac{dp_A}{dx} = \frac{p_{A(L)} - p_{A(0)}}{L}$$

which corresponds to a concentration gradient of

$$\frac{dC_A}{dx} = \frac{p_{A(L)} - p_{A(0)}}{RTL}$$

and hence the molar flux of A due to diffusion is

$$J_A^* = -D\frac{dC_A}{dx} = \frac{D(p_{A(0)} - p_{A(L)})}{RTL} \tag{11.33}$$

11.6 One-Dimensional Steady-State Diffusion of Gas *A* Through Stationary Gas *B*

In this case $\mathbf{v}^* \neq 0$, and hence for one-dimensional steady-state diffusion in which $\partial C_A/\partial t$, v_y^*, v_z^*, and $\partial v_x^*/\partial x$ are all zero, Eqs. (11.23) and (11.27) give

$$v_x^*\frac{dC_A}{dx} = -\frac{dJ_A^*}{dx} = D\frac{d^2C_A}{dx^2} \tag{11.34}$$

As the gas B is stationary, $v_{B,x} = 0$, and hence from Eq. (11.14),

$$J_{B,x}^* = -C_B v_x^*$$

or

$$v_x^* = -\frac{J_{B,x}^*}{C_B}$$

$$= \frac{D}{C_B}\frac{dC_B}{dx} \tag{11.35}$$

substitution of which into Eq. (11.34) gives

$$\frac{1}{C_B}\frac{dC_B}{dx}\frac{dC_A}{dx} = \frac{d^2C_A}{dx^2} \tag{11.36}$$

or from Eq. (11.29),

$$\frac{1}{p_B}\frac{dp_B}{dx}\frac{dp_A}{dx} = \frac{d^2p_A}{dx^2} \tag{11.37}$$

As the total pressure of the gas, P, is constant and $p_A + p_B = P$,

$$dp_A = -dp_B$$

and thus Eq. (11.37) becomes

$$\frac{1}{p_B}\left(\frac{dp_B}{dx}\right)^2 = \frac{d^2p_B}{dx^2} \tag{11.38}$$

From the identity

$$\frac{d}{dx}\left(\frac{1}{y}\frac{dy}{dx}\right) = \frac{1}{y}\frac{d^2y}{dx^2} - \frac{dy}{dx}\left(\frac{1}{y^2}\frac{dy}{dx}\right) = \frac{1}{y}\left[\frac{d^2y}{dx^2} - \frac{1}{y}\left(\frac{dy}{dx}\right)^2\right]$$

Equation (11.38) is equivalent to

$$\frac{d}{dx}\left(\frac{1}{p_B}\frac{dp_B}{dx}\right) = 0$$

The first integration gives

$$\frac{1}{p_B}\frac{dp_B}{dx} = C_1$$

and the second integration gives

$$\ln p_B = C_1 x + C_2$$

For the boundary conditions

$$p_B = \begin{cases} p_{B(0)} & \text{at } x = 0 \\ p_{B(L)} & \text{at } x = L \end{cases}$$

$$C_2 = \ln p_{B(0)}$$

and

$$C_1 = \frac{\ln p_{B(L)} - \ln p_{B(0)}}{L}$$

Thus

$$\ln p_B = \frac{\ln p_{B(L)} - \ln p_{B(0)}}{L} x + \ln p_{B(0)}$$

or

$$\ln \frac{p_B}{p_{B(0)}} = \frac{x}{L} \ln \frac{p_{B(L)}}{p_{B(0)}} \tag{11.39}$$

and with $p_A = P - p_B$,

$$\ln \frac{P - p_A}{P - p_{A(0)}} = \frac{x}{L} \ln \frac{P - p_{A(L)}}{P - p_{A(0)}} \tag{11.40}$$

These gradients are shown in Fig. 11.2 for $p_{B(0)} = 0.1$ and $p_{B(L)} = 0.9$. The partial pressure, and hence concentration of B increase with increasing x, and the concentration gradient of B gives rise to a diffusion flux of B from right to left, $J^*_{B,x}$, which, from Eq. (11.35), is exactly balanced by the transport of B from left to right by the bulk flow of the gas mixture, $C_B v^*_x$.

The absolute molar flux of A is given by Eq. (11.13) as

$$\dot{N}_{A,x} = C_A v_{A,x} = J^*_{A,x} + C_A v^*_x$$

which from Eq. (11.35) can be expressed as

$$\dot{N}_{A,x} = -D\frac{dC_A}{dx} + C_A \left(\frac{D}{C_B}\frac{dC_B}{dx}\right)$$

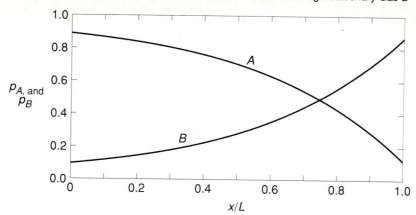

FIGURE 11.2.   Variations, with position, of the partial pressures when gas A diffuses through stationary gas B.

$$= -\frac{D}{RT}\frac{dp_A}{dx} - \frac{p_A}{p_B}\left(\frac{D}{RT}\frac{dp_A}{dx}\right)$$

$$= -\frac{D}{RT}\left(\frac{P}{P - p_A}\right)\frac{dp_A}{dx} \tag{11.41}$$

From Eq. (11.40),

$$\frac{d\ln(P - p_A)}{dx} = \frac{1}{L}\ln\frac{P - p_{A(L)}}{P - p_{A(0)}}$$

or

$$\frac{dp_A}{dx} = -\left(\frac{P - p_A}{L}\right)\ln\frac{P - p_{A(L)}}{P - p_{A(0)}}$$

substitution of which into Eq. (11.41) gives

$$\dot{N}_{A,x} = C_A v_{A,x} = \frac{DP}{RTL}\ln\frac{P - p_{A(L)}}{P - p_{A(0)}} \tag{11.42}$$

which, as is required at steady state, has a constant value. Thus, in Fig. 11.2, as C_A decreases with increasing x, Eq. (11.42) shows that $v_{A,x}$ increases with increasing x.

Diffusion of a gas A through a stationary gas B is realized in Stefan's apparatus shown in Fig. 11.3. In this arrangement liquid A is contained by a vertical tube and a mixture of gas B and vapor A is made to flow through the horizontal tube. The temperature and total pressure within the system are constant. If B does not dissolve in liquid A, the partial pressure of A in the gas mixture at the liquid A–gas interface is the saturated vapor pressure of liquid A at the temperature T, designated $p_{A(0)}$ at z_0, and the partial pressure of A, $p_{A(L)}$, at z_L is determined by the composition of the gas mixture flowing in the horizontal tube. Thus, as $p_A = X_A P$, Eq. (11.40) gives

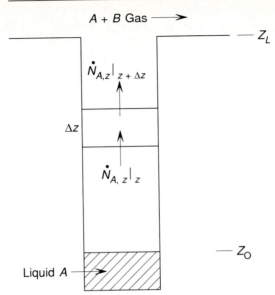

FIGURE 11.3. Stefan's apparatus for measuring the interdiffusion coefficient in the binary gas $A-B$.

the variation of X_A with position in the vertical tube as

$$\ln \frac{1 - X_A}{1 - X_{A(0)}} = \frac{z - z_0}{z_L - z_0} \ln \frac{1 - X_{A(L)}}{1 - X_{A(0)}} \qquad (11.43)$$

when steady state has been achieved and arrangements have been made to maintain the distance $z_L - z_0$ constant. Measurement of the rate of evaporation of liquid A then allows determination of D from Eq. (11.42).

EXAMPLE 11.1

When liquid Mn at 1600°C evaporates and diffuses into pure Ar in an apparatus at which $z_L - z_0$ is maintained constant at 0.02 m, the steady-state rate of loss of mass of liquid Mn is 3.21×10^{-4} kg·m^{-2}·s^{-1}. Calculate D_{Mn-Ar}.

The atomic weight of Mn is 0.05494 kg/g mol, and at 1873 K, the saturated vapor pressure of liquid Mn is 5370 Pa. Thus the flux of Mn vapor from the melt surface is

$$\dot{N}_{Mn,z} = \frac{3.21 \times 10^{-4}}{0.05494} = 5.84 \times 10^{-3} \text{ g mol·m}^{-2}\text{·s}^{-1}$$

and with a total pressure of 101,325 Pa, D_{Mn-Ar} is obtained from Eq. (11.42) as

$$D_{Mn-Ar} = \frac{\dot{N}_{Mn,z} RTL}{P \ln[P/(P - p_{Mn}^\circ)]}$$

$$= \frac{5.84 \times 10^{-3} \times 8.3144 \times 1873 \times 0.02}{101{,}325 \; \ln[101{,}325/(101{,}325 - 5370)]}$$

$$= 3.30 \times 10^{-4} \; m^2/s$$

From Eq. (11.35), with $Mn \equiv A$ and $Ar \equiv B$,

$$v_x^* = \frac{D}{C_{Ar}} \frac{dC_{Ar}}{dx}$$

$$= D \frac{d \ln C_{Ar}}{dx}$$

$$= D \frac{d \ln p_{Ar}}{dx}$$

and from Eq. (11.39),

$$\frac{d \ln p_{Ar}}{dx} = \frac{1}{L} \ln \frac{p_{Ar(L)}}{p_{Ar(0)}}$$

Thus

$$v_x^* = \frac{D}{L} \ln \frac{p_{Ar(L)}}{p_{Ar(0)}}$$

$$= \frac{3.3 \times 10^{-4}}{0.02} \ln \frac{101{,}325}{101{,}325 - 5370}$$

$$= 9.0 \times 10^{-4} \; m/s \tag{i}$$

From Eqs. (11.12) and (11.13),

$$\dot{N}_{Mn,x} = -D \frac{dC_{Mn}}{dx} + C_{Mn} v_x^*$$

The highest value of C_{Mn} is

$$C_{Mn} = \frac{p_{Mn(0)}}{RT} = \frac{5370}{8.3144 \times 1873} = 0.345 \; g \; mol/m^3$$

and thus the highest value of $C_{Mn}v_x^*$ is $9.0 \times 10^{-4} \times 0.345 = 3.1 \times 10^{-4}$ $g \; mol \cdot m^{-2} \cdot s^{-1}$, which is only 5.3% of the total flux of 5.84×10^{-3} $g \; mol \cdot m^{-2} \cdot s^{-1}$. The small contribution of bulk flow to the transport of Mn is caused by the fact that the concentration of Mn vapor in the mixture is very much smaller than that of Ar. In Eq. (i), as $p_{Ar(0)}$ approaches P, v_x^* approaches zero. With very small values of v_x^*, an approximate value of D can be obtained from Fick's first law as

$$J_{Mn}^* = -D \frac{dC_{Mn}}{dx} = \frac{-D}{RT} \frac{dp_{Mn}}{dx}$$

which, assuming a linear variation of p_{Mn} with x, gives

$$D_{Mn-Ar} = -\frac{RTLJ^*_{Mn}}{p_{Mn(L)} - p_{Mn(0)}}$$

$$= \frac{8.3144 \times 1873 \times 0.02 \times 5.84 \times 10^{-3}}{5370}$$

$$= 3.39 \times 10^{-4} \text{ m}^2/\text{s}$$

Assume that zinc is evaporating into argon in the same apparatus at 1037 K and a total pressure of 101.325 kPa at the rate of 5.82×10^{-2} g mol·m^{-2}·s^{-1}. At 1037 K, the saturated vapor pressure of liquid zinc is 50.66 kPa (0.5 atm), and hence from Eq. (11.42),

$$D_{Zn-Ar} = \frac{5.82 \times 10^{-2} \times 8.3144 \times 1037 \times 0.02}{101,325 \ln 2}$$

$$= 1.43 \times 10^{-4} \text{ m}^2/\text{s}$$

and

$$v^*_x = \frac{D}{L} \ln \frac{p_{Ar(L)}}{p_{Ar(0)}}$$

$$= \frac{1.43 \times 10^{-4}}{0.02} \ln 2$$

$$= 4.95 \times 10^{-3} \text{ m/s}$$

The highest value of C_{Zn} is

$$C_{Zn} = \frac{p_{Zn(0)}}{RT} = \frac{50,663}{8.3144 \times 1037} = 5.88 \text{ g mol/m}^3$$

and thus the highest value of $C_{Zn}v^*_x$ is $5.88 \times 4.95 \times 10^{-3} = 0.0291$ g mol·m^{-2}·s^{-1}, which is 50% of the total flux. Application of Eq. (11.13) to Zn gives

$$\dot{N}_{Zn,x} = C_{Zn}v_{Zn,x} = J^*_{Zn,x} + C_{Zn}v^*_x$$

From Eq. (11.39), as

$$p_{Ar} = X_{Ar}P = C_{Ar}\frac{P}{C}$$

and

$$C = \frac{P}{RT} = \frac{101,325}{8.3144 \times 1037} = 11.75 \text{ g mol·m}^{-3}$$

$$\ln C_{Ar} = \frac{x}{L} \ln \frac{C_{Ar(L)}}{C_{Ar(0)}} + \ln C_{Ar(0)}$$

$$= \frac{x}{0.02} \ln 2 + \ln \frac{50,660}{8.3144 \times 1037}$$

$$= 34.7x + 1.771 \tag{ii}$$

This variation of C_{Ar} with x, together with the corresponding variation of $C_{Zn} = C - C_{Ar}$, is shown in Fig. 11.4(a). From Eq. (ii),

$$\ln(C - C_{Zn}) = 34.7x + 1.771$$

and thus

$$\frac{dC_{Zn}}{dx} = -34.7(C - C_{Zn}) = -34.7C_{Ar}$$

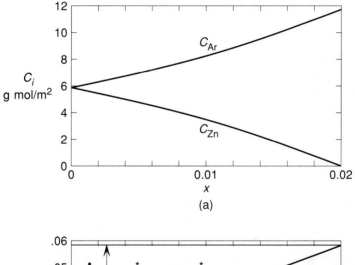

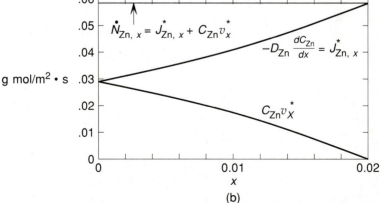

FIGURE 11.4. (a) Concentration profiles of Zn and Ar calculated in Example 11.1; (b) The variations, with position, of the quantities $J^*_{Zn,x}$ and $C_{Zn}v^*_x$ calculated in Example 11.1.

The variation, with x, of the product

$$-D_{Zn} \frac{dC_{Zn}}{dx} = J^*_{Zn,x} = 1.43 \times 10^{-4} \times 34.7 C_{Ar}$$

$$= 4.95 \times 10^{-3} C_{Ar}$$

is shown in Fig. 11.4(b) along with the corresponding variation of the product

$$C_{Zn} v^*_x = 4.95 \times 10^{-3} C_{Zn}$$

The variations of $-D_{Zn}(dC_{Zn}/dx)$ and $C_{Zn} v^*_x$ are such that their sum

$$-D_{Zn} \frac{dC_{Zn}}{dx} + C_{Zn} v^*_x = 4.95 \times 10^{-3} C_{Ar} + 4.95 \times 10^{-3} C_{Zn}$$

$$= 4.95 \times 10^{-3} C$$

$$= 4.95 \times 10^{-3} \times 11.75$$

$$= 0.0582 \text{ g mol·m}^{-2}\text{·s}^{-1}$$

$$= \dot{N}_{Zn}$$

is constant.

11.7 Sublimation of a Sphere into a Stationary Gas

Consider a sphere of solid A of radius a subliming into a stationary gas B. The rate of transport through the stationary B in the radial direction at a radius r, $\dot{W}_A$, in units of g mol/s is

$$\dot{W}_A = 4\pi r^2 \dot{N}_A$$

and hence, from Eq. (11.41),

$$\dot{N}_A = \frac{\dot{W}_A}{4\pi r^2} = -\frac{D}{RT} \left(\frac{P}{P - P_A} \right) \frac{dp_A}{dr} \tag{i}$$

With $p_A = p_{A,2}$ at $r = r_2$ and $p_A = p_{A,1}$ at $r = r_1$, integration of Eq. (i) gives

$$\frac{\dot{W}_A}{4\pi} \left(\frac{1}{r_1} - \frac{1}{r_2} \right) = \frac{DP}{RT} \ln \frac{P - p_{A,2}}{P - p_{A,1}} \tag{ii}$$

With $r_1 = a$ and $r_2 = \infty$, $p_{A,1}$ is the saturated vapor pressure of solid A at the temperature T, p°_A, and $p_{A,2} = 0$, in which case Eq. (ii) gives the molar flux of A from the surface of the sphere as

$$\dot{W}_A = \frac{4\pi a D P}{RT} \ln \frac{P}{P - p^\circ_A} \tag{11.44}$$

However, sublimation causes the radius of the sphere to decrease and equating the molar flux with rate of loss of moles from the sphere of radius a gives

$$\dot{W}_A = -\frac{dn_A}{dt} = -\frac{4}{3} \pi \rho_m \frac{da^3}{dt} = -4\pi \rho_m a^2 \frac{da}{dt} \tag{iii}$$

in which ρ_m is the molar density of solid A. Combination of Eqs. (11.44) and (iii) gives

$$ada = -\frac{DP}{\rho_m RT} \ln \frac{P}{P - p_A^\circ} dt$$

which on integration between $a = a_0$ at $t = 0$ and $a = a$ at $t = t$ gives

$$a_0^2 - a^2 = \frac{2DPt}{\rho_m RT} \ln \frac{P}{P - p_A^\circ} \tag{iv}$$

The time required for sublimation to cause the sphere to disappear is thus

$$t = \frac{a_0^2 \rho_m RT}{2DP \ln [P/(P - p_A^\circ)]} \tag{11.45}$$

EXAMPLE 11.2

Calculate (a) the initial rate of sublimation of a napthalene mothball of diameter 0.01 m into stagnant air at 20°C and a pressure of 101.3 kPa, and (b) the time required for complete sublimation of the mothball. For napthalene,

$$p^\circ \text{ at } 293 \text{ K} = 30 \text{ Pa}$$

$$D = 8 \times 10^{-6} \text{ m}^2/\text{s}$$

$$\rho_m = 5470 \text{ g mol/m}^3$$

(a) From Eq. (11.44),

$$\dot{W} = \frac{4\pi a DP}{RT} \ln \frac{P}{P - p^\circ}$$

$$= \frac{4 \times \pi \times 0.005 \times 8 \times 10^{-6} \times 101,300}{8.3144 \times 293} \ln \frac{101,300}{101,300 - 30}$$

$$= 6.2 \times 10^{-9} \text{ g mol/s} \tag{11.44}$$

(b) From Eq. (11.15) the time required for the ball to disappear due to sublimation is

$$t = \frac{(0.01)^2 \times 5470 \times 8.3144 \times 293}{2 \times 8 \times 10^{-6} \times 101,300 \times \ln[101,300/(101,300 - 30)]}$$

$$= 2.78 \times 10^6 \text{ s}$$

$$= 32.1 \text{ days}$$

The variation of the diameter of the mothball, with time, obtained from Eq. (iv) as

$$a = \sqrt{10^{-4} - 3.60 \times 10^{-11}t}$$

is shown in Fig. 11.5.

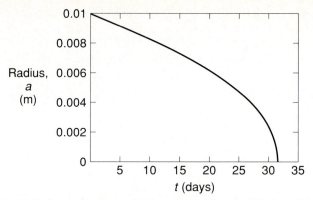

FIGURE 11.5.   Variation, with time, of the radius of the mothball subliming into stagnant air considered in Example 11.2.

11.8 Film Model

Consider Fig. 11.6, which shows a mixture of gas B and vapor A flowing over the surface of solid A at the temperature T. If the partial pressure of B in the mixture is less than the saturated vapor pressure of solid A at the temperature T, then A sublimes from the solid and is transported into the bulk gas. Phase equilibrium between the solid and the gas mixture requires that the partial pressure of A at $y = 0$ be the saturated vapor pressure of solid A at the temperature T, $p_A^\circ(T)$, and thus the concentration of A in the gas phase at $y = 0$ is

$$C_{A,(0)} = \frac{p_A^\circ(T)}{RT}$$

If the concentration of A in the bulk gas phase is $C_{A,b}$, the variation of C_A with distance away from the surface can be expected to be as shown in Fig. 11.7(a), and from Eq. (11.13), the molar flux of A in the y-direction is

$$\dot{N}_A = J_A^* + C_A v_y^* \tag{11.13}$$

In this expression the contribution of bulk flow to the molar flux of A is determined by the magnitudes of C_A and v_y^*. If the product $C_A v_y^*$ is small in comparison with $\dot{N}_A$, treatment of molar transport of A from the solid to the bulk gas is simplified by visualizing a thin film of viscous gas undergoing laminar flow adjacent to the surface

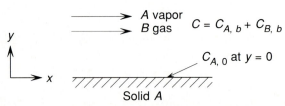

FIGURE 11.6.   Mixture of A vapor and B gas flowing over the surface of solid A.

of the solid through which diffusion is the sole mechanism of mass transport. In Fig. 11.7(b) and molar flux at $y = 0$ is determined by D_A and the tangent to the concentration profile at $y = 0$. The thickness, L, of the imagined film is the value of y at which the tangent line reaches the value $C_A = C_{A,b}$. At $y > L$ the flowing gas is considered to be well mixed, and hence does contain any concentration, temperature, or velocity gradients. From Eq. (11.42) the steady-state molar flux of A through a film of thickness L is

$$\dot{N}_A = \frac{D_A P}{RTL} \ln \frac{p_{B(L)}}{p_{B(0)}}$$

$$= \frac{CD_A}{L} \ln \frac{X_{B(L)}}{X_{B(0)}} \tag{11.46}$$

Defining the logarithmic mean of $X_{B(L)}$ and $X_{A(L)}$ as

$$(X_B)_{lm} = \frac{X_{B(L)} - X_{B(0)}}{\ln (X_{B(L)}/X_{B(0)})}$$

and substituting into Eq. (11.46) gives the molar flux of A through the film as

$$\dot{N}_A = \frac{CD_A}{L} \frac{X_{A(0)} - X_{A(L)}}{(X_B)_{lm}} \tag{11.47}$$

When $X_{A(0)}$ and $X_{A(L)}$ have small values, $(X_B)_{lm}$ has a value close to unity and hence

$$\dot{N}_A \doteq \frac{CD_A}{L} (X_{A(0)} - X_{A(L)}) \tag{11.48}$$

which approaches Fick's first law for diffusion of A down a linear concentration gradient from $y = 0$ to $y = L$.

As L is a function of the velocity and physical properties of the flowing fluid, the difficulty of estimating its value in mass transport systems limits the extent of the practical application of the film model, but it is a useful tool in understanding mass transport in a fluid to and from a condensed phase over which the fluid is flowing. If the temperature of the solid, T_s, differs from the temperature of the bulk gas, T_∞, the temperature profile through the film is as shown in Fig. 11.7(c), which also shows the velocity profile through the film.

11.9 Catalytic Surface Reactions

Figure 11.8 shows a gas mixture $A-B$ flowing over a flat surface on which a heterogeneous catalytic chemical reaction either produces or consumes the species A. It will be assumed that the reaction consumes A by means of a first-order irreversible chemical reaction (one in which the chemical reaction rate, $\dot{r}$, is proportional to the concentration of A at the reaction site). Thus

$$\dot{r} = k_r C_{A(0)} \tag{i}$$

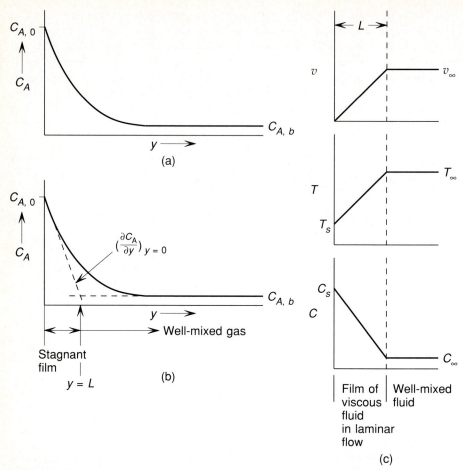

FIGURE 11.7. (a) Concentration profile of A vapor in the gas phase in Fig. 11.6 in a direction normal to the surface of solid A; (b) criterion for calculating the thickness of the hypothetical film of gas on the surface of solid A; (c) presumed variations of local flow velocity, temperature, and concentration with position in the film of gas on the surface of solid A.

in which $\dot{r}$ has the units g mol·m^{-2}·s^{-1}, and k_r, the reaction rate constant, has the units m·s^{-1}. Steady-state conditions require that the rate at which A arrives at $y = 0$ by diffusion through the film of thickness L be equal to the rate at which it is consumed by the chemical reaction at $y = 0$:

$$\dot{r} = -\dot{N}_A|_{y=0}$$

$$= D_A C \left.\frac{dX_A}{dy}\right|_{y=0} \tag{ii}$$

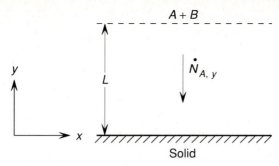

FIGURE 11.8. *A–B* gas mixture flowing over a flat surface on which a catalytic reaction consumes the species *A*.

(The negative sign in the equality arises because the diffusion flux of *A* is in the $-y$-direction.) Steady-state diffusion of *A* in the film requires that

$$\frac{d\dot{N}_A}{dy} = 0$$

or

$$D_A C \frac{d^2 X_A}{dy^2} = 0$$

Thus

$$\frac{dX_A}{dy} = C_1 \tag{iii}$$

and

$$X_A = C_1 y + C_2 \tag{iv}$$

which is the linear variation given by Eq. (11.48). From Eqs. (i), (ii), and (iii),

$$C_1 = \frac{\dot{r}}{D_A C} = \frac{k_r C_{A(0)}}{D_A C} = \frac{k_r X_{A(0)}}{D_A} \tag{v}$$

and as $X_A = X_{A(L)}$ at $y = L$,

$$C_2 = X_{A(L)} - C_1 L$$

$$= X_{A(L)} - \frac{k_r X_{A(0)} L}{D_A} \tag{vi}$$

Substitution of Eqs. (v) and (vi) into Eq. (iv) gives the concentration profile as

$$X_A = X_{A(L)} - \frac{k_r X_{A(0)}}{D_A} (L - y)$$

Thus

$$X_{A(0)} = X_{A(L)} - \frac{k_r X_{A(0)} L}{D_A}$$

or

$$X_{A(0)} = \frac{X_{A(L)}}{1 + k_r L / D_A} \tag{vii}$$

From Eqs. (ii), (iii), and (v), the molar flux of A is

$$\dot{N}_A = -D_A C \frac{dX_A}{dy} = -D_A C \frac{k_r X_{A(0)}}{D_A}$$

or from Eq. (vii),

$$\dot{N}_A = -\frac{k_r C X_{A(L)}}{1 + k_r L / D_A} \tag{viii}$$

The limiting cases are represented by $k_r \to 0$ and $k_r \to \infty$.

Case 1: $k_r \to 0$. From Eq. (vii),

$$X_{A(0)} \to X_{A(L)}$$

and from Eq. (viii),

$$\dot{N}_A \to -k_r C X_{A(L)}$$

Case 2: $k_r \to \infty$. From Eq. (vii),

$$X_{A(0)} \to 0$$

and

$$\dot{N}_A = -\frac{C D X_{A(L)}}{L}$$

In case 1 the reaction rate is controlled by the reaction rate constant and the influence of diffusion is negligible and in case 2 the reaction rate is controlled by diffusion. In the limiting cases the processes are said to be reaction controlled and diffusion controlled, respectively. For finite values of k_r the process is under mixed reaction and diffusion control. The effect of the reaction rate constant on the concentration profile of A in the film is shown in Fig. 11.9.

11.10 Diffusion and Chemical Reaction in Stagnant Film

Figure 11.10 shows gas B flowing over the surface of solid A. In this case A sublimes from the surface and diffuses into the film and B diffuses through the film toward

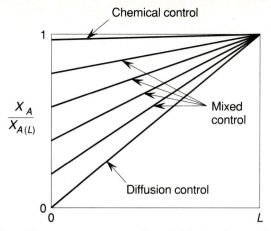

FIGURE 11.9. Possible concentration profiles of A in the film on the surface of solid A shown in Fig. 11.8.

the surface. Gas B and vapor A react to form the solid AB according to

$$A + B = AB$$

which is a second-order irreversible homogeneous reaction for which

$$\dot{r}_A = \dot{r}_B = k_r C_A C_B$$
$$= k_r C^2 X_A X_B \qquad \text{g mol·m}^{-3}\text{·s}^{-1} \qquad \text{(i)}$$

In a control volume of dimensions Δx, Δy, and Δz within the film, the steady-state molar balance requires that the rate at which A (or B) enters minus the rate at which A (or B) leaves be equal to the rate at which A (or B) is consumed by chemical reaction in the control volume. Thus, for A,

$$(\dot{N}_A\big|_y - \dot{N}_A\big|_{y+\Delta y}) \, \Delta x \, \Delta z = \dot{r}_A \, \Delta x \, \Delta y \, \Delta z$$

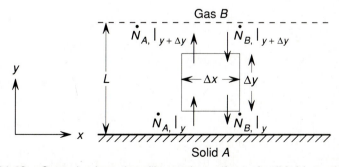

FIGURE 11.10. Control volume in a film on the surface of solid A into which gas B and vapor A are being transported by diffusion.

which leads to

$$-\frac{d\dot{N}_A}{dy} = \dot{r}_A \tag{ii}$$

and for B,

$$(N_B|_{y+\Delta y} - \dot{N}_B|_y)\, \Delta x\, \Delta z = \dot{r}_B\, \Delta x\, \Delta y\, \Delta z$$

or

$$\frac{d\dot{N}_B}{dy} = \dot{r}_B \tag{iii}$$

Thus from Eqs. (ii) and (i),

$$-\frac{d\dot{N}_A}{dy} = D_A C \frac{d^2 X_A}{dy^2} = k_r C^2 X_A X_B \tag{iv}$$

and from Eqs. (iii) and (i),

$$\frac{d\dot{N}_B}{dy} = -D_B C \frac{d^2 X_B}{dy^2} = k_r C^2 X_A X_B \tag{v}$$

The equation

$$D_A \frac{d^2 X_A}{dy^2} = -D_B \frac{d(1 - X_A)^2}{dy^2}$$

is nonlinear and its solution requires a numerical method. However, it was seen in the preceding section that with a large enough reaction rate constant, the concentration of the reactant can be decreased to zero at the reaction site. In the present case, with a large enough value of k_r, A and B are transported by diffusion to the reaction zone at $y = Y$, where fast chemical reaction decreases their concentrations to zero. Thus from Eqs. (iv) and (v),

$$\frac{d^2 X_A}{dy^2} = 0 \qquad \text{for } 0 \le y \le Y \tag{vi}$$

and

$$\frac{d^2 X_B}{dy^2} = 0 \qquad \text{for } Y \le y \le L \tag{vii}$$

Integration of Eq. (vi) with the boundary conditions

$$X_A = \begin{cases} X_{A(0)} & \text{at } y = 0 \\ 0 & \text{at } y = Y \end{cases}$$

gives

$$X_A = X_{A(0)} \left(1 - \frac{y}{Y}\right) \tag{viii}$$

and integration of Eq. (vii) with the boundary conditions

$$X_B = \begin{cases} X_{B(L)} & \text{at } y = L \\ 0 & \text{at } y = Y \end{cases}$$

gives

$$X_B = X_{B(L)} \frac{y - Y}{L - Y} \qquad \text{(ix)}$$

At the reaction site $y = Y$, the ratio of the molar fluxes of A and B must be equal to the ratio of the stoichiometry coefficients in the chemical reaction; that is, as $\dot{r}_A = \dot{r}_B$, Y occurs where $\dot{N}_A = -\dot{N}_B$. From Eq. (viii),

$$\dot{N}_A = -D_A C \frac{dX_A}{dy} = \frac{D_A C X_{A(0)}}{Y} \qquad \text{(x)}$$

and from Eq. (ix),

$$-\dot{N}_B = D_B C \frac{dX_B}{dy} = \frac{D_B C X_{B(L)}}{L - Y}$$

Thus

$$\frac{D_A C X_{A(0)}}{Y} = \frac{D_B C X_{B(L)}}{L - Y}$$

or

$$Y = \frac{L}{1 + D_B X_{B(L)}/D_A X_{A(0)}} \qquad \text{(xi)}$$

substitution of which into Eq. (x) gives the molar flux of A in the film as

$$\dot{N}_A = \frac{D_A C X_{A(0)}}{L} \left(1 + \frac{D_B X_{B(L)}}{D_A X_{A(0)}} \right) \qquad \text{(xii)}$$

In the absence of a chemical reaction and the case in which X_A decreases to zero at $y = L$, the molar flux of A is given by

$$\dot{N}_A = \frac{D_A C X_{A(0)}}{L}$$

and thus when a fast chemical reaction occurs, which increases the concentration gradient of X_A from $-X_{A(0)}/L$ to $-X_{A(0)}/Y$, the flux of A is increased by the factor

$$1 + \frac{D_B X_{B(L)}}{D_A X_{A(0)}}$$

The concentration gradients occurring when a fast reaction occurs in the film are shown in Fig. 11.11.

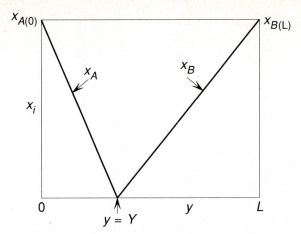

FIGURE 11.11. Concentration profiles in the film shown in Fig. 11.10 when chemical reaction between A and B in the film is rapid enough to decrease the concentrations of both species to zero at $y = Y$.

EXAMPLE 11.3 _____

When air flows over the surface of liquid iron, iron vaporizes, diffuses through the stagnant film and reacts with oxygen gas to form a fume of solid FeO. At what distance from the surface of the liquid iron is the fume formed when the observed rate of formation of the fume is 1.25×10^{-2} kg·m^{-2}·s^{-1}?

On the assumption that the fuming rate is controlled by diffusion of iron in the film, the answer is obtained from Eq. (xii). The diffusion coefficient for iron in nitrogen at 1873 K is 6×10^{-5} m^2/s and the saturated vapor pressure of liquid iron is

$$\ln p_{\text{Fe}}^{\circ} \text{ (Pa)} = \frac{-45{,}390}{T} - 1.27 \ln T + 35.45$$

Thus, at 1873 K, $p_{\text{Fe}}^{\circ} = 5.18$ Pa. The molar density of an ideal gas at 1873 K and 101.3 kPa is

$$C = \frac{n}{V} = \frac{P}{RT} = \frac{101{,}300}{8.3144 \times 1873} = 6.51 \text{ g mol/m}^3$$

and with a molecular weight of FeO of 0.07185 kg/g mole, Fe is consumed by the fuming reaction at the rate

$$\frac{1.25 \times 10^{-2}}{0.07185} = 0.174 \text{ g mol Fe·m}^{-2} \text{·s}^{-1}$$

The mole fraction of iron in the gas phase at the surface of the liquid iron is

$$\frac{5.18}{101{,}300} = 5.11 \times 10^{-5}$$

and thus Y is obtained from Eq. (xii),

$$0.174 = \frac{6 \times 10^{-5} \times 6.51 \times 5.11 \times 10^{-5}}{Y}$$

as $Y = 1.15 \times 10^{-7}$ m.

11.11 Mass Transfer at Large Fluxes and Large Concentrations

In the previous sections it was assumed that the product $C_A v_y^*$ in Eq. (11.13) was small enough to be ignored, in which case the molar flux of A could be expressed by Eq. (11.48). Consider the case in which a significant applied convection flux is superimposed on the diffusion flux. Such a case occurs when the solid A in Fig. 11.6 is porous, and in addition to gas B flowing past the solid in the x-direction, gas B is forced through the pores in the y-direction. When the gas B emerges from the pores it is saturated with A and the molar flux of A in the y-direction is given by Eq. (11.13) as

$$\dot{N}_A = J_A^* + C_A v_y^* \tag{11.13}$$

In this expression

$$C_A v_y^* = \frac{C_A}{C}(C_A v_{A,y} + C_B v_{B,y})$$

$$= X_A(\dot{N}_{A,y} + \dot{N}_{B,y})$$

and as the total molar flux, $\dot{N}_{T,y}$, is the sum of $\dot{N}_{A,y}$ and $\dot{N}_{B,y}$, Eq. (11.13) can be written as

$$\dot{N}_A = J_A^* + X_A \dot{N}_T$$

$$= -D_A C \frac{dX_A}{dy} + X_A \dot{N}_T \tag{i}$$

At steady state in the film, $d\dot{N}_A/dy = 0$, or

$$\frac{d^2 X_A}{dy^2} - \frac{\dot{N}_T}{D_A C}\frac{dX_A}{dy} = 0 \tag{11.49}$$

Letting

$$r = \frac{\dot{N}_T}{D_A C}$$

and defining the differential operators

$$DX_A = \frac{dX_A}{dy}$$

and

$$D^2 X_A = \frac{d^2 X_A}{dy^2}$$

allows Eq. (ii) to be written as

$$D^2 X_A - r D X_A = 0$$

or

$$D(D - r) X_A = 0$$

which has the solutions

$$D X_A = \frac{d X_A}{dy} = 0 \tag{ii}$$

and

$$(D - r) X_A = \frac{d X_A}{dy} - r X_A = 0 \tag{iii}$$

Thus from Eq. (ii),

$$X_A = C_1$$

and from Eq. (iii),

$$X_A = C_2 e^{ry}$$

summation of which gives

$$X_A = C_2 e^{ry} + C_1$$

as the solution to Eq. (11.49). The boundary conditions

$$X_A = \begin{cases} A_{A(0)} & \text{at } y = 0 \\ X_{A(L)} & \text{at } y = L \end{cases}$$

give

$$X_A = X_{A(0)} + \frac{X_{A(L)} - X_{A(0)}}{\exp(\dot{N}_T L / C D_A) - 1} \left[\exp\left(\frac{\dot{N}_T y}{C D_A}\right) - 1 \right] \tag{iv}$$

and thus J_A^* in Eq. (i) is obtained as

$$J_A^* = -D_A C \frac{d X_A}{dy}$$

$$= \frac{\dot{N}_T (X_{A(0)} - X_{A(L)})}{\exp(\dot{N}_T L / C D_A) - 1} \exp\left(\frac{\dot{N}_T y}{C D_A}\right) \tag{v}$$

Substitution of Eqs. (iv) and (v) into Eq. (i) gives

$$\dot{N}_A = \dot{N}_T \left[X_{A(0)} - \frac{X_{A(L)} - X_{A(0)}}{\exp(\dot{N}_T L / C D_A) - 1} \right] \tag{vi}$$

which in the limit of $\dot{N}_T \to 0$ approaches

$$\dot{N}_A = -\frac{CD_A}{L}(X_{A(L)} - X_{A(0)}) \qquad \text{(vii)}$$

and Eq. (vii) is the same as Eq. (11.48).

For a system in which $L = 0.02$ m, $D_A = 3 \times 10^{-5}$ m^2/s, $X_{A(0)} = 0.1$, $X_{A(L)} = 0.01$, and the A–B mixture is an ideal gas at 50°C and 101.3 kPa,

$$C = \frac{P}{RT} = \frac{101,300}{8.3144 \times 323} = 37.72 \text{ g mol/m}^3$$

and Eq. (vi) gives

$$\dot{N}_A = \dot{N}_T \left[0.1 + \frac{0.09}{\exp(17.67\dot{N}_T) - 1} \right]$$

which is shown as a log-log plot in Fig. 11.12. With decreasing $\dot{N}_T$, $\dot{N}_A$ asymptotically approaches the value of 5×10^{-3} g mol·m^{-2}·s^{-1} (log $\dot{N}_A = -2.29$), which is the same as that obtained from Eq. (11.48):

$$\dot{N}_A = -\frac{CD_A}{L}(X_{A(L)} - X_{A(0)})$$

$$= \frac{-37.73 \times 3 \times 10^{-5}}{0.02}(0.01 - 0.1)$$

$$= 5 \times 10^{-3} \text{ g mol·m}^{-2}\text{·s}$$

When the flux of B in the y-direction through the porous A is decreased to zero, the flux of A is given by Eq. (11.42) as

$$\dot{N}_{A,y} = \frac{DP}{RTL} \ln \frac{P - P_{A(L)}}{P - P_{A(0)}}$$

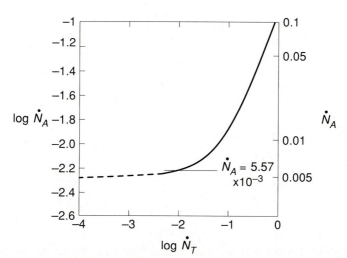

FIGURE 11.12.   Variation of the molar flux of A with total molar flux during mass transfer at large fluxes and large concentrations.

$$= \frac{3 \times 10^{-5} \times 101,300}{8.3144 \times 313 \times 0.02} \ln \frac{1 - 0.01}{1 - 0.1}$$

$$= 5.57 \times 10^{-3} \text{ g mol·m}^{-2}\text{·s}$$

and thus the line drawn in Fig. 11.12 has a physical significance only at values of $\dot{N}_A$ greater than 5.57×10^{-3} g mol·m^{-2}·s^{-1}. The concentration gradients across the film for $\dot{N}_T = 0.1, 0.05$, and 0.01 g mol·m^{-2}·s^{-1} are shown in Fig. 11.13.

11.12 Influence of Mass Transport on Heat Transfer in Stagnant Film

Figure 11.14 shows a mixture of hot gas B and vapor A in contact with the surface of a cooler solid A. The difference in temperature across the film causes the transfer of heat by conduction from the bulk gas to the surface of the solid and the concentration gradient of A in the film causes the mass transport of A by diffusion from the surface to the bulk gas. At finite temperatures the atoms of A have finite enthalpies, and thus the molar flux of A through the film is also an enthalpy flux, which contributes to the total heat flux and thus influences the temperature gradient across the film. The enthalpy flux caused by the molar flux of A is given by $\dot{N}_A H_A$, in which H_A is the enthalpy of A vapor per gram mole. An energy balance at steady state on the control volume shown in Fig. 11.14 is thus

(net heat flux by conduction) + (net enthalphy flux) = 0

or

$$(q'_y|_{y+\Delta y} - q'_y|_y) \, \Delta x \, \Delta z + (\dot{N}_A H_A|_{y+\Delta y} - \dot{N}_A H_A|_y) \, \Delta x \, \Delta z = 0$$

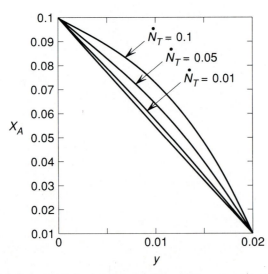

FIGURE 11.13. Influence of the total molar flux on the concentration profile of A in the film on the surface of porous solid A.

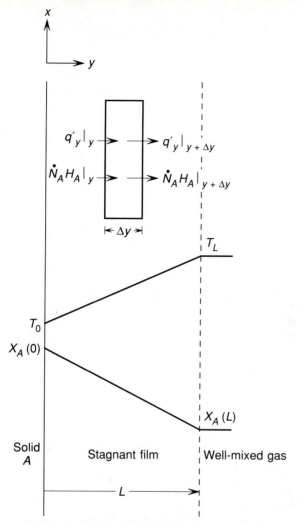

FIGURE 11.14. Temperature and concentration profiles in the film arising from simultaneous mass and heat transport through the film.

Dividing by $\Delta x\,\Delta y\,\Delta z$ and allowing Δz to approach zero gives

$$\frac{dq'_y}{dy} + \dot{N}_A \frac{dH_A}{dy} = 0$$

or for a constant thermal conductivity, k, and a constant heat capacity of A, $C_{p,A}$,

$$-k\frac{d^2T}{dy^2} + \dot{N}_A C_{p,A} \frac{dT}{dy} = 0$$

or

$$\frac{d^2T}{dy^2} - \frac{\dot{N}_A C_{p,A}}{k}\frac{dT}{dy} = 0$$

which is mathematically identical with Eq. (11.49) and has the solution

$$T = C_2 e^{ry} + C_1 \tag{i}$$

in which

$$r = \frac{\dot{N}_A C_{p,A}}{k}$$

Thus with the boundary conditions

$$T = \begin{cases} T_0 & \text{at } y = 0 \\ T_L & \text{at } y = L \end{cases}$$

Eq. (i) becomes

$$T = T_0 + \frac{T_L - T_0}{\exp(rL) - 1}[\exp(ry) - 1] \tag{ii}$$

Thus at $y = 0$, the heat flux due to conduction is

$$q_{\text{cond}} = -k\frac{dT}{dy}\bigg|_{y=0} = \frac{-k(T_L - T_0)}{\exp(rL) - 1}r$$

$$= \frac{-\dot{N}_A C_{p,A}(T_L - T_0)}{\exp(\dot{N}_A C_{p,A}L/k) - 1}$$

and the enthalpy flux due to the molar transport of A across the film from $y = 0$ to $y = L$ is

$$\dot{N}_A C_{p,A}(T_L - T_0)$$

The total heat flux is thus

$$q'_{\text{total}} = \frac{-\dot{N}_A C_{p,A}(T_L - T_0)}{\exp(\dot{N}_A C_{p,A}L/k) - 1} + \dot{N}_A C_{p,A}(T_L - T_0)$$

$$= \dot{N}_A C_{p,A}(T_L - T_0)\left[1 - \frac{1}{\exp(\dot{N}_A C_{p,A}L/k) - 1}\right] \tag{iii}$$

and the heat transfer coefficient, h_x, is identified as

$$h_x = \dot{N}_A C_{p,A}\left[1 - \frac{1}{\exp(\dot{N}_A C_{p,A}L/k) - 1}\right]$$

As $\dot{N}_A$, and hence r, approach zero, Eq. (ii) approaches

$$T = T_0 + (T_L - T_0)\frac{y}{L} \tag{iv}$$

which is the linear variation occurring when heat transport in a stagnant film of constant thermal conductivity, k, is by conduction alone. In comparing Eqs. (ii) and (iv) for a positive value of $\dot{N}_A$ (and hence r),

$$\frac{\exp(ry) - 1}{\exp(rL) - 1} < \frac{y}{L}$$

and thus T_y in Eq. (ii) is less than T_y in Eq. (iv). The enthalpy flux thus causes the temperature profile in the film to be convex downward.

If a steep temperature gradient exists from the surface of a hot liquid A to a cold gas B, the A that evaporates from the surface and diffuses toward the bulk gas can condense within the film, which, by decreasing the value of X_A at the point of condensation, increases the gradient of X_A from the surface and hence increases the diffusion flux of A. Furthermore, if the concentration of A vapor in the film is high enough, the latent heat of condensation of A evolved within the film can influence the shape of the temperature profile in the film.

11.13 Diffusion into a Falling Film of Liquid

Figure 11.15 shows a film of viscous liquid B of thickness δ flowing down a vertical wall. The free surface of the liquid is in contact with a gas A which is slightly soluble in the liquid. The gas dissolves in the liquid and is transported by diffusion and by bulk flow of the viscous liquid. A steady-state molar balance for A in the control

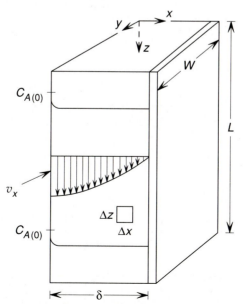

FIGURE 11.15.   Diffusion of gas A into a falling film of viscous liquid B.

volume shown in Fig. 11.15 gives

$$\dot{N}_{A,x}|_x \, \Delta z \, \Delta y - \dot{N}_{A,x}|_{x+\Delta x} \, \Delta z \, \Delta y + \dot{N}_{A,z}|_z \, \Delta x \, \Delta y - \dot{N}_{A,z}|_{z+\Delta z} \, \Delta x \, \Delta y = 0$$

Dividing by the volume $\Delta x \, \Delta y \, \Delta z$ and allowing Δx, Δy, and Δz to approach zero gives

$$\frac{\partial \dot{N}_{A,x}}{\partial x} + \frac{\partial \dot{N}_{A,z}}{\partial z} = 0 \tag{i}$$

From Eq. (11.13),

$$\dot{N}_{A,x} = C_A v_{A,x} = J^*_{A,x} + C_A v^*_x$$

in which $J^*_{A,x}$ is the molar flux of A in the x-direction due to diffusion down the concentration gradient and $C_A v^*_x$ is the molar flux of A due to bulk flow in the x-direction. If the solubility of A is small enough that its presence in the liquid has a negligible influence on the density and the viscosity of the liquid, bulk flow in the x-direction is negligible and hence

$$\dot{N}_{A,x} \doteqdot J^*_{A,x} = -D \frac{\partial C_A}{\partial x} \tag{ii}$$

For molar transport in the z-direction

$$\dot{N}_{A,z} = J^*_{A,z} + C_A v^*_z = -D \frac{\partial C_A}{\partial z} + C_A v^*_z$$

In this case, because of the low solubility, the concentration gradient of A in the z-direction at any x is small enough that the transport of A by diffusion in the z-direction is negligible in comparison with that caused by bulk flow of the liquid. Thus

$$\dot{N}_{A,z} \doteqdot C_A v^*_z \tag{iii}$$

As the presence of A has a negligible influence on the physical properties of B, v^*_z is the local velocity in the bulk flow, given by Eq. (2.15) as

$$v_z = \frac{\rho g}{2\eta} (\delta^2 - x^2) \tag{2.15}$$

from which

$$v_{z,max} = \frac{\rho g}{2\eta} \delta^2$$

and hence

$$v_z = v_{z,max} \left[1 - \left(\frac{x}{\delta} \right)^2 \right] \tag{iv}$$

Substitution of Eqs. (ii), (iii), and (iv) in Eq. (i) then gives

$$-D \frac{\partial^2 C_A}{\partial x^2} = v_{z,\max}\left[1 - \left(\frac{x}{\delta}\right)^2\right]\frac{\partial C_A}{\partial z} \qquad \text{(v)}$$

The solution of Eq. (v) is simplified if the extent of the penetration of A into the liquid is small, in which case A is being transported by bulk flow at a velocity that has a value very close to $v_{z,\max}$. Small penetrations of A are achieved with short times of exposure of the surface of the liquid to the gas phase. The time of this exposure decreases with increasing $v_{z,\max}$ and decreasing L, and hence short exposure times are achieved with low values of $L/v_{z,\max}$, and under this constraint Eq. (v) can be approximated by

$$v_{z,\max}\frac{\partial C_A}{\partial z} = D\frac{\partial^2 C_A}{\partial x^2} \qquad \text{(iv)}$$

Equation (iv) has the mathematical form of Eqs. (8.13) and (10.12) and is dealt with by introducing the function η, given as

$$\eta = \frac{x}{2\sqrt{Dz/v_{z,\max}}} \qquad \text{(v)}$$

which is analogous with $\eta = x/2\sqrt{\alpha t}$ in the solution of Eq. (8.13) and $\eta = x/2\sqrt{Dt}$ in the solution of Eq. (10.12). Then

$$\frac{\partial C_A}{\partial z} = \frac{\partial \eta}{\partial z}\frac{dC_A}{d\eta} = -\frac{x}{4\sqrt{D/v_{z,\max}}\,z^{3/2}}\frac{dC_A}{d\eta} = -\frac{\eta}{2z}\frac{dC_A}{d\eta}$$

and

$$\frac{\partial^2 C_A}{\partial x^2} = \frac{\partial}{\partial x}\left(\frac{\partial \eta}{\partial x}\frac{dC_A}{d\eta}\right) = \frac{\partial}{\partial x}\left(\frac{1}{2\sqrt{Dz/v_{z,\max}}}\frac{dC_A}{d\eta}\right)$$

$$= \frac{d}{d\eta}\left(\frac{1}{2\sqrt{Dz/v_{z,\max}}}\frac{1}{2\sqrt{Dz/v_{z,\max}}}\frac{dC_A}{d\eta}\right)$$

$$= \frac{v_{z,\max}}{4Dz}\frac{d^2 C_A}{d\eta^2}$$

substitution of which into Eq. (iv) gives the ordinary differential equation

$$-2\eta\frac{dC_A}{d\eta} = \frac{d^2 C_A}{d\eta^2} \qquad \text{(vi)}$$

Designating

$$\frac{dC_A}{d\eta} = y$$

gives

$$-2\eta y = \frac{dy}{d\eta}$$

or

$$\frac{dy}{y} = -2\eta \, d\eta$$

integration of which gives

$$\ln y = -\eta^2 + \ln A$$

where $\ln A$ is the integration constant, or

$$\ln \frac{y}{A} = -\eta^2$$

or

$$y = \frac{dC_A}{d\eta} = Ae^{-\eta^2}$$

and integrating again gives

$$\int dC_A = A \int e^{-\eta^2} \, d\eta \qquad \text{(vii)}$$

The boundary conditions are:

1. $C_A = 0$ at $z = 0$ ($\eta = \infty$)
2. $C_A = C_{A(s)}$ at $x = 0$ ($\eta = 0$)
3. $C_A = 0$ at $x = \infty$ ($\eta = \infty$)

Thus integrating Eq. (vii) between the boundary conditions 2 and 3 gives

$$\int_{C_{A(s)}}^{0} dC_A = A \int_{0}^{\infty} e^{-\eta^2} \, d\eta$$

or

$$-C_{A(s)} = A \frac{\sqrt{\pi}}{2}$$

and thus

$$A = \frac{-2C_{A(s)}}{\sqrt{\pi}}$$

and Eq. (vii) becomes

$$\int dC_A = \frac{-2C_{A(s)}}{\sqrt{\pi}} \int e^{-\eta^2} \, d\eta$$

which between the limits $C_A = C_A$ at $\eta = \eta$ and $C_A = C_{A(s)}$ at $\eta = 0$ gives

$$C_A - C_{A(s)} = \frac{-2C_{A(s)}}{\sqrt{\pi}} \frac{\sqrt{\pi}}{2} \, \text{erf}(\eta)$$

$$= -C_{A(s)} \, \text{erf}(\eta)$$

or

$$\frac{C_A}{C_{A(s)}} - 1 = -\text{erf}(\eta)$$

or

$$\frac{C_A}{C_{A(s)}} = 1 - \text{erf}(\eta) = \text{erfc}(\eta)$$

$$= \text{erfc}\,\frac{x}{2\sqrt{Dz/v_{z,\max}}} \qquad \text{(viii)}$$

The molar flux of A into the film at any value of z is

$$\dot{N}_{A,x} = -D\,\frac{\partial C_A}{\partial x}\bigg|_{x=0} \qquad \text{(ix)}$$

From Eq. (viii),

$$\frac{\partial C_A}{\partial x} = C_{A(s)}\,\frac{\partial}{\partial x}\left(1 - \frac{2}{\sqrt{\pi}}\int e^{-\eta^2}\,d\eta\right)$$

$$= -C_{A(s)}\,\frac{1}{2\sqrt{Dz/v_{z,\max}}}\,\frac{2}{\sqrt{\pi}}\,\frac{\partial}{\partial x}\left(\int e^{-\eta^2}\,dx\right)$$

which evaluated at $x = 0$ is

$$\frac{\partial C_A}{\partial x}\bigg|_{x=0} = -C_{A(s)}\sqrt{\frac{v_{z,\max}}{Dz\pi}} \qquad \text{(x)}$$

and thus Eq. (ix) is

$$\dot{N}_{A,x} = C_{A(s)}\sqrt{\frac{Dv_{z,\max}}{z\pi}}$$

The rate at which A is transported into the film is thus

$$\dot{W}_{A,x} = \int_0^W \int_0^L \dot{N}_{Ax|x=0}\,dz\,dy \qquad \text{g mol/s}$$

$$= 2WC_{A(s)}\sqrt{\frac{Dv_{z,\max}L}{\pi}} \qquad \text{(xi)}$$

Application of Eq. (x) to the film model, in which the film has a thickness L, gives

$$\left[\frac{\partial C_A}{\partial x}\right]_{x=0} = \frac{0 - C_{A(s)}}{L} = -C_{A(s)}\sqrt{\frac{v_{z,\max}}{Dz\pi}}$$

or

$$L = 1.77\sqrt{\frac{Dz}{v_{z,\max}}}$$

Alternatively, the value of x at which C_A has decreased to 1% of $C_{A(s)}$ is obtained from Eq. (viii) as

$$\frac{C_A}{C_{A(s)}} = 0.01 = \text{erfc} \frac{x}{2\sqrt{Dz/v_{z,\max}}}$$

or

$$x = 1.82 \left(2\sqrt{\frac{Dz}{v_{z,\max}}} \right)$$

$$= 3.64 \sqrt{\frac{D_z}{v_{z,\max}}}$$

which is twice the calculated value of L.

In the case of dissolved gas A being removed from the falling film by exposure to a vacuum, which decreases the value of C_A at $x = 0$ to zero, the boundary conditions for integration of Eq. (vii) are:

1. $C_A = C_{A,b}$ at $z = 0$ ($\eta = \infty$)
2. $C_A = 0$ at $x = 0$ ($\eta = 0$)
3. $C_A = C_{A,b}$ at $x = 0$ ($\eta = \infty$)

in which $C_{A,b}$ is the concentration of A in the liquid before exposure to the vacuum. Thus integrating Eq. (vii) between boundary conditions (ii) and (iii) gives

$$\int_0^{C_{A,b}} = A \int_0^\infty e^{-\eta^2}\, d\eta$$

or

$$A = \frac{2C_{A,b}}{\sqrt{\pi}}$$

and Eq. (vii) becomes

$$\int dC_A = \frac{2C_{A,b}}{\sqrt{\pi}} \int e^{-\eta^2}\, d\eta$$

which between the limits $C_A = C_A$ at $\eta = \eta$ and $C_A = 0$ at $\eta = 0$ gives

$$C_A = C_{A,b}\, \text{erf}(\eta)$$

$$= C_{A,b}\, \text{erf} \left(\frac{x}{2\sqrt{Dz/v_{z,\max}}} \right) \qquad \text{(xii)}$$

The molar flux of A from the film is thus

$$\dot{N}_{A,x} = -C_{A,b} \sqrt{\frac{Dv_{z,\max}}{\pi z}} \qquad \text{(xiii)}$$

and the total number of moles transported from the film in unit time is

$$2WC_{A,b}\sqrt{\frac{Dv_{z,\max}L}{\pi}} \qquad \text{(xiv)}$$

EXAMPLE 11.4

Hydrogen is removed from a liquid metal by allowing a thin film of the melt to flow down an inclined plane while being exposed to a vacuum. Consider a thin film of metal flowing, at the mass flow rate of 0.15 kg/s, down a plane of length 1 m and width 0.2 m that is inclined at an angle of 1° from the horizontal in which the initial concentration of hydrogen is 40 g mol/m³. Calculate the fraction of hydrogen in the melt that is removed by exposure of the film to a vacuum while it flows down the plane. Data for the metal are $\rho = 7900$ kg/m³, $\eta = 6 \times 10^{-3}$ Pa·s, and $D_H = 1.3 \times 10^{-8}$ m²/s.

From Eq. (2.18) for fluid flow down an inclined plane

$$\dot{M} = \frac{\rho^2 g \cos \theta \, \delta^3 W}{3\eta}$$

$$0.15 = \frac{7900^2 \times 9.81 \times \cos(89) \times \delta^3 \times 0.2}{3 \times 6 \times 10^{-3}} \qquad \text{(2.18)}$$

which gives the film thickness, δ, as 1.08×10^{-3} m. The average linear flow velocity is obtained as

$$\bar{v} = \frac{\dot{M}}{A\rho} = \frac{0.15}{0.2 \times 1.08 \times 10^{-3} \times 7900} = 0.0878 \text{ m/s}$$

and $v_{z,\max}$ is 1.5 times this value,

$$v_{z,\max} = 0.132 \text{ m/s}$$

The rate of removal from

$$1 \times 0.2 \times 1.08 \times 10^{-3} = 2.16 \times 10^{-4} \text{ m}^3 \text{ of metal}$$

is then obtained from Eq. (xiv) as

$$2WC_{H,b}\sqrt{\frac{D_H v_{z,\max}L}{\pi}} = 2 \times 0.2 \times 40 \sqrt{\frac{1.3 \times 10^{-8} \times 0.132}{3.142}}$$

$$= 3.74 \times 10^{-4} \text{ g mol/s}$$

which when multiplied by the time of exposure to the vacuum,

$$\frac{L}{\bar{v}} = \frac{1}{0.087} = 11.4 \text{ s}$$

gives the total number of gram moles of H removed as

$$3.74 \times 10^{-4} \times 11.4 = 4.26 \times 10^{-3}$$

Initially, the number of gram moles of H in 2.16×10^{-4} m³ of metal was

$$40 \times 2.16 \times 10^{-4} = 8.64 \times 10^{-3}$$

and hence the fraction removed by vacuum degassing is

$$\frac{4.26 \times 10^{-3}}{8.64 \times 10^{-3}} = 0.49$$

11.14 Diffusion and the Kinetic Theory of Gases

Application of the kinetic theory of gases to calculation of the viscosity and thermal conductivity of a gas involved consideration of the transfer of momentum and energy, respectively, when spherical atoms of the same size and mass collide with one another, and both cases considered processes occurring in a one-component gas. Consideration of diffusion differs from that of viscosity and thermal conductivity in that interdiffusion down concentration gradients requires the presence of at least two different kinds of atom or molecule, and this complicates the mathematics. Thus, to simplify the mathematics and still show the principles of the application of the kinetic theory of gases to interdiffusion phenomena, an artificially simple system will be examined. The system is the binary gas $A-A^*$, in which the atoms of A and A^* are rigid spheres of the same diameter, d, and the same mass, m. Thus, from kinetic theory, the mean speed of the atoms, $\bar{c}$, is

$$\bar{c} = \left(\frac{8kT}{\pi m}\right)^{1/2}$$

the number of atoms passing through unit area in unit time is

$$\tfrac{1}{6}n\bar{c}$$

and the mean free path of the atoms is

$$\lambda = \frac{1}{\sqrt{2}\,\pi d^2 n}$$

in which n is the number of atoms per unit volume. The concentration of the gas mixture in gram moles per unit volume is thus $C = n/N_0$, and the concentrations of A and A^* are, respectively, $C_A = X_A C$ and $C_A^* = X_A^* C$. Figure 11.16, which is similar to Figs. 2.25 and 6.17, shows three "layers" of gas atoms separated from each other by the mean free path, and a gradient in X_A in the y-direction. In the absence of bulk flow in the gas, the flux of A^* in g mol·m²·s⁻¹ at the center plane in the y-direction is

$$J_{A,*y} = \frac{1}{N_0}\,(\tfrac{1}{6}n\bar{c}X_A^*|_{y-\lambda} - \tfrac{1}{6}n\bar{c}X_A^*|_{y+\lambda})$$

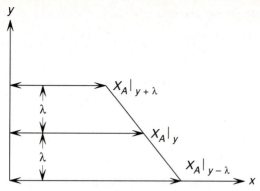

FIGURE 11.16. Construction for application of the kinetic theory of gases to calculation of the interdiffusion coefficient in a binary gas mixture.

in which

$$X_A^*\big|_{y-\lambda} = X_A^*\big|_y - \lambda \frac{dX_A^*}{dy}$$

and

$$X_A^*\big|_{y+\lambda} = X_A^*\big|_y + \lambda \frac{dX_A^*}{dy}$$

Thus

$$
\begin{aligned}
J_{A,*y} &= \frac{1}{N_0}\left(-\frac{1}{6}\,n\bar{c}\lambda\,\frac{dX^*_A}{dy} - \frac{1}{6}\,n\bar{c}\lambda\,\frac{dX^*_A}{dy}\right) \\
&= -\frac{2}{3}\,C\bar{c}\lambda\,\frac{dX^*_A}{dy} \\
&= -\frac{2}{3}\,\bar{c}\lambda\,\frac{dC^*_A}{dy}
\end{aligned}
$$

which on comparison with Fick's first law of diffusion gives the interdiffusion coefficient as

$$
\begin{aligned}
D &= \tfrac{2}{3}\bar{c}\lambda \\
&= \frac{2}{3}\left(\frac{8kT}{\pi m}\right)^{1/2}\frac{1}{\sqrt{2}\,\pi d^2 n}
\end{aligned}
\tag{11.50}
$$

Equation (11.50) shows that in contrast with viscosity and thermal conductivity, the interdiffusion coefficient is a function of n, the concentration of the gas, and hence is a function of pressure. If the gas mixture behaves as an ideal gas,

$$n \text{ (atoms per unit volume)} = \frac{P}{kT}$$

and thus Eq. (11.50) can be written as

$$D = \frac{2}{3}\left(\frac{k^3}{\pi^3 m}\right)^{1/2}\frac{T^{3/2}}{Pd^2} \tag{11.51}$$

which is the interdiffusion coefficient for a mixture of two species of rigid spherical atoms of identical mass and diameter. The result for rigid spherical atoms of A and B with unequal masses and diameters is

$$D = \frac{2}{3}\left(\frac{k^3}{\pi^3}\right)^{1/2}\left(\frac{1}{2m_A} + \frac{1}{2m_B}\right)^{1/2}\frac{T^{3/2}}{P[(d_A + d_B)/2]^2} \tag{11.52}$$

The Chapman–Enskog expression for the interdiffusion coefficient of a binary gas is

$$CD = \frac{2.2646 \times 10^{-5}\sqrt{T(1/M_A + 1/M_B)}}{\sigma_{AB}^2\Omega_D} \tag{11.53}$$

which if the gas is ideal can be written as

$$D = \frac{1.8583 \times 10^{-3}\sqrt{T^3(1/M_A + 1/M_B)}}{P\sigma_{AB}^2\Omega_D} \tag{11.54}$$

The units in Eqs. (11.53) and (11.54) are

$$D,\ cm^2/s$$

$$T,\ K$$

$$M,\ g/g\ mol$$

$$P,\ atm$$

$$C,\ g\ mol/cm^3$$

$$\sigma_{AB},\ angstroms$$

and Ω_D is a dimensionless function of the temperature and the interaction potential between an atom (or molecule) of A and an atom (or molecule) of B. The interaction is taken as being that given by the Lennard-Jones function

$$\phi_{AB} = 4\epsilon_{AB}\left[\left(\frac{\sigma_{AB}}{r}\right)^{12} - \left(\frac{\sigma_{AB}}{r}\right)^{6}\right]$$

in which for nonpolar and nonreacting pairs, it is assumed that the collision diameter, σ_{AB}, is the arithmetic average of σ_A and σ_B,

$$\sigma_{AB} = \frac{\sigma_A + \sigma_B}{2} \tag{11.55}$$

and the interaction energy ϵ_{AB} is the geometric average of ϵ_A and ϵ_B,

$$\epsilon_{AB} = \sqrt{\epsilon_A\epsilon_B} \tag{11.56}$$

The dependence of Ω_D on kT/ϵ is listed in Table 11.1 and it is to be noted that Ω_D

TABLE 11.1

kT/ϵ_{AB}	$\Omega_{D,AB}$	kT/ϵ_{AB}	$\Omega_{D,AB}$
0.30	2.662	2.60	0.9878
0.35	2.476	2.70	0.9770
0.40	2.318	2.80	0.9672
0.45	2.184	2.90	0.9576
0.50	2.066	3.00	0.9490
0.55	1.966	3.10	0.9406
0.60	1.877	3.20	0.9328
0.65	1.798	3.30	0.9256
0.70	1.729	3.40	0.9186
0.75	1.667	3.50	0.9120
0.80	1.612	3.60	0.9058
0.85	1.562	3.70	0.8998
0.90	1.517	3.80	0.8942
0.95	1.476	3.90	0.8888
1.00	1.439	4.00	0.8836
1.05	1.406	4.10	0.8788
1.10	1.375	4.20	0.8740
1.15	1.346	4.30	0.8694
1.20	1.320	4.40	0.8652
1.25	1.296	4.50	0.8610
1.30	1.273	4.60	0.8568
1.35	1.253	4.70	0.8530
1.40	1.233	4.80	0.8492
1.45	1.215	4.90	0.8456
1.50	1.198	5.0	0.8422
1.55	1.182	6.0	0.8124
1.60	1.167	7.0	0.7896
1.65	1.153	8.0	0.7712
1.70	1.140	9.0	0.7556
1.75	1.128	10.0	0.7424
1.80	1.116	20.0	0.6640
1.85	1.105	30.0	0.6232
1.90	1.094	40.0	0.5960
1.95	1.084	50.0	0.5756
2.00	1.075	60.0	0.5596
2.10	1.057	70.0	0.5464
2.20	1.041	80.0	0.5352
2.30	1.026	90.0	0.5256
2.40	1.012	100.0	0.5170
2.50	0.9996		

Source: J. O. Hirschfelder, R. B. Bird, and E. L. Spotz, *Chem. Rev.* (1949), vol. 44, p. 205.

TABLE 11.2

Gas Pair	T (K)	D (m²/s × 10⁵)
Ar–He	273	6.41
Ar–H$_2$	293	7.7
Ar–N$_2$	293	2.0
Ar–O$_2$	293	2.0
Ar–CO$_2$	293	1.4
N$_2$–H$_2$	273	6.74
N$_2$–O$_2$	273	1.81
N$_2$–CO	273	1.92
N$_2$–CO$_2$	273	1.44
H$_2$–O$_2$	273	6.97
H$_2$–CO	273	6.51
H$_2$–CO$_2$	273	5.50
CO–O$_2$	273	1.85
CO$_2$–O$_2$	273	1.39
CO$_2$–CO	273	1.37
H$_2$O–air	298	2.8
CO$_2$–air	298	1.6
H$_2$–air	298	4.1
O$_2$–air	298	2.1

Source: J. O. Hirschfelder, C. F. Curtiss, and R. B. Bird, *Molecular Theory of Gases and Liquids*, Wiley, New York, 1954, p. 579.

is not the same as Ω_η (and Ω_k) listed in Table 2.2. The experimentally measured interdiffusion coefficients for several gas pairs are listed in Table 11.2.

EXAMPLE 11.5

Calculate the interdiffusion coefficient for nitrogen and oxygen at 273.2 K and 1 atm pressure.

From Table 2.1,

	M	σ (Å)	ε/$\bar{k}$ (K)	
O$_2$	32.00	3.433		113
N$_2$	28.02	3.681		91.5

Therefore, from Eq. (11.55),

$$\sigma_{O_2-N_2} = \frac{3.433 + 3.681}{2} = 3.557$$

and from Eq. (11.56),

$$\frac{\epsilon}{\bar{k}} = (113 \times 91.5)^{1/2} = 102$$

or

$$\frac{\bar{k}T}{\epsilon} = \frac{273.2}{102} = 2.68$$

From interpolation in Table 11.1, $\Omega_D = 0.9792$, and thus from Eq. (11.54),

$$D = \frac{1.8583 \times 10^{-3}\sqrt{273.2^3(1/32 + 1/28.02)}}{1 \times 3.557^2 \times 0.9792}$$

$$= 0.175 \text{ cm}^2/\text{s}$$

which is in good agreement with the measured value of 0.181 cm²/s listed in Table 11.2.

Maximum Rate of Evaporation of a Liquid into a Vacuum

In a closed one-component system containing a liquid in equilibrium with its vapor, the vapor pressure is the saturated vapor pressure of the liquid, $p°$, at the temperature of interest and phase equilibrium requires that the rate of evaporation of the liquid equal the rate of condensation of the vapor. On the assumption that any atom in the vapor phase which strikes the surface of the liquid is captured by the liquid and is thus condensed, the rate of condensation equals the rate at which atoms in the vapor phase strike the surface of the liquid. It has been shown that the rate at which atoms in a gas or vapor pass through unit area in unit time is

$$\tfrac{1}{6}n\bar{c} \qquad\qquad\qquad (\text{i})$$

where n is the number of atoms per unit volume and $\bar{c}$ is the mean speed of the atoms. Equation (i) was derived using the simplifying assumption that the motions of all of the atoms in the gas are constrained to the x, y, or z directions. Elimination of this assumption and consideration of the random motions of the atoms gives the correctly calculated rate at which atoms pass through unit area in unit time as

$$\tfrac{1}{4}n\bar{c}$$

At the vapor pressure $p°$ and the temperature T,

$$n = \frac{p°}{\bar{k}T}$$

and thus the rate at which atoms in the vapor phase strike the unit area of the surface of the liquid in unit time $\dot{n}$ is

$$\dot{n} = \frac{1}{4}\frac{p°}{\bar{k}T}\left(\frac{8\bar{k}T}{\pi m}\right)^{1/2}$$

$$= \frac{p^\circ}{\sqrt{2\pi k T m}}$$

The rate of condensation in units of atoms per unit area per unit time is thus

$$\frac{p^\circ N_0}{\sqrt{2\pi RTM}}$$

or in units of gram moles per unit area per unit time $\dot{N}$,

$$\dot{N} = \frac{\dot{n}}{N_0} = \frac{p^\circ}{\sqrt{2\pi RTM}} \tag{11.57}$$

which by virtue of the liquid–vapor phase equilibrium is also the rate of evaporation of the liquid. As the saturated vapor pressure of the liquid is determined by the intrinsic rate of evaporation, Eq. (11.57), which is known as the *Langmuir equation*, gives the maximum rate at which a liquid can evaporate into a vacuum.

EXAMPLE 11.6 —————————————————————————————————————

In Example 11.1 the rate of evaporation of liquid manganese into argon at 1600°C was 5.84×10^{-3} g mol·m^{-2}·s^{-1}. Calculate the maximum rate at which manganese can evaporate into a vacuum at 1600°C.

With a saturated vapor pressure of 5370 Pa at 1600°C and an atomic weight of 0.05494 kg/g mol, Eq. (11.57) gives

$$\dot{N}_{Mn}(max) = \frac{5370}{\sqrt{2 \times \pi \times 1873 \times 0.05494}}$$

$$= 73 \text{ g mol·m}^{-2}\cdot\text{s}^{-1}$$

Hirschfelder et al. (J. O. Hirschfelder, C. F. Curtiss, and R. B. Bird, *Molecular Theory of Gases and Liquids*, Wiley, New York, 1954, p. 244) have shown that, using the force constants for the Lennard-Jones potential, the reduced pressure, defined as

$$P* = \frac{P\sigma^3}{\epsilon}$$

the reduced volume, defined as

$$V* = \frac{V}{N_0\sigma^3}$$

and the reduced temperature, defined as

$$T* = \frac{kT}{\epsilon}$$

are approximately constants when evaluated at the critical points of nonpolar and almost spherical molecules heavier than H_2 and He. This is illustrated in Table 11.3.

TABLE 11.3

Gas	T_{cr} (K)	V_{cr} (cm³)	P_{cr} (atm)	T_{cr}^*	V_{cr}^*	P_{cr}^*
He	5.3	57.8	2.26	0.52	5.75	0.027
H_2	33.3	65.0	12.8	0.90	4.30	0.064
Ne	44.5	41.7	25.9	1.25	3.33	0.111
Ar	151	75.2	48	1.26	3.16	0.116
Xe	289.81	120.2	57.89	1.31	2.90	0.132
N_2	126.1	90.1	33.5	1.33	2.96	0.131
O_2	154.4	74.4	49.7	1.31	2.69	0.142
CH_4	190.7	99.0	45.8	1.29	2.96	0.126

Source: J. O. Hirschfelder, C. F. Curtiss, and R. B. Bird, *Molecular Theory of Gases and Liquids*, Wiley, New York, 1954, p. 245.

The approximation permits the estimation of the force constants from knowledge of the critical point

$$P_{cr}^* = \frac{P_{cr}\sigma^3}{\epsilon} = 0.126 \tag{i}$$

$$V_{cr}^* = \frac{V_{cr}}{N_0\sigma^3} = 3.00 \tag{ii}$$

$$T_{cr}^* = \frac{kT_{cr}}{\epsilon} = 1.29 \tag{iii}$$

Thus from Eq. (iii),

$$\frac{\epsilon}{k} = \frac{T_{cr}}{T_{cr}^*} = 0.77T_{cr} \tag{11.58}$$

and from Eq. (ii),

$$\sigma = \left(\frac{V_{cr}}{3N_0}\right)^{1/3}$$

$$= \left(\frac{10^8 \times 3}{3 \times 6.0232 \times 10^{23}}\right)^{1/3} V_{cr}^{1/3}$$

$$= 0.821 V_{cr}^{1/3} \tag{11.59}$$

in which σ is in units of angstroms and V_{cr} is cm³/g mol. Other empirical relations include

$$T_{cr} = 1.5T_b = 2.5T_m \tag{iv}$$

in which T_m and T_b are the normal melting and boiling temperatures of the chemical compound or element, and

$$V_{cr} = 2.66V_{b,\text{liq}} = 3.12V_{m,\text{sol}} \tag{v}$$

in which $V_{b,\text{liq}}$ is the molar volume of the liquid state at the normal boiling temperature and $V_{m,\text{sol}}$ is the molar volume of the solid state at the normal melting temperature. Combination of Eqs. (11.58), (11.59), (iv), and (v) give

$$\frac{\epsilon}{k} = 0.77T_{\text{cr}} = 1.92T_m = 1.115T_b \qquad (11.60)$$

and

$$\sigma = 0.821V_{\text{cr}}^{1/3} = 1.14V_{b,\text{liq}}^{1/3} = 1.20V_{m,\text{sol}}^{1/3} \qquad (11.61)$$

The molar volumes of some metals at melting and boiling temperatures and the estimated force constants are listed in Table 11.4. Although agreement between the values of σ calculated from $V_{m,\text{sol}}$ and $V_{b,\text{liq}}$ is generally good, the agreement between the values of ϵ/k calculated from T_m and T_b can be poor, as is the case with Ga, In, Li, Pu, and Sn.

TABLE 11.4

Metal	T_m (°C)	T_b (°C)	Molar Volume (cm³/g mol)		σ (Å)		ϵ/k	
			$V_{m,\text{sol}}$	$V_{b,\text{liq}}$	From $V_{m,\text{sol}}$	From $V_{b,\text{liq}}$	From T_m	From T_b
Ag	960.8	2163	11.02	13.06	2.67	2.68	2375	2816
Al	660	2057	10.49	13.26	2.63	2.69	1796	2691
Bi	271	1477	—	24.07	—	3.28	1047	2021
Cd	321	765	13.34	14.89	2.85	2.80	1143	1199
Co	1493	2877	7.14	8.98	2.31	2.36	3400	3638
Cu	1083	2570	7.58	9.15	2.36	2.38	2610	3284
Fe	1535	2833	7.60	9.55	2.36	2.41	3480	3587
Ga	29.8	1983	—	13.87	—	2.73	583	2606
Hg	−38.9	357	14.09	15.71	2.90	2.85	451	728
In	156.4	2087	15.94	19.98	3.02	3.09	827	2726
K	63.7	760	46.0	59.36	4.30	4.44	648	1193
Li	186	1317	13.27	16.76	2.84	2.91	884	1836
Mg	651	1103	14.66	20.30	2.94	3.10	1779	1589
Na	97.9	883	24.04	31.07	3.46	3.57	714	1335
Ni	1453	2816	7.11	9.02	2.31	2.37	3326	3568
Pb	327.4	1717	18.35	22.47	3.17	3.21	1156	2298
Pu	637	3300	14.20	18.78	2.91	3.02	1752	4127
Sb	630.5	1440	18.59	20.29	3.18	3.10	1739	1979
Sn	231.9	2770	16.51	21.63	3.06	3.17	972	3515
Tl	303	1457	—	20.64	—	3.12	1109	1998
Zn	419.5	906	9.56	10.19	2.55	2.47	1333	1362

Source: Adapted from E. T. Turkdogan, *Steelmaking: The Chipman Conference*, MIT Press, Cambridge, Mass., 1965, p. 77.

EXAMPLE 11.7 _____

Estimate the interdiffusion coefficient for mixtures of argon and zinc vapor at 1037 K and 1 atm pressure.

From Table 11.4, for zinc (of $M = 65.4$ g/g mol)

$$\sigma = \frac{2.55 + 2.47}{2} = 2.51 \qquad \text{and} \qquad \frac{\epsilon}{k} = \frac{1333 + 1362}{2} = 1348 \text{ K}$$

and from Table 2.2, for argon,

$$\sigma = 3.418 \text{ Å} \qquad \text{and} \qquad \frac{\epsilon}{k} = 124 \text{ K}$$

Therefore,

$$\sigma_{Zn-Ar} = \frac{2.51 + 3.418}{2} = 2.96$$

and

$$\frac{\epsilon}{k} = (124 \times 1348)^{1/2} = 409$$

$$\frac{kT}{\epsilon} = \frac{1037}{409} = 2.54$$

From Table 11.1, $\Omega_D = 0.9945$, and thus from Eq. (11.54),

$$D = \frac{1.8583 \times 10^{-3}\sqrt{1037^3(1/39.9 + 1/65.4)}}{1 \times 2.96^2 \times 0.9945}$$

$$= 1.43 \text{ cm}^2/\text{s}$$

11.15 Mass Transfer Coefficient and Concentration Boundary Layer on a Flat Plate

Consideration of heat transfer from the surface of a solid at the temperature T_s to a fluid at the temperature T_∞ flowing over surface lead to the definition of the local heat transfer coefficient, h_x, at the location x on the surfaces as

$$q_x' = h_x(T_s - T_\infty)$$

The no-slip condition at the interface between the solid and fluid at $y = 0$, which requires that both v_x and v_y be zero at $y = 0$, requires that heat transfer across the layer of stagnant fluid in contact with the surface be by conduction, in which case

$$q_x' = -k\left(\frac{\partial T}{\partial y}\right)_{y=0}$$

and thus the local heat transfer coefficient was defined as

$$h_x = \frac{-k(\partial T/\partial y)_{y=0}}{T_s - T_\infty}$$

Consider now a fluid B flowing past the surface of solid A which is slightly soluble in the fluid B. If the fluid is not saturated with A, mass transport occurs as the dissolution of A into the liquid B or the sublimation of A into the gas B. If phase equilibrium is maintained at the interface between the solid and the fluid, the molar concentration of A at $x = 0$, $C_{A,s}$, is the solubility limit of A in B, and if the molar concentration of A in the bulk fluid is $C_{A,\infty}$, a local mass transfer coefficient, $h_{m,x}$, at the location x on the surface is defined as

$$\dot{N}_{A,y} = h_{m,x}(C_{A,s} - C_{A,\infty}) \tag{11.62a}$$

Alternatively, in terms of mass concentration of A,

$$\dot{n}_{A,y} = h_{m,x}(\rho_{A,s} - \rho_{A,\infty}) \tag{11.62b}$$

and $h_{m,x}$ has the units m/s. This analogy between heat and mass transfer is not exact because although heat can be transferred by pure conduction through the stagnant layer of fluid at $y = 0$, mass transport of A from the surface cannot occur if v_y (and hence v_x) are zero. However, the analogy is forced by assuming that v_y is zero at $y = 0$, which requires the assumption that the mass concentration of A in the momentum boundary layer on the surface is small enough that it does not influence the physical properties of the fluid and hence does not influence the characteristics of the momentum boundary layer. With this forced analogy mass transport at $y = 0$ occurs by diffusion, in which case

$$\dot{n}_{A,y} = j_{A,y}|_{y=0} = -D_A \left(\frac{\partial \rho_A}{\partial y}\right)_{y=0}$$

and thus

$$h_{m,x} = \frac{-D_A(\partial \rho_A/\partial y)_{y=0}}{\rho_{A,s} - \rho_{A,\infty}} \tag{11.63}$$

In heat transfer the Nusselt number is defined as the ratio of the contributions of convection and conduction to heat transport through the thermal boundary layer on the surface,

$$Nu_x = \frac{q'(\text{conv.})}{q'(\text{cond.})} = \frac{h_x x}{k}$$

in which x is a characteristic distance. The mass transfer equivalent of the Nusselt number is the Sherwood number, Sh_x, which is defined as the ratio of the contributions of convection and diffusion to mass transport through the concentration boundary layer. Considering diffusion of A through a stagnant film of thickness L, the mass flux is

$$j_{A,y} = \frac{D_A(\rho_{A,s} - \rho_{A,\infty})}{L}$$

and the mass flux by convection is

$$\dot{n}_{A,y} = h_{m,x}(\rho_{A,s} - \rho_{A,\infty})$$

The ratio of $\dot{n}_{A,y}$ to $j_{A,y}$ defines the mass transfer Nusselt number, or the Sherwood number, Sh, as

$$Sh = \frac{h_{m,x}L}{D_A}$$

The local Sherwood number, Sh_x, is defined as

$$Sh_x = \frac{h_{m,x}x}{D} \qquad (11.64)$$

in which x is a characteristic distance.

The dissolution of A into the flowing fluid B causes the formation of a concentration boundary layer, shown in Fig. 11.17, which is mathematically similar to the thermal boundary layer shown in Fig. 7.1. In Fig. 11.17 the concentration of A in the fluid B at any value of y near the plate increases with increasing x due to dissolution of the plate. Consequently, the gradients $(\partial \rho_A / \partial y)$ at the fixed value of y decrease with increasing x and also the gradients $(\partial \rho_A / \partial y)_{y=0}$ decrease with increasing x. The concentration boundary layer that develops over the plate has its upper

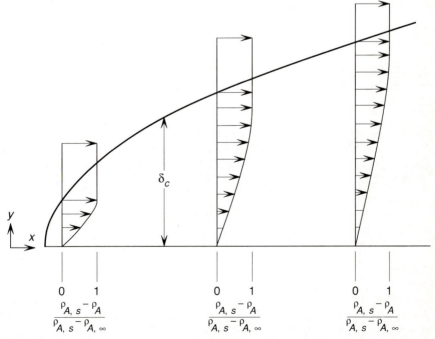

FIGURE 11.17. Normalized concentration profiles of A in the concentration boundary layer on the surface of solid A over which fluid B is flowing, when A is slightly soluble in B.

surface at values of $(\rho_{A,s} - \rho_A)/(\rho_{A,s} - \rho_{A,\infty}) = 0.99$, and as $(\partial\rho_A/\partial y)_{y=0}$ decreases with increasing x, the local rate of mass transport of A from the plate, and hence $h_{m,x}$ decrease with increasing x.

The steady-state mass balance for two-dimensional molar transport of A through a control volume in the concentration boundary is obtained from Eq. (11.23) as

$$v_x^* \frac{\partial C_A}{\partial x} + v_y^* \frac{\partial C_A}{\partial y} = -\frac{\partial J_{A,x}^*}{\partial x} - \frac{\partial J_{A,y}^*}{\partial y} - C_A \left(\frac{\partial v_x^*}{\partial x} + \frac{\partial v_y^*}{\partial y} \right)$$

The equivalent expression for mass transport is

$$v_x \frac{\partial \rho_A}{\partial x} + v_y \frac{\partial \rho_A}{\partial y} = -\frac{\partial j_{A,x}}{\partial x} - \frac{\partial j_{A,y}}{\partial y} - \rho_A \left(\frac{\partial v_x}{\partial x} + \frac{\partial v_y}{\partial y} \right)$$

For a fluid of constant density, the equation of continuity, Eq. (3.3), makes the term in parentheses zero, and thus for a constant value of D_A, the mass balance is

$$v_x \frac{\partial \rho_A}{\partial x} + v_y \frac{\partial \rho_A}{\partial y} = D_A \left(\frac{\partial^2 \rho_A}{\partial x^2} + \frac{\partial^2 \rho_A}{\partial y} \right) \tag{11.65}$$

which is the mass transport equivalent of the momentum boundary equation

$$v_x \frac{\partial v_x}{\partial x} + v_y \frac{\partial v_x}{\partial y} = \nu \left(\frac{\partial^2 v_x}{\partial x^2} + \frac{\partial^2 v_x}{\partial y^2} \right)$$

and the thermal boundary equation, Eq. (7.3),

$$v_x \frac{\partial T}{\partial x} + v_y \frac{\partial T}{\partial y} = \alpha \left(\frac{\partial^2 T}{\partial x^2} + \frac{\partial^2 T}{\partial z^2} \right) \tag{7.3}$$

In eq. (11.65) v_x and v_y are the x and y components of the local mass average velocity of the fluid $\mathbf{v}$, defined in Eq. (11.1) as

$$\rho \mathbf{v} = \rho_A \mathbf{v}_A + \rho_B \mathbf{v}_B \tag{11.1}$$

However, as the concentration of A in the boundary layer is low enough that its presence has a negligible influence on the physical properties of B, $\mathbf{v}$ is virtually the same as $\mathbf{v}_B$, the averaged value of the absolute velocities of the particles of B in the fluid, which, in turn, is the velocity of the fluid. In Eq. (11.1) as ρ_A approaches zero, $\mathbf{v}$ approaches $\mathbf{v}_B$.

Comparison of the momentum and thermal boundary layers was facilitated by introduction of the Prandtl number

$$Pr = \frac{\nu}{\alpha} \tag{7.4}$$

In a similar manner, comparison of the momentum and concentration boundary layers is facilitated by the Schmidt number, Sc, which is the ratio of viscous momentum transport and diffusion transfer, defined as

$$Sc = \frac{\nu}{D} \tag{11.66}$$

Comparison of the thermal and concentration boundary layers is facilitated by the Lewis number, Le:

$$Le = \frac{\alpha}{D} = \frac{Sc}{Pr} \qquad (11.67)$$

The exact solution of the momentum and concentration boundary equations for laminar flow of fluids of $(0.6 \leq Sc \leq 50)$ gives the local mass transfer coefficient, $h_{m,x}$, as

$$h_{m,x} = 0.332 D_{AB} Sc^{0.343} \sqrt{\frac{v_\infty}{\nu x}} \qquad (11.68)$$

and hence a local Sherwood number of

$$Sh_x = \frac{h_{m,x}}{D_{AB}} = 0.332\ Sc^{0.343} Re_x^{1/2} \qquad (11.69)$$

Equations (11.68) and (11.69) are equivalent to Eqs. (7.5) and (7.9), respectively. The average mass transfer coefficient $\overline{h}_{m,L}$ is obtained by averaging the values of $h_{m,x}$ between $x = 0$ and $x = L$ as

$$\overline{h}_{m,L} = \frac{1}{L} \int_0^L h_{m,x}\ dx$$

$$= 0.664\ D_{AB} Sc^{0.343} \sqrt{\frac{v_\infty}{\eta L}} \qquad (11.70)$$

and hence the average Sherwood number from $x = 0$ to $x = L$ is

$$\overline{Sh}_L = 0.664 Sc^{0.343} Re_L^{1/2} \qquad (11.71)$$

11.16 Approximate Integral Method

The application of the approximate integral method to mass transport in the concentration boundary layer is analogous to its application to heat transport in the thermal boundary layer. The control volume of length Δx and unit width has the concentration boundary layer as its upper surface, as shown in Fig. 11.18. At steady state A enters and leaves the control volume at equal rates. A enters the front face due to fluid flow at the rate

$$\int_0^{\delta_C} \rho_A v_x\ dy$$

and leaves through the rear face at the rate

$$\int_0^{\delta_C} \rho_A v_x\ dy + \frac{d}{dx} \int_0^{\delta_C} \rho_A v_x\ dy\ \Delta x$$

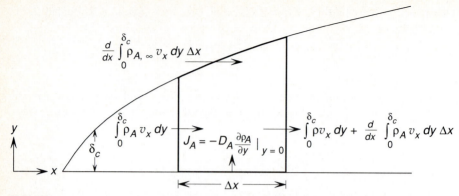

FIGURE 11.18. Control volume considered in the approximate integral method of determining the properties of the concentration boundary layer.

Thus A enters through the upper surface due to fluid flow at the rate

$$\frac{d}{dx} \int_0^{\delta_c} \rho_{A,\infty} v_x \, dy \, \Delta x$$

and A is transported into the control volume by dissolution of A at $y = 0$ at the rate

$$-D_A \left(\frac{\partial \rho_A}{\partial y} \right)_{y=0} \Delta x$$

The mass balance is thus

$$\frac{d}{dx} \int_0^{\delta_c} (\rho_A - \rho_{A,\infty}) v_x \, dy = -D_A \left(\frac{\partial \rho_A}{\partial y} \right)_{y=0} \tag{11.72}$$

which is equivalent to Eq. (7.11). Following the assumption made with respect to the shape of the velocity profile in the momentum boundary layer and the shape of the temperature profile in the thermal boundary layer, it is assumed that the concentration profile in the concentration boundary layer is of the form

$$\rho_{A,s} - \rho_A = ay + by^3$$

which with the boundary conditions

$$\rho_A = \rho_{A,\infty} \qquad \text{at } y = \delta_C$$

and

$$\frac{\partial \rho_A}{\partial y} = 0 \qquad \text{at } y = \delta_C$$

gives the concentration profile as

$$\frac{\rho_{A,s} - \rho_A}{\rho_{A,s} - \rho_{A,\infty}} = \frac{3}{2} \left(\frac{y}{\delta_C} \right) - \frac{1}{2} \left(\frac{y}{\delta_C} \right)^3 \tag{11.73}$$

which is equivalent to Eq. (7.14). From Eq. (11.73),

$$\rho_{A,\infty} - \rho_A = (\rho_{A,\infty} - \rho_{A,s}) \left[1 - \frac{3}{2} \left(\frac{y}{\delta_c} \right) + \frac{1}{2} \left(\frac{y}{\delta_c} \right)^3 \right]$$

and

$$\left(\frac{\partial \rho_A}{\partial y} \right)_{y=0} = (\rho_{A,\infty} - \rho_{A,s}) \frac{3}{2} \left(\frac{1}{\delta_c} \right) \qquad (11.74)$$

which when substituted along with

$$\frac{v_x}{v_\infty} = \frac{3}{2} \left(\frac{y}{\delta} \right) - \frac{1}{2} \left(\frac{y}{\delta} \right)^3 \qquad (3.19)$$

into Eq. (11.72) yields

$$\frac{d}{dx} \left\{ (\rho_A - \rho_{A,s}) v_\infty \delta \left[0.15 \left(\frac{\delta_c}{\delta} \right)^2 - 0.0107 \left(\frac{\delta_c}{\delta} \right)^4 \right] \right\} = D_A \left(\frac{\partial \rho_A}{\partial y} \right)_{y=0} \qquad (11.75)$$

which is the equivalent of Eq. (7.16). Manipulation of Eq. (11.75) in a manner identical with that used for Eq. (7.16) yields

$$\frac{\delta_c}{\delta} = \frac{1}{1.025 Sc^{1/3}} \qquad (11.76)$$

For $\delta_c = \delta$, $Sc = 0.929$, and hence elimination of the fourth-power term in Eq. (11.75) restricts the application of Eq. (11.76) to fluids of Schmidt number greater than 0.929. In the exact solution $\delta_c = \delta$ when $\nu = D_A$ (i.e., when $Sc = 1$) and the exact solution gives

$$\frac{\delta_c}{\delta} = \frac{1}{Sc^{1/3}} \qquad (11.77)$$

Application of Eq. (11.74) to the film model, in which the film has a thickness of L, gives

$$\left(\frac{\partial \rho_A}{\partial y} \right)_{y=0} = \frac{\rho_{A,\infty} - \rho_{A,s}}{L} = (\rho_{A,\infty} - \rho_{A,s}) \frac{3}{2} \left(\frac{1}{\delta_c} \right)$$

or

$$L = \tfrac{2}{3} \delta_c$$

The exact solution is limited to fluids of $0.6 \le Sc \le 50$. From Eq. (11.62),

$$h_{m,x} = \frac{-D_A (\partial \rho_A / \partial y)_{y=0}}{\rho_{A,s} - \rho_{A,\infty}}$$

which with Eq. (11.74) gives

$$
\begin{aligned}
h_{m,x} &= \frac{1.5D_A}{\delta_C} \\
&= \frac{1.5D_A \times 1.025Sc^{1/3}}{\delta} \\
&= 1.5D_A \times 1.025Sc^{1/3} \times \frac{1}{4.64}\left(\frac{v_\infty}{vx}\right)^{1/2} \\
&= 0.331D_A Sc^{1/3}\left(\frac{v_\infty}{vx}\right)^{1/2}
\end{aligned}
\tag{11.78}
$$

which is virtually identical with Eq. (11.68). It thus follows that

$$
\bar{h}_{m,L} = 0.662D_A Sc^{1/3}\left(\frac{v_\infty}{vL}\right)^{1/2}
\tag{11.79}
$$

$$
Sh_x = \frac{h_{m,x}x}{D_A} = 0.331Sc^{1/3}Re_x^{1/2}
\tag{11.80}
$$

and

$$
\overline{Sh}_L = \frac{\bar{h}_{m,L}L}{D_A} = 0.662Sc^{1/3}Re_L^{1/2}
\tag{11.81}
$$

which are the equivalents of Eqs. (7.24), (7.25) and (7.26).

From the Chapman–Enskog equation for the viscosities of gases, Eq. (2.35), the kinematic viscosity of a gas of molecular weight M is obtained as

$$
v = \frac{\eta}{\rho} = \frac{\eta}{CM} = \frac{2.6693 \times 10^{-5}}{C\sigma^2\Omega}\sqrt{\frac{T}{M}} \qquad g\cdot cm^{-1}\cdot s^{-1}
\tag{i}
$$

and division of Eq. (i) by the Chapman–Enskog equation, Eq. (11.53), for the interdiffusion coefficient of a binary gas in which $M_A = M_B = M$ and $\sigma_A = \sigma_B$ gives the Schmidt number as

$$
\begin{aligned}
Sc &= \frac{v}{D} = \frac{2.6693 \times 10^{-5}}{2.2646 \times 10^{-5}}\sqrt{\frac{1}{2}}\frac{\Omega_D}{\Omega} \\
&= 0.833\frac{\Omega_D}{\Omega}
\end{aligned}
\tag{ii}
$$

From comparison of Tables 2.2 and 11.1, the ratio Ω_D/Ω varies from 0.96 for $kT/\epsilon = 0.3$ to 0.91 for $kT/\epsilon = 100$, and thus Sc calculated from Eq. (ii) has values in the range 0.76 to 0.8. The Schmidt numbers for all real gases have values close to unity and the Sc, Pr, and Le numbers for the system air–water vapor are shown in Fig. 11.19. Conversely, the low diffusivities in liquids give rise to high Schmidt numbers and the Schmidt numbers for liquids lie in the range 500 to 1000.

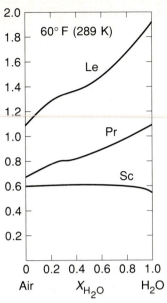

FIGURE 11.19. Variations of the Lewis number, the Prandtl number, and the Schmid number with composition in the system air–water vapor at 289 K.

In considering heat transfer, the Stanton number was defined as

$$St = \frac{Nu}{Re \cdot Pr} \tag{7.27}$$

and in mass transport, the mass transport Stanton number is defined as

$$St_m = \frac{Sh}{Re \cdot Sc} \tag{11.82}$$

and, for laminar flow, the Reynolds analogy, Eq. (7.28), gives

$$St_{m,x} = 0.5 f_x Sc^{-2/3} \tag{11.83}$$

in which f_x is the local friction factor

$$f_x = 0.646 Re_x^{1/2} \tag{4.23a}$$

In the film model $\delta = \delta_T = \delta_c = L$, and consideration of Fig. 11.7(c) gives rise to the following relationships:

$$h = \frac{k}{L} \tag{iii}$$

$$h_m = \frac{D}{L} \tag{iv}$$

$$\tau_0 = \eta \frac{v_\infty}{L} \tag{v}$$

Eliminating the film thickness, L, from Eqs. (iii) to (v) gives

$$\frac{h}{k} = \frac{h_m}{D} = \frac{\tau_0}{v_\infty \eta}$$ (vi)

and multiplying Eq. (vi) by the distance x gives

$$\frac{hx}{k} = \frac{h_m x}{D} = \frac{\tau_0 x}{v_\infty \eta}$$ (vii)

The first term in Eq. (vii) is Nu_x, the second is Sh_x, and the third can be rearranged as

$$\frac{\tau_0 x}{v_\infty \eta} = \frac{1}{2} \frac{\tau_0}{\frac{1}{2}\rho v_\infty^2} \frac{\rho v_\infty x}{\eta} = \frac{1}{2} f_x Re_x$$

Therefore, Eq. (vii) becomes

$$Nu_x = Sh_x = \tfrac{1}{2} f_x Re_x$$ (viii)

which in view of Eqs. (7.27) and (11.82) gives

$$St \cdot Pr = St_m Sc = \tfrac{1}{2} f_x$$ (ix)

However, as in the film model, $\delta = \delta_T = \delta_C$,

$$\alpha = D = v$$

and thus as both Pr and Sc are unity, Eq. (ix) is

$$St = St_m = \tfrac{1}{2} f_x$$ (x)

In turbulent flow, at $Re_x > 5 \times 10^5$, the mass transport equivalent of Eq. (7.28) is

$$Sh_x = St_{m,x} Re_x Sc = 0.0296 Sc^{1/3} Re_x^{4/5}$$ (11.84)

and the average mass transfer coefficient for a plate of length L on which a transition from laminar to turbulent flow occurs is thus

$$\bar{h}_{m,L} = \frac{1}{L} \left(\int_0^{x_{trans}} h_{m,x lam} \, dx + \int_{x_{trans}}^{L} h_{m,x turb} \, dx \right)$$

$$= \frac{1}{L} [0.662 DSc^{1/3} Re_{trans}^{1/2} + \tfrac{5}{4} \times 0.0296 DSc^{1/3} (Re_L^{4/5} - Re_{trans}^{4/5})]$$

$$= \frac{DSc^{1/3}}{L} (0.037 Re^{4/5} - 873)$$ (11.85)

Equation (11.85) is valid in the range $0.6 < Sc < 3000$.

EXAMPLE 11.8

Cooling water for a power plant is stored in a pond 1000 m in length and 500 m wide. Calculate the rate of evaporation from the pond when a dry wind at 300 K blows in a horizontal direction parallel to the 1000-m side of the pond at a velocity of 2 m/s. The cooling water is also at 300 K.

For air at 300 K, $\nu = 1.67 \times 10^{-5}$ m^2/s.

The transition from laminar to turbulent boundary conditions occurs at x_{trans}, where $\text{Re}_x = 500{,}000$:

$$x_{\text{trans}} = \frac{500{,}000\nu}{v_\infty} = \frac{500{,}000 \times 1.57 \times 10^{-5}}{2} = 3.9 \text{ m}$$

As only 0.39% of the boundary layer is laminar, it will be considered that the entire layer is turbulent. From Eq. (11.84),

$$\text{Sh}_x = \frac{h_{m,x}x}{D} = 0.0296\text{Sc}^{1/3}\text{Re}_x^{4/5}$$

or the local mass transfer coefficient is

$$h_{m,x} = 0.0296D\text{Sc}^{1/3}\left(\frac{v_\infty}{\nu}\right)^{4/5}x^{-1/5}$$

The average mass transfer coefficient is thus

$$\bar{h}_{m,L} = \frac{1}{L}0.0296D\text{Sc}^{1/3}\left(\frac{v_\infty}{\nu}\right)^{4/5}\int_0^L x^{-1/5}\, dx$$

$$= \frac{5}{4} \times 0.0296D\text{Sc}^{1/3}\left(\frac{v_\infty}{\nu}\right)^{4/5}L^{-1/5}$$

From Table 11.2, $D_{\text{H}_2\text{O}-\text{air}} = 2.6 \times 10^{-5}$ m^2/s and thus

$$\bar{h}_{m,L} = \frac{5}{4} \times 0.0296 \times 2.6 \times 10^{-5} \times \left(\frac{1.67 \times 10^{-5}}{2.6 \times 10^{-5}}\right)^{1/3} \times \left(\frac{2}{1.67 \times 10^{-5}}\right)^{4/5}$$

$$\times\; 1000^{-1/5}$$

$$= 2.48 \times 10^{-3} \text{ m/s}$$

The mass flux from the surface of the pond is thus

$$\dot{M} = \bar{h}_{m,L}A(\rho_{\text{H}_2\text{O}(s)} - \rho_{\text{H}_2\text{O}(\infty)})$$

which with

$$\rho = \frac{pM}{RT}$$

gives

$$\dot{M} = \frac{\bar{h}_{m,L}AM_{\text{H}_2\text{O}}}{R}\left(\frac{p_{\text{H}_2\text{O}(s)}^\circ}{T_s} - \frac{p_{\text{H}_2\text{O}(\infty)}}{T_\infty}\right)$$

The variation, with temperature, of the saturated vapor pressure of water is

$$\ln p_{\text{H}_2\text{O}}^\circ(\text{Pa}) = \frac{-6679}{T} - 4.65 \ln T + 56.97$$

and hence at 300 K, $p^\circ_{H_2O(s)} = 3580$ Pa, which with $p_{H_2O,\infty} = 0$ gives

$$\dot{M} = 2.48 \times 10^{-3} \times 500 \times 1000 \times \frac{0.018 \times 3580}{8.3144 \times 300}$$

$$= 32 \text{ kg/s}$$

Maintenance of a constant level in the storage pond thus requires that water be added at the rate of 2.77×10^6 kg per day. This calculation does not consider the influence of the absorption of the latent heat of evaporation, which tends to cool the water.

In the discussion of diffusion into a falling film of liquid, in Section 11.5, the local molar flux of gas A to the film was obtained as

$$\dot{N}_{A,x} = C_{A(s)} \sqrt{\frac{Dv_{z,\max}}{z\pi}}$$

The local mass transfer coefficient is thus

$$h_{m,z} = \frac{\dot{N}_x|_{x=0}}{C_{A(s)} - C_{A(\infty)}}$$

which with $C_{A(\infty)} = 0$ becomes

$$h_{m,z} = \frac{\dot{N}_x|_{x=0}}{C_{A(s)}} = \sqrt{\frac{Dv_{z,\max}}{z\pi}}$$

In the laminar flow of the film of liquid

$$v_{z,\max} = \tfrac{3}{2}\bar{v}_z$$

and hence

$$h_{m,z} = \sqrt{\frac{3\bar{v}_z D}{2z\pi}}$$

which can be rearranged as

$$\frac{h_{m,z}z}{D} = \sqrt{\frac{3}{2\pi}} \sqrt{\frac{\bar{v}_z z}{\nu}} \sqrt{\frac{\nu}{D}}$$

in which the term on the left-hand side is the local Sherwood number, Sh_z, the second term on the right-hand side is the local Reynolds number, Re_z, and the third term on the right-hand side is the Schmid number, Sc. Thus

$$Sh_z = \sqrt{\frac{3}{2\pi}} Re_z^{1/2} Sc^{1/2}$$

The average flux to the film over the length $z = 0$ to $z = L$ is obtained as

$$\bar{\dot{N}}_{A,L} = \frac{1}{L} \int_0^L \dot{N}_{A,z}\, dz$$

$$= \frac{1}{L} C_{A(s)} \sqrt{\frac{3D\bar{v}_z}{2\pi}} \int_0^L \frac{dz}{z^{1/2}}$$

$$= 2C_{A(s)} \sqrt{\frac{3D\bar{v}_z}{2\pi L}}$$

and hence the average Sherwood number is

$$\overline{Sh}_L = 2Sh_{x=L} = \sqrt{\frac{6}{\pi}} \, Re_L^{1/2} Sc^{1/2}$$

EXAMPLE 11.9 _____

In Example 11.4 a thin film of liquid metal was vacuum degassed as it flowed down an inclined plane of length 1 m and width 0.2 m. The film had an initial hydrogen content of 40 g mol/m^3, flowed with an average linear velocity of 0.0878 m/s, and was in contact with the vacuum for 11.4 s. The diffusion coefficient for hydrogen in the metal was $D_H = 1.3 \times 10^{-8}$ m^2/s.

The local mass transfer coefficient is

$$h_{m,z} = \sqrt{\frac{D v_{z,max}}{z\pi}}$$

and thus the average mass transport coefficient is

$$\bar{h}_m = \frac{1}{L} \int_0^L h_{m,z} \, dz$$

$$= 2 \sqrt{\frac{D v_{z,max}}{\pi L}}$$

$$= 2 \sqrt{\frac{1.3 \times 10^{-8} \times 1.5 \times 0.0878}{\pi \times 1}}$$

$$= 4.67 \times 10^{-5} \text{ m/s}$$

The molar flux from the surface of the film is thus

$$\dot{W}_H = \bar{h}_m A (C_{H,b} - C_{H(s)})$$
$$= 4.67 \times 10^{-5} \times 1 \times 0.2 \times 40$$
$$= 3.73 \times 10^{-4} \text{ g mol of H/s}$$

which when multiplied by the time of contact, 11.4 s, gives the total number of gram moles of H removed from the film as

$$3.37 \times 10^{-4} \times 11.39 = 4.26 \times 10^{-3}$$

The correlation for heat transfer by forced convection from a sphere was given by Eq. (7.64) as

$$\mathrm{Nu}_D = 2 + (0.4\mathrm{Re}_D^{0.5} + 0.06\mathrm{Re}_D^{0.667})\mathrm{Pr}^{0.4}\left(\frac{\eta_\infty}{\eta_s}\right)^{0.25} \tag{7.64}$$

The analog for isothermal mass transport from a sphere is thus

$$\mathrm{Sh}_D = 2 + (0.4\mathrm{Re}_D^{0.5} + 0.06\mathrm{Re}_D^{0.667})\mathrm{Sc}^{0.4} \tag{11.86}$$

EXAMPLE 11.10

In Example 11.2 it was calculated that the rate of sublimation of a naphthalene mothball of diameter 0.01 m into stagnant air at 20°C and 101.3 kPa was 6.2×10^{-9} g mol/s. Using the mass transfer analogy given by Eq. (11.86), calculate the rate of sublimation of the mothball (a) into still air at 20°C and 101.3 kPa, and (b) into an airstream that is flowing past the mothball at a velocity of 10 m/s.

For naphthalene at 20°C,

$$p° = 30 \text{ Pa}$$

$$D = 8 \times 10^{-6} \text{ m}^2/\text{s}$$

and for air at 20°C,

$$v = 1.54 \times 10^{-5} \text{ m}^2/\text{s}$$

With $v_\infty = 0$, $\mathrm{Re}_D = 0$ and hence

$$\mathrm{Sh}_D = 2 = \frac{h_m 2R}{D}$$

Thus

$$h_m = \frac{2 \times 8 \times 10^{-6}}{0.01} = 0.0016 \text{ m/s}$$

and the rate of mass transport to the still air by sublimation is

$$h_m 4\pi R^2 \frac{p°}{RT} = \frac{0.0016 \times 4 \times \pi \times 0.005^2 \times 30}{8.3144 \times 293}$$

$$= 6.2 \times 10^{-9} \text{ g mol/s}$$

which is in exact agreement with the value obtained in Example 11.2.
With $v_\infty = 10$ m/s,

$$\mathrm{Re}_D = \frac{v_\infty 2R}{v} = \frac{10 \times 0.01}{1.54 \times 10^{-5}} = 6493$$

and thus in Eq. (11.86),

$$\mathrm{Sh}_D = 2 + (0.4 \times 6493^{0.5} + 0.06 \times 6493^{0.667})\left(\frac{1.54 \times 10^{-5}}{8 \times 10^{-6}}\right)^{0.4}$$

$$= 2 + 69 = 71$$

Therefore,

$$h_m = \frac{71 \times 8 \times 10^{-6}}{0.01} = 0.0568 \text{ m/s}$$

and hence the rate of sublimation is

$$6.2 \times 10^{-9} \times \frac{0.0568}{0.0016} = 2.2 \times 10^{-7} \text{ g mol/s}$$

11.17 Mass Transfer by Free Convection

In Section 7.8 it was seen that in heat transfer by free convection in a fluid at a vertical plate, the motion of the fluid is caused by differences in the density of the fluid, which, in turn, are caused by the differences in temperature giving rise to the heat transfer. Similarly, if a vertical plate of solid A is soluble in the fluid B, mass transfer to A to the fluid gives rise to concentration gradients in the fluid adjacent to the plate, which, in turn, establish a variation in the density of the fluid. In a gravitational field the variation in density gives rise to buoyancy forces, and at steady state, velocity and concentration boundary layers, as shown in Fig. 11.20, are established on the surface of the plate, which are analogous with the boundary layers established by heat transfer from a vertical plate by free convection. The equation of conservation of x-directed momentum in the control volume within the velocity boundary layer in Fig. 11.20 is given by Eq. (7.90):

$$\rho \left(v_x \frac{\partial v_x}{\partial x} + v_y \frac{\partial v_x}{\partial y} \right) = \eta \frac{\partial v_x}{\partial y^2} - \frac{\partial P}{\partial x} + \rho g_x \qquad (7.90)$$

and in this expression, the last two terms on the right side were given by Eq. (7.91) as

$$-\frac{\partial P}{\partial x} + \rho g_x = (\rho - \rho_\infty)g_x \qquad (7.91)$$

which with an isobaric thermal expansivity of the fluid β,

$$\beta = -\frac{1}{\rho} \left(\frac{\partial \rho}{\partial T} \right)_P$$

gave

$$-\frac{\partial P}{\partial x} + \rho g_x = g_x \beta \rho (T_\infty - T)$$

and hence an x-directed momentum balance of

$$v_x \frac{\partial v_x}{\partial x} + v_y \frac{\partial v_x}{\partial y} = \nu \frac{\partial^2 v_x}{\partial y^2} + g_x \beta (T_\infty - T) \qquad (7.93)$$

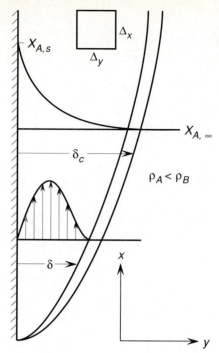

FIGURE 11.20. Concentration and velocity profiles in the boundary layers on a vertical plate undergoing mass transfer from the plate to the fluid by natural convection.

In considering mass transfer by natural convection, the coefficient of the change in density of a binary fluid, $A-B$, with composition is

$$\beta' = -\frac{1}{\rho} \left(\frac{\partial \rho}{\partial X_A} \right)_{T,P}$$

which for small changes in ρ and X_A gives

$$\beta' \rho = -\frac{\rho - \rho_\infty}{X_A - X_{A,\infty}}$$

and thus the x-directed momentum balance in the control volume is

$$v_x \frac{\partial v_x}{\partial x} + v_y \frac{\partial v_x}{\partial y} = \nu \frac{\partial^2 v_x}{\partial y^2} + g_x \beta'(X_{A,\infty} - X_A) \tag{11.87}$$

Equation (11.87) is coupled with the equation for conservation of mass

$$v_x \frac{\partial X_A}{\partial x} + v_y \frac{\partial X_A}{\partial y} = D \frac{\partial^2 X_A}{\partial y^2} \tag{11.88}$$

and Eqs. (11.87) and (11.88) must be solved simultaneously. The results, shown in Fig. 11.21(a) and (b), are identical with the solutions to the heat transfer equations

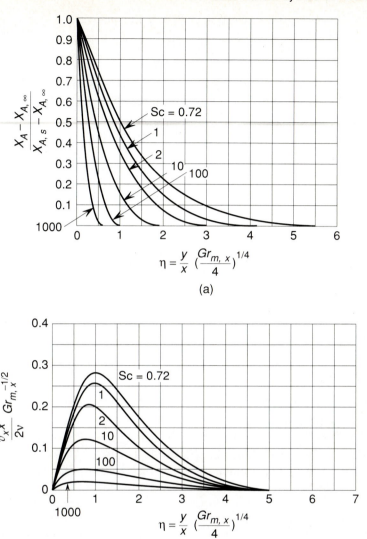

FIGURE 11.21. (a) Normalized concentration profiles in the concentration boundary layer on a vertical plate undergoing mass transport from the plate to the fluid by natural convection; (b) velocity profiles in the momentum boundary layer on a vertical plate undergoing mass transport from the plate to the fluid by natural convection.

shown in Fig. 7.15(a) and (b). Examination of Fig. 11.21(a) and (b) shows that the high Schmidt numbers of liquids, which are in the range 500 to 1000, cause the concentration boundary layers in liquids to be much smaller than the corresponding velocity boundary layers.

In heat transfer by free convection the Grashof number

$$Gr_x = \frac{g\beta(T_s - T_\infty)x^3}{\nu^2}$$

was recognized as an important variable and the Nusselt number was found to be a function of Gr_x and Pr, being given for laminar flow by Eq. (7.113) as

$$Nu_x = \frac{2Pr^{1/2}Gr_x^{1/4}}{[336(Pr + 0.556)]^{1/4}}$$

In mass transfer by free convection the mass transfer Grashof number, $Gr_{m,x}$, is defined as

$$Gr_{m,x} = \frac{g\beta'(X_{A,s} - X_{A,\infty})x^3}{\nu^2} \tag{11.89}$$

and the Sherwood number is a function of $Gr_{m,x}$ and Sc. The approximate integral method presented in Section 7.8 and applied to mass transfer by free convection and laminar flow gives

$$Sh_x = \frac{2Sc^{1/2}Gr_{m,x}^{1/4}}{[336(Pr + 0.556)]^{1/4}} \tag{11.90}$$

11.18 Simultaneous Heat and Mass Transfer: Evaporative Cooling

Figure 11.22 shows gas B at the temperature T_∞ flowing over the surface of liquid A. If the vapor pressure of A in the gas is less than the saturated vapor pressure of A at the temperature of the surface, T_s, A evaporates into the flowing gas. However, as evaporation is an endothermic process, it causes a decrease in the enthalpy of the liquid A and hence a decrease in its temperature. In turn, the difference between T_∞ and T_s causes the transfer of heat from the gas to the liquid, and a steady state is reached when the rate of heat transfer from the gas to the liquid equals the rate at which the latent heat of evaporation is absorbed. The rate of heat transfer from the gas to the liquid is

$$q' = h(T_\infty - T_s)$$

and with a mass evaporation rate of $\dot{n}_A$ and a latent heat of evaporation of $\Delta H_{A,evap}$ J/kg, latent heat is supplied at the rate

$$q'_{evap} = \dot{n}_A \, \Delta H_{A,evap} = h_m(\rho_{A,T_s} - \rho_{A,T_\infty}) \, \Delta H_{A,evap}$$

Thus at steady state,

$$T_\infty - T_s = \frac{h_m}{h} \Delta H_{A,evap}(\rho_{A,T_s} - \rho_{A,T_\infty}) \tag{11.91}$$

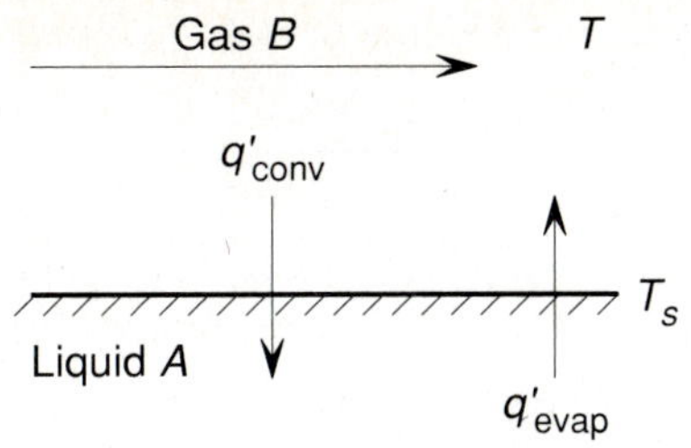

FIGURE 11.22. Gas B at the temperature T_∞ flowing over the surface of liquid A at the temperature T_s and evaporation of A into the gas stream.

From Eqs. (7.23) and (11.77),

$$\frac{h_m}{h} = \frac{D_A \text{Sc}^{1/3}}{k_B \text{Pr}^{1/3}} = \frac{D_A}{k_B} \left(\frac{\nu_B}{D_A}\right)^{1/3} \left(\frac{\alpha_B}{\nu_B}\right)^{1/3}$$

$$= \frac{D_A^{2/3} \alpha_B^{1/3}}{k_B}$$

or as $k = \alpha \rho C_p$,

$$\frac{h_m}{h} = \frac{1}{\rho_B C_{p,B}} \left(\frac{D_A}{\alpha_B}\right)^{2/3} = \frac{1}{\rho_B C_{p,B} \text{Le}^{2/3}} \tag{11.92}$$

Furthermore, if vapor A and gas B mix ideally,

$$\rho_A = \frac{p_A M_A}{RT} \tag{11.93}$$

and substitution of Eqs. (11.92) and (11.93) in Eq. (11.91) gives

$$T_\infty - T_s = \frac{M_A \Delta H_{A,\text{evap}}}{R \rho_B C_{p,B}} \left(\frac{1}{\text{Le}}\right)^{2/3} \left(\frac{p_{A,T_s}^\circ}{T_s} - \frac{p_{A,T_\infty}}{T_\infty}\right) \tag{11.94}$$

in which p_{A,T_∞} is the vapor pressure of A in the bulk gas stream. In Eq. (11.94) the properties ρ_B, $C_{p,B}$, and Le should be evaluated at the mean temperature $(T_\infty + T_s)/2$.

EXAMPLE 11.11

Calculate the steady-state temperature of liquid water when dry air at 40°C flows over it.

As T_s is unknown, the mean temperature cannot be calculated, and hence for the first calculation, the properties of dry air at 40°C will be used. The data are

$$M_{H_2O} = 0.018 \text{ kg/g mol}$$

$$\Delta H_{H_2O,\text{evap}} = 3.338 \times 10^5 \text{ J/kg}$$

$$C_{p,\text{air}} = 1005 \text{ J/kg·K}$$

$$\rho_{air} = 1.133 \text{ kg/m}^3 \text{ (at } P_{total} = 101.3 \text{ kPa)}$$

$$\ln p_{H_2O}^{\circ}(\text{Pa}) = -\frac{6679}{T} - 4.65 \ln T + 56.97 \tag{11.95}$$

From Fig. 11.19 the Lewis number for moist air is approximately 1.1, in which case Eq. (11.94) gives

$$313 - T_s = \frac{0.018 \times 3.338 \times 10^5}{8.3144 \times 1.133 \times 1005} \left(\frac{1}{1.1}\right)^{2/3} \frac{p_{H_2O,T_s}^{\circ}}{T_s}$$

$$= \frac{0.596}{T_s} \exp\left(-\frac{6679}{T_s} - 4.65 \ln T_s + 56.97\right) \tag{i}$$

which has the solution $T_s = 304$ K. Recalculation using a mean temperature of $(304 + 313)/2 = 309$ K changes the 0.596 in Eq. (i) to 0.584, which also gives a solution of 304 K.

The steady-state temperature of the liquid is independent of the velocity of the gas, because from the heat transfer analogy, the mass transfer coefficient, h_m, is the same function of the Reynolds number as is the heat transfer coefficient, and thus the ratio h_m/h in Eq. (11.91) is independent of the Reynolds number. At a film temperature of $(313 + 304)/2 = 309$ K, the Prandtl number and thermal conductivity of air are, respectively, 0.711 and 0.0266 W/m·K and thus at a Reynolds number of 5×10^5,

$$h = 0.332k\text{Pr}^{1/3}\text{Re}^{1/2}$$
$$= 0.332 \times 0.0266 \times 0.711^{1/3} \times (5 \times 10^5)^{1/2}$$
$$= 5.57 \text{ W/m}^2\cdot\text{K}$$

and hence from Eq. (11.90),

$$\dot{n}_A = \frac{h(T_\infty - T_s)}{\Delta H_{evap}}$$
$$= \frac{5.57 \times (313 - 300)}{3.338 \times 10^5}$$
$$= 1 \times 10^{-4} \text{ kg·m}^{-2}\cdot\text{s}^{-1}$$

Wet Bulb Psychrometer

In Example 11.10, Eq. (11.94) was used to obtain the difference between T_s and T_∞ when p_{H_2O,T_s}° and p_{H_2O,T_∞} were known (or could be expressed as functions of T_s and T_∞). Equation (11.94) can also be used to calculate p_{H_2O,T_∞} if T_s, T_∞, and p_{H_2O,T_s}° are known, and this is the basis for the wet bulb psychrometer, which facilitates determination of the relative humidity of moist air. The wet bulb psychrometer is a thermometer whose bulb is covered with a cloth wick soaked with water. When moist air whose humidity is to be measured flows over the wick, a steady state is reached in which heat is transferred from the moist air at the temperature T_{db} (which is measured at a second dry bulb thermometer) to the wet bulb thermometer at the temperature T_{wb} at the rate at which the latent heat of evaporation of water from the wick is absorbed. Measurement of T_{db} and T_{wb} thus allows calculation of $p_{H_2O,T_{db}}$ and hence the relative humidity of the moist air.

EXAMPLE 11.12

Calculate the relative humidity of moist air if $T_{wb} = 295$ K and $T_{db} = 298$ K. With Le $= 1.1$, rearrangement of Eq. (11.94) gives

$$\frac{p_{A,T_{wb}}^\circ}{T_{wb}} - \frac{p_{A,T_{db}}}{T_{db}} = \frac{R\rho_{air}C_{p,air}Le^{2/3}}{M_{H_2O}\ \Delta H_{evap,H_2O}}(T_{db} - T_{wb})$$

Thus, for a film temperature of 296.5 K,

$$\frac{2656}{295} - \frac{p_{A,T_{db}}}{298} = \frac{8.3144 \times 1.177 \times 1005 \times 1.1^{2/3} \times 3}{0.018 \times 3.338 \times 10^5}$$

or

$$p_{A,T_{db}} = 1123 \text{ Pa}$$

From Eq. (11.95) the saturated water vapor pressure at $T_{db} = 298$ K is 3183 Pa, and thus the relative humidity RH, of the moist air is

$$\text{RH} = \frac{1123}{3183} \times 100 = 35\%$$

11.19 Chemical Reaction and Mass Transfer: Mixed Control

Consider a gas A_2 which when adsorbed on the surface of liquid B dissociates to form adsorbed atoms of A according to the overall reaction

$$A_{2(gas)} \rightarrow 2A_{(ad)} \tag{11.96}$$

The rate at which adsorbed A atoms are produced on the surface is

$$\dot{n}_A = k_f P_{A_2} - k_f \rho_{A,s}^2$$

in which P_{A_2} is the pressure of the gaseous A_2, $\rho_{A,s}$ is the concentration of A on the surface, k_f (in $kg \cdot m^{-2} \cdot s^{-1} \cdot Pa^{-1}$) is the reaction rate constant for the forward reaction $A_{2(gas)} \rightarrow 2A_{(ad)}$, and k_b (in $m^4 \cdot kg^{-1} \cdot s^{-1}$) is the reaction rate constant for the backward reaction $2A_{(ad)} \rightarrow A_{2(gas)}$. At thermodynamic equilibrium the forward and backward reactions proceed at the same rate, in which case the net rate of production of adsorbed A is zero and hence

$$k_f P_{A_2} = k_b \rho_{A,s}^2$$

or

$$\frac{k_f}{k_b} = \frac{\rho_{A,s}^2}{P_{A2}}$$ (11.97)

However, the equilibrium constant, K, for the reaction given by Eq. (11.96) is

$$K = \frac{\rho_{A,s}^2}{P_{A2}}$$ (11.98)

in which $\rho_{A,s}$ is the concentration of A in B in equilibrium with A_2 gas at the pressure P_{A_2}. Thus from Eqs. (11.97) and (11.98),

$$\dot{n}_A = k_f \left(P_{A2} - \frac{\rho_{A,s}^2}{K} \right)$$ (11.99)

which is the net rate at which adsorbed A is produced on the surface of B when the pressure of A_2 gas is P_{A2} and the concentration on the surface is $\rho_{A,s}$. If A is soluble in B, the atoms of A adsorbed on the surface are transported into the bulk liquid at the rate

$$\dot{n}_A = h_m(\rho_{A,s} - \rho_{A,b})$$ (11.100)

in which $\rho_{A,b}$ is the mass concentration of A in the bulk B. At steady state the rate of dissociation of A_2 on the surface equals the mass flux of A from the surface into the bulk,

$$k_f \left(P_{A2} - \frac{\rho_{A,s}^2}{K} \right) = h_m(\rho_{A,s} - \rho_{A,b})$$

or

$$\frac{k_f}{K} \rho_{A,s}^2 + h_m \rho_{A,s} = k_f P_{A2} + h_m \rho_{A,b}$$ (11.101)

and hence the concentration of A at the surface and the rate at which A is transported from the gas phase to the bulk liquid are determined by the relative magnitudes of K, k_f, and h_m. Designating

$$\frac{h_m}{k_f} = r$$

in Eq. (11.101) gives $\rho_{A,s}$ as

$$\rho_{A,s} = \frac{-r + \sqrt{r^2 + 4(P_{A2} + r\rho_{A,b})/K}}{2/K}$$ (11.102)

For a system in which $K = 1.02 \times 10^{-4}$, the variation of $\rho_{A,s}$ with r for $P_{A2} = 101.3$ kPa and $\rho_{A,b} = 0$ are as shown in Fig. 11.23.

The two processes that occur in series are:

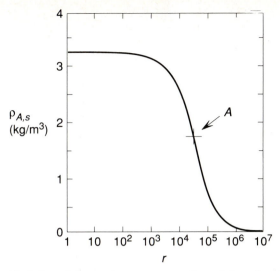

FIGURE 11.23. Variation of the surface concentration of A on liquid B with $r = h_m/k_f$ during a process involving the adsorption of gas A_2 on the surface of liquid B, the dissociation of A_2 to form adsorbed A and mass transport of A from the surface to the bulk liquid.

1. Adsorption of A_2 on the surface followed by dissociation to form adsorbed A.
2. Mass transport of A from the surface into the bulk liquid.

If the rate of dissociation is high and h_m is small, the surface concentration approaches the value for equilibrium with the gas phase and the overall process is controlled by mass transport of A in the liquid. Conversely, if the rate of dissociation is small and h_m is large, mass transport from the surface to the bulk can be fast enough that $\rho_{A,s}$ is decreased to a small fraction of its equilibrium value, and in this case the overall process is controlled by the rate of dissociation of adsorbed A. In intermediate cases the process is under mixed control in which the values of $\rho_{A,s}$ and j_A are determined by both the rate of dissociation and the rate of mass transport.

Consider systems in which $K = 1.02 \times 10^{-4}$ and $k_f = 6.9 \times 10^{-8}$, $k_f = 6.9 \times 10^{-9}$, and $k_f = 6.9 \times 10^{-10}$. The variations of $\rho_{A,s}$ with h_m at fixed k_f and $P_{A_2} = 101.3$ kPa, $\rho_{A,b} = 0$ are shown in Fig. 11.24 and the corresponding variations of j_A with h_m at fixed k_f are shown in Fig. 11.25, in which j_A is obtained as

$$j_A = h_m \rho_{A,s}$$

In Fig. 11.24 at $h_m = 0$, $\rho_{A,s} = 3.22$ kg/m³, which is the equilibrium value for contact with A_2 gas at 101.3 kPa. Increasing h_m at constant k_f decreases $\rho_{A,s}$, but from Fig. 11.24, as the decrease in $\rho_{A,s}$ is less than the increase in h_m, j_A increases with increasing h_m. At constant h_m, increasing k_f increases $\rho_{A,s}$ and hence increases j_A.

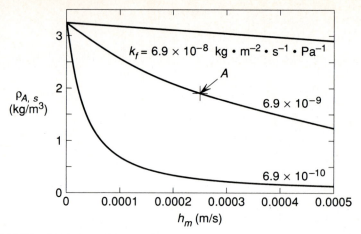

FIGURE 11.24. Variation of the surface concentration of A on liquid B with h_m at constant k_f.

EXAMPLE 11.13

The concentration of nitrogen in solution in liquid iron which is in equilibrium with nitrogen gas at 1600°C and $P = 101.3$ kPa is 0.045 wt %. The density of liquid iron at 1600°C is 7160 kg/m³, and hence 0.045 wt % corresponds to

$$\frac{0.045 \times 7160}{100} = 3.22 \text{ kg/m}^3$$

and thus for the equilibrium,

$$N_{2(\text{gas})} = 2[N]_{\text{dissolved in Fe}}$$

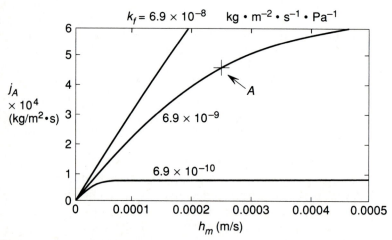

FIGURE 11.25. Variation of the mass flux of A with h_m at constant k_f corresponding to Fig. 11.24.

$$K = \frac{\rho_N^2}{P_{N_2}} = \frac{(3.22)^2}{101{,}325} = 1.02 \times 10^{-4}$$

For the dissociation of adsorbed nitrogen on liquid iron according to

$$N_{2(gas)} \rightarrow 2N_{ad}$$

at 1600°C, $k_f = 6.9 \times 10^{-9} \ \text{kg·m}^{-2}\text{·s}^{-1}\text{·Pa}^{-1}$ and the mass transfer coefficient of nitrogen in liquid iron is $h_m = 2.5 \times 10^{-4} \ \text{m/s}$. Insertion of these values of K, k_f, and h_m, with $P_{N_2} = 101.3 \ \text{kPa}$ and $p_{N,b} = 0$, into Eq. (11.101) gives

$$\rho_{N,s} = 1.86 \ \text{kg/m}^3$$

and

$$j_N = 2.5 \times 10^{-4} \times 1.86$$
$$= 4.65 \times 10^{-4} \ \text{kg·m}^{-2}\text{·s}^{-1}$$

which are points *A* in Figs. 11.23, 11.24, and 11.25. The transport of nitrogen from the gas phase to the bulk liquid iron phase is thus under mixed chemical kinetics and mass transport control.

11.20 Dissolution of Pure Metal *A* in Liquid *B*: Mixed Control

The phase diagram for the system $A-B$ is shown in Fig. 11.26. When a piece of pure solid *A* is immersed in pure liquid *B* at the temperature *T*, it begins to dissolve in the melt to form a liquid $A-B$ alloy whose composition eventually reaches the liquidus composition C_s^o. At this point thermodynamic equilibrium is established between pure solid *A* and the saturated melt and dissolution ceases. The dissolution

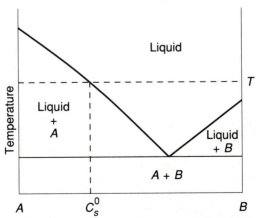

FIGURE 11.26. Phase diagram for the system $A-B$ showing the limit of solubility of *A* in liquid *B* at the temperature *T*.

process involves two steps: (1) a spontaneous escape mechanism by means of which an atom of A is transferred from the surface of the solid to the melt in contact with the solid, and (2) the diffusion or mass transport of A from the surface to the bulk melt. If the spontaneous escape mechanism is not rapid enough for equilibrium to be maintained at the solid–liquid interface, the concentration of A in the melt at $x = 0$, C_s, is less than the equilibrium value, C_s°, and the concentration profile of A into the bulk melt is as shown in Fig. 11.27, in which C_b is the concentration of A in the bulk melt.

The rate of the spontaneous escape process is

$$\dot{r} = \frac{dn_A}{dt} = k_1 A C_s^\circ - k_2 A C_s \qquad \text{g mol } A\cdot\text{s}^{-1} \qquad (11.103)$$

where n_A is g mol of A
$\quad$ A is the surface area of the immersed solid A, m^2
$\quad$ k_1 is the forward rate constant, m·s^{-1}
$\quad$ k_2 is the backward rate constant, m·s^{-1}

At equilibrium

$$\dot{r} = 0 = k_1 A C_s^\circ - k_2 A C_s^\circ$$

and thus $k_1 = k_2 = k$ and Eq. (i) becomes

$$\frac{dn_A}{dt} = kA(C_s^\circ - C_s)$$

Dividing by the volume of the melt, V, gives the rate of change of concentration of A in the melt as

$$\frac{dC}{dt} = \frac{kA(C_s^\circ - C_s)}{V} \qquad (11.104) \quad 2$$

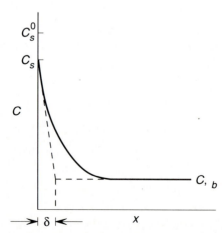

FIGURE 11.27. Concentration profile of A dissolving in liquid B when phase equilibrium is not maintained at the interface between solid A and the liquid.

The rate of transport of A by diffusion away from the interface is given by

$$\frac{dn_A}{dt} = \frac{AD(C_s - C_b)}{\delta}$$

in which the diffusion thickness, δ, is defined in Fig. 11.27. Or again dividing by V,

$$\frac{dC}{dt} = \frac{AD(C_s - C_b)}{\delta V} \tag{11.105}$$

Combining Eqs. (11.104) and (11.105) gives

$$C_s = \frac{C_s^\circ[k + (C_b/C_s^\circ)(D/\delta)]}{k + D/\delta} \tag{11.106}$$

substitution of which into Eq. (11.104) gives

$$\frac{dC}{dt} = \frac{kA}{V}\frac{C_s^\circ - C_b}{1 + k\delta/D} \tag{11.107}$$

Integration of Eq. (11.107) from

$$C_b = 0 \qquad \text{at } t = 0$$

to

$$C_b = C_b \qquad \text{at } t = t$$

gives

$$-\ln\frac{C_s^\circ - C_b}{C_s^\circ} = \frac{kAt}{V(1 + k\delta/D)}$$

or

$$C_b = C_s^\circ\left\{1 - \exp\left[\frac{-kAt}{V(1 + k\delta/D)}\right]\right\} \tag{11.108}$$

as the variation of the concentration of A in the bulk melt with time. In Eq. (11.108) if $k \gg D/\delta$, C_s approaches C_s° and the dissolution process is controlled by diffusion, in which case Eq. (11.108) gives

$$C_b = C_s^\circ\left[1 - \exp\left(-\frac{ADt}{V\delta}\right)\right] \tag{11.109}$$

Conversely, if $k \ll D/\delta$, C_s approaches C_b and the dissolution process is controlled by spontaneous escape, in which case the concentration gradient in the melt is virtually eliminated and Eq. (11.108) gives

$$C_b = C_s^\circ\left[1 - \exp\left(-\frac{kAt}{V}\right)\right] \tag{11.110}$$

If either diffusion or spontaneous escape is the rate-limiting step in the dissolution process, the variation of C_b with time, from Eq. (11.109) or Eq. (11.110), is of the form

$$C_b = C_s^\circ \left[1 - \exp \left(-\frac{KAt}{V} \right) \right]$$

or

$$\ln \frac{C_s^\circ}{C_s^\circ - C_b} = \frac{KAt}{V} \tag{11.111}$$

where K is the dissolution rate constant, which is equal to k for a process controlled by spontaneous escape and is equal to D/δ for a process controlled by diffusion. The velocity of the melt relative to the solid has a significant influence on the magnitude of δ, and hence on K, if the process is controlled by diffusion. Conversely, if the process is controlled by spontaneous escape the dissolution rate is independent of the velocity of the melt. Stevenson and Wulff [D. A. Stevenson and J. Wulff, *Trans. Met. Soc. AIME* (1961), vol. 221, p. 279] have measured the rates of solution when cylinders of copper and nickel were rotated in liquid lead, and analyzed their results in terms of Eq. (11.111). Their experimental results for copper are shown in Fig. 11.28(a) and (b). In Fig. 11.28(a) the dependence of the slope of the straight line on the rate of rotation of the cylinder, at lower rates of rotation, indicates diffusion control and the increase in K with rotation rate shows that δ decreases with increasing velocity of the liquid. At the highest velocity studied, deviation from linearity at high percentages of saturation of the liquid lead with copper indicates a transition from diffusion control to mixed control. The results at the higher temperature of 727°C, shown in Fig. 11.28(b), show deviation from linearity at all rates of rotation, indicating that the factors contributing to transition from diffusion control to mixed control are high rotation rates, high percentages of saturation, and high temperature.

11.21 Summary

Measurement of a mass transport flux in fluids is determined by the relative motion of the observer of the flux. An observer moving with the velocity of the bulk fluid sees mass transport due to diffusion down concentration gradients in the fluid and a stationary observer sees mass transport due to bulk fluid flow and diffusion.

Consideration of mass transport from the surface of a solid to a flowing fluid is facilitated by the film model in which it is envisioned that mass transport in a thin film of fluid adjacent to the surface occurs solely by diffusion, and that at distances from the surface greater than the thickness of the film, the fluid is well mixed and hence does not contain any concentration gradients. When mass transport occurs from a cool solid to a warm fluid, the mass flux is also an enthalpy flux, and hence the mass flux can influence the temperature gradient in the fluid and hence the heat transfer process.

The mathematics of mass transport in gases can be derived from the kinetic theory of gases in a manner similar to that used to derive the equations describing heat transfer and momentum transport in gases. The interdiffusion coefficient in a given system is proportional to the 3/2 power of the temperature and is inversely proportional to the pressure of the gas.

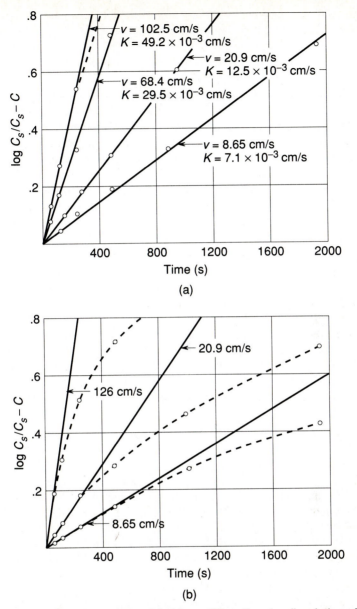

FIGURE 11.28. Dependence of $\log[C_s/(C_s - C)]$ on time for dissolution of copper into lead at (a) 527°C and (b) 727°C.

The local mass transfer coefficient at the interface between a solid and a fluid is defined in a manner analogous to that used to define the local heat transfer coefficient, and the mass flux is proportional to the difference between the mass density in the fluid in contact with the surface and the mass density in the bulk fluid, with the mass transfer coefficient being the proportionality constant. The variations of the local mass transfer coefficient with position on the surface, the averaged mass transfer coefficient, and the thickness of the concentration boundary layer are derived by analogy with heat transfer by convection from the solid to the fluid. This analogy leads to the definition of the Sherwood number and the Schmid number, which are the mass transport analogs of the Nusselt number and the Prandtl numbers. In forced convection the Sherwood number is the same function of the Reynolds number and the Schmid number, as is the Nusselt number of the Reynolds number and the Prandtl number. In natural convection the buoyancy arises from concentration (and hence density) differences in the fluid, and the Sherwood number is a function of the Schmid number and the mass transport equivalent of the Grashof number.

When mass transport involves more than one mechanism and no single mechanism is rate controlling, mixed control of the process occurs. Mixed control is common in reactions occurring at interfaces in which neither the resistance to transport across the interface nor resistance to transport by diffusion to the interface predominates.

Problems

PROBLEM 11.1

Two large vessels containing N_2-CO_2 mixtures at 273 K and a total pressure of 101.3 kPa are connected by a tube of length 1 m and 0.05 m diameter. If the partial pressure of CO_2 in one tank is maintained constant at 1×10^4 Pa and in the other tank is maintained constant at 5000 Pa, calculate the mass flow rate of CO_2 between the tanks under conditions of equimolar diffusion. The interdiffusion coefficient is listed in Table 11.2.

PROBLEM 11.2

A container is cooled by evaporation of a volatile liquid A from a cloth wrapped around the container which is saturated with the liquid. The evaporation occurs into air at 50°C, and at steady state, the temperature of the container is 8°C. Calculate the interdiffusion coefficient D_{A-air} given the following properties for A.

$$\ln p^{\circ}_{A,\text{liq}} \text{ (Pa)} = -\frac{1200}{T} + 12.79$$

$$MW_A = 0.2 \text{ kg/g mol}$$

$$\Delta H_{\text{evap},A} = 50,000 \text{ J/kg}$$

Assume a film temperature of 300 K.

PROBLEM 11.3

Consider a thin layer of water lying on the ground under a clear night sky. The effective temperature of the sky is -50°C, the emissivity of the surface of the water is 1.0 and the heat transfer coefficient for transfer of heat from the air to the water is 30 $\text{W}\cdot\text{m}^{-2}\cdot\text{K}^{-1}$.

(a) Estimate the mass transfer coefficient for evaporation of the water.
(b) Calculate the air temperature below which the ground water will freeze.

It can be assumed that there is no transfer of heat by conduction from the water to the ground below it and that the night air is dry. The saturated water vapor pressure at 0°C is 614 Pa and the latent heat of freezing of water is 3.338×10^5 J/kg.

PROBLEM 11.4

A circular disk of diameter 0.05 m supports a thin film of water. When 0.95 W of electric power is supplied, the water reaches a steady-state temperature of 305 K and water evaporates at the rate of 5×10^{-3} kg/h into the surrounding dry air, which is at a temperature of 295 K.

(a) Calculate the heat transfer coefficient for transfer of heat by convection from the water to the air.
(b) If the relative humidity of the air is increased to 50% at 295 K and the power input is maintained at 0.95 W, does the water temperature increase or decrease?
(c) Does the rate of evaporation increase or decrease?
(d) Does the mass transport–heat transport analogy hold?

PROBLEM 11.5

Calculate the interdiffusion coefficient for air and CO_2, D_{CO_2-air}, at 1000 K and 1100 K at 1 atm pressure. Use the data in Tables 2.1 and 11.1.

PROBLEM 11.6

During the calcination of limestone, the rate of decomposition of the $CaCO_3$ is controlled by mass transport of CO_2 from the surface of the solid to the surrounding atmosphere. On the assumptions that the CO_2 diffuses through a stagnant film of air of thickness 0.001 m above the surface, and the concentration of CO_2 in the calcining furnace is zero, calculate the rates of decomposition of $CaCO_3$ at 1000 K and at 1100 K. Use the diffusion coefficients calculated in Problem 11.5. The standard free energy for the reaction

$$CaO_{(s)} + CO_{2(g)} = CaCO_{3(s)}$$

based on 1 atm pressure as the standard state is

$$\Delta G° = -161{,}300 + 137T \text{ J/g mol}$$

PROBLEM 11.7

Hydrogen, present in solution in liquid aluminum, is removed by exposing the surface of a thin film of the liquid metal to a vacuum while the film flows down an inclined plane. The flowing film is 1 mm deep and the plane, which is inclined at an angle of 2° to the horizontal, is 1 m long and 0.5 m wide. The initial concentration of hydrogen in the aluminum is 0.6 g mol H/m^3. Calculate the percentage of hydrogen removed by the procedure. Data for aluminum at 1000 K are

$$\rho = 2500 \text{ kg/m}^3$$

$$\eta = 3 \times 10^{-3} \text{ Pa·s}$$

$$D_H = 1 \times 10^{-7} \text{ m}^2/\text{s}$$

PROBLEM 11.8

Three rectangular shallow trays each of length 0.5 m and containing water are laid end to end in contact with one another and a dry airstream at 300 K flows over the three trays in the lengthwise direction. Calculate the electrical power inputs to each of the three trays required to maintain the water at a temperature of 300 K when the free-stream velocity of the air is (a) 5 m/s, and (b) 10.47 m/s. (c) When the free-stream velocity is 10.47 m/s, what would the relative humidity of the air have to be to require a power input of 50 W to the middle tray in order to maintain the water in the tray at 300 K? The saturated water vapor pressure at 300 K is 3580 Pa and the latent heat of evaporation of water is 3.338×10^5 J/kg.

PROBLEM 11.9

Fine particles of coal are combusted at 1450 K in pure oxygen gas at a pressure of 101.3 kPa. If the reaction

$$C_{(s)} + O_{(g)} = CO_{2(g)}$$

is sufficiently rapid that the concentration of oxygen in the gas phase at the surface of the particle is zero, and the reaction rate is controlled by diffusion of oxygen in the gas to the surface of the particle, calculate:

(a) The rate of combustion of a spherical particle of initial radius 10^{-4} m
(b) The time for a particle of initial radius 10^{-4} m to be completely combusted

Consider the particle to be pure carbon of density 2000 kg/m³. The interdiffusion coefficient $D_{O_2-CO_2}$ is 1.7×10^{-4} m/s at 1450 K, and for steady-state diffusion of A in spherical coordinates,

$$\frac{d}{dr}\left(r^2 D \frac{dC_A}{dr}\right) = 0$$

PROBLEM 11.10

Repeat Problem 11.9 assuming that the combustion reaction obeys the rate law

$$\dot{r} = k_r C_{O_2}(r_0)$$

where the rate constant is $k_r = 0.1$ m/s, $\dot{r}$ is the rate of consumption of oxygen in g mol·m^{-2}·s^{-1}, and $C_{O_2}(r_0)$ is the concentration of oxygen gas at the surface of the particle.

Elementary and Derived SI Units and Symbols

	Unit	Formula	Symbol
Elementary units			
Length	meter	—	m
Mass	kilogram	—	kg
Time	second	—	s
Electric current	ampere	—	A
Temperature	kelvin	—	K
Luminous intensity	candela	—	cd
Plane angle	radian	—	rad
Solid angle	steradian	—	sr
Derived units			
Acceleration	meter per second squared	$m \cdot s^{-2}$	
Area	square meter	m^2	
Capacitance	farad	$A \cdot s \cdot V^{-1}$	F
Charge	coulomb	$A \cdot s$	C
Density	kilogram per cubic meter	$kg \cdot m^{-3}$	
Electric field strength	volt per meter	$V \cdot m^{-1}$	
Energy	joule	$N \cdot m$	J
Force	newton	$kg \cdot m \cdot s^{-2}$	N
Frequency	hertz	s^{-1}	Hz
Illumination	lux	$lm \cdot m^{-2}$	lx
Inductance	henry	$V \cdot s \cdot A^{-1}$	H
Kinematic viscosity	meters squared per second	$m^2 \cdot s^{-1}$	
Luminance	lumen	$cd \cdot sr$	lm
Magnetic field strength	ampere per meter	$A \cdot m^{-1}$	
Magnetic flux	weber	$V \cdot s$	Wb

	Unit	Formula	Symbol
Magnetic flux density	telsa	$Wb{\cdot}m^{-2}$	T
Power	watt	$J{\cdot}s^{-1}$	W
Pressure	pascal	$N{\cdot}m^{-2}$	Pa
Resistance	ohm	$V{\cdot}A^{-1}$	Ω
Stress	pascal	$N{\cdot}m^{-2}$	Pa
Velocity	meter per second	$m{\cdot}s^{-1}$	
Viscosity	pascal second	$Pa{\cdot}s$	
Voltage	volt	$W{\cdot}A^{-1}$	V
Volume	cubic meter	m^3	

Prefixes and Symbols for Multiples and Submultiples of SI Units

Multiple or Submultiple	Prefix	Symbol
10^{12}	tera	T
10^{9}	giga	G
10^{6}	mega	M
10^{3}	kilo	k
10^{2}	hecto	h
10	deka	da
10^{-1}	deci	d
10^{-2}	centi	c
10^{-3}	milli	m
10^{-6}	micro	μ
10^{-9}	nano	n
10^{-12}	pico	p
10^{-15}	femto	f
10^{-18}	atto	a

Conversion from British and U.S. Units to SI Units

Acceleration	$1 \text{ ft/s}^2 = 0.3048 \text{ m/s}^2$
Area	$1 \text{ ft}^2 = 0.0929 \text{ m}^2$
Density	$1 \text{ lb}_m/\text{ft}^2 = 16.02 \text{ kg/m}^3$
Energy	$1 \text{ Btu} = 1055 \text{ J}$
	$1 \text{ ft·lb}_f = 1.356 \text{ J}$
Force	$1 \text{ lb}_f = 4.448 \text{ N}$
Heat flux	$1 \text{ Btu/h·ft}^2 = 3.155 \text{ W/m}^2$
Heat generation rate	$1 \text{ Btu/h·ft}^3 = 10.35 \text{ W/m}^3$
Heat transfer coefficient	$1 \text{ Btu/h·ft}^2\text{·°F} = 5.678 \text{ W/m}^2\text{·K}$
Heat transfer rate	$1 \text{ Btu/h} = 0.2931 \text{ W}$
Kinematic viscosity	$1 \text{ ft}^2/\text{s} = 0.0929 \text{ m}^2/\text{s}$
Latent heat	$1 \text{ Btu/lb}_m = 2326 \text{ J/kg}$
Length	$1 \text{ ft} = 0.3048 \text{ m}$
	$1 \text{ in.} = 2.54 \text{ cm}$
	$1 \text{ mile} = 1609 \text{ m}$
Mass	$1 \text{ lb}_m = 0.4536 \text{ kg}$
Mass flow rate	$1 \text{ lb}_m/\text{s} = 0.4536 \text{ kg/s}$
Power	$1 \text{ hp} = 745.7 \text{ W}$
	$1 \text{ ft·lb}_f/\text{s} = 1.356 \text{ W}$
	$1 \text{ Btu/h} = 0.2931 \text{ W}$
Pressure	$1 \text{ lb}_f/\text{in.}^2 = 6895 \text{ Pa}$
	$1 \text{ atm} = 101,325 \text{ Pa}$
	$1 \text{ bar} = 10^5 \text{ Pa}$
Specific heat	$1 \text{ Btu/lb}_m\text{·°F} = 4187 \text{ J/kg·K}$
Thermal conductivity	$1 \text{ Btu/h·ft·°F} = 1.731 \text{ W/m·K}$
Thermal diffusivity	$1 \text{ ft}^2/\text{s} = 0.0929 \text{ m}^2/\text{s}$
Thermal resistance	$1 \text{ h·°F/Btu} = 1.896 \text{ K/W}$

Velocity	$1\ \text{ft/s} = 0.3048\ \text{m/s}$
Viscosity	$1\ \text{lb}_m/\text{ft·s} = 1.488\ \text{Pa·s}$
Volume	$1\ \text{ft}^3 = 0.02832\ \text{m}^3$
	$1\ \text{U.S. gal} = 0.003785\ \text{m}^3$
Volume flow rate	$1\ \text{ft}^3/\text{s} = 0.02832\ \text{m}^3/\text{s}$
	$1\ \text{U.S. gal/min} = 6.309 \times 10^{-5}\ \text{m}^3/\text{s}$

Properties of Solid Metals

Properties at 20°C   Thermal Conductivity, k (W/m·K)

Material	Melting Point (K)	ρ (kg/m³)	C_p (J/kg·K)	α (m²/s)	0°C	200°C	400°C	600°C	800°C	1000°C
Aluminum	933	2702	902	9.7×10^{-5}	236	238	228	215		
Beryllium	1550	1850	1809	6.3×10^{-5}	218	144	118	100	83	69
Boron	2573	2500	1090	1.1×10^{-5}	31.7	18.3	10.0	7.4	5.9	4.8
Cadmium	594	8650	231	5.2×10^{-5}	104	99				
Chromium	2118	7160	451	2.9×10^{-5}	95	86	77	69	64	62
Cobalt	1765	8862	419	2.7×10^{-5}	104	77				
Copper										
Pure	1356	8933	385	11.6×10^{-5}	401	389	378	366	352	336
Al Bronze (90% Cu, 10% Al)	1293	8800	420	1.4×10^{-5}	49	54	61			
Brass (70% Cu, 30% Zn)	1188	8526	382	3.4×10^{-5}	110	142	151			
Constantan (55% Cu, 45% Ni)	1493	8921	410	6.3×10^{-6}	22	26				
Germanium	1211	5360	318	3.6×10^{-5}	67	40	19	17		
Gold	1336	19300	129	12.7×10^{-5}	318	309	299	286	273	254
Iron										
Pure	1810	7870	440	2.3×10^{-5}	83	66	53	41	32	30
Wrought		7850	460	1.6×10^{-5}	59	52	45	33		
Armco		7870	447	2.1×10^{-5}	75	62	49	38	29	29
Cast (4% C)		7272	420	1.7×10^{-5}	52	40	34	24	21	
Steels										
1% carbon		7801	473	1.2×10^{-5}	43	42	36	29	28	
1% chrome		7913	448	1.7×10^{-5}	62	52	42	36	33	
304 stainless (18% Cr, 8% Ni)		7900	477	4.0×10^{-6}	14	18	21	24	26	

Material	Melting Point (K)	Properties at 20°C			Thermal Conductivity, k (W/m·K)					
		ρ (kg/m³)	C_p (J/kg·K)	α (m²/s)	0°C	200°C	400°C	600°C	800°C	1000°C
Lead	601	11340	129	2.4×10^{-5}	36	33				
Lithium	454	534	3560	4.0×10^{-5}	79					
Magnesium	923	1740	1017	8.8×10^{-5}	157	151	147	145		
Manganese	1517	7430	477	2.2×10^{-5}	7.7					
Molybdenum	2883	10240	255	5.3×10^{-5}	139	131	123	116	109	103
Nickel										
Pure	1726	8900	442	2.3×10^{-5}	94	74	65	69	73	78
Nichrome (80% Ni, 20% Cr)	1672	8360	430	3.5×10^{-6}	12	15	18	23		
Platinum										
Pure	2042	21450	130	2.6×10^{-5}	72	72	74	76	80	84
60% Pt, 40% Rh	1800	16600	162	1.7×10^{-5}	46	54	62	66	70	74
Rhenium	3453	21100	138	1.7×10^{-5}	49	45	44	44	45	46
Rhodium	2233	12450	247	4.9×10^{-5}	151	141	132	125	119	113
Silicon	1685	2330	691	9.5×10^{-5}	168	93	59	41	30	25
Silver	1234	10500	235	17.3×10^{-5}	428	415	399	383	368	355
Sodium	371	971	1220	11.2×10^{-5}	135					
Tin	505	5750	227	5.1×10^{-5}	68	60				
Titanium	1953	4500	510	9.6×10^{-6}	22	20	19	20	21	23
Tungsten	3653	19300	133	6.9×10^{-5}	182	152	134	125	118	114
Uranium	1406	19070	116	1.2×10^{-5}	27	30	35	39	45	
Vanadium	2192	6100	486	1.1×10^{-5}	31	33	35	37	40	44
Zinc	693	7140	388	4.4×10^{-5}	122	112	102			
Zirconium	2125	6570	280	1.2×10^{-5}	23	21	21	22	24	26

Source: Y. S. Touloukian et al. *Thermophysical Properties of Matter,* 13 volumes, IFI/Plenum, New York, 1970–1977.

Properties of
Nonmetallic Solids

Properties at 20°C — Thermal Conductivity, k (W/m·K)

Material	Melting Point (K)	ρ (kg/m³)	c_p (J/kg·K)	α (m²/s)	0°C	200°C	400°C	600°C	800°C	1000°C
Aluminum oxide	2323	3970	765	1.3×10^{-5}	40	22	13	9.3	7.3	6.2
Asbestos		383	816	3.6×10^{-6}	0.11					
Beryllium oxide	2725	3000	1030	9.2×10^{-5}	302	159	93	60	41	30
Bricks										
Common		1600	840	5.2×10^{-7}	0.7					
Chrome		3000	840	8.7×10^{-7}	2.2	2.3	2.4	2.4	2.1	
Fireclay		2000	960	5.2×10^{-7}	1.0	1.0	1.0	1.1	1.1	1.1
Magnesite			1130		4.0	3.6	2.8	2.4	2.1	1.8
Masonry		1700	837	4.6×10^{-7}	0.66					
Silica		1900			1.1					
Carbon	4073	1950			1.6	1.9	2.2	2.4	2.6	2.8
Cement mortar		1860	780	6.2×10^{-7}	0.9					
Concrete		2300	880	4.9×10^{-7}	1.0					
Coal		1370	1260	1.4×10^{-7}	0.24					
Diamonds										
Type I					1000.0	300.0				
Type IIa		3500	510	1.4×10^{-3}	2650.0	1300.0				
Type IIb					1510.0	780.0				
Earths										
Clay		1500	880	1.1×10^{-6}	1.4					
Diatomaceous					1.3					
Sand		1500	800	2.5×10^{-7}	0.3					
Glasses										
Pyroceram		2600	810	1.9×10^{-6}	4.1	3.6	3.2	3.0	2.9	2.8
Window		2700	800	3.9×10^{-7}	0.84					
Ice		920	2000	1.2×10^{-6}	2.2					

Properties at 20°C | Thermal Conductivity, k (W/m·K)

Material	Melting Point (K)	ρ (kg/m³)	C_p (J/kg·K)	α (m²/s)	0°C	200°C	400°C	600°C	800°C	1000°C
Insulations										
Cork, granular		45–120	1900	2–5×10^{-7}	0.045					
Corkboard		160	1900	1.4×10^{-7}	0.043					
Cellulose, loose		45			0.038					
Feltboard		50–125			0.035					
Glass fiber		220			0.035					
Glass wool		40	700	1.4×10^{-6}	0.038					
Kapok					0.035					
Magnesia, 85%		270			0.065	0.085				
Polystyrene		50			0.025					
Rock wool		160			0.040					
Rubber, foam		70			0.030					
Sawdust					0.059					
Vermiculite, loose		80			0.058					
Magnesium oxide					53	29	18	12	8.8	7.3
Plaster, gypsum		1600	1000	3.8×10^{-7}	0.5					
Quartz, fused					1.3	1.6	1.9	2.3	3.2	3.8
Rocks										
Granite		2640	800	1.4×10^{-6}	3.0					
Limestone		2400	860	1.0×10^{-6}	2.0					
Marble		2650	1000	1.0×10^{-6}	2.7					
Sandstone		2200	740	1.7×10^{-6}	2.8					
Rubber, hard		1170	2000	6.8×10^{-8}	0.16					
Skin, human					0.37					
Snow										
Loose		110			0.05					

Properties at 20°C

Material	Melting Point (K)	ρ (kg/m³)	C_p (J/kg·K)	α (m²/s)	Thermal Conductivity, k (W/m·K) 0°C	200°C	400°C	600°C	800°C	1000°C
Packed	500				0.19					
Teflon	2200				0.35					
Woods										
Balsa		140			0.055					
Cypress		460			0.097					
Fir		420	2700	9.7×10^{-8}	0.11					
Maple or oak		600	2400	1.2×10^{-7}	0.17					
Pine, white		440			0.11					
Pine, yellow		640	2800	8.4×10^{-8}	0.15					
Plywood		550	1200	1.8×10^{-7}	0.12					
Wool		200			0.038					

Source: Y. S. Touloukian et al. *Thermophysical Properties of Matter*, 13 volumes, IFI/Plenum, New York, 1970–1977.

Properties of Gases at 1 Atm Pressure

T (K)	ρ (kg/m³)	C_p (J/kg·K)	k (W/m·K)	α (m²/s)	η (kg/m·s)	ν (m²/s)	Pr	$g\beta/\nu^2$ (m⁻³·K⁻¹)
Air								
200	1.766	1003	0.0181	1.02×10^{-5}	1.34×10^{-5}	0.76×10^{-5}	0.740	85700×10^4
250	1.413	1003	0.0223	1.57×10^{-5}	1.61×10^{-5}	1.14×10^{-5}	0.724	30200×10^4
300	1.177	1005	0.0261	2.21×10^{-5}	1.85×10^{-5}	1.57×10^{-5}	0.712	13300×10^4
350	1.009	1008	0.0297	2.92×10^{-5}	2.08×10^{-5}	2.06×10^{-5}	0.706	6600×10^4
400	0.883	1013	0.0331	3.70×10^{-5}	2.29×10^{-5}	2.60×10^{-5}	0.703	3630×10^4
450	0.785	1020	0.0363	4.54×10^{-5}	2.49×10^{-5}	3.18×10^{-5}	0.700	2160×10^4
500	0.706	1029	0.0395	5.44×10^{-5}	2.68×10^{-5}	3.80×10^{-5}	0.699	1360×10^4
550	0.642	1039	0.0426	6.39×10^{-5}	2.86×10^{-5}	4.45×10^{-5}	0.698	900×10^4
600	0.589	1051	0.0456	7.37×10^{-5}	3.03×10^{-5}	5.15×10^{-5}	0.698	616×10^4
700	0.504	1075	0.0513	9.46×10^{-5}	3.35×10^{-5}	6.64×10^{-5}	0.702	318×10^4
800	0.441	1099	0.0569	11.7×10^{-5}	3.64×10^{-5}	8.25×10^{-5}	0.704	180×10^4
900	0.392	1120	0.0625	14.2×10^{-5}	3.92×10^{-5}	9.99×10^{-5}	0.705	109×10^4
1000	0.353	1141	0.0672	16.7×10^{-5}	4.18×10^{-5}	11.8×10^{-5}	0.709	70×10^4
1200	0.294	1175	0.0759	22.2×10^{-5}	4.65×10^{-5}	15.8×10^{-5}	0.720	33×10^4
1400	0.252	1201	0.0835	27.6×10^{-5}	5.09×10^{-5}	20.2×10^{-5}	0.732	17.2×10^4
1600	0.221	1240	0.0904	33.0×10^{-5}	5.49×10^{-5}	24.9×10^{-5}	0.753	9.9×10^4
1800	0.196	1276	0.0970	38.8×10^{-5}	5.87×10^{-5}	29.9×10^{-5}	0.772	6.1×10^4
2000	0.177	1327	0.1032	44.1×10^{-5}	6.23×10^{-5}	35.3×10^{-5}	0.801	3.9×10^4
Ammonia								
200	1.038	2199	0.0153	0.67×10^{-5}	6.89×10^{-6}	0.66×10^{-5}	0.990	113000×10^4
250	0.831	2248	0.0197	1.05×10^{-5}	8.53×10^{-6}	1.03×10^{-5}	0.973	37000×10^4
300	0.692	2298	0.0246	1.55×10^{-5}	10.27×10^{-6}	1.48×10^{-5}	0.959	14900×10^4
350	0.593	2349	0.0302	2.17×10^{-5}	12.06×10^{-6}	2.03×10^{-5}	0.938	6800×10^4
400	0.519	2402	0.0364	2.92×10^{-5}	13.90×10^{-6}	2.68×10^{-5}	0.917	3400×10^4
450	0.461	2455	0.0433	3.82×10^{-5}	15.76×10^{-6}	3.42×10^{-5}	0.894	1860×10^4

T (K)	ρ (kg/m³)	C_p (J/kg·K)	k (W/m·K)	α (m²/s)	η (kg/m·s)	ν (m²/s)	Pr	$g\beta/\nu^2$ (m⁻³·K⁻¹)
500	0.415	2507	0.0506	4.86×10^{-5}	17.63×10^{-6}	4.25×10^{-5}	0.873	1090×10^{4}
550	0.378	2559	0.0580	6.00×10^{-5}	19.5×10^{-6}	5.16×10^{-5}	0.860	670×10^{4}
600	0.346	2611	0.0656	7.26×10^{-5}	21.4×10^{-6}	6.18×10^{-5}	0.852	430×10^{4}
700	0.297	2710	0.0811	10.1×10^{-5}	25.1×10^{-6}	8.45×10^{-5}	0.839	196×10^{4}
800	0.260	2810	0.0977	13.4×10^{-5}	28.8×10^{-6}	1.11×10^{-5}	0.828	100×10^{4}
900	0.231	2907	0.1146	17.1×10^{-5}	32.4×10^{-6}	14.0×10^{-5}	0.822	56×10^{4}
1000	0.208	3001	0.1317	21.1×10^{-5}	35.9×10^{-6}	17.3×10^{-5}	0.818	33×10^{4}
Argon								
200	2.435	523.6	0.0124	0.98×10^{-5}	1.60×10^{-5}	0.66×10^{-5}	0.674	113000×10^{4}
250	1.948	522.2	0.0152	1.49×10^{-5}	1.95×10^{-5}	1.00×10^{-5}	0.672	39200×10^{4}
300	1.623	521.6	0.0177	2.09×10^{-5}	2.27×10^{-5}	1.40×10^{-5}	0.669	16700×10^{4}
350	1.392	521.2	0.0201	2.78×10^{-5}	2.57×10^{-5}	1.85×10^{-5}	0.666	8200×10^{4}
400	1.218	521.0	0.0223	3.52×10^{-5}	2.85×10^{-5}	2.34×10^{-5}	0.665	4480×10^{4}
450	1.082	520.9	0.0244	4.33×10^{-5}	3.12×10^{-5}	2.88×10^{-5}	0.665	2630×10^{4}
500	0.974	520.8	0.0264	5.20×10^{-5}	3.37×10^{-5}	3.45×10^{-5}	0.664	1640×10^{4}
550	0.886	520.7	0.0283	6.14×10^{-5}	3.60×10^{-5}	4.07×10^{-5}	0.662	1080×10^{4}
600	0.812	520.6	0.0301	7.12×10^{-5}	3.83×10^{-5}	4.72×10^{-5}	0.662	730×10^{4}
700	0.696	520.6	0.0336	9.28×10^{-5}	4.25×10^{-5}	6.11×10^{-5}	0.658	375×10^{4}
800	0.609	520.5	0.0369	11.6×10^{-5}	4.64×10^{-5}	7.62×10^{-5}	0.655	211×10^{4}
900	0.541	520.5	0.0398	14.1×10^{-5}	5.01×10^{-5}	9.26×10^{-5}	0.654	127×10^{4}
1000	0.487	520.5	0.0427	16.8×10^{-5}	5.35×10^{-5}	11.0×10^{-5}	0.652	81×10^{4}
1200	0.406	520.5	0.0481	22.8×10^{-5}	5.99×10^{-5}	14.8×10^{-5}	0.648	38×10^{4}
1400	0.348	520.4	0.0535	29.6×10^{-5}	6.56×10^{-5}	18.9×10^{-5}	0.638	19.7×10^{4}
1600	0.304	520.4	0.0588	37.1×10^{-5}	7.10×10^{-5}	23.2×10^{-5}	0.628	11.3×10^{4}
1800	0.271	520.4	0.0641	45.5×10^{-5}	7.60×10^{-5}	28.1×10^{-5}	0.617	6.9×10^{4}
2000	0.244	520.4	0.0692	54.6×10^{-5}	8.07×10^{-5}	33.1×10^{-5}	0.607	4.5×10^{4}

T (K)	ρ (kg/m³)	C_p (J/kg·K)	k (W/m·K)	α (m²/s)	η (kg/m·s)	ν (m²/s)	Pr	$g\beta/\nu^2$ (m⁻³·K⁻¹)
Carbon Dioxide								
200	2.683	759	0.0095	0.47×10^{-5}	1.02×10^{-5}	0.38×10^{-5}	0.814	338000×10^4
250	2.146	806	0.0129	0.75×10^{-5}	1.26×10^{-5}	0.59×10^{-5}	0.790	113000×10^4
300	1.789	852	0.0166	1.09×10^{-5}	1.50×10^{-5}	0.84×10^{-5}	0.768	46500×10^4
350	1.533	897	0.0205	1.49×10^{-5}	1.73×10^{-5}	1.13×10^{-5}	0.755	22100×10^4
400	1.341	939	0.0244	1.94×10^{-5}	1.94×10^{-5}	1.45×10^{-5}	0.747	16900×10^4
450	1.192	979	0.0283	2.43×10^{-5}	2.15×10^{-5}	1.80×10^{-5}	0.743	6700×10^4
500	1.073	1017	0.0323	2.96×10^{-5}	2.35×10^{-5}	2.19×10^{-5}	0.740	4100×10^4
550	0.976	1049	0.0363	3.55×10^{-5}	2.54×10^{-5}	2.60×10^{-5}	0.734	2630×10^4
600	0.894	1077	0.0403	4.18×10^{-5}	2.72×10^{-5}	3.04×10^{-5}	0.727	1770×10^4
700	0.767	1126	0.0487	5.64×10^{-5}	3.06×10^{-5}	3.99×10^{-5}	0.708	880×10^4
800	0.671	1169	0.0560	7.14×10^{-5}	3.39×10^{-5}	5.05×10^{-5}	0.708	480×10^4
900	0.596	1205	0.0621	8.65×10^{-5}	3.69×10^{-5}	6.19×10^{-5}	0.716	284×10^4
1000	0.537	1235	0.0680	10.25×10^{-5}	3.97×10^{-5}	7.40×10^{-5}	0.721	179×10^4
1200	0.447	1283	0.0780	13.6×10^{-5}	4.49×10^{-5}	10.04×10^{-5}	0.739	81×10^4
1400	0.383	1315	0.0867	17.2×10^{-5}	4.97×10^{-5}	13.0×10^{-5}	0.754	42×10^4
Carbon Monoxide								
200	1.708	1045	0.0175	0.98×10^{-5}	1.27×10^{-5}	0.75×10^{-5}	0.763	88100×10^4
250	1.366	1048	0.0214	1.50×10^{-5}	1.54×10^{-5}	1.13×10^{-5}	0.753	30900×10^4
300	1.138	1051	0.0252	2.11×10^{-5}	1.78×10^{-5}	1.56×10^{-5}	0.743	13400×10^4
350	0.976	1056	0.0288	2.80×10^{-5}	2.01×10^{-5}	2.50×10^{-5}	0.735	6640×10^4
400	0.854	1060	0.0323	3.57×10^{-5}	2.21×10^{-5}	2.59×10^{-5}	0.727	3650×10^4
450	0.759	1065	0.0355	4.39×10^{-5}	2.41×10^{-5}	3.18×10^{-5}	0.723	2160×10^4
500	0.683	1071	0.0386	5.28×10^{-5}	2.60×10^{-5}	3.80×10^{-5}	0.720	1360×10^4
550	0.621	1077	0.0416	6.22×10^{-5}	2.77×10^{-5}	4.46×10^{-5}	0.717	896×10^4

T (K)	ρ (kg/m³)	C_p (J/kg·K)	k (W/m·K)	α (m²/s)	η (kg/m·s)	ν (m²/s)	Pr	$g\beta/\nu^2$ (m⁻³·K⁻¹)
600	0.569	1084	0.0444	7.20×10^{-5}	2.94×10^{-5}	5.17×10^{-5}	0.718	613×10^4
700	0.488	1099	0.0497	9.27×10^{-5}	3.25×10^{-5}	6.66×10^{-5}	0.718	316×10^4
800	0.427	1114	0.0549	11.5×10^{-5}	3.54×10^{-5}	8.29×10^{-5}	0.718	178×10^4
900	0.379	1128	0.0596	13.9×10^{-5}	3.81×10^{-5}	10.04×10^{-5}	0.721	108×10^4
1000	0.342	1142	0.0644	16.5×10^{-5}	4.06×10^{-5}	11.9×10^{-5}	0.720	69×10^4
1100	0.310	1155	0.0692	19.3×10^{-5}	4.30×10^{-5}	13.9×10^{-5}	0.718	47×10^4
1200	0.285	1168	0.0738	22.2×10^{-5}	4.53×10^{-5}	15.9×10^{-5}	0.717	32×10^4
Helium								
200	0.2440	5197	0.115	0.91×10^{-4}	1.50×10^{-5}	0.61×10^{-4}	0.676	1320×10^4
250	0.1952	5197	0.134	1.54×10^{-4}	1.75×10^{-5}	0.90×10^{-4}	0.680	448×10^4
300	0.1627	5197	0.150	1.77×10^{-4}	1.99×10^{-5}	1.22×10^{-4}	0.690	219×10^4
350	0.1394	5197	0.165	2.28×10^{-4}	2.21×10^{-5}	1.59×10^{-4}	0.698	111×10^4
400	0.1220	5197	0.180	2.83×10^{-4}	2.43×10^{-5}	1.99×10^{-4}	0.703	61.9×10^4
450	0.1085	5197	0.195	3.45×10^{-4}	2.63×10^{-5}	2.43×10^{-4}	0.702	37×10^4
500	0.0976	5197	0.211	4.17×10^{-4}	2.83×10^{-5}	2.90×10^{-4}	0.695	23.4×10^4
550	0.0887	5197	0.229	4.97×10^{-4}	3.02×10^{-5}	3.40×10^{-4}	0.684	15.4×10^4
600	0.0813	5197	0.247	5.84×10^{-4}	3.20×10^{-5}	3.93×10^{-4}	0.673	10.6×10^4
700	0.0697	5197	0.278	7.67×10^{-4}	3.55×10^{-5}	5.09×10^{-4}	0.663	5.41×10^4
800	0.0610	5197	0.307	9.68×10^{-4}	3.88×10^{-5}	6.37×10^{-4}	0.657	3.02×10^4
900	0.0542	5197	0.335	11.9×10^{-4}	4.20×10^{-5}	7.75×10^{-4}	0.652	1.82×10^4
1000	0.0488	5197	0.363	14.3×10^{-4}	4.50×10^{-5}	9.23×10^{-4}	0.645	1.15×10^4
1200	0.0407	5197	0.416	19.7×10^{-4}	5.08×10^{-5}	12.5×10^{-4}	0.635	5240
1400	0.0349	5197	0.469	25.9×10^{-4}	5.61×10^{-5}	16.1×10^{-4}	0.622	2700
1600	0.0305	5197	0.521	32.9×10^{-4}	6.10×10^{-5}	20.0×10^{-4}	0.608	1530
1800	0.0271	5197	0.570	40.4×10^{-4}	6.57×10^{-5}	24.2×10^{-4}	0.599	930
2000	0.0244	5197	0.620	48.9×10^{-4}	7.00×10^{-5}	28.7×10^{-4}	0.587	595

T (K)	ρ (kg/m^3)	C_p (J/kg·K)	k (W/m·K)	α (m^2/s)	η (kg/m·s)	ν (m^2/s)	Pr	$g\beta/\nu^2$ (m^{-3}·K^{-1})
Hydrogen								
200	0.1229	13,540	0.128	0.77×10^{-4}	0.68×10^{-5}	0.55×10^{-4}	0.717	1610×10^4
250	0.0983	14,070	0.156	1.13×10^{-4}	0.79×10^{-5}	0.80×10^{-4}	0.713	607×10^4
300	0.0819	14,320	0.182	1.55×10^{-4}	0.89×10^{-5}	1.09×10^{-4}	0.705	275×10^4
350	0.0702	14,420	0.203	2.01×10^{-4}	0.99×10^{-5}	1.42×10^{-4}	0.705	140×10^4
400	0.0614	14,480	0.221	2.49×10^{-4}	1.09×10^{-5}	1.78×10^{-4}	0.714	77.8×10^4
450	0.0546	14,500	0.239	3.02×10^{-4}	1.18×10^{-5}	2.17×10^{-4}	0.719	46.4×10^4
500	0.0492	14,510	0.256	3.59×10^{-4}	1.27×10^{-5}	2.59×10^{-4}	0.721	29.2×10^4
550	0.0447	14,520	0.274	4.22×10^{-4}	1.36×10^{-5}	3.04×10^{-4}	0.722	19.3×10^4
600	0.0410	14,540	0.291	4.89×10^{-4}	1.45×10^{-5}	3.54×10^{-4}	0.724	13.0×10^4
700	0.0351	14,610	0.325	6.34×10^{-4}	1.61×10^{-5}	4.59×10^{-4}	0.724	6.66×10^4
800	0.0307	14,710	0.360	7.97×10^{-4}	1.77×10^{-5}	5.76×10^{-4}	0.723	3.69×10^4
900	0.0273	14,840	0.394	10.8×10^{-4}	1.92×10^{-5}	7.03×10^{-4}	0.723	2.2×10^4
1000	0.0246	14,990	0.428	11.6×10^{-4}	2.07×10^{-5}	8.42×10^{-4}	0.724	1.38×10^4
1200	0.0205	15,370	0.495	15.7×10^{-4}	2.36×10^{-5}	11.5×10^{-4}	0.733	6150
Nitrogen								
200	1.708	1043	0.0183	1.02×10^{-5}	1.29×10^{-5}	0.75×10^{-5}	0.734	86500×10^4
250	1.367	1042	0.0222	1.56×10^{-5}	1.55×10^{-5}	1.13×10^{-5}	0.725	30700×10^4
300	1.139	1040	0.0260	2.19×10^{-5}	1.79×10^{-5}	1.57×10^{-5}	0.715	13300×10^4
350	0.967	1041	0.0294	2.92×10^{-5}	2.01×10^{-5}	2.08×10^{-5}	0.711	6500×10^4
400	0.854	1045	0.0325	3.64×10^{-5}	2.21×10^{-5}	2.59×10^{-5}	0.710	3650×10^4
450	0.759	1050	0.0356	4.47×10^{-5}	2.41×10^{-5}	3.17×10^{-5}	0.709	2170×10^4
500	0.683	1057	0.0387	5.36×10^{-5}	2.59×10^{-5}	3.79×10^{-5}	0.708	1370×10^4
550	0.621	1065	0.0414	6.26×10^{-5}	2.76×10^{-5}	4.45×10^{-5}	0.711	900×10^4

T (K)	ρ (kg/m³)	C_p (J/kg·K)	k (W/m·K)	α (m²/s)	η (kg/m·s)	ν (m²/s)	Pr	$g\beta/\nu^2$ (m⁻³·K⁻¹)
600	0.569	1075	0.0441	7.20×10^{-5}	2.93×10^{-5}	5.14×10^{-5}	0.713	620×10^4
700	0.488	1098	0.0493	9.20×10^{-5}	3.24×10^{-5}	6.63×10^{-5}	0.720	319×10^4
800	0.427	1122	0.0541	11.3×10^{-5}	3.52×10^{-5}	8.24×10^{-5}	0.730	181×10^4
900	0.380	1146	0.0587	13.5×10^{-5}	3.79×10^{-5}	9.97×10^{-5}	0.739	110×10^4
1000	0.342	1168	0.0631	15.8×10^{-5}	4.04×10^{-5}	11.8×10^{-5}	0.747	70×10^4
1200	0.285	1205	0.0713	20.8×10^{-5}	4.50×10^{-5}	15.8×10^{-5}	0.761	33×10^4
1400	0.244	1233	0.0797	26.5×10^{-5}	4.92×10^{-5}	20.2×10^{-5}	0.761	17×10^4
Oxygen								
200	1.951	906	0.0182	1.03×10^{-5}	1.47×10^{-5}	0.75×10^{-5}	0.728	87000×10^4
250	1.561	914	0.0225	1.58×10^{-5}	1.78×10^{-5}	1.14×10^{-5}	0.721	30300×10^4
300	1.301	920	0.0267	2.23×10^{-5}	2.07×10^{-5}	1.59×10^{-5}	0.711	12900×10^4
350	1.115	929	0.0306	2.95×10^{-5}	2.34×10^{-5}	2.10×10^{-5}	0.710	6380×10^4
400	0.976	942	0.0342	3.72×10^{-5}	2.59×10^{-5}	2.65×10^{-5}	0.713	3480×10^4
450	0.867	956	0.0377	4.55×10^{-5}	2.83×10^{-5}	3.26×10^{-5}	0.717	2050×10^4
500	0.780	971	0.0412	5.44×10^{-5}	3.05×10^{-5}	3.91×10^{-5}	0.720	1280×10^4
550	0.709	987	0.0447	6.38×10^{-5}	3.27×10^{-5}	4.61×10^{-5}	0.722	840×10^4
600	0.650	1003	0.0480	7.36×10^{-5}	3.47×10^{-5}	5.34×10^{-5}	0.725	574×10^4
700	0.557	1032	0.0544	9.46×10^{-5}	3.85×10^{-5}	6.91×10^{-5}	0.730	294×10^4
800	0.488	1054	0.0603	11.7×10^{-5}	4.21×10^{-5}	8.63×10^{-5}	0.736	165×10^4
900	0.434	1074	0.0661	14.2×10^{-5}	4.54×10^{-5}	10.5×10^{-5}	0.738	99×10^4
1000	0.390	1091	0.0717	16.8×10^{-5}	4.85×10^{-5}	12.4×10^{-5}	0.738	63×10^4
1200	0.325	1116	0.0821	22.6×10^{-5}	5.42×10^{-5}	16.7×10^{-5}	0.737	29×10^4
1400	0.278	1136	0.0921	29.1×10^{-5}	5.95×10^{-5}	21.3×10^{-5}	0.734	15×10^4

Water Vapor (Steam)

T (K)	ρ (kg/m³)	C_p (J/kg·K)	k (W/m·K)	α (m²/s)	η (kg/m·s)	ν (m²/s)	Pr	$g\beta/\nu^2$ (m⁻³·K⁻¹)
300	0.0253[a]	2041	0.0181	35.1×10^{-5}[a]	0.91×10^{-5}	36.1×10^{-5}[a]	1.03	25×10^{4}[a]
350	0.258[a]	2037	0.0222	4.22×10^{-5}[a]	1.12×10^{-5}	4.33×10^{-5}[a]	1.02	1490×10^{4}[a]
400	0.555	2000	0.0264	2.38×10^{-5}	1.32×10^{-5}	2.38×10^{-5}	1.00	4330×10^{4}
450	0.491	1968	0.0307	3.17×10^{-5}	1.52×10^{-5}	3.10×10^{-5}	0.98	2270×10^{4}
500	0.441	1977	0.0357	4.09×10^{-5}	1.73×10^{-5}	3.92×10^{-5}	0.96	1280×10^{4}
550	0.401	1994	0.0411	5.15×10^{-5}	1.93×10^{-5}	4.82×10^{-5}	0.94	770×10^{4}
600	0.367	2022	0.0464	6.25×10^{-5}	2.13×10^{-5}	5.82×10^{-5}	0.93	480×10^{4}
700	0.314	2083	0.0572	8.74×10^{-5}	2.54×10^{-5}	8.09×10^{-5}	0.93	214×10^{4}
800	0.275	2148	0.0686	11.6×10^{-5}	2.95×10^{-5}	10.7×10^{-5}	0.92	106×10^{4}
900	0.244	2217	0.078	14.4×10^{-5}	3.36×10^{-5}	13.7×10^{-5}	0.95	58×10^{4}
1000	0.220	2288	0.087	17.3×10^{-5}	3.76×10^{-5}	17.1×10^{-5}	0.99	33×10^{4}

Source: Y. S. Touloukian et al. *Thermophysical Properties of Matter*, 13 volumes, IFI/Plenum, New York, 1970–1977.
[a]Saturation pressure is less than 1 atm.

APPENDIX **G**

Properties of Saturated Liquids

T (°C)	ρ (kg/m³)	C_p (J/kg·K)	k (W/m·K)	α (m²/s)	η (kg/m·s)	ν (m²/s)	Pr	$g\beta/\nu^2$ (m⁻³·K⁻¹)
Ammonia								
−40	692	4467	0.546	1.78×10^{-7}	2.81×10^{-4}	4.06×10^{-7}	2.28	1.05×10^{11}
−20	667	4509	0.546	1.82×10^{-7}	2.54×10^{-4}	3.81×10^{-7}	2.09	1.31×10^{11}
0	640	4635	0.540	1.82×10^{-7}	2.39×10^{-4}	3.73×10^{-7}	2.05	1.51×10^{11}
20	612	4789	0.521	1.78×10^{-7}	2.20×10^{-4}	3.59×10^{-7}	2.02	1.81×10^{11}
40	581	4999	0.493	1.70×10^{-7}	1.98×10^{-4}	3.40×10^{-7}	2.00	2.34×10^{11}
Ethyl Alcohol (CH₃CH₂OH)								
−40	823	2037	0.186	1.11×10^{-7}	4.81×10^{-3}	5.84×10^{-6}	52.7	0.29×10^{9}
−20	815	2124	0.179	1.03×10^{-7}	2.83×10^{-3}	3.47×10^{-6}	33.6	0.84×10^{9}
0	806	2249	0.174	0.960×10^{-7}	1.77×10^{-3}	2.20×10^{-6}	22.9	2.12×10^{9}
20	789	2395	0.168	0.889×10^{-7}	1.20×10^{-3}	1.52×10^{-6}	17.0	4.54×10^{9}
40	772	2572	0.162	0.816×10^{-7}	0.834×10^{-3}	1.08×10^{-6}	13.2	9.31×10^{9}
60	755	2781	0.156	0.743×10^{-7}	0.592×10^{-3}	0.784×10^{-6}	10.6	18.1×10^{9}
80	738	3026	0.150	0.672×10^{-7}	0.430×10^{-3}	0.583×10^{-6}	8.7	33.5×10^{9}
Ethyl Glycol (CH₂OHCH₂OH)								
0	1131	2295	0.254	9.79×10^{-8}	65.1×10^{-3}	57.5×10^{-6}	588	0.0192×10^{8}
20	1117	2386	0.257	9.64×10^{-8}	21.4×10^{-3}	19.2×10^{-6}	199	0.173×10^{8}
40	1101	2476	0.259	9.50×10^{-8}	9.57×10^{-3}	8.69×10^{-6}	91	0.844×10^{8}
60	1088	2565	0.262	9.39×10^{-8}	5.17×10^{-3}	4.75×10^{-6}	51	2.82×10^{8}
80	1078	2656	0.265	9.26×10^{-8}	3.21×10^{-3}	2.98×10^{-6}	32	7.18×10^{8}
100	1059	2750	0.267	9.17×10^{-8}	2.15×10^{-3}	2.03×10^{-6}	22	15.5×10^{8}

T (°C)	ρ (kg/m³)	C_p (J/kg·K)	k (W/m·K)	α (m²/s)	η (kg/m·s)	ν (m²/s)	Pr	$g\beta/\nu^2$ (m⁻³·K⁻¹)
Freon-12 (CCl_2F_2)								
−40	1515	885	0.069	5.14×10^{-8}	4.24×10^{-4}	2.80×10^{-7}	5.4	2.52×10^{11}
−20	1457	907	0.071	5.38×10^{-8}	3.43×10^{-4}	2.35×10^{-7}	4.4	3.73×10^{11}
0	1393	935	0.073	5.59×10^{-8}	2.98×10^{-4}	2.14×10^{-7}	3.8	5.04×10^{11}
20	1327	966	0.073	5.66×10^{-8}	2.62×10^{-4}	1.97×10^{-7}	3.5	6.54×10^{11}
40	1254	1002	0.069	5.46×10^{-8}	2.40×10^{-4}	1.91×10^{-7}	3.5	8.64×10^{11}
Glycerin ($HOCH_2CH(OH)CH_2OH$)								
−20	1288	2143	0.282	1.02×10^{-7}	134	104×10^{-3}	1020×10^{3}	0.42
0	1276	2261	0.284	0.98×10^{-7}	12.1	9.5×10^{-3}	96×10^{3}	50
20	1264	2386	0.287	0.95×10^{-7}	1.49	1.2×10^{-3}	12.4×10^{3}	3200
40	1252	2513	0.290	0.92×10^{-7}	0.27	0.2×10^{-3}	2.3×10^{3}	101000
Unused Engine Oil								
0	899	1796	0.147	9.11×10^{-8}	3850×10^{-3}	4280×10^{-6}	47100	350
20	888	1880	0.145	8.72×10^{-8}	800×10^{-3}	901×10^{-6}	10400	79000
40	876	1964	0.144	8.34×10^{-8}	212×10^{-3}	242×10^{-6}	2870	111000
60	864	2047	0.140	8.00×10^{-8}	72.5×10^{-3}	83.9×10^{-6}	1050	939000
80	852	2131	0.138	7.69×10^{-8}	32.0×10^{-3}	37.5×10^{-6}	490	4.77×10^{6}
100	840	2219	0.137	7.38×10^{-8}	17.1×10^{-3}	20.3×10^{-6}	276	16.5×10^{6}
120	829	2307	0.135	7.10×10^{-8}	10.2×10^{-3}	12.4×10^{-6}	175	44.8×10^{6}
140	817	2395	0.133	6.86×10^{-8}	6.53×10^{-3}	8.0×10^{-6}	116	109×10^{6}
160	806	2483	0.132	6.63×10^{-8}	4.49×10^{-3}	5.6×10^{-6}	84	226×10^{6}

T (°C)	ρ (kg/m^3)	C_p (J/kg·K)	k (W/m·K)	α (m^2/s)	η (kg/m·s)	ν (m^2/s)	Pr	$g\beta/\nu^2$ (m^{-3}·K^{-1})
Water								
273.2	1000	4205	0.564	1.34×10^{-7}	1.79×10^{-3}	1.79×10^{-6}	13.4	-21×10^7
280	1000	4197	0.582	1.39×10^{-7}	1.44×10^{-3}	1.44×10^{-6}	10.4	$+22 \times 10^7$
300	997	4177	0.608	1.46×10^{-7}	0.857×10^{-3}	0.86×10^{-6}	5.88	366×10^7
320	989	4176	0.637	1.54×10^{-7}	0.579×10^{-3}	0.59×10^{-6}	3.79	1250×10^7
340	980	4187	0.659	1.61×10^{-7}	0.423×10^{-3}	0.43×10^{-6}	2.69	2980×10^7
360	967	4204	0.674	1.66×10^{-7}	0.320×10^{-3}	0.33×10^{-6}	2.00	6250×10^7
373.2	958	4220	0.681	1.68×10^{-7}	0.282×10^{-3}	0.29×10^{-6}	1.75	8500×10^7
400	937	4241	0.686	1.73×10^{-7}	0.219×10^{-3}	0.23×10^{-6}	1.35	16100×10^7
450	890	4419	0.673	1.71×10^{-7}	0.153×10^{-3}	0.17×10^{-6}	1.01	40200×10^7
500	832	4647	0.635	1.64×10^{-7}	0.118×10^{-3}	0.14×10^{-6}	0.86	77100×10^7
550	756	5272	0.571	1.43×10^{-7}	0.095×10^{-3}	0.13×10^{-6}	0.88	144000×10^7
600	650	6691	0.481	1.11×10^{-7}	0.076×10^{-3}	0.12×10^{-6}	1.05	295000×10^7
647.3[a]	315							

[a]Critical point.

Properties of
Liquid Metals

T (K)	ρ (kg/m³)	C_p (J/kg·K)	k (W/m·K)	α (m²/s)	η (kg/m·s)	ν (m²/s)	Pr	$g\beta/\nu^2$ (m⁻³·K⁻¹)
Bismuth								
545[a]	10069	143	16.8	1.17×10^{-5}	1.75×10^{-3}	1.74×10^{-7}	0.0148	43×10^{8}
600	9997	145	16.4	1.13×10^{-5}	1.61×10^{-3}	1.61×10^{-7}	0.0142	49×10^{8}
700	9867	150	15.6	1.06×10^{-5}	1.34×10^{-3}	1.36×10^{-7}	0.0128	68×10^{8}
800	9752	154	15.6	1.04×10^{-5}	1.12×10^{-3}	1.15×10^{-7}	0.0111	88×10^{8}
900	9636	159	15.6	1.02×10^{-5}	0.96×10^{-3}	0.99×10^{-7}	0.0098	126×10^{8}
1000	9510	163	15.6	1.01×10^{-5}	0.83×10^{-3}	0.87×10^{-7}	0.0087	177×10^{8}
Lead								
601[a]	10588	161	15.5	0.91×10^{-5}	2.62×10^{-3}	2.47×10^{-7}	0.0272	145×10^{8}
700	10476	157	17.4	1.06×10^{-5}	2.15×10^{-3}	2.05×10^{-7}	0.0194	257×10^{8}
800	10359	153	19.0	1.20×10^{-5}	2.05×10^{-3}	1.98×10^{-7}	0.0165	289×10^{8}
900	10237	149	20.3	1.33×10^{-5}	1.54×10^{-3}	1.50×10^{-7}	0.0113	528×10^{8}
1000	10111	145	21.5	1.47×10^{-5}	1.32×10^{-3}	1.30×10^{-7}	0.0089	736×10^{8}
Mercury								
234[a]	13723	142	7.3	3.8×10^{-6}	2.00×10^{-3}	1.46×10^{-7}	0.0389	82×10^{9}
273	13628	140	8.2	4.3×10^{-6}	1.69×10^{-3}	1.24×10^{-7}	0.0289	115×10^{9}
300	13562	139	8.9	4.7×10^{-6}	1.51×10^{-3}	1.11×10^{-7}	0.0237	143×10^{9}
350	13441	138	10.0	5.4×10^{-6}	1.31×10^{-3}	0.98×10^{-7}	0.0181	185×10^{9}
400	13320	137	11.0	6.1×10^{-6}	1.18×10^{-3}	0.89×10^{-7}	0.0147	227×10^{9}
500	13081	136	12.7	7.1×10^{-6}	1.02×10^{-3}	0.78×10^{-7}	0.0109	292×10^{9}
600	12816	134	14.2	8.3×10^{-6}	0.84×10^{-3}	0.66×10^{-7}	0.0080	480×10^{9}
Lithium								
454[a]	512	4190	43	2.0×10^{-5}	6.1×10^{-4}	1.18×10^{-6}	0.059	134×10^{7}
500	508	4190	44	2.1×10^{-5}	5.9×10^{-4}	1.17×10^{-6}	0.056	136×10^{7}

T (K)	ρ (kg/m³)	C_p (J/kg·K)	k (W/m·K)	α (m²/s)	η (kg/m·s)	ν (m²/s)	Pr	$g\beta/\nu^2$ (m⁻³·K⁻¹)
600	498	4190	48	2.3×10^{-5}	5.7×10^{-4}	1.14×10^{-6}	0.050	143×10^{7}
700	489	4190	51	2.4×10^{-5}	5.4×10^{-4}	1.11×10^{-6}	0.045	151×10^{7}
800	480	4190	54	2.7×10^{-5}	5.2×10^{-4}	1.08×10^{-6}	0.040	160×10^{7}
900	471	4190	57	2.9×10^{-5}	4.9×10^{-4}	1.05×10^{-6}	0.036	169×10^{7}
1000	462	4190	60	3.1×10^{-5}	4.7×10^{-4}	1.02×10^{-6}	0.033	179×10^{7}
Potassium								
337[a]	827	802	55	8.3×10^{-5}	4.7×10^{-4}	5.6×10^{-7}	0.0068	86×10^{8}
400	812	798	52	8.0×10^{-5}	3.9×10^{-4}	4.9×10^{-7}	0.0061	119×10^{8}
500	789	790	48	7.7×10^{-5}	3.0×10^{-4}	3.8×10^{-7}	0.0050	199×10^{8}
600	766	783	44	7.3×10^{-5}	2.3×10^{-4}	3.0×10^{-7}	0.0041	331×10^{8}
700	742	775	40	7.0×10^{-5}	1.8×10^{-4}	2.4×10^{-7}	0.0034	550×10^{8}
800	718	767	37	6.7×10^{-5}	1.6×10^{-4}	2.2×10^{-7}	0.0033	683×10^{8}
900	693	760	34	6.5×10^{-5}	1.4×10^{-4}	2.0×10^{-7}	0.0032	840×10^{8}
1000	669	752	31	6.2×10^{-5}	1.3×10^{-4}	1.9×10^{-7}	0.0030	1040×10^{8}
Sodium								
371[a]	929	1382	88	6.9×10^{-5}	7.0×10^{-4}	7.5×10^{-7}	0.0110	51×10^{8}
400	922	1371	87	6.9×10^{-5}	6.1×10^{-4}	6.7×10^{-7}	0.0097	66×10^{8}
500	896	1334	82	6.8×10^{-5}	4.1×10^{-4}	4.6×10^{-7}	0.0067	145×10^{8}
600	871	1309	76	6.7×10^{-5}	3.2×10^{-4}	3.6×10^{-7}	0.0054	238×10^{8}
700	846	1284	72	6.6×10^{-5}	2.6×10^{-4}	3.0×10^{-7}	0.0046	356×10^{8}
800	822	1259	67	6.5×10^{-5}	2.1×10^{-4}	2.6×10^{-7}	0.0040	507×10^{8}
900	797	1256	63	6.2×10^{-5}	1.9×10^{-4}	2.4×10^{-7}	0.0039	604×10^{8}
1000	773	1256	58	6.0×10^{-5}	1.8×10^{-4}	2.3×10^{-7}	0.0038	708×10^{8}

[a]Melting point.

Recommended Reading

1. *Fluid Dynamics and Heat Transfer*
   J. G. Knudsen and D. L. Katz
   University of Michigan, Engineering Research Institute
   Bulletin 37, 1954

2. *Transport Phenomena*
   R. B. Bird, W. E. Stewart, and E. N. Lightfoot
   Wiley, New York, 1960

3. *Rate Phenomena in Process Metallurgy*
   J. Szekely and N. J. Themelis
   Wiley-Interscience, New York, 1971

4. *Transport Phenomena in Metallurgy*
   G. H. Geiger and D. R. Poirier
   Addison-Wesley, Reading, Mass., 1973

5. *Viscous Fluid Flow*
   F. M. White
   McGraw-Hill, New York, 1974

6. *Turbulence*, 2nd Ed.
   J. O. Hinze
   McGraw-Hill, New York, 1975

7. *Mass Transfer*
   T. K. Sherwood, R. L. Pigford, and C. R. Wilke
   McGraw-Hill, New York, 1975

8. *Boundary Layer Theory*, 7th Ed.
   H. Schlichting
   McGraw-Hill, New York, 1979

9. *Momentum, Heat and Mass Transfer*, 3rd Ed.
   C. O. Bennett and J. E. Meyers
   McGraw-Hill, New York, 1982

10. *Heat Transfer*
    F. M. White
    Addison-Wesley, Reading, Mass., 1984

11. *Fluid Mechanics and Transfer Processes*
    J. M. Kay and R. M. Nedderman
    Cambridge University Press, Cambridge, 1985

12. *Heat Transfer, A Basic Approach*
    M.N. Ozisik
    McGraw-Hill, New York, 1985

13. *Fundamentals of Fluid Mechanics*
    J. B. Evett and C. Liu
    McGraw-Hill, New York, 1987

14. *Engineering in Process Metallurgy*
    R. I. L. Guthrie
    Clarendon Press, New York, 1989

15. *Fundamentals of Heat and Mass Transfer*, 3rd Ed.
    F. P. Incropera and D. P. DeWitt
    Wiley, New York, 1990

16. *Introduction to Heat Transfer*, 2nd Ed.
    F. P. Incropera and D. P. DeWitt
    Wiley, New York, 1990

Answers to Problems

Chapter One

1.1   88,000 N
1.2   23,050 N
1.3   3840 N
1.4   0.34 m/s
1.5   (a) 5.08 cm^3 (b) 2967 kg/m^3
1.6   (a) 2004 kg/m^3 (b) 830 kg/m^3
1.7   (a) 800 kg/m^3 (b) 20760 Pa
1.8   3104 m
1.9   100 kg
1.10   94,380 Pa

Chapter Two

2.1 3.86×10^{-2} N·m per unit length, 4.04 W per unit length
2.2   3.93 N
2.3   0.49 Pa·s
2.4   (a) 0.403 kg/m·s   (b) 5.61×10^{-3} m   (c) 8.03×10^{-2} m/s
2.5   (a) 4.77×10^{-3} m/s   (b) 0.116 m/s   (c) 43 s   (d) 0.423
2.6   0.1 m
2.7   (a) 4.63 Pa/m   (b) 0.78 m/s   (c) 497   (d) 0.0116 Pa
2.9   (a) $2^{1/2}R$ (b) $2^{1/4}R$
2.10   Procedure 2 [in (1) $\dot{V}$ increases by the factor 1.33; in (2) $\dot{V}$ increases by the factor 1.88; in (3) $\dot{V}$ increases by the factor 1.6]
2.11 4.63×10^{-4} P (4.63×10^{-5} Pa·s)
2.12 3.11×10^{-4} P (3.11×10^{-5} Pa·s)
2.13 5.8×10^{-3} Pa·s
2.14   0.5 m/s
2.15   (a) 213 Pa   (b) 5690 Pa/m   (d) $v_x = 10 - 1778r^2$ m/s
2.16   (a) $v_x = \dfrac{V}{\ln(b/a)} \ln \dfrac{r}{a}$

(b) $\dot{V} = \dfrac{\pi V}{\ln(b/a)} \left(b^2 \ln \dfrac{a}{b} - \dfrac{a^2 - b^2}{2} \right)$

2.17 Upper layer is fluid 1, lower layer is fluid 2

(a) $v_{x,1} = \dfrac{\Delta P}{L} \left(\dfrac{\delta^2}{\eta_1 + \eta_2} - \dfrac{\delta y}{2\eta_1} \dfrac{\eta_1 - \eta_2}{\eta_1 + \eta_2} - \dfrac{y^2}{2\eta_1} \right)$

$v_{x,2} = \dfrac{\Delta P}{L} \left(\dfrac{\delta^2}{\eta_1 + \eta_2} - \dfrac{\delta y}{2\eta_2} \dfrac{\eta_1 - \eta_2}{\eta_1 + \eta_2} - \dfrac{y^2}{2\eta_2} \right)$

(b) $\tau_{yx,1} = \tau_{yx,2} = \dfrac{\Delta P}{L} \left(y - \dfrac{\delta}{2} \dfrac{\eta_2 - \eta_1}{\eta_1 + \eta_2} \right)$

$\tau_{yx} = 0$ at $y = \dfrac{\delta}{2} \dfrac{\eta_2 - \eta_1}{\eta_1 + \eta_2}$

which is positive and thus occurs in layer 1.

(c) In layer 1, v_{max} occurs at $y = \dfrac{\delta}{2} \dfrac{\eta_2 - \eta_1}{\eta_1 + \eta_2}$

In layer 2, v_{max} occurs at $y = 0$

2.18 7.55×10^{-4} m

Chapter Three

3.1 14.96 Pa·s
3.2 (a) 6.53×10^{-3} m (b) 0.2 m/s (c) 0.57 m (d) v_x (m/s) = $3.65y - 180{,}700y^3$ (y in m)
3.3 (a) 6.23×10^{-2} Pa (b) 8.81×10^{-2} N (c) 0.0467 Pa (d) 42.3 Pa
3.4 (a) 2.34×10^{-5} m (b) $v_t = 1.28 \times 10^{-3}$ m/s
3.5 (a) 7.27×10^{-3} m²/s (b) 1.07×10^{-2} m (c) 0.034 m/s
3.6 109.7 W

Chapter Four

4.1 (a) 1.84 m/s (b) 1.79 m/s (c) 0.019 m
4.2 (a) 72.05 kPa (b) 0.0182 m
4.3 33.3 kPa
4.4 0.828 m/s
4.5 (a) 4.12×10^{-2} N (b) 29.3 m/s (c) 0.96 m/s
4.6 2.02 m/s
4.7 7.13 m/s
4.8 21.7 kPa
4.9 8.48×10^{-2} m/s
4.10 2.35
4.11 When entire layer is laminar $F = 0.614v_\infty^{3/2}$
When transition occurs $F = 0.312v_\infty + 2.263v_\infty^{9/5}(1 - 0.338/v_\infty^{4/5})$
4.12 6.79 m³/s
4.13 1.98 m/s in (1), 2.18 m/s in (2), and 2.17 m/s in (3)
4.14 (a) 2.71 m/s in (1) and 1.86 m/s in (2) (b) 153 kPa

4.15   8.76 N
4.16   1.3 to 38.9 m/s

Chapter Five

5.1   (a) 6.9 J/kg   (b) 6.88 kPa
5.2   (a) 1038 J/kg   (b) 7350 W
5.3   40 kW
5.4   77%
5.5   197 kg/s
5.6   2.21 m/s
5.7   112 kW
5.8   (a) 67.8 s   (b) 83.9 kg/s   (c) 59.3 kg/s
5.9   0.24 m
5.10   12.7 s
5.11   157 s
5.12   (a) 0.091 m^3/s   (b) 427 kPa   (c) 2.63 h
5.13   (a) 2.36 m/s   (b) 74.6 kPa
5.14   450 kPa
5.15 2.66×10^{-2} kg/s

Chapter Six

6.1   (a) 61.1 W   (b) 37.4 W   (c) 0.30 W   (d) 0.25 W   (e) 0.013 W
       (f) 0.33 m
6.2   (a) 2.83 W/m·K   (b) 2830 W/m^2 (c) $T = 100 \ln(2.718 - 17.16x)$
       (d) 62.1°C   (e) 50°C   (f) -923°C/m   (g) -1000°C/m
6.3   (a) 125 W/m^2 (b) increased to 341 W/m^2 (c) 6.9×10^{-3} m
6.4   (a) 0.39 m   (b) 504 W/m^2
6.5 1.57×10^{-4} kg/s
6.6   0.056 m
6.7   35.4°C
6.8   (a) 13.5 W/m   (b) 235°C   (c) 2.3×10^{-2} Celsius degrees   (d) 115°C
       and 92°C   (e) 11.7 A
6.9   (a) 142°C   (b) 64°C
6.10   260 Celsius degrees
6.11   (a) 9.24%   (b) 250 W/m^2
6.12   (a) $T = -81{,}000x^2 + 3550x + 25$($x$ in m)   (b) 0.0219 m
       (c) 0.715 W and -0.558 W
6.13   0.243 m
6.14   0.0242 W/m·K
6.15 $T_1 = T_2 = 171$°C, $T_3 = T_4 = 64$°C, $T_5 = T_6 = 21$°C
6.16 $T_1 = T_7 = 171$°C, $T_2 = T_8 = 219$°C, $T_3 = T_9 = 279$°C, $T_4 = 167$°C,
       $T_5 = 225$°C, $T_6 = 296$°C
6.17 $T_1 = 66.7$°C, $T_2 = T_3 = 50$°C, $T_4 = 33.3$°C

Chapter Seven

7.1   1575 W
7.2   19.7 m/s

7.3   12.1 kW

7.4   313 K

7.5   Maximum power of 1184 W to the seventh strip

7.6   (a) 4.67 m

$$\text{(b) } L = 0.1115 \left[\log \frac{1.564 \times 10^{-4}}{\bar{v}} \right]^2 \left[1 - \frac{4.325}{\log[(1.564 \times 10^{-4})/\bar{v}]} \right]$$

7.7 1.04×10^{-2} m/s

7.8   (a) 8×10^{-3} kg/s (b) 72°C

7.9   204°C

7.10   1056 W/m

7.11   1.93 m/s

7.12   (a) 89.3°C (b) 0.0016 m (c) $-14,300$ W and 12,400 W

7.13   (a) 0.589 m (b) 1.35×10^{-2} m (c) 0.48 m/s (d) 3.46×10^{-3} m

7.14   (a) 12.7 kW/m² from each side (b) 0.01 m

7.15   (a) 164 W/m (b) 0.03 m

7.16   7930 W

7.17   (a) 89°C (b) T calculated as 1884°C

Chapter Eight

8.1   28.8 W/m²·K

8.2   0.53 m/min

8.3   (a) 0.096 m (b) 10.1 kW/m² (c) 6060 kW/m²

8.4   5.21 min

8.5   (a) 479°C (b) 497°C

8.6   53.9 A

8.7 $T_s = 75$°C, $T(x = 0.05$ m$) = 80$°C, $T(x = 0.1$ m$) = 83$°C, $T(x = 0.15$ m$) = 84$°C.

8.8 $T_s = 78$°C, $T(x = 0.05$ m$) = 83$°C, $T(x = 0.1$ m$) = 87$°C, $T(x = 0.15$ m$) = 91$°C, $T(x = 0.20$ m$) = 93$°C, $T(x = 0.25$ m$) = 94$°C, $T_s = 93$°C

8.9 $T_s = 20$°C, $T(x = 0.01$ m$) = 58$°C, $T(x = 0.02$ m$) = 83$°C, $T(x = 0.03$ m$) = 94$°C

8.10   (a) $T_s = 37$°C, $T(x = 0.01$ m$) = 35$°C, $T(x = 0.02$ m$) = 33$°C, $T(x = 0.03$ m$) = 31$°C, $T(x = 0.04$ m$) = 31$°C (b) 11,500 W/m²
       (c) 1.7×10^7 J/m

8.11

	(a) steady state, $\dot{q} = 10^7$ W/m³	(b) at $t = 2$s with $\dot{q} = 10^8$ W/m³
T_s	341°C	395°C
$T(x = 0.002$ m$)$	347°C	404°C
$T(x = 0.004$ m$)$	352°C	410°C
$T(x = 0.006$ m$)$	355°C	414°C
$T(x = 0.008$ m$)$	357°C	417°C
$T(x = 0.01$ m$)$	358°C	417°C

(c) Boils after 33.2 s

8.12   3.26 h
8.13   In 5 min T at $x = 0.1$ m is 31.8°C
8.14 1.55×10^7 J/m^2
8.15   (a) 2.3 h   (b) 55 s
8.16   No. $T_s = 788$°C is higher than $T_{m,Al} = 660$°C
8.17 3.9×10^{-3} m

Chapter Nine

9.1   (a) 2.87×10^5 W/m^2 (b) 2.74 μm   (c) 98,080 W/m^2·μm at $\lambda = 193$ μm
9.2   276.7 K
9.3   (a) $\epsilon = 0.176$, $\alpha = 0.833$, J $= 1000$ W/m^2 (b) 1500 W/m^2
9.4   347 K
9.5   (a) 0.6   (b) 540 W/m^2 to the surface   (c) zero
9.6   (a) 1.03×10^5 W   (b) 4.12×10^5 W
9.7   (a) 189 W   (b) 893 W
9.8   100°C
9.9   292 kW/m
9.10   (a) 60.2 W/m   (b) 21.2 W/m   (c) Yes   (d) No   (e) 0.0031 m
       (f) 632 K   (g) 0.0052 m
9.11   323 K
9.12 $q_1' = 8560$ W/m, $q_2' = -4560$ W/m, $q_3' = -4000$ W/m
9.13   (a) 6087 W/m   (b) 725 K
9.14   796 K
9.15   552 K

Chapter Ten

10.1   (a) 4.0 hrs   (b) 1366 K
10.2   0.8 cm
10.3   (a) 6.67×10^{-6} g mol C/m^2·s   (b) 9.29×10^{-4} m
10.4   (a) 413 s   (b) 1.15×10^{-5} g mol H/m^2·s   (c) 9.46×10^{-3} g mol H/m^2
10.5 1.8×10^{-4} m

Chapter Eleven

11.1 9.87×10^{-6} kg/h
11.2 7.79×10^{-5} m^2/s
11.3   (a) 4.45×10^{-2} m/s   (b) 8.2°C
11.4   (a) 24.8 W/m^2·K   (b) T increases to 307.5 K   (c) Evaporation rate
       decreases   (d) Yes
11.5 D_{CO_2-air} at 1000 K $= 4.71 \times 10^{-4}$ m^2/s,
       D_{CO_2-air} at 1100 K $= 5.54 \times 10^{-4}$ m^2/s
11.6   At 1000 K, 0.318 g mol/m^2·s; at 1100 K, 2.31 g mol/m^2·s
11.7   71%
11.8   (a) 50.0 W/m to first, 20.7 W/m to second, 15.9 W/m to third
       (b) 72.4 W/m to first, 68.6 W/m to second, 123 W/m to third
       (c) 27.1%
11.9   (a) 1.8×10^{-6} g mol C/s   (b) 0.22 s
11.10 (a) 9.97×10^{-8} g mol C/s   (b) 0.25 s

Index